Neuroproteomics

Neuroproteomics

Special Issue Editors

Angus C. Nairn
Kenneth R. Williams

MDPI • Basel • Beijing • Wuhan • Barcelona • Belgrade

Special Issue Editors
Angus C. Nairn
Yale University School of Medicine
USA

Kenneth R. Williams
Yale University School of Medicine
USA

Editorial Office
MDPI
St. Alban-Anlage 66 4052 Basel,
Switzerland

This is a reprint of articles from the Special Issue published online in the open access journal *Proteomes* (ISSN 2227-7382) from 2018 to 2019 (available at: https://www.mdpi.com/journal/proteomes/special_issues/neuroproteomics).

For citation purposes, cite each article independently as indicated on the article page online and as indicated below:

LastName, A.A.; LastName, B.B.; LastName, C.C. Article Title. *Journal Name* **Year**, *Article Number*, Page Range.

ISBN 978-3-03928-106-0 (Pbk)
ISBN 978-3-03928-107-7 (PDF)

The upper and lower images on the front cover are monochrome versions of Figure 3H (synaptome diversity map) and Figure 3E (synaptome dominant subtype map) respectively from Zhu, F., Cizeron, M., Qiu, Z., Franse, E., Komiyama, N.H., Grant, S.G.N. (2018) Architecture of the Mouse Brain Synaptome, Neuron 99(4):781-799 (PMID:30078578, PMCID:PMC6117470). The original versions of these images were prepared by Dr. Zhen Qiu in the laboratory of Dr. Seth Grant, University of Edinburgh.

Contents

About the Special Issue Editors vii

Kenneth R. Williams and Angus C. Nairn
Editorial for Special Issue: Neuroproteomics
Reprinted from: *Proteomes* **2019**, *7*, 24, doi:10.3390/proteomes7020024 1

Rashaun S. Wilson and Angus C. Nairn
Cell-Type-Specific Proteomics: A Neuroscience Perspective
Reprinted from: *Proteomes* **2018**, *6*, 51, doi:10.3390/proteomes6040051 10

Yi-Zhi Wang and Jeffrey N. Savas
Uncovering Discrete Synaptic Proteomes to Understand Neurological Disorders
Reprinted from: *Proteomes* **2018**, *6*, 30, doi:10.3390/proteomes6030030 33

Marcia Roy, Oksana Sorokina, Colin McLean, Silvia Tapia-González, Javier DeFelipe, J. Douglas Armstrong and Seth G. N. Grant
Regional Diversity in the Postsynaptic Proteome of the Mouse Brain
Reprinted from: *Proteomes* **2018**, *6*, 31, doi:10.3390/proteomes6030031 45

Tony Cijsouw, Austin M. Ramsey, TuKiet T. Lam, Beatrice E. Carbone, Thomas A. Blanpied and Thomas Biederer
Mapping the Proteome of the Synaptic Cleft through Proximity Labeling Reveals New Cleft Proteins
Reprinted from: *Proteomes* **2018**, *6*, 48, doi:10.3390/proteomes6040048 63

Rashaun S. Wilson, Navin Rauniyar, Fumika Sakaue, TuKiet T. Lam, Kenneth R. Williams and Angus C. Nairn
Development of Targeted Mass Spectrometry-Based Approaches for Quantitation of Proteins Enriched in the Postsynaptic Density (PSD)
Reprinted from: *Proteomes* **2019**, *7*, 12, doi:10.3390/proteomes7020012 90

Becky C. Carlyle, Bianca A. Trombetta and Steven E. Arnold
Proteomic Approaches for the Discovery of Biofluid Biomarkers of Neurodegenerative Dementias
Reprinted from: *Proteomes* **2018**, *6*, 32, doi:10.3390/proteomes6030032 112

Brianna M. Lutz and Junmin Peng
Deep Profiling of the Aggregated Proteome in Alzheimer's Disease: From Pathology to Disease Mechanisms
Reprinted from: *Proteomes* **2018**, *6*, 46, doi:10.3390/proteomes6040046 133

Luis A. Natividad, Matthew W. Buczynski, Daniel B. McClatchy and John R. Yates III
From Synapse to Function: A Perspective on the Role of Neuroproteomics in Elucidating Mechanisms of Drug Addiction
Reprinted from: *Proteomes* **2018**, *6*, 50, doi:10.3390/proteomes6040050 145

Darlene A. Pena, Mariana Lemos Duarte, Dimitrius T. Pramio, Lakshmi A. Devi and Deborah Schechtman
Exploring Morphine-Triggered PKC-Targets and Their Interaction with Signaling Pathways Leading to Pain via TrkA
Reprinted from: *Proteomes* **2018**, *6*, 39, doi:10.3390/proteomes6040039 163

About the Special Issue Editors

Angus C. Nairn completed his undergraduate training in biochemistry at the University of Edinburgh, Scotland, and received his Ph.D. in 1979 in Muscle Biochemistry for his work in the laboratory of Professor Sam Perry at Birmingham University, England. He then carried out postdoctoral research in Molecular Neuroscience with Professor Paul Greengard at Yale, and moved with Professor Greengard to Rockefeller University in 1983 as a faculty member. He moved back to Yale University in 2001, where he is currently the Charles B.G. Murphy Professor of Psychiatry. He also holds a joint appointment in the Department of Pharmacology and is Co-Director of the Yale/National Institute of Drug Abuse Neuroproteomics Center at the Yale School of Medicine. Dr. Nairn has very extensive experience in the enzymology, protein chemistry, and molecular biology of signal transduction, particularly with respect to the role of protein phosphorylation in the nervous system. With more than 400 publications in the area, Dr. Nairn has identified, purified, and characterized a variety of neuronal phosphoproteins that are important in mediating the actions of the neurotransmitter dopamine in the CNS. Dr. Nairn has also carried out detailed studies of the structure and function of many protein kinases and protein phosphatases that play critical roles in neuronal function. Recent studies by Dr. Nairn and his colleagues have focused on identifying long-term adaptive changes in signal transduction processes that might be involved in mediating the actions of psychomotor stimulants, and other drugs of abuse.

Kenneth R. Williams received his Ph.D. degree in 1976 from the University of Vermont and then held a postdoctoral position in the Department of Molecular Biophysics and Biochemistry at Yale University where he advanced up the research track until he was appointed in 1989 to Professor (Adjunct) Research. In 1980, he founded the Keck Foundation Biotechnology Laboratory (http://keck.med.yale.edu/). As Director/Co-Director of the Keck Lab from 1980 through 2014, Dr. Williams wrote or helped Keck staff write 25 successful NIH/NSF Shared Instrumentation Grants—with the resulting instrumentation bringing state-of-the art biotechnologies within reach of the >1000 investigators at >200 institutions who annually use the Keck Lab. In 1986, Dr. Williams was one of six founding members of the Association of Biomolecular Resource Facilities (ABRF, http://www.abrf.org/). In 2000, he was PI on one of ten NIH/NIDDK Microarray Biotechnology Center grants, and in 2002, he was PI on one of ten contracts that established the Yale/NHLBI Proteomics Center. In 2004, Dr. Williams was PI on one of two grants to establish NIH/NIDA Neuroproteomics Centers. From 2004 through 2015, he was Director, and since then, he has been the Co-Director of the Yale/NIDA Neuroproteomics Center (http://medicine.yale.edu/keck/nida/). As Founder of the Keck Laboratory, his focus is on bringing advanced mass spectrometry technologies to biomedical research, the results of which are described in 179 publications and that recently includes uncovering protein biomarkers for delayed recovery from kidney transplants and for the early detection of ovarian cancer.

proteomes

Editorial

Editorial for Special Issue: Neuroproteomics

Kenneth R. Williams [1,2,]* and Angus C. Nairn [1,3,]*

1 Yale/NIDA Neuroproteomics Center, New Haven, CT 06511, USA
2 Molecular Biophysics and Biochemistry, Yale University School of Medicine, New Haven, CT 06511, USA
3 Department of Psychiatry, Yale School of Medicine, Connecticut Mental Health Center, New Haven, CT 06511, USA
* Correspondence: kenneth.williams@yale.edu (K.R.W.); angus.nairn@yale.edu (A.C.N.)

Received: 24 May 2019; Accepted: 27 May 2019; Published: 31 May 2019

Recent advances in mass spectrometry (MS) instrumentation [1,2], especially in MS resolution and scan rate enable the quantitation of expression of more than 15,000 proteins (>12,000 genes) from mammalian tissue samples [3,4]. These advances have opened the door to the proteome and already are having an impact that extends from biology to clinical proteomics. With no theoretical limits in sight—with regard to further improvements in MS instrumentation and improved peptide identification algorithms and bioinformatics—the future of MS-based, quantitative proteomics is incredibly promising and exciting. Indeed, new chemical labeling technologies that incorporate multiple isobaric tags now enable concurrent analyses of up to 11 different samples using commercially available reagents [5]. While these methods are beginning to be applied to neuroproteomics, the central nervous system (CNS) poses unique challenges to quantitative proteomics that begin with the immense level of cellular and sub-cellular heterogeneity. The human CNS has ~100 billion neurons, each with 10,000 to 100,000 synaptic connections; and even larger numbers of glial cells. Moreover, there is a large variety in cell morphology with individual neurons typically being intermingled in close contact with several different types of neurons and with axonal projections from an individual neuron often projecting over relatively long distances. Given that it is now clear that each of the ~500–1000 individual types of nerve cells exhibit distinct patterns of gene expression [6,7], it is critically important to develop and publish the technologies and methodologies needed to enable quantitative MS/proteomic analyses of specific neuronal cell types and their organelles. This topic is reviewed by Wilson and Nairn [8], and Wang and Savas [9], who highlight that cell-type-specific analysis has become a major focus for many neuroscience investigators. While the whole brain or large regions of brain tissue can be used for proteomic analysis, the useful data that can be gathered is limited because of cellular and sub-cellular heterogeneity. Analysis of mixed populations of distinct cell types not only limits our understanding of where a particular protein expression change might have occurred, it also minimizes our ability to detect significant changes in protein expression and/or modification levels due to issues related to dilution effects and low signal to high noise. Moreover, isolation of specific cell types can be challenging due to their nonuniformity and complex projections to different brain regions. In addition, many analytical techniques used for protein detection and quantitation remain insensitive to the low amounts of protein extracted from specific cell populations. Despite these challenges, methods to improve the proteomic yield and increase resolution continue to develop at a rapid rate.

The review by Wang and Savas [9], and the article by Roy et al. [10], show that proteomic heterogeneity in the brain extends beyond the cell type to synaptic and postsynaptic density (PSD) proteomes, respectively. Different types of synapses in the brain have highly specialized neuronal cell-cell junctions, with both common and distinct functional features that arise from their individual synaptic protein compositions. Even a single neuron can have several different types of synapses that each contain hundreds or even thousands of different proteins. While MS/proteomic analyses provide a powerful approach for characterizing different types of synapses and to potentially identify disease-causing alterations in synaptic proteomes, the value of most synaptic proteomic analyses that have been published are also limited by the molecular averaging of proteins from the multiple types of neurons and synapses that often have been analyzed together. In their review, Wang and Savas [9] summarize a wide range of currently available technologies for analyzing neuron-type specific and synapse-type specific proteomes and discuss strengths and limitations of each of these technologies for successfully addressing the "averaging problem".

The study by Roy et al. [10] was designed to determine if the synaptic proteome differs across anatomically distinct brain regions. Postsynaptic protein extracts were isolated from seven forebrain and hindbrain regions in mice and their compositions were determined using MS/proteomics. Across these regions 74% of proteins showed differential expression with each region having a distinctive composition. These compositions correlated with the anatomical regions of the brain and their embryological origins. Proteins in biochemical pathways controlling plasticity and disease, protein interaction networks, and individual proteins involved with cognition all showed differential regional expression. In toto, the Roy et al. [10] study showed that interconnected regions have characteristic proteome signatures and that diversity in synaptic proteome composition is an important feature of mouse and human brain structure.

Both Wilson and Nairn [8], and Wang and Savas [9], described the use of in situ proximity labeling methods to identify protein-protein interactions within discrete cellular compartments. As an example of the use of this technology, the Cijsouw et al. [11] article describes the use of this approach to map the proteome of the synaptic cleft, which is the space between two neurons at a nerve synapse. Cijsouw et al. [11] used a peroxidase-mediated proximity labeling approach with the excitatory-specific synaptic cell adhesion protein SynCAM 1 fused to horseradish peroxidase (HRP) as a reporter in cultured cortical neurons. This reporter marked excitatory synapses, as detected by confocal microcopy, and was localized in the edge zone of the synaptic cleft, as determined using 3D dSTORM super-resolution imaging. Proximity labeling with a membrane-impermeant biotin-phenol compound limited labeling to the cell surface, and label-free quantitation (LFQ) MS combined with ratiometric HRP tagging of membrane vs. synaptic surface proteins was used to determine the protein composition of excitatory clefts. Novel cleft proteins were identified and one of these, Receptor-type tyrosine-protein phosphatase zeta, was independently validated using immunostaining. The Cijsouw et al. [11] study supports the use of peroxidase-mediated proximity labeling for quantifying changes in the synaptic cleft proteome that may occur in diseases such as psychiatric disorders and addiction.

The ability of targeted mass spectrometry technologies to quantify the same proteins in multiple samples with the highest possible sensitivity, quantification precision, and accuracy [12] makes these technologies ideal for analyzing the small amounts of protein that result from the use of fluorescence-activated cell sorting (FACS), laser capture microdissection (LCM), and other technologies described by Wilson and Nairn [8] and Wang and Savas [9] to analyze single cell types and region-specific synaptic proteomes. In regard to the latter, there is increasing interest especially in understanding the functions of proteins in the PSD because of their potential involvement in a wide variety of neuropsychiatric disorders including autism spectrum disorder (ASD) [13–15] and schizophrenia [16]. As described in the Wilson et al. [17] article, the PSD is an electron-dense region located just beneath the postsynaptic membrane of excitatory glutamatergic synapses, which is involved in a wide range of cellular and signaling processes in neurons. Biochemical fractionation combined with MS/proteomics analyses has enabled cataloging of the PSD proteome. However, since the PSD composition may change rapidly in response to stimuli, robust and reproducible technologies are needed to quantify changes in PSD protein abundance. Using a data-independent acquisition (DIA) approach on PSD fractions isolated from mouse cortical brain tissue and a pre-determined spectral library, Wilson et al. [17] quantified over 2,100 proteins. In addition, Wilson et al. [17] designed a targeted, parallel reaction monitoring (PRM) assay with heavy-labeled, synthetic internal peptide standards to rigorously quantify 50 PSD proteins. Wilson et al. [17] suggest that the PSD/PRM assay is particularly appropriate for validating differentially expressed proteins identified by the DIA assay.

Despite the challenges in carrying out quantitative MS/proteomics analyses on neural tissues, sufficient progress has been made that neuroproteomics is increasingly being used to improve diagnosis and staging, and to help develop better treatments for a broad range of neurological diseases. With the number of Americans with Alzheimer's disease (AD) expected to increase from an estimated 5 million in 2014 to nearly 14 million in 2060 [18] and with the costs of treating this disease expected to increase from $190 billion in 2019 to between $379 and $500 billion annually in 2040 [19]; there is considerable interest in finding more sensitive and specific diagnostic tools for this devastating disease that is now the 5th leading cause of death among adults aged 65 years or older [20]. As described in the review article by Carlyle et al. [21], neurodegenerative dementias like AD are highly complex diseases. While most can be diagnosed by pathological analyses of the postmortem brain, clinical disease symptoms often involve overlapping cognitive, behavioral, and functional impairments that pose diagnostic challenges in living patients. As global demographics shift towards an aging population, especially in developed countries, clinicians need more sensitive and specific assays that can be carried out on readily available bodily fluids, such as sera or plasma to diagnose, monitor, and treat neurodegenerative diseases. The Carlyle et al. [21] review provides an overview of how contemporary MS/proteomic and state of the art capture-based technologies can contribute to the discovery of improved biofluid biomarkers for neurodegenerative diseases, and the limitations of these technologies. The Carlyle et al. [21] review also discusses technical considerations and data processing approaches for achieving accurate and reproducible findings and reporting requirements to help improve our ability to compare data from different laboratories.

As reviewed in the Lutz and Peng [22] article, characteristic features of AD include protein aggregates such as amyloid beta plaques and tau neurofibrillary tangles in the patient's brain. Determining the complete composition and structure of the protein aggregates in AD can increase our understanding of the underlying mechanisms of AD development and progression. The Lutz and Peng [22] review summarizes the use of LCM—which was also reviewed in the Wilson and Nairn [8], and Wang and Savas [9] articles—and the differential extraction approaches needed to achieve deep profiling of the aggregated proteomes in AD samples, and discusses the resulting novel insights from these analyses that may contribute to AD pathogenesis.

A number of articles in this Special Issue are focused on addictive diseases. To grasp the importance of this area of research one has only to glance at data in the Surgeon General's Report [23] for 2015 that states that 66.7 million people in the U.S. reported binge drinking in the past month and 27.1 million people were current users of illicit drugs or misused prescription drugs. While the accumulated costs of addiction to the individual, family, and the community are staggering, with the economic burden of prescription opioid misuse alone in the U.S. amounting to $78.5 billion annually [24], the most devastating consequences are the tens of thousands of fatalities each year as a result of substance abuse. In this regard, alcohol misuse contributes to 88,000 deaths annually in the U.S. In addition, in 2014 there were 47,055 drug overdose deaths, including 28,647 people who died from an opioid overdose—more than in any previous year. As reviewed by Natividad et al. [25], drug addiction is a complex disease caused by abnormally regulated molecular signaling across several brain reward regions. Due to our incomplete understanding of the molecular pathways that underlie addiction, there currently are only a few treatment options. Recent research suggests that addiction results from the overall impact of many small changes in molecular signaling networks that include neuropeptides (neuropeptidome), protein-protein interactions (interactome), and protein post-translational modifications (PTMs) such as protein phosphorylation (phosphoproteome). Advances in MS/proteomics instrumentation and technologies are increasingly able to identify the molecular changes that occur in the reward regions of the addicted brain and to translate these findings into new treatments. In their review Natividad et al. [25] provide an overview of MS/proteomics approaches for addressing critical questions in addiction neuroscience and they highlight recent innovative studies that demonstrate how analyses of the neuroproteome can increase our understanding of the molecular mechanisms that underlie drug addiction.

As discussed by Pena et al. [26], the treatment of chronic pain has been challenging as the most effective treatment that uses opiates has many unwanted side effects. For example, treatment with morphine quickly leads to μ opioid receptor (MOR) desensitization and the development of morphine tolerance. MOR activation by the peptide agonist, [D-Ala2, N-MePhe4, Gly-ol]-enkephalin (DAMGO), leads to G protein receptor kinase activation, β-arrestin recruitment, and subsequent receptor endocytosis, which does not occur with morphine. However, MOR activation by morphine induces receptor desensitization in a protein kinase C (PKC)-dependent manner. While PKC inhibitors decrease receptor desensitization, reduce opiate tolerance, and increase analgesia; the mechanism of action of PKC in these processes is not well understood. The challenges in establishing a role for PKC result, in part, from the inability to identify PKC targets. To meet this challenge Pena et al. [26] generated a conformation state-specific anti-PKC antibody that preferentially recognizes the active state of this kinase. Using this antibody to isolate PKC substrates and MS/proteomics to identify the resulting proteins, Pena et al. [26] determined the effect of morphine treatment on PKC targets. They found that morphine strengthens the interactions of several proteins with active PKC. Pena et al. [26] describe the role of these proteins in PKC-mediated MOR desensitization and analgesia, and they propose a role for some of these proteins in mediating pain by tropomyosin receptor kinase A (TrKA) activation. Finally, Pena et al. [26] discuss how these PKC interacting proteins and pathways might be targeted for more effective pain treatment.

As described by Mervosh et al. [27], there is increasing interest in the role that neuroimmune interactions play in the development of psychiatric illness, including addiction. This raises the possibility that targeting neuroimmune signaling pathways may be a viable treatment for substance use disorders. Calipari et al. [28] recently determined that granulocyte-colony stimulating factor (G-CSF), which is a cytokine, is up-regulated following chronic cocaine use [11]. Peripheral injections of G-CSF potentiated the development of locomotor sensitization, conditioned place preference, and self-administration of cocaine, and blocking G-CSF function in the mesolimbic dopamine system abrogated the formation of conditioned place preference. Despite these effects on behavior and neurophysiology, the molecular mechanisms by which G-CSF brings about these changes in brain function are unclear. In the Mervosh et al. [27] study, mice were treated with repeated injections of G-CSF, cocaine, or both, and changes in protein expression in the ventral tegmental area (VTA) were examined using 10-plex tandem mass tag (TMT) labeling coupled with LC-MS/MS analyses. Repeated G-CSF treatment resulted in differential expression of 475 proteins in multiple synaptic plasticity and neuronal morphology signaling pathways. While there was significant overlap in the proteins that were differentially expressed in each of the three treatment groups, injections of cocaine and the combination of cocaine and G-CSF also resulted in subsets of differentially expressed proteins that were unique to each treatment group. This study identified proteins and pathways that were differentially regulated by G-CSF in an important limbic brain region and will help guide further study of G-CSF function and its evaluation as a possible therapeutic target for the treatment of drug addiction.

As summarized by Natividad et al. [25], MS/phosphoproteomics has provided addiction researchers with a useful tool for measuring changes in activated states that may be devoid of changes in the corresponding protein levels. The phosphorylation of serine, threonine and tyrosine residues is one of the most common post-translational modifications (PTMs) that can act as a molecular switch and modulate a wide range of biological activity including signal transduction, cell differentiation/proliferation, protein-protein and protein-gene interactions, and subcellular localization. Natividad et al. [25] note that many hypotheses invoke differential protein phosphorylation to control the activities of key regulators of gene transcription (e.g., the cAMP response element-binding protein, delta fosB), membrane receptors (e.g., GluA1) and other important binding partners (e.g., transmembrane α-amino-3-hydroxy-5-methyl-4-isoxazolepropionic acid (AMPA) receptor regulatory proteins as summarized by Park [29]) that modulate neuroplasticity. Indeed, there are several hundred eukaryotic kinases and phosphatases that have a broad range of substrate targets [30]. Since a substantial component of receptor-mediated neuronal signaling involves modulation of the activities of kinases and phosphatases, large-scale phosphoproteome profiling is a key technology that can provide unique information into the roles of protein phosphorylation in addiction.

As summarized by Park [29], strengthening and weakening of synaptic transmission (i.e., synaptic plasticity) provides a critical mechanism for many brain functions including learning, memory, and drug addiction. Long-term potentiation (LTP) and depression (LTD) are well-characterized models of synaptic plasticity that can be regulated by changes at presynaptic (e.g., changes in the release of neurotransmitters) and postsynaptic (e.g., changes in the number and properties of neurotransmitter receptors) sites. As shown in cellular models of synaptic plasticity, changes in the post-synaptic activity of the AMPA receptor (AMPAR) complex mediates these phenomena. In particular, Park [29] notes that protein phosphorylation plays a key role in controlling synaptic plasticity, for example, Ca^{2+}/CaM-dependent protein kinase II (CaMKII) in hippocampal LTP. The Park [29] review summarizes studies on phosphorylation of the AMPAR pore-forming subunits and auxiliary proteins including transmembrane AMPA receptor regulatory proteins (TARPs) and discusses its role in synaptic plasticity.

Just as protein phosphorylation plays a key role in the molecular mechanisms underlying drug addiction, the articles by Bertholomey et al. [31] and Miller et al. [32] indicate that this PTM also plays an important role in alcohol use disorders (AUDS) and nicotine addiction, respectively. Bertholomey et al. [31] describe how early life stress is associated with an increased risk of developing AUDs. Although the neurobiological mechanisms underlying this effect are not well understood, abnormal glucocorticoid and noradrenergic system functioning may play a role. Bertholomey et al. [31] studied the impact of chronic exposure during adolescence to elevated levels of the glucocorticoid stress hormone corticosterone (CORT) on amygdalar function and on the risk of developing AUDS. Adolescent CORT exposure increased alcohol, but not sucrose self-administration, and enhanced stress-induced reinstatement with yohimbine in adulthood. LFQ phosphoproteomic analyses revealed that adolescent CORT exposure resulted in 16 changes in protein phosphorylation in the amygdala, which provided a list of potential novel mechanisms involved in increasing the risk of alcohol drinking. Of particular interest, Bertholomey et al. [31] found that adolescent CORT exposure resulted in increased phosphorylation of the α_{2A} adrenergic receptor (α_{2A}AR) mediated by G protein-coupled receptor kinase 2 (GRK2). Bertholomey et al. [31] also found that intra-amygdala infusion of a peptidergic GRK2 inhibitor reduced alcohol seeking, suggesting that GRK2 may provide a novel target for treating stress-induced AUDS.

As described by Miller et al [32], high-affinity nicotinic acetylcholine receptors containing α4 and β2 subunits (α4/β2* nAChRs, where * denotes other, potentially unidentified subunits) are essential for the rewarding and reinforcing properties of nicotine. α4/β2* nAChRs are ion channel-containing proteins that flux positive ions, including calcium, in response to nicotine or the endogenous neurotransmitter acetylcholine. Activation of α4/β2* nAChRs in the mammalian brain results in the depolarization of neurons on which they are expressed, leading to changes in intracellular signaling, such as the activation of calcium-dependent kinases. Interactions have previously been identified between α4/β2* nAChRs and calcium/calmodulin-dependent protein kinase II (CaMKII) in mouse and human brains [33,34]. Following co-expression of α4/β2 nAChR subunits with CaMKII in human embryonic kidney (HEK) cells, MS/proteomic analyses described by Miller et al. [32] identified eight phosphorylation sites in the α4 subunit. One of these sites and an additional site were identified when α4/β2* nAChRs were dephosphorylated and then incubated with CaMKII in vitro, while three phosphorylation sites were identified following incubation with protein kinase A (PKA) in vitro. Miller et al. [32] then isolated native α4/β2* nAChRs from mouse brain following acute or chronic exposure to nicotine. Two CaMKII sites identified in HEK cells were phosphorylated, and one PKA site was dephosphorylated following acute nicotine administration in vivo, whereas phosphorylation of the PKA site was increased back to baseline levels following repeated nicotine exposure. Although significant changes in β2 nAChR subunit phosphorylation were not observed under these conditions, two novel sites were identified on this subunit, one in HEK cells and one in vitro.

As described in the Watkins et al. [35] article, reversible protein phosphorylation that modulates neuronal signaling, communication, and synaptic plasticity is controlled by competing kinase and phosphatase activities. Glutamatergic projections from the cortex and dopaminergic projections from the substantia nigra or ventral tegmental area synapse on dendritic spines of specific gamma-aminobutyric acid (GABA)ergic medium spiny neurons (MSNs) in the striatum. Direct pathway MSNs (dMSNs) are positively coupled to PKA signaling and the activation of these neurons enhance specific motor programs, whereas indirect pathway MSNs (iMSNs) are negatively coupled to PKA and inhibit competing motor programs. Psychostimulant drugs increase dopamine signaling and cause an imbalance in the activities of these two programs. While changes in specific kinases, such as PKA, regulate different effects in the two MSN populations, alterations in the specific activity of serine/threonine phosphatases, such as protein phosphatase 1 (PP1), are less well understood. This lack of knowledge partly results from unknown, cell-specific changes in PP1 targeting proteins. Spinophilin is the major PP1-targeting protein in striatal postsynaptic densities. Using MS/proteomics and immunoblotting together with a transgenic mouse expressing hemagglutinin (HA)-tagged spinophilin in dMSNs or iMSNs, Watkins et al. [35] identified novel spinophilin interactions modulated by amphetamine in the different striatal cell types. These results increase our understanding of cell type-specific, phosphatase-dependent signaling pathways that are altered by the use of psychostimulants.

As described by Luxmi et al. [36], identification of enkephalins as endogenous ligands for opioid receptors led to the identification of hundreds of additional bioactive peptides in the nervous systems of species as diverse as *Drosophila* and *Hydra*. The precursors to these neuropeptides have N-terminal signal sequences with multiple potential paired basic amino acid endoproteolytic cleavage sites. Genomic and transcriptomic data from a diverse array of organisms indicated that neuropeptide precursors were present in species lacking neurons or endocrine cells. The enzymes involved in converting neuropeptide precursors into bioactive peptides are highly conserved. The identification of catalytically active peptidylglycine α-amidating monooxygenase (PAM) in *Chlamydomonas reinhardtii*, a unicellular green alga, suggested the presence of a PAM-like gene and peptidergic signaling in the last eukaryotic common ancestor (LECA). Luxmi et al. [36] identified prototypical neuropeptide precursors and essential peptide processing enzymes in the *C. reinhardtii* genome. Positing that sexual reproduction by *C. reinhardtii* requires communication between cells, they used MS to identify proteins in the soluble secretome of mating gametes, and searched for evidence that the putative peptidergic processing enzymes were functional. After fractionation by SDS-PAGE, they identified intact signal peptide-containing proteins as well as those that had been cleaved. The *C. reinhardtii* mating secretome contained multiple matrix metalloproteinases, cysteine endopeptidases, and serine carboxypeptidases, along with one subtilisin-like proteinase. Transcriptomic studies suggest these proteases are involved in sexual reproduction. Multiple extracellular matrix proteins (ECM) were identified in the secretome. Several pherophorins and ECM glycoproteins were present, with most containing typical peptide processing sites, and many had been cleaved, generating stable N- or C-terminal fragments. The Luxmi et al. [36] study suggests that subtilisin endoproteases and matrix metalloproteinases similar to those involved in vertebrate peptidergic and growth factor signaling play an important role in stage transitions during the life cycle of *C. reinhardtii*. Moreover, this study [36] further suggests that endoproteolytic activation of proneuropeptides and growth factors originated in unicellular organisms. The complex endomembrane system in LECA presumably gave rise to the evolution of the preproneuropeptides and growth factors essential for nervous system development and function well before the appearance of neurons.

Despite its low prevalence in the U.S. of ~0.25% [37], schizophrenia (SZ) results in significant health, social, and economic concerns and is one of the 15 leading causes of disability worldwide [38]. Individuals with SZ have an increased risk of premature death with the estimated potential life lost for SZ patients in the U.S. being 28.5 years [39]. As described in the Sowers et al. [40] article, male mice lacking fibroblast growth factor 14 (FGF14) (i.e., *Fgf14*$^{-/-}$) recapitulate key features of SZ, including loss of parvalbumin-positive GABAergic interneurons in the hippocampus, disrupted gamma frequency, and reduced working memory. FGF14 is one of the intracellular FGF proteins that are involved in neuronal ion channel regulation and synaptic transmission. As the molecular basis of SZ and its sex-specific onset are not well understood, the *Fgf14*$^{-/-}$ model may provide a valuable tool to interrogate pathways related to SZ disease mechanisms. Sowers et al. [40] performed LFQ MS to identify enriched pathways in both male and female hippocampi from *Fgf14*$^{+/+}$ and *Fgf14*$^{-/-}$ mice. They found that all of the differentially expressed proteins in *Fgf14*$^{-/-}$ animals, relative to their same-sex wild type counterparts, are associated with SZ, based on genome-wide association data. In addition, differentially expressed proteins were predominantly sex-specific, with male *Fgf14*$^{-/-}$ mice having increased expression of proteins in pathways associated with neuropsychiatric disorders. The Sowers et al. [40] article increases our understanding of the role of FGF14, confirms that the *Fgf14*$^{-/-}$ mouse provides a valuable and experimentally accessible model for studying the molecular basis and gender-specificity of SZ, and also highlights the importance of sex-specific biomedical research.

The articles in the Neuroproteomics Special Issue provide an overview of the unique challenges that must be addressed to carry out meaningful MS/proteomics analyses on neural tissues and the tools and technologies that are available to meet these challenges. The several articles that cover Alzheimer's disease, addiction, and schizophrenia illustrate how MS/proteomics technologies can be used to help improve our ability to diagnose and understand the molecular basis for neurological diseases. We believe that several of the articles in this Special Issue will be of interest to investigators beyond the field of neurological disorders. In particular, the review by Carlyle et al. [21], "Proteomic Approaches for the Discovery of Biofluid Biomarkers of Neurodegenerative Dementias", may be of interest to investigators searching for blood and cerebrospinal fluid (CSF) biomarkers for virtually any disease. Similarly, the review by Natividad et al. [25], "From Synapse to Function, A Perspective on the Role of Neuroproteomics in Elucidating Mechanisms of Drug Addiction", provides a general overview of the utility of MS/proteomics approaches for addressing critical questions in addiction neuroscience that should be equally applicable to investigators involved in virtually any area of biomedical research. Likewise, the article by Wilson et al. [17], "Development of Targeted Mass Spectrometry-Based Approaches for Quantitation of Proteins Enriched in the Postsynaptic Density", may be useful for any investigator who wishes to design and validate DIA and/or PRM assays for virtually any proteins. Finally, the peroxidase-mediated proximity labeling technology described in the article by Cijsouw et al. [11], "Mapping the Proteome of the Synaptic Cleft through Proximity Labeling Reveals New Cleft Proteins", may be of interest to investigators interested in mapping many other spatially restricted proteomes.

Author Contributions: The initial draft of this manuscript was written by K.R.W., which was then edited by A.C.N.; K.R.W. and A.C.N. had equal responsibility for overseeing the selection and review of the 16 articles in this Special Issue.

Funding: This work was supported by the NIH/NIDA grant DA018343 that supports the Yale/NIDA Neuroproteomics Center.

Conflicts of Interest: The authors declare no conflict of interest.

References

1. Aebersold, R.; Mann, M. Mass-spectrometric exploration of proteome structure and function. *Nature* **2016**, *537*, 347–355. [CrossRef]
2. Zhang, Y.; Fonslow, B.R.; Shan, B.; Baek, M.C.; Yates, J.R., 3rd. Protein analysis by shotgun/bottom-up proteomics. *Chem. Rev.* **2013**, *113*, 2343–2394. [CrossRef]
3. Mertins, P.; Mani, D.R.; Ruggles, K.V.; Gillette, M.A.; Clauser, K.R.; Wang, P.; Wang, X.; Qiao, J.W.; Cao, S.; Petralia, F.; et al. Proteogenomics connects somatic mutations to signalling in breast cancer. *Nature* **2016**, *534*, 55–62. [CrossRef]

4. Stewart, E.; McEvoy, J.; Wang, H.; Chen, X.; Honnell, V.; Ocarz, M.; Gordon, B.; Dapper, J.; Blankenship, K.; Yang, Y.L.; et al. Identification of therapeutic targets in rhabdomyosarcoma through integrated genomic, epigenomic, and proteomic analyses. *Cancer Cell* **2018**, *34*, 411–426. [CrossRef]
5. Rauniyar, N.; Yates, J.R., 3rd. Isobaric labeling-based relative quantification in shotgun proteomics. *J. Proteome Res.* **2014**, *13*, 5293–5309. [CrossRef]
6. Zeisel, A.; Hochgerner, H.; Lonnerberg, P.; Johnsson, A.; Memic, F.; van der Zwan, J.; Haring, M.; Braun, E.; Borm, L.E.; La Manno, G.; et al. Molecular architecture of the mouse nervous system. *Cell* **2018**, *174*, 999–1014. [CrossRef] [PubMed]
7. Saunders, A.; Macosko, E.Z.; Wysoker, A.; Goldman, M.; Krienen, F.M.; de Rivera, H.; Bien, E.; Baum, M.; Bortolin, L.; Wang, S.; et al. Molecular diversity and specializations among the cells of the adult mouse brain. *Cell* **2018**, *174*, 1015–1030. [CrossRef]
8. Wilson, R.S.; Nairn, A.C. Cell-type-specific proteomics: A neuroscience perspective. *Proteomes* **2018**, *6*, 51. [CrossRef] [PubMed]
9. Wang, Y.Z.; Savas, J.N. Uncovering discrete synaptic proteomes to understand neurological disorders. *Proteomes* **2018**, *6*, 30. [CrossRef] [PubMed]
10. Roy, M.; Sorokina, O.; McLean, C.; Tapia-Gonzalez, S.; DeFelipe, J.; Armstrong, J.D.; Grant, S.G.N. Regional diversity in the postsynaptic proteome of the mouse brain. *Proteomes* **2018**, *6*, 31. [CrossRef]
11. Cijsouw, T.; Ramsey, A.M.; Lam, T.T.; Carbone, B.E.; Blanpied, T.A.; Biederer, T. Mapping the proteome of the synaptic cleft through proximity labeling reveals new cleft proteins. *Proteomes* **2018**, *6*, 48. [CrossRef]
12. Gillette, M.A.; Carr, S.A. Quantitative analysis of peptides and proteins in biomedicine by targeted mass spectrometry. *Nat. Methods* **2013**, *10*, 28–34. [CrossRef]
13. Peca, J.; Feliciano, C.; Ting, J.T.; Wang, W.; Wells, M.F.; Venkatraman, T.N.; Lascola, C.D.; Fu, Z.; Feng, G. Shank3 mutant mice display autistic-like behaviours and striatal dysfunction. *Nature* **2011**, *472*, 437–442. [CrossRef]
14. Dhamne, S.C.; Silverman, J.L.; Super, C.E.; Lammers, S.H.T.; Hameed, M.Q.; Modi, M.E.; Copping, N.A.; Pride, M.C.; Smith, D.G.; Rotenberg, A.; et al. Replicable in vivo physiological and behavioral phenotypes of the shank3b null mutant mouse model of autism. *Mol. Autism* **2017**, *8*, 26. [CrossRef]
15. Peixoto, R.T.; Wang, W.; Croney, D.M.; Kozorovitskiy, Y.; Sabatini, B.L. Early hyperactivity and precocious maturation of corticostriatal circuits in shank3b(-/-) mice. *Nat. Neurosci.* **2016**, *19*, 716–724. [CrossRef] [PubMed]
16. Fernandez, E.; Collins, M.O.; Uren, R.T.; Kopanitsa, M.V.; Komiyama, N.H.; Croning, M.D.R.; Zografos, L.; Armstrong, J.D.; Choudhary, J.S.; Grant, S.G.N. Targeted tandem affinity purification of psd-95 recovers core postsynaptic complexes and schizophrenia susceptibility proteins. *Mol. Syst. Biol.* **2009**, *5*, 269. [CrossRef] [PubMed]
17. Wilson, R.S.; Rauniyar, N.; Sakaue, F.; Lam, T.T.; Williams, K.R.; Nairn, A.C. Development of targeted mass spectrometry-based approaches for quantitation of proteins enriched in the postsynaptic density (psd). *Proteomes* **2019**, *7*, 12. [CrossRef]
18. Matthews, K.A.; Xu, W.; Gaglioti, A.H.; Holt, J.B.; Croft, J.B.; Mack, D.; McGuire, L.C. Racial and ethnic estimates of alzheimer's disease and related dementias in the united states (2015–2060) in adults aged ≥ 65 years. *Alzheimers Dement.* **2019**, *15*, 17–24. [CrossRef]
19. Hurd, M.D.; Martorell, P.; Delavande, A.; Mullen, K.J.; Langa, K.M. Monetary costs of dementia in the united states. *New Eng. J. Med.* **2013**, *368*, 1326–1334. [CrossRef]
20. Heron, M. Deaths: Leading causes for 2010. *Natl. Vital Stat. Rep.* **2013**, *62*, 1–96.
21. Carlyle, B.C.; Trombetta, B.A.; Arnold, S.E. Proteomic approaches for the discovery of biofluid biomarkers of neurodegenerative dementias. *Proteomes* **2018**, *6*, 32. [CrossRef] [PubMed]
22. Lutz, B.M.; Peng, J. Deep profiling of the aggregated proteome in alzheimer's disease: From pathology to disease mechanisms. *Proteomes* **2018**, *6*, 46. [CrossRef]
23. United States Department of Health and Human Services. *Facing Addiction in America: The Surgeon General's Report on Alcohol, DRUGS and Health*; Department of Health & Human Services: Washington, DC, USA, 2016; p. 1.
24. Florence, C.S.; Zhou, C.; Luo, F.; Xu, L. The economic burden of prescription opioid overdose, abuse, and dependence in the united states, 2013. *Med. Care* **2016**, *54*, 901–906. [CrossRef]

25. Natividad, L.A.; Buczynski, M.W.; McClatchy, D.B.; Yates, J.R., 3rd. From synapse to function: A perspective on the role of neuroproteomics in elucidating mechanisms of drug addiction. *Proteomes* **2018**, *6*, 50. [CrossRef] [PubMed]
26. Pena, D.A.; Duarte, M.L.; Pramio, D.T.; Devi, L.A.; Schechtman, D. Exploring morphine-triggered pkc-targets and their interaction with signaling pathways leading to pain via trka. *Proteomes* **2018**, *6*, 39. [CrossRef]
27. Mervosh, N.L.; Wilson, R.; Rauniyar, N.; Hofford, R.S.; Kutlu, M.G.; Calipari, E.S.; Lam, T.T.; Kiraly, D.D. Granulocyte-colony-stimulating factor alters the proteomic landscape of the ventral tegmental area. *Proteomes* **2018**, *6*, 35. [CrossRef] [PubMed]
28. Calipari, E.S.; Godino, A.; Peck, E.G.; Salery, M.; Mervosh, N.L.; Landry, J.A.; Russo, S.J.; Hurd, Y.L.; Nestler, E.J.; Kiraly, D.D. Granulocyte-colony stimulating factor controls neural and behavioral plasticity in response to cocaine. *Nat. Commun.* **2018**, *9*, 9. [CrossRef] [PubMed]
29. Park, J. Phosphorylation of the ampar-tarp complex in synaptic plasticity. *Proteomes* **2018**, *6*, 40. [CrossRef] [PubMed]
30. Ardito, F.; Giuliani, M.; Perrone, D.; Troiano, G.; Lo Muzio, L. The crucial role of protein phosphorylation in cell signaling and its use as targeted therapy (review). *Int. J. Mol. Med.* **2017**, *40*, 271–280. [CrossRef]
31. Bertholomey, M.L.; Stone, K.; Lam, T.T.; Bang, S.; Wu, W.; Nairn, A.C.; Taylor, J.R.; Torregrossa, M.M. Phosphoproteomic analysis of the amygdala response to adolescent glucocorticoid exposure reveals g-protein coupled receptor kinase 2 as a target for reducing motivation for alcohol. *Proteomes* **2018**, *6*, 41. [CrossRef] [PubMed]
32. Miller, M.B.; Wilson, R.S.; Lam, T.T.; Nairn, A.C.; Picciotto, M.R. Evaluation of the phosphoproteome of mouse alpha 4/beta 2-containing nicotinic acetylcholine receptors in vitro and in vivo. *Proteomes* **2018**, *6*, 42. [CrossRef]
33. McClure-Begley, T.D.; Stone, K.L.; Marks, M.J.; Grady, S.R.; Colangelo, C.M.; Lindstrom, J.M.; Picciotto, M.R. Exploring the nicotinic acetylcholine receptor-associated proteome with itraq and transgenic mice. *Genom. Proteom. Bioinform.* **2013**, *11*, 207–218. [CrossRef]
34. McClure-Begley, T.D.; Esterlis, I.; Stone, K.L.; Lam, T.T.; Grady, S.R.; Colangelo, C.M.; Lindstrom, J.M.; Marks, M.J.; Picciotto, M.R. Evaluation of the nicotinic acetylcholine receptor-associated proteome at baseline and following nicotine exposure in human and mouse cortex. *eNeuro* **2016**, *3*. [CrossRef]
35. Watkins, D.S.; True, J.D.; Mosley, A.L.; Baucum, A.J., 2nd. Proteomic analysis of the spinophilin interactome in rodent striatum following psychostimulant sensitization. *Proteomes* **2018**, *6*, 53. [CrossRef]
36. Luxmi, R.; Blaby-Haas, C.; Kumar, D.; Rauniyar, N.; King, S.M.; Mains, R.E.; Eipper, B.A. Proteases shape the chlamydomonas secretome: Comparison to classical neuropeptide processing machinery. *Proteomes* **2018**, *6*, 53. [CrossRef]
37. Desai, P.R.; Lawson, K.A.; Barner, J.C.; Rascati, K.L. Estimating the direct and indirect costs forcommunity-dwelling patients with schizophrenia. *J. Pharm. Health Serv. Res.* **2013**, *4*, 187–194. [CrossRef]
38. Vos, T.; Abajobir, A.; Abate, K.; Abbafati, C.; Abbas, K.M.; Abd-Allah, F.; Abdulkader, R.S.; Abdulle, A.M.; Abebo, T.A.; Abera, S.F.; et al. Global, regional, and national incidence, prevalence, and years lived with disability for 328 diseases and injuries for 195 countries, 1990-2016: A systematic analysis for the global burden of disease study 2016. *Lancet* **2017**, *390*, 1211–1259. [CrossRef]
39. Olfson, M.; Gerhard, T.; Huang, C.; Crystal, S.; Stroup, T.S. Premature mortality among adults with schizophrenia in the united states. *JAMA Psychiatry* **2015**, *72*, 1172–1181. [CrossRef]
40. Sowers, M.L.; Re, J.D.; Wadsworth, P.A.; Shavkunov, A.S.; Lichti, C.; Zhang, K.; Laezza, F. Sex-specific proteomic changes induced by genetic deletion of fibroblast growth factor 14 (fgf14), a regulator of neuronal ion channels. *Proteomes* **2019**, *7*, 5. [CrossRef]

proteomes

Review

Cell-Type-Specific Proteomics: A Neuroscience Perspective

Rashaun S. Wilson [1] and Angus C. Nairn [1,2,*]

[1] Yale/NIDA Neuroproteomics Center, 300 George St., New Haven, CT 06511, USA; rashaun.wilson@yale.edu
[2] Department of Psychiatry, Yale School of Medicine, Connecticut Mental Health Center, New Haven, CT 06511, USA
* Correspondence: angus.nairn@yale.edu; Tel.: +1-203-974-7725

Received: 13 November 2018; Accepted: 5 December 2018; Published: 9 December 2018

Abstract: Cell-type-specific analysis has become a major focus for many investigators in the field of neuroscience, particularly because of the large number of different cell populations found in brain tissue that play roles in a variety of developmental and behavioral disorders. However, isolation of these specific cell types can be challenging due to their nonuniformity and complex projections to different brain regions. Moreover, many analytical techniques used for protein detection and quantitation remain insensitive to the low amounts of protein extracted from specific cell populations. Despite these challenges, methods to improve proteomic yield and increase resolution continue to develop at a rapid rate. In this review, we highlight the importance of cell-type-specific proteomics in neuroscience and the technical difficulties associated. Furthermore, current progress and technological advancements in cell-type-specific proteomics research are discussed with an emphasis in neuroscience.

Keywords: cell type; neuroscience; proteomics; mass spectrometry; neuron; proximity labeling; affinity chromatography; neuroproteomics; biotinylation

1. Introduction

Novel methods for proteomic analysis of biological tissues have developed rapidly in the past decade; however, neuroproteomics remains a challenging field of study. The mammalian central nervous system (CNS) is far different from any other organ in the mammalian system, primarily because it is made up of several hundred different cell types [1]. Each cell type has unique characteristics, and distinct populations of cells are present in different brain regions. For instance, although 40% of all cells in the brain are astrocytes, neurons outnumber astrocytes in the cerebellum, whereas there is an inverse correlation in the cortex [2]. Furthermore, Herculano-Houzel et al. [3] determined that almost 70% of the two billion neurons found in the adult rat brain are located in the cerebellum, and five-fold less are present in the cortex. Brain cells also possess region-specific identities and biomarkers that have proven useful in cell-type-specific studies but can also complicate analyses [4,5]. In addition, neural cells lack uniformity and make projections to different brain regions, resulting in spatiotemporal regulation of many signaling processes within the brain. Consequently, these factors make separation and isolation of specific cell types from brain challenging.

A second issue is that proteomic analysis of brain cells has lagged behind in comparison to its transcriptomic counterpart, which continues to make rapid advances. The facile method of RNA amplification has enabled over 500 single-cell transcript expression analyses [6]. In a few years, the field has moved from the use of quantitative reverse transcription-polymerase chain reaction (qRT-PCR) to quantify globin gene expression in human erythroleukemic cells [7] or measure expression levels of five genes in single cells isolated from mouse pancreatic islets [8], to methods with greater scope and scale. For example, RNA sequencing (RNA-seq) methods have been used to successfully analyze gene expression in single cells [9–14]. One study classified 3005 cells in the mouse cortex and hippocampal CA1 region using single-cell RNA-seq, revealing 47 subclasses from nine known cell types [13]. A later report used single-nuclei RNA-seq to identify 16 neuronal subtypes from 3227 single-neuron datasets isolated from six different regions of the postmortem human brain [14]. Recently, a study successfully profiled gene expression in 4347 single cells from mutant human oligodendrogliomas [10]. Variations of the RNA-seq method have been developed to enable more high-throughput, comprehensive analyses [15,16], including a recent study that profiled over 400,000 single-cell transcriptomes from more than 800 mouse cell types using a method termed Microwell-seq [15]. This rapid, cost-effective method uses an agarose microwell system for single-cell isolation and barcoded magnetic beads for mRNA capture. Drop-seq uses a similar concept but isolates and lyses single cells in nanoliter droplets of liquid prior to barcode labeling [16–19]. This method enabled isolation and characterization of over 44,000 transcriptomes from mouse retinal cells, which were ultimately grouped into 39 different cell types [16]. Drop-seq has also been used to analyze RNA expression levels in 690,000 cells from 9 different adult mouse brain regions [18]. Though comprehensive transcriptomic analyses have proven useful in the characterization of specific cell types, these methods do not account for differential control of protein synthesis and degradation. Therefore, mRNA expression often does not correlate with protein abundance and may not be reliably used as a predictive tool for proteomics [20].

Large-scale proteomic studies use mass spectrometry, an approach that continues to improve in terms of accuracy and sensitivity [21–24]. However, one major difference between transcriptomic and proteomic profiling is that protein abundance cannot be amplified in the same way that nucleic acids can. Therefore, the protein quantity isolated from a cell population must be above the threshold of detection for mass spectrometry analysis. While highly abundant proteins can be analyzed by mass spectrometry at the single cell level (see below), the protein yields obtained from a single cell are often below the levels necessary for reliable quantitation and therefore do not allow the depth of coverage observed in transcriptomic analyses. Moreover, past and current cell isolation techniques are often inefficient and collect small quantities of cells in a given experiment, which in turn results in low protein yields. Specific to neurons and other CNS cells, due to their non-uniformity of size and subcellular organization, many of the current separation techniques are incapable of retaining cellular structure, often resulting in leakage of cellular contents or loss of cell integrity entirely. Furthermore, protein/peptide loss can occur during sample preparation, either through peptide adsorption to sample tubes and/or during transfer of sample to and from multiple tubes [25,26]. Mass spectrometry analysis itself can also influence the number of proteins identified, which can often be attributed to ionization efficiency and instrument sensitivity [26].

Overcoming the challenges facing cell-type-specific proteomics is of critical importance, as many types of psychiatric, developmental, and neurodegenerative disorders are associated with specific cell types in the brain. Drug addiction is one of these psychiatric disorders in which specific neuronal cell types are implicated. For instance, the psychostimulant, cocaine, regulates the reuptake of the neurotransmitter, dopamine, leading to aberrant signaling in specific sub-types of striatal medium spiny neuron (MSN) in the dorsal and ventral striatum [27]. While morphologically similar, MSNs can be separated into at least two large subtypes that differentially express D1- or D2-classes of dopamine receptors that are in turn differentially coupled to either increased or decreased cAMP signaling, respectively [28,29]. Thus, exposure to cocaine results in opposite patterns of phosphorylation of important intracellular targets such as DARPP-32 in intermixed sub-populations of MSNs [29]. Biochemical analysis of striatum, in the absence of separation of different MSN cell types, leads to an averaging of the increased or decreased signals, and a loss of important information.

In addition to drug addiction, neurodegenerative disorders like Alzheimer's disease (AD) and Down syndrome (DS) are associated with specific cell types in the brain [30,31]. For instance, pathology of both AD and DS patients involves overproduction of amyloid beta peptide, and the development of neurofibrillary tangles and amyloid plaques. Astrocytes, which are a type of glial brain cell, also play active roles in pathogenesis of AD brain tissue [5,32]. In mice overexpressing amyloid beta, plaques are surrounded by reactive astrocytes and activated microglia [33,34]. Furthermore, brain inflammation caused by glial and microglial activation is observed in brain tissue of AD patients [33,35,36].

Other cell-type-associated disorders include Parkinson's disease (PD), Amyotrophic lateral sclerosis (ALS), and Huntington's disease (HD). In PD subjects, pathology within the *substantia nigra* revealed a loss of a sub-population of dopaminergic neurons, followed by an increase in Lewy body structures within the retained neurons [5,37,38]. The subsequent DA depletion causes cell-specific effects such as hyper- and hypoactivation of D2 and D1 MSNs, respectively [39–41]. Astrocytes are also implicated in PD in many animal-based studies [5]. ALS is a degenerative disease that affects the motor cortex, brain stem, and spinal cord and ultimately results in motor neuron death [5,42,43]. Patients with HD exhibit a preferential loss of D2 MSNs, and an accumulation of the mutant form of Huntingtin (HTT) protein occurs in human neurons and astrocytes [5,44,45].

It is clear from the ongoing list of disorders that a greater focus needs to be placed on biochemical characterization of neural cell types. Though many technologies have advanced in recent years to address the issues of cell separation and isolation as well as increasing the depth of proteomic coverage for cell-type-specific analyses, there are still many aspects that need to be improved. This review will outline the different methods available, while also noting the benefits and limitations of each. Studies which have employed these techniques will also be highlighted, and potential improvements for these methods will be discussed.

2. Cell-Type-Specific Isolation Methods

The nonuniformity and complex networks of different cell populations within the brain often require the use of cell-type-specific markers to improve the accuracy of isolation. This can be accomplished through promoter-directed expression of a reporter protein either through viral transduction (transient) or generation of a transgenic animal (stable). While viral transduction can be useful for some experimental applications (See Proteome labeling methods), expression levels may be variable when compared to transgenic animals, which may ultimately affect proteomic analyses. Though generation of transgenic animals can be time- and resource-intensive, many groups have now successfully developed transgenic tools for characterization of brain cell types [46,47]. One of these tools was developed by taking advantage of a bacterial artificial chromosome (BAC) to express a green fluorescent protein (GFP) marker in specific neural cell types [46]. The same BAC approach was used to generate Ribo-tagged transgenic mice expressing an enhanced green fluorescence protein (EGFP)-L10a ribosomal protein under the control of cell-type-specific promoters [47]. Along with cell-type-specific visualization, this design has the added advantage of enabling translating ribosome affinity purification (TRAP) to isolate ribosomes from target cell types. Emergence of these tools coupled to cell isolation techniques is useful for proteomic analysis of CNS cell types.

One frequently-used method to isolate specific cell types is fluorescence-activated cell sorting (FACS) (Figure 1A), which relies on a fluorescent cellular marker that can be endogenously-expressed or immunolabeled for detection. In an early study, 5000–10,000 striatal MSNs were isolated via FACS from fluorescently-labeled neurons expressing EGFP under the *Drd1*, *Drd2*, or *Chrm4* promoter (BAC transgenic mice) [48]. FACS of tissue from transgenic mice expressing GFP under the control of the parvalbumin-expressing interneuron (*Pvalb*) promoter was later used to isolate approximately 5000 and 10,000 GFP-positive nuclei from striatal and hippocampal tissue, respectively [49]. Nuclei from different sub-populations of MSNs were also subjected to FACS after acute or chronic cocaine treatment to observe cell-type-specific differential post-translational modification of histones [50]. FACS has also been used for glutamatergic synaptosomal enrichment by expressing fluorescent VGLUT1 protein in mice, which resulted in identification of 163 enriched proteins after mass spectrometry analysis [51]. Recently, FACS and subsequent LC-MS/MS was performed on sensory inner ear hair cells, enabling identification of 6333 proteins [52].

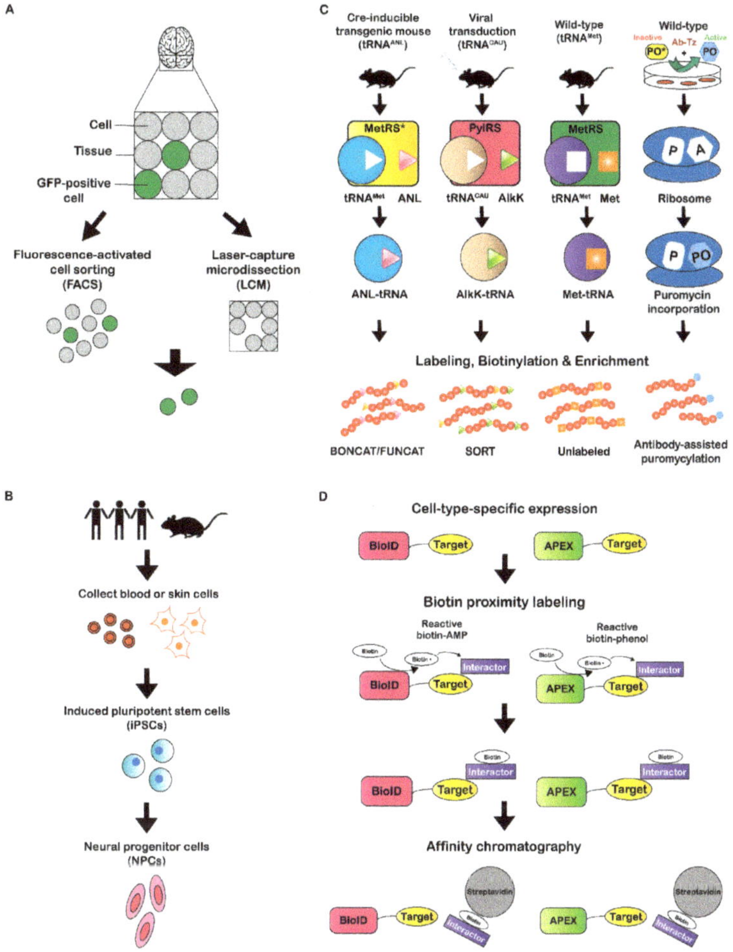

Figure 1. Methods for cell-type-specific isolation and proteome enrichment. (**A**) Two methods for specific cell isolation from a total cell population. Animal models can be generated that express fluorescent markers in a cell type of interest. Fluorescent cells can be detected and isolated using fluorescence-activated cell sorting (FACS) or laser capture microdissection (LCM). FACS requires homogenization of tissue prior to cell sorting, while LCM enables cells isolation from intact tissue slices. (**B**) Basic workflow of induced pluripotent stem cell (iPSC) differentiation. Skin or blood cells are collected from a biological organism of interest and used to generate induced pluripotent stem cells (iPSCs). Factors are then added to iPSCs for differentiation into neural progenitor cells (NPCs). (**C**) Cell-type-specific labeling methods enable stochastic incorporation of a non-canonical amino acid or puromycin into the target proteome. The cell-type-specific expression of a tRNA synthetase is accomplished either by genetic engineering of a Cre-dependent transgenic mouse (BONCAT/FUNCAT) or via viral transduction (SORT). The incorporated amino acid can be further biotinylated for enrichment prior to LC-MS/MS analysis (BONCAT/SORT) or modified with a fluorescent probe for visualization (FUNCAT). Puromycin labeling occurs through introduction of a cell-type-specific enzyme-tagged antibody (Ab-Tz) followed by an inactive puromycin analog. Activation of puromycin occurs after Tz reacts with the inactive puromycin analog. (**D**) Experimental workflow for BioID and APEX proximity labeling techniques. BioID or APEX fusion target proteins are expressed in a specific cell type. Reactive biotin is supplemented, and target interacting proteins are biotinylated via BioID or APEX. Biotinylated interactors can be enriched using affinity chromatography techniques with a stationary phase such as streptavidin prior to LC-MS/MS.

An alternative single-cell isolation method is termed laser capture microdissection (LCM) (Figure 1A), which uses a microscope equipped with a high-precision laser to dissect small areas within a tissue slice (>100 μm^2). Imaging and dissection can be performed in fluorescence or bright-field modes, enabling a variety of experimental applications. For instance, Drummund et al. [53] performed LCM on neurons isolated from formalin-fixed, paraffin-embedded (FFPE) AD cortical brain tissue, which yielded more than 400 proteins identified by LC-MS/MS analysis. In this study, extensive sample treatment optimization was also performed on tissue isolated via LCM from the temporal cortex. Results from this optimization ranged from 202 to over 1700 proteins identified from approximately 4000–80,000 neurons. Another study identified 1000 proteins from tissue sections of neuromelanin granules isolated from the human *substantia nigra* [54]. Furthermore, mass spectrometry analysis of four different compartments in FFPE fetal human brain tissue identified a total of 3041 proteins [55]. Two recent reports isolated cells from human post-mortem tissue using LCM to identify a small number of potential biomarkers from AD [56] and ischemic stroke [57] patients via mass spectrometry. LCM was also recently used to quantify approximately 1000 proteins from 10–18 cells (100-μm-diameter) isolated from different rat brain regions [26]. For these analyses, optimization was first performed with 50 μm (2–6 cells), 100 μm (10–18 cells), and 200 μm (30–50 cells) diameter tissue sections from rat brain cortex, where 180, 695, and 1827 protein groups were identified, respectively.

While LCM clearly offers precision for a variety of experimental workflows, it does have limitations. If an endogenously-expressed fluorescent protein is used as a cell-type-specific marker in the tissue of interest, it must be expressed at an intensity above the threshold of detection for the microscope to accurately dissect. Furthermore, most LCM microscopes are not capable of cooling the tissue specimen during dissection. Therefore, the user must work rapidly to prevent altered protein expression and/or degradation, particularly when using fresh tissue. Moreover, dissection of the tissue can be more tedious and time-consuming than many other isolation methods, which could result in a lower number of cells (and protein) isolated in a given amount of time. Finally, if the tissue must be immunolabeled, the antibody is often processed with the rest of the cellular protein extract. This could ultimately affect proteomic results depending on the amount of antibody used. Despite these potential issues, LCM is clearly a powerful method that can be useful for many types of cell-type-specific applications.

Although animal models are useful for investigative research in neuroscience, results and treatments do not always translate to the human system. It is difficult to obtain brain tissue from human subjects, particularly over a range of development with age-matched controls and within a post-mortem interval short enough to avoid protein degradation and variations in post-translational modifications (PTMs) [58–61]. In an effort to address these challenges, researchers have turned to developing specific neuron cell types from induced pluripotent stem cells (iPSCs) (Figure 1B) [62,63]. A major benefit of using iPSCs is that they can be produced from human somatic cells such as dermal fibroblasts (HDF) instead of embryonic stem cells, which have ethical conflicts associated. Furthermore, these iPSCs can be directly reprogrammed to differentiate into virtually any cell type with patient- or disease-specificity [62]. Many studies have already demonstrated successful production of a variety of region-specific neuronal cell types including ventral forebrain cholinergic, ventral midbrain dopaminergic, cortical glutamatergic, and cholinergic motor neurons [64–68]. Recently, iPSCs have undergone proteomic characterization for numerous experimental applications [69–74]. For instance, Yamana et al. [69] compared lysates of iPSCs and fibroblast cells to identify a total of 9510 proteins via mass spectrometry analysis. A later study used quantitative mass spectrometry to identify 2217 total proteins in spinal muscular atrophy (SMA) patient-derived and healthy control motor neurons differentiated from iPSCs [73]. A comparison of the two groups indicated that 63 and 30 proteins were up-regulated in control and SMA motor neurons, respectively. Recently, three-dimensional neuron-spheroids were derived from AD and control patient iPSCs and subjected to tandem mass tag (TMT) LC-MS/MS analysis [74], which is a quantitative mass spectrometry approach that uses reporter ions generated during MS/MS fragmentation for quantitation [75]. Collectively,

1855 proteins were identified in the 3D neuro-spheroid samples that were differentiated from a total of ten iPSC lines between both the AD and control subjects. Furthermore, 8 proteins were found to be up-regulated in AD subjects, while 13 proteins were down-regulated. Another recent study profiled the proteomes of iPSCs, neural progenitor cells (NPCs), and differentiated neurons in cell culture to identify a total of 2875 proteins among all three groups [55]. Notably, 90, 33, and 126 proteins were unique to iPSCs, NPCs, and neurons, respectively. Although differentiation of iPSCs has demonstrated significant promise for moving closer to a human model system while also improving protein yield, these analyses are still being performed in vitro. It therefore becomes difficult to maintain true neural connectivity, which could ultimately result in altered protein expression compared to what would normally be observed in the human brain. Nevertheless, this approach still has potential for a variety of neurological applications in the future.

3. Proteome Labeling Methods

Cell-type-specific proteome labeling is a technique that can be used to circumvent the issue of maintaining cellular integrity during isolation. Until recent years, proteome labeling studies were performed primarily using Stable Isotope Labeling with Amino acids in Cell culture (SILAC) [76–83]. The obvious caveat to SILAC, however, is that experiments must be performed in cell culture. A variation termed Stable Isotope Labeling with Amino acids in Mammals (SILAM) can be used for quantitation of protein expression in vivo, however, labeling times are long (~25 d) and it cannot be performed in a cell-type-specific manner. Recent efforts have attempted to make in vivo labeling methods compatible with cell-type-specific applications. One of the first studies to perform in situ proteome labeling over a short, 2 h time course, was termed BioOrthogonal Non-Canonical Amino acid Tagging (BONCAT) [84]. BONCAT takes advantage of a cell's protein synthesis machinery and enables incorporation of a noncanonical amino acid into the proteome of interest (Figure 1C). Recently, this method has transitioned to cell-type-specific labeling of proteomes through generation of transgenic mice that express a mutated methionyl-tRNA synthase (MetRS*) with an expanded amino acid binding site that recognizes the noncanonical amino acid ANL [85]. Expression of MetRS* is driven by a cell-specific promoter and enables charging of supplemented ANL onto an endogenous $tRNA^{Met}$, which is then stochastically incorporated into the target cell proteome. After labeling, click-chemistry can be performed to biotinylate ANL residues, followed by enrichment via streptavidin affinity chromatography. Mass spectrometry analysis of ANL-labeled, enriched proteins in hippocampal neurons and Purkinje cells resulted in 2384 and 1687 proteins identified, respectively [85]. Furthermore, a hippocampal proteome analysis of mice exposed to standard (SC) or enriched (EE) housing environments identified 2384 and 2365 proteins, respectively, of which 225 were significantly regulated after statistical comparison. Not only can click-chemistry be used for biotinylation, but fluorescent probes can be added to the ANL residues, which Dietrich et al. [86], termed FlUorescent Non-Canonical Amino acid Tagging (FUNCAT) (Figure 1C). This method can be used for temporal visualization of newly-synthesized proteins, while also enabling post-visualization enrichment by methods such as immunoaffinity chromatography.

A similar technique called Stochastic Orthogonal Recoding of Translation (SORT) has also recently been established to label proteomes in vivo [87,88]. Instead of requiring generation of a transgenic animal, SORT uses targeted, viral-mediated expression of an orthogonal pyrrolysyl-tRNA synthetase-$tRNA_{xxx}$ pair that recognizes and incorporates a non-canonical amino acid AlkK into the target proteome of interest (Figure 1C). Click-chemistry can then be performed in the same way as BONCAT/FUNCAT. Recently, SORT was used to label, biotinylate, and enrich proteins in mouse striatal MSNs prior to mass spectrometry analysis, which resulted in identification of 1780 cell-type specific proteins [89].

While these methods of cell-type-specific proteome labeling seem advantageous for future studies in neuroproteomics, there are still associated challenges and extensive optimization required for each experiment. For BONCAT/FUNCAT, transgenic animals must be generated and characterized, which is not only time-consuming, but costly. Furthermore, the MetRS* expression levels may vary depending on the cell-type-specific promoter used, which could result in low labeling efficiency and ultimately low protein yield for mass spectrometry analysis. Similarly, low expression levels of the pyrrolysyl-tRNA synthetase-tRNA$_{xxx}$ pair could also be observed for the SORT method for a variety of reasons including promoter selection, transduction efficiency, and accuracy of injection. Both methods also require supplementation of the non-canonical amino acid, either through drinking water intake or injection. This supplementation also needs to be optimized to ensure equivalent dosages and labeling efficiencies occur between animals. Moreover, the proteomics results from the aforementioned studies [85,89] indicate that improvements need to be made to reach a greater depth of proteomic coverage. The observed number of protein identifications is far below the known upper limit of detection (~12,000 proteins) [90,91] and could potentially be improved by a variety of factors such as increasing the number of animals used and/or selecting a promoter that labels at a level above the limit of detection for the assay but does not label proteins at a level that could interfere with cellular processes.

Another labeling approach that takes advantage of the cell's native protein synthesis machinery uses a puromycin analog tag [92–95]. The puromycin analog binds the acceptor (A) site of the ribosome and is then incorporated into the nascent polypeptide chain prior to inhibition of protein synthesis. The incorporated puromycin analog can then be chemically modified to enrich for newly synthesized proteins. This method was first demonstrated in cultured cells and mice using *O*-propargyl-puromycin (OP-puro), where newly-synthesized proteins were visualized via fluorescence microscopy after a copper(I)-catalyzed azide-alkyne cycloaddition (CuAAC) reaction with a fluorescent azide [92]. Recently, a similar technique was modified for cell-type-specific labeling of proteomes in vivo [94]. This modification involves introduction of a cell-type-specific antibody bearing a tetrazing (Tz) tag and a "caged" form of puromycin (TCO-PO), which is unable to be incorporated into the proteome. When the Tz-tagged antibody and a TCO-PO molecule come in contact, a reaction occurs which results in conjugation of TCO to the antibody, rendering the PO molecule "uncaged" and free to incorporate into the proteome of the target cell. From this study, more than 1200 proteins were identified via LC-MS/MS when this method was employed in A431 cells. An earlier study performed a similar type of experiment with cell-type-specific, viral-mediated expression of an enzyme capable of activating a "caged" puromycin analog in mouse pancreatic islets and HEK 293T cells [95]. Mass spectrometry analysis of the HEK 293T cell proteome resulted in identification of 1165 proteins enriched puromycin-incorporated, enzyme-expressing proteome.

There are several advantages to using a puromycin labeling strategy over the biorthogonal labeling methods. First, the functional concentration of puromycin is much lower than that of noncanonical amino acids, reducing the likelihood of unwanted side-effects [92,94–96]. Furthermore, unlike noncanonical amino acids, methionine does not directly compete with puromycin for incorporation into the proteome. Therefore, animals that undergo puromycin labeling do not require the low-methionine diet which may be necessary for biorthogonal labeling methods and are not subject to potential bias toward proteins with higher methionine content [92,93]. Another advantage is that puromycin incorporation may not require use of a genetically modified organism, which does not always represent a true native biological environment [94]. Moreover, puromycin incorporation displays higher temporal resolution than biorthogonal labeling, which requires charging of the non-canonical amino acid to the tRNA prior to incorporation [92–94]. Despite the advantages of in vivo puromycin incorporation, cell-type-specific variations have only been demonstrated in cultured cells to date [94,95].

Not only are specific cellular proteomes being labeling for general protein identification, but in situ proximity labeling methods have recently emerged to identify protein-protein interactors within discrete cellular compartments. In general, these methods rely on expression of a promiscuous biotin protein ligase fused to a target protein whose interacting proteins are being investigated. After biotin supplementation, the target interacting proteins are biotinylated by the ligase and can then be enriched and identified using proteomic analysis (Figure 1D). One of these methods has been termed BioID, which was originally developed by Roux et al. [97] and used to identify lamin-A (LaA) interacting proteins. In this study, an *E. coli* biotin protein ligase BirA was fused to LaA and expressed in HEK293 cells to identify 122 proteins unique to BioID-LaA via LC-MS/MS. A more recent study used the BioID method to identify interacting proteins of excitatory and inhibitory postsynaptic protein complexes [98]. Viral-mediated expression of BirA, PSD-95-BirA, or BirA-gephyrin, BirA-collybistin, and BirA-InSyn1 was performed in mouse brain tissue prior to enrichment of biotinylated proteins and subsequent mass spectrometry analysis. For the PSD analysis, PSD-95-BirA interacting proteins were compared to those of the BirA control. In total, 2183 proteins were identified, 121 of which were enriched at least two-fold in PSD-95-BirA samples compared to the BirA control. For the inhibitory protein complexes, gephyrin-, collybistin-, and InSyn1-BirA interacting proteins were compared to those of the BirA control. Mass spectrometry analysis of the samples identified 2533 total proteins with a combined 181 proteins significantly enriched in the three target interactomes compared to the BirA control. More recently, BioID2 was developed, which is a similar method that employs a smaller promiscuous biotin ligase [99]. This improved method has several advantages to traditional BioID, including increased selectivity of targeting fusion proteins, a reduced amount of biotin required, and enhanced labeling of proximal proteins. TurboID is a similar approach developed recently that takes advantage of a different mutated form of biotin ligase, which is capable of proximity labeling within 10 min [100]. In this study, TurboID displayed a significantly higher biotin labeling efficiency and a similar proteome coverage of subcellular compartments within HEK293T cells after quantitative LC-MS/MS when compared to BioID.

A second method termed APEX (short for Enhanced APX) uses an engineered ascorbate peroxidase fusion protein for biotin labeling of target interacting proteins. This method was first demonstrated in HEK293 cells, where APEX was targeted to the mitochondrial matrix, and biotinylated interacting proteins were enriched and subjected to LC-MS/MS [101]. In total, 495 proteins were identified in the mitochondrial matrix proteome. Recently, APEX was used in *C. elegans* to identify tissue-specific and subcellular-localized proteomes [102]. APEX was targeted to the nucleus or cytoplasm of intestine, epidermis, body wall muscle, or pharyngeal muscle tissues, from which 3180 interacting proteins were collectively identified. A separate study used APEX to identify spatiotemporal interacting proteins of the delta opioid receptor (DOR) in HEK cells [103]. This study observed changes in DOR interactions over an activation time course of 1–30 min as well as different subcellular compartments, including the plasma membrane (PM) and endosome (Endo). Recently, a modified APEX strategy was used to map proteins at excitatory and inhibitory synaptic clefts of rat cortical neurons, resulting in identification of 199 and 42 proteins, respectively [104].

Like the other labeling techniques, extensive optimization of these proximity labeling assays is required for optimal performance. Moreover, the amount of starting material needed for adequate protein enrichment for LC-MS/MS analysis is substantial and not feasible for small amounts of tissue or certain cell types. Furthermore, standardization and reproducibility of labeling methods becomes difficult since protein output is often not provided (See Table A1) and can vary between organisms. Though these proximity labeling methods are similar in practice, APEX labeling times are much faster (~1 min) compared to the 24 h labeling time of the BioID method, which could significantly impact proteomics results. Notably, however, APEX has limited stability in heated or reducing environments compared to BioID, and the presence of H_2O_2 in the cell can lead to toxicity. Nevertheless, APEX does have great appeal, particularly for those interested in rapid proteomic changes such as altered subcellular localization or metabolic regulation.

4. Mass Spectrometry Methods

One of the major challenges in workflows related to cell-type-specific proteomics is loss of protein during sample handling, which occurs at various steps between isolation of the single or multiple cell and peptide injection onto the mass spectrometer. Furthermore, enzymatic cleavage is necessary to generate peptides for bottom-up proteomics, but this can result in partial or incomplete digestion depending on the amino acid composition of the protein. Peptides generated from poor cleavage are often too large for ionization and detection via LC-MS/MS, ultimately resulting in loss of information for these specific regions of the protein. Instrument issues also include sensitivity and accuracy as well as chromatographic and spectral reproducibility between sample runs.

Efforts to overcome some of these issues have utilized alternative workflows in an attempt to obtain cell-type level proteome or metabolome analysis (Figure 2). One such method termed mass spectrometry imaging (MSI) can analyze tissue sections with high spatial resolution to determine relative abundances and distribution of proteins [105–111]. Of the MS ionization sources available, matrix-assisted laser desorption/ionization (MALDI) and secondary ion mass spectrometry (SIMS) microprobes are most commonly used for imaging mass spectrometry due to their softer, non-destructive qualities that enable ionization of intact biomolecules at micro- and nanometer resolutions, respectively [105,112,113]. MALDI uses a laser light for desorption and ionization of the sample, and SIMS uses a more focused, accelerated primary ion beam to ionize analytes from the surface of cells. Furthermore, MALDI is particularly useful for detecting higher molecular weight species (2–70 kDa), while SIMS offers detection of molecules below 1 kDa or 2000 *m/z* [112,114–116].

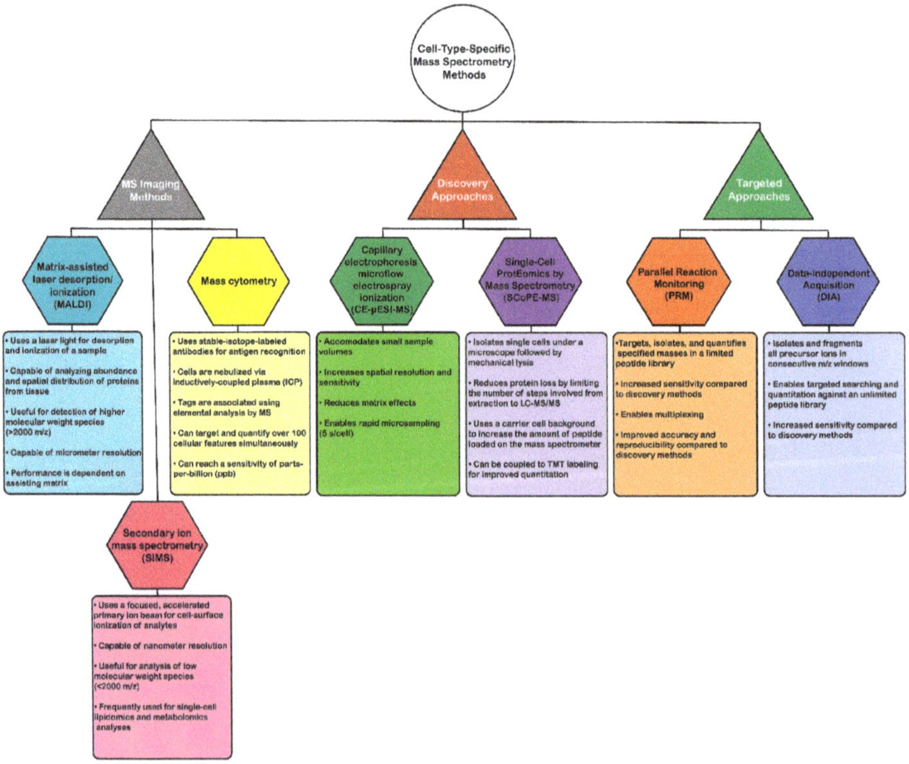

Figure 2. Overview of common mass spectrometry-based methods that are currently used for cell-type-specific analyses. Tree includes method type (triangles), name (hexagon) and a list of features associated with each method (rectangle).

These methods have been used for a range of experimental cell-type-specific applications [106,107,117–121]. For instance, MALDI-MSI was performed in mouse pituitary gland samples at a spatial resolution of 5 μm to identify ten neuropeptides at up to 2500 *m/z* [117].

An earlier study identified proteins in over 82 mass ranges in different mouse brain regions as well as 150 proteins in human glioblastoma tissue using MALDI-MSI [107]. One of the most recent MALDI-MSI applications demonstrated proteomic profiling of over 1000 rat dorsal root ganglia cells, which were classified into three separate groups on a peptide and lipid data basis [118]. SIMS has also been used for identification of single-cell metabolites, however, the majority of these studies focus on lipidomic analyses [120,121]. One study also used both SIMS and MALDI-MSI approaches to investigate the biomolecular and spatial composition of rat spinal cord tissue [116].

Mass cytometry is another type of MSI method that uses inductively coupled plasma (ICP) as an ionization source. This method is viewed as a targeted approach to MSI and uses metal-conjugated antibodies to enable antigen localization within the tissue or cell of interest, ultimately improving the limits of detection for target proteins. This multiplexing method enables quantitation of 100 target features, simultaneously without spectral overlap [122–124]. Bandura et al. [122] developed a 20-antigen targeted mass cytometry expression assay using lanthanide-tagged antibodies. This assay was then used to label cell lines from human leukemia patients (monoblastic M5 AML and monocytic M5 AML) and model cell lines (KG1a and Ramos) and subsequently map the isotope tag intensity profiles for an average of 15,000–20,000 cells [122]. A later report used bone marrow aspirates from a total of 46 leukemia and healthy patients to quantify 20 target biomarkers via mass cytometry [125]. Recently, tissue preparation techniques were compared for mass cytometry analysis of single-cell suspensions of human glioma, melanoma, and tonsil tissues [124]. A variation on this method was later developed, termed multiplexed ion beam imaging (MIBI), which images metal isotope-labeled antibodies using SIMS [123]. This method is also capable of imaging up to 100 features simultaneously at a parts-per-billion (ppb) sensitivity and is compatible with fixed tissue. Angelo et al. [123] used MIBI to quantify 10 biomarker targets in breast cancer biopsy tissue, which performed at the same level or better than other quantitative clinical immunohistochemistry (IHC) methods.

While there are clear advantages associated with MSI methods for single-cell proteomic and metabolic analyses, including sensitivity and multiplexing capabilities, there are still several drawbacks to these methods. As previously mentioned, MALDI-MSI is limited to higher molecular weight species (>2 kDa), while SIMS is limited to low molecular weight species (<2 kDa). Furthermore, MALDI is only capable of micrometer resolution and performance is dependent on the assisting matrix [105,112,113,126]. Mass cytometry is limited by the number of available metal-isotope-labeled antibodies and the specificity of the antibodies to the target antigen(s). Despite the possible disadvantages, advances in these mass spectrometry techniques have enormous potential to significantly improve the quality of data obtained from cell-type-specific proteomic analyses.

5. Future Perspectives

Cell-type-specific proteomics has undoubtedly made considerable progress in recent years, particularly in the field of neuroscience. Not only have cell isolation methods improved, but the instrumentation used for proteomic analysis has significantly advanced regarding sensitivity and reproducibility. Based on many of the neural cell-type-specific datasets available, however, the average number of proteins identified continues to fall far below the acceptable threshold of previous neural proteomics reports (Table A1) [90,91]. As discussed, there are several possible reasons for the discrepancy in protein identifications found in brain tissue versus single-cell datasets. One is the lack of organism- and tissue-specific standardization to determine the threshold of cellular material necessary for adequate proteomic analysis. As displayed in Table A1, the number of proteins identified in each of the listed techniques varies drastically between studies. Moreover, many of the results listed are lacking experimental information that is necessary for reproduction. For instance, several reports provide the number of cells and/or tissue quantity isolated but do not include the amount of protein extracted from this material or injected onto the mass spectrometer. This calls attention to the benefit of better standardization methods for cell-type-specific proteomics, in order to improve overall reproducibility and quality of datasets. Furthermore, method development for cell-type-specific proteomics in neuroscience needs to continue with increased focus placed on factors such as improving the efficiencies of cell isolation methods and reducing protein loss during sample preparation.

Recent efforts have also been made to improve these issues in the context of FACS for proteomic analysis. For instance, Zhu et al. [25] identified an average of 670 protein groups from single HeLa cells after integrating FACS and a novel method called nanoPOTS (nano-droplet processing in one-pot for trace samples). After cells are sorted via FACS, the nanoPOTS method relies on robotic liquid handling to perform sample processing in nanoliter volumes to help minimize sample loss. In this study, FACS was noted to have several advantages in a single-cell proteomic workflow such as precise cell counting and enabling removal of unwanted background contamination through cell dilution in PBS [25].

In addition to FACS-based approaches, development of mass spectrometry-based methods that combine different analytical features have made considerable progress in the advancement of single-cell proteomics. Capillary electrophoresis (CE) is one feature that has been recently coupled to mass spectrometry methods for single-cell analysis [127–136]. Benefits of using CE for single-cell analyses include small sample volume accommodation, increased spatial resolution and sensitivity, and reduced matrix effects [131,137–139]. One group recently coupled CE to microflow electrospray ionization mass spectrometry (CE-μESI-MS) to identify metabolites in different cell types of South African clawed frog (*Xenopus laevis*) embryos in three consecutive studies [129–131]. In the first of these studies, CE-μESI-MS was used to compare metabolites in three different *Xenopus* blastomere cell types dissected from the dorsal-ventral and animal-vegetal regions of the 16-cell embryo [130]. In total, 40 metabolites were significantly altered among the three cell types, indicating both specificity and metabolic interconnection. A year later, this group used a similar method to identify 55 unique small molecules in left and right D1 cells isolated from 8-cell *Xenopus* embryos [131]. After multivariate and statistical analyses, an equal number of five metabolites were found to be significantly enriched in the left and right D1 cells. Recently, this group was able to use CE-μESI-MS for direct analysis of live *Xenopus* embryo cells [129]. In this study, approximately 230 different molecular features were identified during mass spectrometry analysis of dorsal and ventral 8–32-cell-embryos. Not only has this group identified metabolites using CE-μESI-MS, but they have also performed proteomic analyses. In one report, they identified a total of 438 proteins from 16 ng of protein digest from a single blastomere of a *Xenopus* 16-cell embryo [132]. In the same year, they also reported identification of a total of 1709 protein groups from 20 ng of *Xenopus* protein digest from three cell types of the 16-cell embryo [133]. In addition to electrophoresis, capillaries have recently been used for microsampling of biomolecules from single neurons [140]. This study integrated this technique with downstream ESI-IMS-MS, which had only previously been performed in human carcinoma cells [141] and *Arabidopsis thaliana* epidermal cells [142]. Another study developed a neuron-in-capillary method to culture and isolate single *Aplysia californica* bag cell neurons prior to LC-MS/MS analysis [143].

Recently, a mass spectrometry-based approach called Single Cell ProtEomics by Mass Spectrometry (SCoPE-MS) was developed to address two of the major challenges facing cell-type-specific proteomic analysis: minimizing protein loss that can occur from protein extraction to mass spectrometry analysis and improving quantitation of low-abundant peptides identified from single cells [144]. To achieve these goals, live single mouse embryonic stem cells were isolated under a microscope prior to mechanical lysis and protein extraction. Next, single-cell protein was added to that of carrier cells to further reduce sample loss and increase the amount of protein injected on the mass spectrometer. To improve quantitation, tryptic peptides were then subjected to TMT labeling prior to LC-MS/MS, which resulted in quantitation of over 1000 proteins.

Despite the many advantages discovery mass spectrometry has to offer, more quantitative MS approaches have become increasingly popular in recent years. Targeted methods such as parallel reaction monitoring (PRM) and data-independent acquisition (DIA) have emerged in recent years in efforts to improve sensitive, accurate, and reproducible peptide quantitation. Though PRM is limited by the number of peptides that can be quantified in a given assay, it enables multiplexing, which can result in quantitation of multiple peptides in a single run for a more high-throughput analysis. Recently, Wan et al. [145] used PRM to quantify phosphorylation of PINK1 substrates in human and mouse cortical neurons. Data-independent acquisition (DIA) is not as sensitive as PRM, however, it has a much greater assay capacity. For instance, DIA analysis of fractionated mouse hippocampal neurons resulted in identification of 4558 proteins among all fractions [146]. A similar method to DIA was recently reported termed "BoxCar" which enabled identification of more than 10,000 proteins from mouse brain tissue [147]. Finally, label-based quantitation is another method that is becoming increasingly popular for neuroproteomic analyses. Recently, 11,840 protein groups were identified across two brain regions of control, AD, PD, and AD/PD human patients using TMT 10-plex labeling [148]. While these and other results mentioned above using LCM together with fixed tissue or MALDI-MSI are encouraging, there is a need for systematic and comprehensive cell-type-specific LC-MS-MS analyses in human tissue.

Targeted mass spectrometry is also useful for quantitation of protein isoforms, which can have cell-type- and tissue-specific expression profiles. Since the majority of isoform sequences are highly conserved, they can only be distinguished by isoform-specific peptides, which are often lower in abundance than peptides within the conserved regions. If these specific peptides are not detected via discovery LC-MS/MS, the isoforms cannot be distinguished and are consequently grouped by the mass spectrometry search software. This ultimately results in loss of isoform-specific expression profiles. Using a more sensitive targeted approach drastically improves the probability that isoform-specific peptides will be detected and quantifiable. Depending on the protein sequence, however, it may not be possible to identify specific peptides for all isoforms using the targeted mass spectrometry approach. One of the remaining ways to elucidate isoform-specific expression patterns is through mRNA sequencing. mRNA is alternatively spliced prior to protein translation and is therefore a blueprint for the protein sequence. By integrating the mRNA and protein datasets, a more complete picture of the proteome can be generated. Tools to achieve this type of data integration have already been developed, and continue to improve, which could prove useful for future cell-type-specific analyses [149,150].

In summary, there is an overwhelming demand for comprehensive and consistent cell-type-specific data in neuroscience, and novel techniques have been evolving rapidly in attempts to fill this gap. This review has outlined methods and technical challenges present in this area of research as well as potential improvements for these analyses. Collectively, these methods are making substantial progress to increase the sensitivity, reproducibility and depth of proteome coverage necessary for future cell-type-specific studies.

Author Contributions: All authors had equal contribution in writing and preparing the manuscript.

Funding: We acknowledge support from the NIH (Yale/NIDA Neuroproteomics Center DA018343; DA040454; MH106934; MH16488).

Acknowledgments: Support was also obtained from the State of Connecticut, Department of Mental Health and Addiction Services.

Conflicts of Interest: The authors declare no conflict of interest.

Appendix A

Table A1. List of cell-type-specific methods for isolation, enrichment, and detection of proteins. The advantages and disadvantages of each technique are listed in Columns 1–3. References (Ref.) which have demonstrated the corresponding technique for a cell-type-specific application are listed in Column 4. Columns 5–8 contain the cell source (5), isolated cell/tissue quantity (6), protein quantity used for MS analysis (7), and number of proteins identified in the MS analysis (8) for each of the listed references. N/A indicates that information was not provided in the reference text.

Technique	Advantages	Disadvantages	Ref.	Cell Source	# Cells/Tissue Quantity Isolated	Protein Quantity for MS Analysis	# Proteins Identified from MS Analysis
Fluorescence-activated cell sorting (FACS)	• Can purify functionally homogenous cell populations • Offers precise cell counting • Enables removal of background contamination	• Cellular integrity can be compromised • Low cell/protein yield if fluorescent expression/signal is low	[151]	Human neuronal nuclei	1 g starting tissue, >5 × 10^6 nuclei	25 µg	1755
			[52]	Mouse inner ear hair cells	199,894 cells	3 µg	6333
			[51]	Mouse glutamatergic synapses	485 synapses	8 µg	2044 total, 163 enriched
			[25]	HeLa cells	1 cell	N/A	670
Laser-capture microdissection (LCM)	• High-precision laser enables isolation of neurons (<100 µm²) • Imaging and dissection can be performed in fluorescence or • bright-field modes • Compatible with fixed tissue	• Tissue cannot be kept cold and may endure heat damage from the laser, potentially causing changes in protein expression or post-translational modifications • Limited by the number of cells that can be analyzed per tissue slice • Endogenous expression of fluorescent marker may not be adequate to visualize and dissect	[53]	Human cortical neurons from AD patients	4000–80,000 neurons	N/A	202 (4k neurons), 1773 (80k neurons)
			[54]	Human *substantia nigra*	550,000 µm² neuromelanin (NM) granules	200 ng	1000
			[57]	Human neurons and blood brain barrier (BBB) structures	2500 neurons and 4000 BBB units	N/A	365 (Neurons), 539 (BBB)
			[26]	Rat cortical cells	2–6, 10–18, and 30–50 cells	N/A	180 (2–6 cells), 695 (10–18 cells), 1827 (30–50 cells)
			[152]	Human pancreatic islets	18 islets	N/A	3219
			[55]	FFPE fetal human brain tissue	36 samples (4 compartments, 8–15 mm²/compartment)	10 µg	3041
Induced pluripotent stem cells (iPSCs)	• Resembles the human model more than commonly-used rodent models • Can be differentiated into any cell type • Less ethical challenges than embryonic cells	• All analyses are in vitro • Neural connectivity is lost	[70]	iPSCs	10^8 cells	N/A	7952
			[69]		N/A	4 µg	9510
			[72]		N/A	40 µg	673
			[73]		6 × 10^4 cells	240 µg	2217
			[74]		N/A	100 µg	1855
			[55]		2 × 10^7 cells	10 µg	2875
BioOrthogonal Non-Canonical Amino Acid Tagging (BONCAT)	• Enables in situ proteome labeling • Enables time-dependent profiling of protein synthesis • Non-canonical amino acid administration through drinking water • or via injection • Can be performed in fixed tissue	• Metabolic incorporation needs to be performed in Met-free media or animals on a low Met diet • Temporal resolution is limited by conversion of non-canonical amino acid into aminoacyl-tRNA prior to protein synthesis • Labeled peptides are poorly detected with mass spectrometry • Requires optimization of labeling efficiency	[84]	HEK293T cells	N/A	1.95–2.1 mg input	195
			[153]	HEK293T cells	N/A	N/A	138
			[85]	Excitatory hippocampal neurons, cerebellar Purkinje cells	130–200 k neurons (Purkinje)	N/A	2384 (hippocampal), 1687 (Purkinje)

Table A1. *Cont.*

Technique	Advantages	Disadvantages	Ref.	Cell Source	# Cells/Tissue Quantity Isolated	Protein Quantity for MS Analysis	# Proteins Identified from MS Analysis
Stochastic Orthogonal Recoding of Translation (SORT)	• Viral-mediated expression of a modified tRNA • (Does not require generation of a transgenic mouse) • Can be performed in fixed tissue	• Viral expression could be variable depending on the promoter used • Optimization is required to determine time-dependent expression levels of tRNA synthase and labeling efficiency	[87]	Fly germ cells	500 ovaries	7 mg	299
			[89]	Mouse striatal medium spiny neurons (MSNs)	N/A	N/A	1780
Antibody-assisted cell-type-specific puromycylation	• Does not require use of transgenic animal • Displays high temporal resolution • Functions at lower concentrations than noncanonical amino acids	• Relies on antibody specificity	[94]	A431 cells	N/A	N/A	>1200
			[95]	HEK293T cells	2×10^7 cells	N/A	1165 enriched
BioID	• Enables screening of proximal protein interactors in situ	• Time-consuming (Need to generate and characterize transgenic mice) • Extensive assay optimization required • Labeling times are slow (~24 h)	[97]	HEK293T cells	4×10^7 cells	N/A	122
			[154]	*Toxoplasma gondii* parasite	N/A	N/A	19
			[98]	Mouse cortical and hippocampal neurons	N/A	N/A	121 (ePSD), 181 (iPSD)
BioID2	• Uses a smaller biotin ligase than BioID • Enables more selective targeting of fusion proteins than BioID • Requires less biotin supplementation than BioID • Displays enhanced labeling of proximal interacting proteins than BioID	• Time-consuming (Need to generate and characterize transgenic mice) • Extensive assay optimization required • Labeling times are moderately slow (~16 h)	[99]	HEK293T cells	4×10^7 cells	100 µg	260
TurboID	• Efficient labeling time (~10 min) • Compatible with TMT labeling • Enables labeling of organelle-specific proteomes	• Can sequester endogenous biotin and cause toxicity • Long labeling times can cause toxicity	[100]	HEK293T cells	N/A	3 mg input	314 (mito), 186 (ER), 1455 (nuclear)
Engineered ascorbate peroxidase (APEX)	• Enables screening of proximal protein interactors in situ • Labeling is very rapid (~1 min) • Applicable for labeling of subcellular compartments	• Limited stability in heated or reducing environments • Generating a transgenic organism is necessary • H2O2 can cause cellular toxicity	[101]	HEK293T cells	7–8 million cells	4 mg input	495
			[155]	*Drosophila melanogaster*	N/A	N/A	389
			[102]	*C. elegans* L4 larvae	30,000 larval cells	450–500 µg input	3180
Matrix-assisted laser desorption/ionization MS imaging (MALDI-MSI)	• Enables spatial quantitation of proteins in tissue sections • Non-destructive method	• Low spatial resolution (µm) • Broad mass range (~500–100 kDa)	[107]	APP23 transgenic mouse tissue	50 µm resolution	N/A	5 Aβ peptides
			[116]	Rat spinal cord	20 µm tissue sections	N/A	27 peptides
			[117]	Mouse pituitary gland	1.5 mm × 2.5 mm tissue sections	N/A	10 neuropeptides
			[118]	Rat dorsal root ganglia	>1000 cells	N/A	26 peptides
Secondary ion mass spectrometry (SIMS)	• Enables spatial quantitation of proteins in tissue sections • Non-destructive method	• High spatial resolution (nm) • Low mass range (<1000 Da)	[121]	Benign prostatic hyperplasia (BPH), HeLa,and human cheek cells	25–30 µm diameter tissue (BPH: 180×180 µm^2, HeLa: 88×108 µm^2, cheek cells: 150×175 µm^2)	N/A	<10 biomolecule ions
			[116]	Rat spinal cord	2.3 µm spatial resolution	N/A	18 biomolecule ions
			[120]	*Aplysia californica* neurons	0.39–2.3 µm resolution	N/A	3 biomolecule ions

Table A1. *Cont.*

Technique	Advantages	Disadvantages	Ref.	Cell Source	# Cells/Tissue Quantity Isolated	Protein Quantity for MS Analysis	# Proteins Identified from MS Analysis
Mass cytometry	• Enables multiplexed targeting of 100 target features without spectral overlap	• Limited by the number and specificity of available metal-isotope-labeled antibodies	[122]	Human leukemia cells (monoblastic M5 AML, monocytic M5 AML) and model cell lines (KG1a, Ramos)	15,000–20,000 cells	N/A	20 target antigens
			[123]	Human breast tumor cells	N/A	N/A	10 target antigens
			[125]	Human bone marrow aspirates	480,000 cells	N/A	28 target antigens
			[124]	Human glioma, melanoma, and tonsil tissue cells	N/A	N/A	8 target antigens
Capillary electrophoresis microflow electrospray ionization mass spectrometry (CE-μESI-MS)	• Accommodates small sample volumes • High spatial resolution and sensitivity • Low matrix effects • Can be temperature controlled to avoid sample heating	• Extensive optimization required	[143]	*Aplysia californica* neurons	1 neuron	N/A	>300 metabolites
			[127]	*Aplysia californica* neurons	25 B1 and B2 buccal neurons	N/A	>300 metabolites
			[130]	*Xenopus laevis* 16-cell embryo	15 blastomeres	N/A	40 metabolites
			[131]	*Xenopus laevis* 8-cell embryo	D1 blastomere	N/A	55 small molecules
			[132]	*Xenopus laevis* 16-cell embryo	1 blastomere	16 ng	438
			[133]	*Xenopus laevis* 16-cell embryo	1 blastomere	20 ng	500–800
			[129]	*Xenopus laevis* 8–32-cell embryo	1 blastomere	N/A	230 molecular features
Single Cell ProtEomics by Mass Spectrometry (SCoPE-MS)	• Minimizes protein loss from protein extraction to LC-MS/MS • Quantitative MS approach (TMT labeling)	• Has been demonstrated in few organisms	[144]	Mouse embryonic stem cells	1 cell		>1000 proteins

References

1. Kitchen, R.R.; Rozowsky, J.S.; Gerstein, M.B.; Nairn, A.C. Decoding neuroproteomics: Integrating the genome, translatome and functional anatomy. *Nat. Neurosci.* **2014**, *17*, 1491–1499. [CrossRef] [PubMed]
2. Herculano-Houzel, S. The glia/neuron ratio: How it varies uniformly across brain structures and species and what that means for brain physiology and evolution. *Glia* **2014**, *62*, 1377–1391. [CrossRef] [PubMed]
3. Herculano-Houzel, S.; Lent, R. Isotropic Fractionator: A Simple, Rapid Method for the Quantification of Total Cell and Neuron Numbers in the Brain. *J. Neurosci.* **2005**, *25*, 2518–2521. [CrossRef] [PubMed]
4. Castelo-Branco, G.; Sousa, K.M.; Bryja, V.; Pinto, L.; Wagner, J.; Arenas, E. Ventral midbrain glia express region-specific transcription factors and regulate dopaminergic neurogenesis through Wnt-5a secretion. *Mol. Cell. Neurosci.* **2005**, *31*, 251–262. [CrossRef] [PubMed]
5. Crompton, L.A.; Cordero-Llana, O.; Caldwell, M.A. Astrocytes in a dish: Using pluripotent stem cells to model neurodegenerative and neurodevelopmental disorders. *Brain Pathol.* **2017**, *27*, 530–544. [CrossRef] [PubMed]
6. Angerer, P.; Simon, L.; Tritschler, S.; Wolf, F.A.; Fischer, D.; Theis, F.J. Single cells make big data: New challenges and opportunities in transcriptomics. *Curr. Opin. Syst. Biol.* **2017**, *4*, 85–91. [CrossRef]
7. Smith, R.D.; Malley, J.D.; Schechter, A.N. Quantitative analysis of globin gene induction in single human erythroleukemic cells. *Nucleic Acids Res.* **2000**, *28*, 4998–5004. [CrossRef] [PubMed]
8. Bengtsson, M.; Ståhlberg, A.; Rorsman, P.; Kubista, M. Gene expression profiling in single cells from the pancreatic islets of Langerhans reveals lognormal distribution of mRNA levels. *Genome Res.* **2005**, *15*, 1388–1392. [CrossRef]
9. Levsky, J.M.; Shenoy, S.M.; Pezo, R.C.; Singer, R.H. Single-Cell Gene Expression Profiling. *Science* **2002**, *297*, 836–840. [CrossRef]
10. Tirosh, I.; Venteicher, A.S.; Hebert, C.; Escalante, L.E.; Patel, A.P.; Yizhak, K.; Fisher, J.M.; Rodman, C.; Mount, C.; Filbin, M.G.; et al. Single-cell RNA-seq supports a developmental hierarchy in human oligodendroglioma. *Nature* **2016**, *539*, 309–313. [CrossRef]
11. Hashimshony, T.; Wagner, F.; Sher, N.; Yanai, I. CEL-Seq: Single-Cell RNA-Seq by Multiplexed Linear Amplification. *Cell Rep.* **2012**, *2*, 666–673. [CrossRef] [PubMed]
12. Ziegenhain, C.; Vieth, B.; Parekh, S.; Reinius, B.; Guillaumet-Adkins, A.; Smets, M.; Leonhardt, H.; Heyn, H.; Hellmann, I.; Enard, W. Comparative Analysis of Single-Cell RNA Sequencing Methods. *Mol. Cell* **2017**, *65*, 631–643.e4. [CrossRef] [PubMed]
13. Zeisel, A.; Muñoz-Manchado, A.B.; Codeluppi, S.; Lönnerberg, P.; La Manno, G.; Juréus, A.; Marques, S.; Munguba, H.; He, L.; Betsholtz, C.; et al. Cell types in the mouse cortex and hippocampus revealed by single-cell RNA-seq. *Science* **2015**, *347*, 1138–1142. [CrossRef]
14. Lake, B.; Shen, R.; Ronaghi, M.; Fan, J.; Wang, W.; Zhang, K. Neuronal subtypes and diverstiy revealed by single-nucleus RNA sequencing of human brain. *Science* **2016**, *35*, 1586–1590. [CrossRef]
15. Han, X.; Wang, R.; Zhou, Y.; Fei, L.; Sun, H.; Lai, S.; Saadatpour, A.; Zhou, Z.; Chen, H.; Ye, F.; et al. Mapping the Mouse Cell Atlas by Microwell-Seq. *Cell* **2018**, *172*, 1091–1097.e17. [CrossRef] [PubMed]
16. Macosko, E.Z.; Basu, A.; Satija, R.; Nemesh, J.; Shekhar, K.; Goldman, M.; Tirosh, I.; Bialas, A.R.; Kamitaki, N.; Martersteck, E.M.; et al. Highly Parallel Genome-wide Expression Profiling of Individual Cells Using Nanoliter Droplets. *Cell* **2015**, *161*, 1202–1214. [CrossRef] [PubMed]
17. Klein, A.M.; Mazutis, L.; Akartuna, I.; Tallapragada, N.; Veres, A.; Li, V.; Peshkin, L.; Weitz, D.A.; Kirschner, M.W. Droplet barcoding for single-cell transcriptomics applied to embryonic stem cells. *Cell* **2015**, *161*, 1187–1201. [CrossRef] [PubMed]
18. Saunders, A.; Macosko, E.Z.; Wysoker, A.; Goldman, M.; Krienen, F.M.; de Rivera, H.; Bien, E.; Baum, M.; Bortolin, L.; Wang, S.; et al. Molecular Diversity and Specializations among the Cells of the Adult Mouse Brain. *Cell* **2018**, *174*, 1015–1030.e16. [CrossRef] [PubMed]
19. Zeisel, A.; Hochgerner, H.; Lönnerberg, P.; Johnsson, A.; Memic, F.; van der Zwan, J.; Häring, M.; Braun, E.; Borm, L.E.; La Manno, G.; et al. Molecular Architecture of the Mouse Nervous System. *Cell* **2018**, *174*, 999–1014.e22. [CrossRef] [PubMed]
20. Liu, Y.; Beyer, A.; Aebersold, R. On the Dependency of Cellular Protein Levels on mRNA Abundance. *Cell* **2016**, *165*, 535–550. [CrossRef]

21. Vidova, V.; Spacil, Z. A review on mass spectrometry-based quantitative proteomics: Targeted and data independent acquisition. *Anal. Chim. Acta* **2017**, *964*, 7–23. [CrossRef] [PubMed]
22. Eliuk, S.; Makarov, A. Evolution of Orbitrap Mass Spectrometry Instrumentation. *Annu. Rev. Anal. Chem.* **2015**, *8*, 61–80. [CrossRef] [PubMed]
23. Sinitcyn, P.; Daniel Rudolph, J.; Cox, J. Computational Methods for Understanding Mass Spectrometry-Based Shotgun Proteomics Data. *Annu. Rev. Biomed. Data Sci.* **2018**, *1*, 207–234. [CrossRef]
24. O'Connell, J.D.; Paulo, J.A.; O'Brien, J.J.; Gygi, S.P. Proteome-Wide Evaluation of Two Common Protein Quantification Methods. *J. Proteome Res.* **2018**, *17*, 1934–1942. [CrossRef]
25. Zhu, Y.; Clair, G.; Chrisler, W.B.; Shen, Y.; Zhao, R.; Shukla, A.K.; Moore, R.J.; Misra, R.S.; Pryhuber, G.S.; Smith, R.D.; et al. Proteomic analysis of single mammalian cells enabled by microfluidic nanodroplet sample preparation and ultrasensitive nanoLC-MS. *Angew. Chem. Int. Ed.* **2018**, *57*, 1–6. [CrossRef] [PubMed]
26. Zhu, Y.; Dou, M.; Piehowski, P.D.; Liang, Y.; Wang, F.; Chu, R.K.; Chrisler, W.B.; Smith, J.N.; Schwarz, K.C.; Shen, Y.; et al. Spatially resolved proteome mapping of laser capture microdissected tissue with automated sample transfer to nanodroplets Running Title: Spatially-resolved proteomics using nanoPOTS platform. *Mol. Cell. Proteom.* **2018**. [CrossRef] [PubMed]
27. Bertran-Gonzalez, J.; Bosch, C.; Maroteaux, M.; Matamales, M.; Hervé, D.; Valjent, E.; Girault, J.-A. Opposing Patterns of Signaling Activation in Dopamine D1 and D2 Receptor-Expressing Striatal Neurons in Response to Cocaine and Haloperidol. *J. Neurosci.* **2008**, *28*, 5671–5685. [CrossRef]
28. Clark, D.; White, F.J. D1 dopamine receptor - the search for a function: A critical evaluation of the D1/D2 dopamine classification and its functional implications. *Synapse* **1997**, *1*, 347–388. [CrossRef]
29. Bateup, H.S.; Svenningsson, P.; Kuroiwa, M.; Gong, S.; Nishi, A.; Heintz, N.; Greengard, P. Cell type-specific regulation of DARPP-32 phosphorylation by psychostimulant and antipsychotic drugs. *Nat. Neurosci.* **2008**, *11*, 932–939. [CrossRef]
30. Braak, H.; Braak, E. Staging of alzheimer's disease-related neurofibrillary changes. *Neurobiol. Aging* **1995**, *16*, 271–278. [CrossRef]
31. Braak, H.; Braak, E. Neuropathological stageing of Alzheimer-related changes. *Acta Neuropathol.* **1991**, *82*, 239–259. [CrossRef] [PubMed]
32. Wyss-Coray, T.; Loike, J.D.; Brionne, T.C.; Lu, E.; Anankov, R.; Yan, F.; Silverstein, S.C.; Husemann, J. Adult mouse astrocytes degrade amyloid-β in vitro and in situ. *Nat. Med.* **2003**, *9*, 453–457. [CrossRef] [PubMed]
33. Chun, H.; Lee, C.J. Reactive astrocytes in Alzheimer's disease: A double-edged sword. *Neurosci. Res.* **2018**, *126*, 44–52. [CrossRef] [PubMed]
34. Jo, S.; Yarishkin, O.; Hwang, Y.J.; Chun, Y.E.; Park, M.; Woo, D.H.; Bae, J.Y.; Kim, T.; Lee, J.; Chun, H.; et al. GABA from reactive astrocytes impairs memory in mouse models of Alzheimer's disease. *Nat. Med.* **2014**, *20*, 886–896. [CrossRef] [PubMed]
35. Serrano-Pozo, A.; Muzikansky, A.; Gómez-Isla, T.; Growdon, J.H.; Betensky, R.A.; Frosch, M.P.; Hyman, B.T. Differential Relationships of Reactive Astrocytes and Microglia to Fibrillar Amyloid Deposits in Alzheimer Disease. *J. Neuropathol. Exp. Neurol.* **2013**, *72*, 462–471. [CrossRef]
36. Itagaki, S.; Mcgeer, P.L.; Akiyama, H.; Zhu, S.; Selkoe, D. Relationship of microglia and astrocytes to amyloid deposits of Alzheimer disease. *J. Neuroimmunol.* **1989**, *24*, 173–182. [CrossRef]
37. Spillantini, M.G.; Schmidt, M.L.; Lee, V.M.-Y.; Trojanowski, J.Q.; Jakes, R.; Goedert, M. alpha-Synuclein in Lewy bodies. *Nature* **1997**, *388*, 839–840. [CrossRef]
38. Brichta, L.; Shin, W.; Jackson-Lewis, V.; Blesa, J.; Yap, E.L.; Walker, Z.; Zhang, J.; Roussarie, J.P.; Alvarez, M.J.; Califano, A.; et al. Identification of neurodegenerative factors using translatome-regulatory network analysis. *Nat. Neurosci.* **2015**, *18*, 1325–1333. [CrossRef]
39. Zhai, S.; Tanimura, A.; Graves, S.M.; Shen, W.; Surmeier, D.J. Striatal synapses, circuits, and Parkinson's disease. *Curr. Opin. Neurobiol.* **2018**, *48*, 9–16. [CrossRef]
40. Mallet, N.; Ballion, B.; Le Moine, C.; Gonon, F. Cortical Inputs and GABA Interneurons Imbalance Projection Neurons in the Striatum of Parkinsonian Rats. *J. Neurosci.* **2006**, *26*, 3875–3884. [CrossRef]
41. Kravitz, A.V.; Freeze, B.S.; Parker, P.R.L.; Kay, K.; Thwin, M.T.; Deisseroth, K.; Kreitzer, A.C. Regulation of parkinsonian motor behaviours by optogenetic control of basal ganglia circuitry. *Nature* **2010**, *466*, 622–626. [CrossRef] [PubMed]

42. Kiernan, M.C.; Vucic, S.; Cheah, B.C.; Turner, M.R.; Eisen, A.; Hardiman, O.; Burrell, J.R.; Zoing, M.C. Amyotrophic lateral sclerosis. *Lancet* **2011**, *377*, 942–955. [CrossRef]
43. Rowland, L.P.; Shneider, N.A. Amyotrophic Lateral Sclerosis. *N. Engl. J. Med.* **2001**, *344*, 1688–1700. [CrossRef] [PubMed]
44. Faideau, M.; Kim, J.; Cormier, K.; Gilmore, R.; Welch, M.; Auregan, G.; Dufour, N.; Guillermier, M.; Brouillet, E.; Hantraye, P.; et al. In vivo expression of polyglutamine-expanded huntingtin by mouse striatal astrocytes impairs glutamate transport: A correlation with Huntington's disease subjects. *Hum. Mol. Genet.* **2010**, *19*, 3053–3067. [CrossRef] [PubMed]
45. Santhakumar, V.; Jones, R.T.; Mody, I. Developmental regulation and neuroprotective effects of striatal tonic GABAA currents. *Neuroscience* **2010**, *167*, 644–655. [CrossRef] [PubMed]
46. Gong, S.; Zheng, C.; Doughty, M.L.; Losos, K.; Didkovsky, N.; Schambra, U.B.; Nowak, N.J.; Joyner, A.; Leblanc, G.; Hatten, M.E.; et al. A gene expression atlas of the central nervous system based on artificial chromosomes. *Nature* **2003**, *425*, 917–925. [CrossRef] [PubMed]
47. Heiman, M.; Schaefer, A.; Gong, S.; Peterson, J.D.; Day, M.; Ramsey, K.E.; Suá Rez-Fariñ, M.; Schwarz, C.; Stephan, D.A.; Surmeier, D.J.; et al. A Translational Profiling Approach for the Molecular Characterization of CNS Cell Types. *Cell* **2008**, *135*, 438–748. [CrossRef] [PubMed]
48. Lobo, M.K.; Karsten, S.L.; Gray, M.; Geschwind, D.H.; Yang, X.W. FACS-array profiling of striatal projection neuron subtypes in juvenile and adult mouse brains. *Nat. Neurosci.* **2006**, *9*, 443–452. [CrossRef] [PubMed]
49. Marion-Poll, L.; Montalban, E.; Munier, A.; Hervé, D.; Girault, J.A. Fluorescence-activated sorting of fixed nuclei: A general method for studying nuclei from specific cell populations that preserves post-translational modifications. *Eur. J. Neurosci.* **2014**, *39*, 1234–1244. [CrossRef]
50. Jordi, E.; Heiman, M.; Marion-Poll, L.; Guermonprez, P.; Cheng, S.K.; Nairn, A.C.; Greengard, P.; Girault, J.-A. Differential effects of cocaine on histone posttranslational modifications in identified populations of striatal neurons. *Proc. Natl. Acad. Sci. USA* **2013**, *110*, 9511–9516. [CrossRef]
51. Biesemann, C.; Grønborg, M.; Luquet, E.; Wichert, S.P.; Eronique Bernard, V.; Bungers, S.R.; Cooper, B.; Ed Erique Varoqueaux, F.; Li, L.; Byrne, J.A.; et al. Proteomic screening of glutamatergic mouse brain synaptosomes isolated by fluorescence activated sorting. *EMBO J.* **2014**, *33*, 157–170. [CrossRef] [PubMed]
52. Hickox, A.E.; Wong, A.C.Y.; Pak, K.; Strojny, C.; Ramirez, M.; Yates, J.R.; Ryan, A.F.; Savas, J.N. Global Analysis of Protein Expression of Inner Ear Hair Cells. *J. Neurosci.* **2016**, *37*, 1320–1339. [CrossRef] [PubMed]
53. Drummond, E.S.; Nayak, S.; Ueberheide, B.; Wisniewski, T.; Huang, T.T. Proteomic analysis of neurons microdissected from formalin-fixed, paraffin-embedded Alzheimer's disease brain tissue. *Sci. Rep.* **2015**, *5*, 15456. [CrossRef] [PubMed]
54. Plum, S.; Steinbach, S.; Attems, J.; Keers, S.; Riederer, P.; Gerlach, M.; May, C.; Marcus, K. Proteomic characterization of neuromelanin granules isolated from human substantia nigra by laser-microdissection. *Sci. Rep.* **2016**, *6*, 4–11. [CrossRef] [PubMed]
55. Djuric, U.; Rodrigues, D.C.; Batruch, I.; Ellis, J.; Shannon, P.; Diamandis, P. Spatiotemporal proteomic profiling of human cerebral development. *Mol. Cell. Proteom.* **2017**, *16*, 1548–1562. [CrossRef] [PubMed]
56. Hondius, D.C.; Eigenhuis, K.N.; Morrema, T.H.J.; Van Der Schors, R.C.; Van Nierop, P.; Bugiani, M.; Li, K.W.; Hoozemans, J.J.M.; Smit, A.B.; Rozemuller, A.J.M. Proteomics analysis identifies new markers associated with capillary cerebral amyloid angiopathy in Alzheimer's disease. *Acta Neuropathol. Commun.* **2018**, *6*, 1–19. [CrossRef] [PubMed]
57. García-Berrocoso, T.; Llombart, V.; Colàs-Campàs, L.; Hainard, A.; Licker, V.; Penalba, A.; Ramiro, L.; Simats, A.; Bustamante, A.; Martínez-Saez, E.; et al. Single Cell Immuno-Laser Microdissection Coupled to Label-Free Proteomics to Reveal the Proteotypes of Human Brain Cells After Ischemia. *Mol. Cell. Proteom.* **2018**, *17*, 175–189. [CrossRef] [PubMed]
58. Tagawa, K.; Homma, H.; Saito, A.; Fujita, K.; Chen, X.; Imoto, S.; Oka, T.; Ito, H.; Motoki, K.; Yoshida, C.; et al. Comprehensive phosphoproteome analysis unravels the core signaling network that initiates the earliest synapse pathology in preclinical Alzheimer's disease brain. *Hum. Mol. Genet.* **2015**, *24*, 540–558. [CrossRef]
59. Oka, T.; Tagawa, K.; Ito, H.; Okazawa, H. Dynamic changes of the phosphoproteome in postmortem mouse brains. *PLoS ONE* **2011**, *6*, e21405. [CrossRef]
60. Li, J.; Gould, T.D.; Yuan, P.; Manji, H.K.; Chen, G. Post-mortem Interval Effects on the Phosphorylation of Signaling Proteins. *Neuropsychopharmacology* **2003**, *28*, 1017–1025. [CrossRef]

61. O'Callaghan, J.P.; Sriram, K. Focused microwave irradiation of the brain preserves in vivo protein phosphorylation: Comparison with other methods of sacrifice and analysis of multiple phosphoproteins. *J. Neurosci. Methods* **2004**, *135*, 159–168. [CrossRef] [PubMed]
62. Takahashi, K.; Tanabe, K.; Ohnuki, M.; Narita, M.; Ichisaka, T.; Tomoda, K.; Yamanaka, S. Induction of Pluripotent Stem Cells from Adult Human Fibroblasts by Defined Factors. *Cell* **2007**, *131*, 861–872. [CrossRef] [PubMed]
63. Takahashi, K.; Yamanaka, S. Induction of Pluripotent Stem Cells from Mouse Embryonic and Adult Fibroblast Cultures by Defined Factors. *Cell* **2006**, *126*, 663–676. [CrossRef] [PubMed]
64. Paolo, F.; Giorgio, D.; Boulting, G.L.; Bobrowicz, S.; Eggan, K.C. Cell Stem Cell Human Embryonic Stem Cell-Derived Motor Neurons Are Sensitive to the Toxic Effect of Glial Cells Carrying an ALS-Causing Mutation. *Stem Cell* **2008**, *3*, 637–648. [CrossRef]
65. Krencik, R.; Weick, J.P.; Liu, Y.; Zhang, Z.-J.; Zhang, S.-C. Specification of transplantable astroglial subtypes from human pluripotent stem cells. *Nat. Biotechnol.* **2011**, *29*, 528–535. [CrossRef]
66. Kriks, S.; Shim, J.-W.; Piao, J.; Ganat, Y.M.; Wakeman, D.R.; Xie, Z.; Carrillo-Reid, L.; Auyeung, G.; Antonacci, C.; Buch, A.; et al. Dopamine neurons derived from human ES cells efficiently engraft in animal models of Parkinson's disease. *Nature* **2011**, *480*, 547–551. [CrossRef]
67. Liu, H.; Zhang, S.-C. Specification of neuronal and glial subtypes from human pluripotent stem cells. *Cell. Mol. Life Sci.* **2011**, *68*, 3995–4008. [CrossRef]
68. Shi, Y.; Kirwan, P.; Smith, J.; Maclean, G.; Orkin, S.H.; Livesey, F.J. A human stem cell model of early Alzheimer's disease pathology in Down syndrome. *Sci. Transl. Med.* **2012**, *4*, 124–129. [CrossRef]
69. Yamana, R.; Iwasaki, M.; Wakabayashi, M.; Nakagawa, M.; Yamanaka, S.; Ishihama, Y. Rapid and Deep Profiling of Human Induced Pluripotent Stem Cell Proteome by One-shot NanoLC−MS/MS Analysis with Meter-scale Monolithic Silica Columns. *J. Proteome Res.* **2013**, *12*, 214–221. [CrossRef]
70. Phanstiel, D.H.; Brumbaugh, J.; Wenger, C.D.; Tian, S.; Bolin, J.M.; Ruotti, V.; Stewart, R.; Thomson, J.A.; Coon, J.J. Proteomic and phosphoproteomic comparison of human ES and iPS cells. *Nat. Methods* **2012**, *8*, 821–827. [CrossRef]
71. Chae, J.-I.; Kim, D.-W.; Lee, N.; Jeon, Y.-J.; Jeon, I.; Kwon, J.; Kim, J.; Soh, Y.; Lee, D.-S.; Kang, S.; et al. Quantitative proteomic analysis of induced pluripotent stem cells derived from a human Huntington's disease patient. *Biochem. J* **2012**, *446*, 359–371. [CrossRef] [PubMed]
72. Hao, J.; Li, W.; Dan, J.; Ye, X.; Wang, F.; Zeng, X.; Wang, L.; Wang, H.; Cheng, Y.; Liu, L.; et al. Reprogramming- and pluripotency-associated membrane proteins in mouse stem cells revealed by label-free quantitative proteomics. *J. Proteom.* **2013**, *86*, 70–84. [CrossRef]
73. Fuller, H.R.; Mandefro, B.; Shirran, S.L.; Gross, A.R.; Kaus, A.S.; Botting, C.H.; Morris, G.E.; Sareen, D. Spinal Muscular Atrophy Patient iPSC-Derived Motor Neurons Have Reduced Expression of Proteins Important in Neuronal Development. *Front. Cell. Neurosci.* **2016**, *9*, 506. [CrossRef] [PubMed]
74. Chen, M.; Lee, H.-K.; Moo, L.; Hanlon, E.; Stein, T.; Xia, W. Common proteomic profiles of induced pluripotent stem cell-derived three-dimensional neurons and brain tissue from Alzheimer patients. *J. Proteom.* **2018**, *182*, 21–33. [CrossRef] [PubMed]
75. Thompson, A.; Schäfer, J.; Kuhn, K.; Kienle, S.; Schwarz, J.; Schmidt, G.; Neumann, T.; Hamon, C. Tandem mass tags: A novel quantification strategy for comparative analysis of complex protein mixtures by MS/MS. *Anal. Chem.* **2003**, *75*, 1895–1904. [CrossRef] [PubMed]
76. Ong, S.-E.; Blagoev, B.; Kratchmarova, I.; Kristensen, D.B.; Steen, H.; Pandey, A.; Mann, M. Stable Isotope Labeling by Amino Acids in Cell Culture, SILAC, as a Simple and Accurate Approach to Expression Proteomics. *Mol. Cell. Proteom.* **2002**, *1*, 376–386. [CrossRef] [PubMed]
77. Mann, M. Functional and quantitative proteomics using SILAC. *Nat. Rev. Mol. Cell Biol.* **2006**, *7*, 952–958. [CrossRef] [PubMed]
78. Schwanhäusser, B.; Gossen, M.; Dittmar, G.; Selbach, M. Global analysis of cellular protein translation by pulsed SILAC. *Proteomics* **2009**, *9*, 205–209. [CrossRef] [PubMed]
79. de Godoy, L.M.F.; Olsen, J.V.; de Souza, G.A.; Li, G.; Mortensen, P.; Mann, M. Status of complete proteome analysis by mass spectrometry: SILAC labeled yeast as a model system. *Genome Biol.* **2006**, *7*, 1–15. [CrossRef] [PubMed]
80. Zhang, A.; Uaesoontrachoon, K.; Shaughnessy, C.; Das, J.R.; Rayavarapu, S.; Brown, K.J.; Ray, P.E.; Nagaraju, K.; van den Anker, J.N.; Hoffman, E.P.; et al. The use of urinary and kidney SILAM proteomics to

monitor kidney response to high dose morpholino oligonucleotides in the mdx mouse. *Toxicol. Rep.* **2015**, *2*, 838–849. [CrossRef]
81. McClatchy, D.B.; Liao, L.; Park, S.K.; Xu, T.; Lu, B.; Yates, J.R. Differential proteomic analysis of mammalian tissues using SILAM. *PLoS ONE* **2011**, *6*, 1–10. [CrossRef] [PubMed]
82. Mcclatchy, D.B.; Liao, L.; Park, S.K.; Venable, J.D.; Yates, J.R. Quantification of the synaptosomal proteome of the rat cerebellum during post-natal development. *Genome Res.* **2007**, *17*, 1–12. [CrossRef] [PubMed]
83. Rauniyar, N.; McClatchy, D.B.; Yates, J.R. Stable isotope labeling of mammals (SILAM) for in vivo quantitative proteomic analysis. *Methods* **2013**, *61*, 260–268. [CrossRef] [PubMed]
84. Dieterich, D.C.; Link, A.J.; Graumann, J.; Tirrell, D.A.; Schuman, E.M.; Sharpless, K.B. Selective identification of newly synthesized proteins in mammalian cells using bioorthogonal noncanonical amino acid tagging (BONCAT). *Proc. Natl. Acad. Sci. USA* **2006**, *103*, 9482–9487. [CrossRef] [PubMed]
85. Alvarez-Castelao, B.; Schanzenbächer, C.T.; Hanus, C.; Glock, C.; Tom Dieck, S.; Dörrbaum, A.R.; Bartnik, I.; Nassim-Assir, B.; Ciirdaeva, E.; Mueller, A.; et al. Cell-type-specific metabolic labeling of nascent proteomes in vivo. *Nat. Biotechnol.* **2017**, *35*, 1196–1201. [CrossRef] [PubMed]
86. Dieterich, D.C.; Hodas, J.J.L.; Gouzer, G.; Shadrin, I.Y.; Ngo, J.T.; Triller, A.; Tirrell, D.A.; Schuman, E.M. In situ visualization and dynamics of newly synthesized proteins in rat hippocampal neurons. *Nat. Neurosci.* **2010**, *13*, 897–905. [CrossRef] [PubMed]
87. Elliott, T.S.; Bianco, A.; Townsley, F.M.; Fried, S.D.; Chin, J.W. Tagging and Enriching Proteins Enables Cell-Specific Proteomics. *Cell Chem. Biol.* **2016**, *23*, 805–815. [CrossRef] [PubMed]
88. Elliott, T.S.; Townsley, F.M.; Bianco, A.; Ernst, R.J.; Sachdeva, A.; Elsässer, S.J.; Davis, L.; Lang, K.; Pisa, R.; Greiss, S.; et al. Proteome labeling and protein identification in specific tissues and at specific developmental stages in an animal. *Nat. Biotechnol.* **2014**, *32*, 465–472. [CrossRef] [PubMed]
89. Krogager, T.P.; Ernst, R.J.; Elliott, T.S.; Calo, L.; Beránek, V.; Ciabatti, E.; Spillantini, M.G.; Tripodi, M.; Hastings, M.H.; Chin, J.W. Labeling and identifying cell-specific proteomes in the mouse brain. *Nat. Biotechnol.* **2018**, *36*, 156–159. [CrossRef] [PubMed]
90. Sharma, K.; Schmitt, S.; Bergner, C.G.; Tyanova, S.; Kannaiyan, N.; Manrique-Hoyos, N.; Kongi, K.; Cantuti, L.; Hanisch, U.-K.; Philips, M.-A.; et al. Cell type-and brain region-resolved mouse brain proteome. *Nat. Neurosci.* **2015**, *18*, 1–16. [CrossRef] [PubMed]
91. Carlyle, B.C.; Kitchen, R.R.; Kanyo, J.E.; Voss, E.Z.; Pletikos, M.; Sousa, A.M.M.; Lam, T.T.; Gerstein, M.B.; Sestan, N.; Nairn, A.C. A Multiregional Proteomic Survey of the Postnatal Human Brain. *Nat. Neurosci.* **2017**, *20*, 1787–1795. [CrossRef] [PubMed]
92. Liu, J.; Xu, Y.; Stoleru, D.; Salic, A. Imaging protein synthesis in cells and tissues with an alkyne analog of puromycin. *Proc. Natl. Acad. Sci. USA* **2012**, *109*, 413–418. [CrossRef] [PubMed]
93. Ge, J.; Zhang, C.W.; Ng, X.W.; Peng, B.; Pan, S.; Du, S.; Wang, D.; Li, L.; Lim, K.L.; Wohland, T.; et al. Puromycin Analogues Capable of Multiplexed Imaging and Profiling of Protein Synthesis and Dynamics in Live Cells and Neurons. *Angew. Chem. Int. Ed.* **2016**, *55*, 4933–4937. [CrossRef] [PubMed]
94. Du, S.; Wang, D.; Lee, J.-S.; Peng, B.; Ge, J.; Yao, S. Cell Type-Selective Imaging and Profiling of Newly Synthesized Proteomes by Using Puromycin Analogues. *Chem. Commun.* **2017**, *53*, 8443–8446. [CrossRef] [PubMed]
95. Barrett, R.M.; Liu, H.W.; Jin, H.; Goodman, R.H.; Cohen, M.S. Cell-specific Profiling of Nascent Proteomes Using Orthogonal Enzyme-mediated Puromycin Incorporation. *ACS Chem. Biol.* **2016**, *11*, 1532–1536. [CrossRef] [PubMed]
96. Li, Z.; Zhu, Y.; Sun, Y.; Qin, K.; Liu, W.; Zhou, W.; Chen, X. Nitrilase-Activatable Noncanonical Amino Acid Precursors for Cell-Selective Metabolic Labeling of Proteomes. *ACS Chem. Biol.* **2016**, *11*, 3273–3277. [CrossRef] [PubMed]
97. Roux, K.J.; Kim, D.I.; Raida, M.; Burke, B. A promiscuous biotin ligase fusion protein identifies proximal and interacting proteins in mammalian cells. *J. Cell Biol.* **2012**, *196*, 801–810. [CrossRef]
98. Uezu, A.; Kanak, D.J.; Bradshaw, T.W.A.; Soderblom, E.J.; Catavero, C.M.; Burette, A.C.; Weinberg, R.J.; Soderling, S.H. Identification of an elaborate complex mediating postsynaptic inhibition. *Science* **2016**, *353*, 1123–1129. [CrossRef]
99. Kim, D.I.; Jensen, S.C.; Noble, K.A.; KC, B.; Roux, K.H.; Motamedchaboki, K.; Roux, K.J. An improved smaller biotin ligase for BioID proximity labeling. *Mol. Biol. Cell* **2016**, *27*, 1188–1196. [CrossRef]

100. Branon, T.C.; Bosch, J.A.; Sanchez, A.D.; Udeshi, N.D.; Svinkina, T.; Carr, S.A.; Feldman, J.L.; Perrimon, N.; Ting, A.Y. Efficient proximity labeling in living cells and organisms with TurboID. *Nat. Biotechnol.* **2018**, *36*, 880–898. [CrossRef]
101. Rhee, H.-W.; Zou, P.; Udeshi, N.D.; Martell, J.D.; Mootha, V.K.; Carr, S.A.; Ting, A.Y. Proteomic Mapping of Mitochondria in Living Cells via Spatially- Restricted Enzymatic Tagging. *Science* **2013**, *339*, 1328–1331. [CrossRef]
102. Reinke, A.W.; Mak, R.; Troemel, E.R.; Bennett, E.J. In vivo mapping of tissue-and subcellular-specific proteomes in Caenorhabditis elegans. *Sci. Adv.* **2017**, *3*, e1602426. [CrossRef]
103. Lobingier, B.T.; Hüttenhain, R.; Eichel, K.; Miller, K.B.; Ting, A.Y.; von Zastrow, M.; Krogan, N.J. An Approach to Spatiotemporally Resolve Protein Interaction Networks in Living Cells. *Cell* **2017**, *169*, 350–360.e12. [CrossRef] [PubMed]
104. Loh, K.H.; Stawski, P.S.; Draycott, A.S.; Udeshi, N.D.; Lehrman, E.K.; Wilton, D.K.; Svinkina, T.; Deerinck, T.J.; Ellisman, M.H.; Stevens, B.; et al. HHS Public Access. *Cell* **2015**, *359*, 1018–1026. [CrossRef]
105. Comi, T.J.; Do, T.D.; Rubakhin, S.S.; Sweedler, J.V. Categorizing Cells on the Basis of their Chemical Profiles: Progress in Single-Cell Mass Spectrometry. *J. Am. Chem. Soc.* **2017**, *139*, 3920–3929. [CrossRef] [PubMed]
106. Stoeckli, M.; Chaurand, P.; Hallahan, D.E.; Caprioli, R.M. Imaging mass spectrometry: A new technology for the analysis of protein expression in mammalian tissues. *Nat. Med.* **2001**, *7*, 493–496. [CrossRef]
107. Stoeckli, M.; Staab, D.; Staufenbiel, M.; Wiederhold, K.-H.; Signor, L. Molecular imaging of amyloid b peptides in mouse brain sections using mass spectrometry. *Anal. Biochem.* **2002**, *311*, 33–39. [CrossRef]
108. Schwamborn, K.; Caprioli, R.M. Molecular imaging by mass spectrometry—Looking beyond classical histology. *Nat. Rev. Cancer* **2010**, *10*, 639–646. [CrossRef]
109. Bozdon-Kulakowska, A.; Suder, P. Imaging mass specrometry: Instrumentation, applications, and combination with other visualization techniques. *Mass Spectrom. Rev.* **2016**, *35*, 147–169. [CrossRef]
110. Rocha, B.; Ruiz-Romero, C.; Blanco, F.J. Mass spectrometry imaging: A novel technology in rheumatology. *Nat. Rev. Rheumatol.* **2016**, *13*, 52–63. [CrossRef]
111. Spengler, B. Mass Spectrometry Imaging of Biomolecular Information. *Anal. Chem.* **2015**, *87*, 64–82. [CrossRef] [PubMed]
112. Reyzer, M.L.; Caprioli, R.M. MALDI Mass Spectrometry for Direct Tissue Analysis: A New Tool for Biomarker Discovery. *J. Proteome Res.* **2005**, *4*, 1138–1142. [CrossRef] [PubMed]
113. Giesen, C.; Wang, H.A.O.; Schapiro, D.; Zivanovic, N.; Jacobs, A.; Hattendorf, B.; Schüffler, P.J.; Grolimund, D.; Buhmann, J.M.; Brandt, S.; et al. Highly multiplexed imaging of tumor tissues with subcellular resolution by mass cytometry. *Nat. Methods* **2014**, *11*, 417–422. [CrossRef] [PubMed]
114. Karas, M.; Hillenkamp, F. Laser desorption ionization of proteins with molecular masses exceeding 10,000 daltons. *Anal. Chem.* **1988**, *60*, 2299–2301. [CrossRef] [PubMed]
115. Zhang, L.; Vertes, A. Single-Cell Mass Spectrometry Approaches to Explore Cellular Heterogeneity. *Angew. Chem. Int. Ed.* **2018**, *57*, 4466–4477. [CrossRef] [PubMed]
116. Monroe, E.B.; Annangudi, S.P.; Hatcher, N.G.; Gutstein, H.B.; Rubakhin, S.S.; Sweedler, J.V. SIMS and MALDI MS imaging of the spinal cord. *Proteomics* **2008**, *8*, 3746–3754. [CrossRef]
117. Guenther, S.; Römpp, A.; Kummer, W.; Spengler, B. AP-MALDI imaging of neuropeptides in mouse pituitary gland with 5μm spatial resolution and high mass accuracy. *Int. J. Mass Spectrom.* **2011**, *305*, 228–237. [CrossRef]
118. Do, T.D.; Ellis, J.F.; Neumann, E.K.; Comi, T.J.; Tillmaand, E.G.; Lenhart, A.E.; Rubakhin, S.S.; Sweedler, J.V. Optically Guided Single Cell Mass Spectrometry of Rat Dorsal Root Ganglia to Profile Lipids, Peptides and Proteins. *ChemPhysChem* **2018**, *19*, 1180–1191. [CrossRef] [PubMed]
119. Do, T.D.; Comi, T.J.; Dunham, S.J.B.; Rubakhin, S.S.; Sweedler, J.V. Single Cell Profiling Using Ionic Liquid Matrix-Enhanced Secondary Ion Mass Spectrometry for Neuronal Cell Type Differentiation. *Anal. Chem.* **2017**, *89*, 3078–3086. [CrossRef] [PubMed]
120. Tucker, K.R.; Li, Z.; Rubakhin, S.S.; Sweedler, J.V. Secondary Ion Mass Spectrometry Imaging of Molecular Distributions in Cultured Neurons and Their Processes: Comparative Analysis of Sample Preparation. *J. Am. Soc. Mass Spectrom.* **2012**, *23*, 1931–1938. [CrossRef] [PubMed]
121. Fletcher, J.S.; Rabbani, S.; Henderson, A.; Blenkinsopp, P.; Thompson, S.P.; Lockyer, N.P.; Vickerman, J.C. A New Dynamic in Mass Spectral Imaging of Single Biological Cells. *Anal. Chem.* **2008**, *80*, 9058–9064. [CrossRef] [PubMed]

122. Bandura, D.R.; Baranov, V.I.; Ornatsky, O.I.; Antonov, A.; Kinach, R.; Lou, X.; Pavlov, S.; Vorobiev, S.; Dick, J.E.; Tanner, S.D. Mass Cytometry: Technique for Real Time Single Cell Multitarget Immunoassay Based on Inductively Coupled Plasma Time-of-Flight Mass Spectrometry. *Anal. Chem.* **2009**, *81*, 6813–6822. [CrossRef] [PubMed]
123. Angelo, M.; Bendall, S.C.; Finck, R.; Hale, M.B.; Hitzman, C.; Borowsky, A.D.; Levenson, R.M.; Lowe, J.B.; Liu, S.D.; Zhao, S.; et al. Multiplexed ion beam imaging of human breast tumors. *Nat. Med.* **2014**, *20*, 436–442. [CrossRef] [PubMed]
124. Leelatian, N.; Doxie, D.B.; Greenplate, A.R.; Mobley, B.C.; Lehman, J.M.; Sinnaeve, J.; Kauffmann, R.M.; Werkhaven, J.A.; Mistry, A.M.; Weaver, K.D.; et al. Single cell analysis of human tissues and solid tumors with mass cytometry. *Cytom. Part B Clin. Cytom.* **2017**, *92B*, 68–78. [CrossRef] [PubMed]
125. Behbehani, G.K.; Samusik, N.; Bjornson, Z.B.; Fantl, W.J.; Medeiros, B.C.; Nolan, G.P. Mass cytometric functional profiling of acute myeloid leukemia defines cell-cycle and immunophenotypic properties that correlate with known responses to therapy. *Cancer Discov.* **2015**, *5*, 988–1003. [CrossRef] [PubMed]
126. Alexander, G.M.; Huang, Y.Z.; Soderblom, E.J.; He, X.P.; Moseley, M.A.; McNamara, J.O. Vagal nerve stimulation modifies neuronal activity and the proteome of excitatory synapses of amygdala/piriform cortex. *J. Neurochem.* **2017**, *140*, 629–644. [CrossRef] [PubMed]
127. Nemes, P.; Knolhoff, A.M.; Rubakhin, S.S.; Sweedler, J.V. Metabolic differentiation of neuronal phenotypes by single-cell capillary electrophoresis-electrospray ionization-mass spectrometry. *Anal. Chem.* **2011**, *83*, 6810–6817. [CrossRef]
128. Nemes, P.; Knolhoff, A.M.; Rubakhin, S.S.; Sweedler, J.V. Single-cell metabolomics: Changes in the metabolome of freshly isolated and cultured neurons. *ACS Chem. Neurosci.* **2012**, *3*, 782–792. [CrossRef]
129. Onjiko, R.M.; Portero, E.P.; Moody, S.A.; Nemes, P. In Situ Microprobe Single-Cell Capillary Electrophoresis Mass Spectrometry: Metabolic Reorganization in Single Differentiating Cells in the Live Vertebrate (Xenopus laevis) Embryo. *Anal. Chem.* **2017**, *89*, 7069–7076. [CrossRef]
130. Onjiko, R.M.; Moody, S.A.; Nemes, P. Single-cell mass spectrometry reveals small molecules that affect cell fates in the 16-cell embryo. *Proc. Natl. Acad. Sci. USA* **2015**, *112*, 6545–6550. [CrossRef]
131. Onjiko, R.M.; Morris, S.E.; Moody, S.A.; Nemes, P. Single-cell mass spectrometry with multi-solvent extraction identifies metabolic differences between left and right blastomeres in the 8-cell frog (Xenopus) embryo. *Analyst* **2016**, *141*, 3648–3656. [CrossRef] [PubMed]
132. Lombard-Banek, C.; Reddy, S.; Moody, S.A.; Nemes, P. Label-free Quantification of Proteins in Single Embryonic Cells with Neural Fate in the Cleavage-Stage Frog (*Xenopus laevis*) Embryo using Capillary Electrophoresis Electrospray Ionization High-Resolution Mass Spectrometry (CE-ESI-HRMS). *Mol. Cell. Proteom.* **2016**, *15*, 2756–2768. [CrossRef] [PubMed]
133. Lombard-Banek, C.; Moody, S.A.; Nemes, P. Single-Cell Mass Spectrometry for Discovery Proteomics: Quantifying Translational Cell Heterogeneity in the 16-Cell Frog (Xenopus) Embryo. *Angew. Chem. Int. Ed.* **2016**, *55*, 2454–2458. [CrossRef] [PubMed]
134. Hofstadler, S.A.; Swanek, F.D.; Gale, D.C.; Ewing, A.G.; Smith, R.D. Capillary Electrophoresis-Electrospray Ionization Fourier Transform Ion Cyclotron Resonance Mass Spectrometry for Direct Analysis of Cellular Proteins. *J. Neurosci. Methods* **1995**, *67*, 1477–1480. [CrossRef]
135. Mellors, J.S.; Jorabchi, K.; Smith, L.M.; Ramsey, J.M. Integrated microfluidic device for automated single cell analysis using electrophoretic separation and electrospray ionization mass spectrometry. *Anal. Chem.* **2010**, *82*, 967–973. [CrossRef] [PubMed]
136. Valaskovic, G.A.; Kelleher, N.L.; McLafferty, F.W. Attomole Protein Characterization by Capillary Electrophoresis-Mass Spectrometry. *Science* **1996**, *273*, 1199–1202. [CrossRef]
137. Smith, R.D.; Shen, Y.; Tang, K. Ultrasensitive and Quantitative Analyses from Combined Separations-Mass Spectrometry for the Characterization of Proteomes. *Acc. Chem. Res.* **2004**, *37*, 269–278. [CrossRef]
138. Cecala, C.; Sweedler, J.V. Sampling techniques for single-cell electrophoresis. *Analyst* **2013**, *137*, 2922–2929. [CrossRef]
139. Zhu, Y.; Zhao, R.; Piehowski, P.D.; Moore, R.J.; Lim, S.; Orphan, V.J.; Paša-Tolić, L.; Qian, W.J.; Smith, R.D.; Kelly, R.T. Subnanogram proteomics: Impact of LC column selection, MS instrumentation and data analysis strategy on proteome coverage for trace samples. *Int. J. Mass Spectrom.* **2018**, *427*, 4–10. [CrossRef]

140. Zhang, L.; Khattar, N.; Kemenes, I.; Kemenes, G.; Zrinyi, Z.; Pirger, Z.; Vertes, A. Subcellular Peptide Localization in Single Identified Neurons by Capillary Microsampling Mass Spectrometry. *Sci. Rep.* **2018**, *8*, 12227. [CrossRef]
141. Zhang, L.; Vertes, A. Energy Charge, Redox State, and Metabolite Turnover in Single Human Hepatocytes Revealed by Capillary Microsampling Mass Spectrometry. *Anal. Chem.* **2015**, *87*, 10397–10405. [CrossRef] [PubMed]
142. Zhang, L.; Foreman, D.P.; Grant, P.A.; Shrestha, B.; Moody, S.A.; Villiers, F.; Kwak, J.M.; Vertes, A. In Situ metabolic analysis of single plant cells by capillary microsampling and electrospray ionization mass spectrometry with ion mobility separation. *Analyst* **2014**, *139*, 5079–5085. [CrossRef] [PubMed]
143. Lee, C.Y.; Fan, Y.; Rubakhin, S.S.; Yoon, S.; Sweedler, J.V. A neuron-in-capillary platform for facile collection and mass spectrometric characterization of a secreted neuropeptide. *Sci. Rep.* **2016**, *6*, 1–8. [CrossRef] [PubMed]
144. Budnik, B.; Levy, E.; Slavov, N. Mass-spectrometry of single mammalian cells quantifies proteome heterogeneity during cell differentiation. *bioRxiv* **2017**. [CrossRef] [PubMed]
145. Wan, H.; Tang, B.; Liao, X.; Zeng, Q.; Zhang, Z.; Liao, L. Analysis of neuronal phosphoproteome reveals PINK1 regulation of BAD function and cell death. *Cell Death Differ.* **2018**, *25*, 904–917. [CrossRef] [PubMed]
146. Distler, U.; Schmeisser, M.J.; Pelosi, A.; Reim, D.; Kuharev, J.; Weiczner, R.; Baumgart, J.; Boeckers, T.M.; Nitsch, R.; Vogt, J.; et al. In-depth protein profiling of the postsynaptic density from mouse hippocampus using data-independent acquisition proteomics. *Proteomics* **2014**, *14*, 2607–2613. [CrossRef] [PubMed]
147. Meier, F.; Geyer, P.E.; Virreira Winter, S.; Cox, J.; Mann, M. BoxCar acquisition method enables single-shot proteomics at a depth of 10,000 proteins in 100 minutes. *Nat. Methods* **2018**, *15*, 440–448. [CrossRef]
148. Ping, L.; Duong, D.M.; Yin, L.; Gearing, M.; Lah, J.J.; Levey, A.I.; Seyfried, N.T. Data Descriptor: Global quantitative analysis of the human brain proteome in Alzheimer's and Parkinson's Disease. *Nat. Publ. Gr.* **2018**, *5*, 1–12. [CrossRef]
149. Carlyle, B.C.; Kitchen, R.R.; Zhang, J.; Wilson, R.; Lam, T.; Rozowsky, J.S.; Williams, K.R.; Sestan, N.; Gerstein, M.; Nairn, A.C. Isoform level interpretation of high-throughput proteomic data enabled by deep integration with RNA-seq. *J. Proteome Res* **2018**, *17*, 3431–3444. [CrossRef]
150. Menschaert, G.; Van Criekinge, W.; Notelaers, T.; Koch, A.; Crappé, J.; Gevaert, K.; Van Damme, P. Deep proteome coverage based on ribosome profiling aids MS-based protein and peptide discovery and provides evidence of alternative translation products and near-cognate translation initiation events. *Mol. Cell. Proteom.* **2013**, *17*, 1–41. [CrossRef]
151. Dammer, E.B.; Duong, D.M.; Diner, I.; Gearing, M.; Feng, Y.; Lah, J.J.; Levey, A.I.; Seyfried, N.T. Neuron Enriched Nuclear Proteome Isolated from Human Brain. *J. Proteome Res* **2013**, *12*, 3193–3206. [CrossRef] [PubMed]
152. Zhu, Y.; Piehowski, P.D.; Zhao, R.; Chen, J.; Shen, Y.; Moore, R.J.; Shukla, A.K.; Petyuk, V.A.; Campbell-Thompson, M.; Mathews, C.E.; et al. Nanodroplet processing platform for deep and quantitative proteome profiling of 10-100 mammalian cells. *Nat. Commun.* **2018**, *9*, 1–10. [CrossRef] [PubMed]
153. Tcherkezian, J.; Brittis, P.A.; Thomas, F.; Roux, P.P.; Flanagan, J.G. Transmembrane Receptor DCC Associates with Protein Synthesis Machinery and Regulates Translation. *Cell* **2010**, *141*, 632–644. [CrossRef] [PubMed]
154. Chen, A.L.; Kim, E.W.; Toh, J.Y.; Vashisht, A.A.; Rashoff, A.Q.; Van, C.; Huang, A.S.; Moon, A.S.; Bell, H.N.; Bentolila, L.A.; et al. Novel components of the toxoplasma inner membrane complex revealed by BioID. *MBio* **2015**, *6*, 1–12. [CrossRef] [PubMed]
155. Chen, C.-L.; Hu, Y.; Udeshi, N.D.; Lau, T.Y.; Wirtz-Peitz, F.; He, L.; Ting, A.Y.; Carr, S.A.; Perrimon, N.; Axelrod, J.D.; et al. Proteomic mapping in live Drosophila tissues using an engineered ascorbate peroxidase. *Proc. Natl. Acad. Sci. USA* **2015**, *112*, 12093–12098. [CrossRef] [PubMed]

proteomes

MDPI

Review

Uncovering Discrete Synaptic Proteomes to Understand Neurological Disorders

Yi-Zhi Wang and Jeffrey N. Savas *

Department of Neurology, Northwestern University Feinberg School of Medicine, Chicago, IL 60611, USA; yi-zhi.wang@northwestern.edu

* Correspondence: jeffrey.savas@northwestern.edu; Tel.: +1-312-503-3089

Academic Editors: Angus C. Nairn and Kenneth R. Williams
Received: 2 June 2018; Accepted: 13 July 2018; Published: 19 July 2018

Abstract: The mammalian nervous system is an immensely heterogeneous organ composed of a diverse collection of neuronal types that interconnect in complex patterns. Synapses are highly specialized neuronal cell-cell junctions with common and distinct functional characteristics that are governed by their protein composition or synaptic proteomes. Even a single neuron can possess a wide-range of different synapse types and each synapse contains hundreds or even thousands of proteins. Many neurological disorders and diseases are caused by synaptic dysfunction within discrete neuronal populations. Mass spectrometry (MS)-based proteomic analysis has emerged as a powerful strategy to characterize synaptic proteomes and potentially identify disease driving synaptic alterations. However, most traditional synaptic proteomic analyses have been limited by molecular averaging of proteins from multiple types of neurons and synapses. Recently, several new strategies have emerged to tackle the 'averaging problem'. In this review, we summarize recent advancements in our ability to characterize neuron-type specific and synapse-type specific proteomes and discuss strengths and limitations of these emerging analysis strategies.

Keywords: proteomics; basal ganglia; synapses; synapse specificity; neuronal circuits; axons; dendrites; neurodegeneration

1. Introduction

The mammalian nervous system is a complex organ assembled from millions of neurons in complex circuits arranged into elaborate networks. Neuronal networks provide information-processing capabilities and also facilitate the transfer of large amounts of data between brain regions and the rest of the organism [1]. To establish proper neuronal circuits during development, axons must innervate the appropriate target regions and form precise connections with the proper neurons [2]. Axons form short-range neuronal connections with nearby neuronal dendrites within the same brain region and long-range nerve fibers connect distant territories of the brain [3]. Chemical synapses represent specialized cell junctions that allow information, in the form of neurotransmitters, to flow from presynaptic axons to postsynaptic dendrites. Synaptic proteomes mature during development and are refined through activity dependent changes [4]. Proper synaptic communication is essential for many physiological functions from breathing to learning and memory [5,6].

The disruption of synaptic transmission from long-range projection neurons plays a key role in a variety of neurological disorders. For example, impaired neurotransmission of projection neurons is a defining characteristic of neurodegeneration in basal ganglia diseases including Parkinson's disease (PD), Huntington's disease (HD), and dystonia [7]. Death of substantia nigra dopaminergic neurons in PD patients leads to loss of a major afferent basal ganglia projection. Importantly, administration of levodopa mimicking dopamine is a widely used therapeutic for PD [8]. Furthermore, in HD, discrete projections are (i.e., indirect pathway) impaired and responsible for the core aspects of pathology [9]. Neurodevelopmental disorders including autism and schizophrenia are also believed to be caused by the impairment of multiple neuronal circuits [10,11]. Specific brain regions, neurons, synapses, and synaptic proteins are also impaired in Alzheimer's disease (AD) [12,13]. Hippocampal synapses are the first to be affected in AD, but later pathology spreads to cortical regions and beyond. However, some brain regions such as the cerebellum seem to be resistant to the AD-related synaptic dysfunction even in late stages of AD [14]. Therefore, a deep understanding of the mechanisms and proteins regulating discrete synapses formed locally or between long-range axonal projections and their postsynaptic counterparts is a key step towards developing effective treatments for a wide range of neurological disorders.

Systematic characterization of neuron-neuron connections or the "connectome" has recently received significant biomedical research attention, since altered connections may play a causative role in neurological dysfunction. The goal is to map the complete set of anatomical and functional connectivity within the healthy mammalian brain. This is no small challenge since there are billions of neurons and 10^{12-14} synapses in the mammalian brain [15,16]. Clearly, this is an immense challenge that will require decades of research to complete [17]. While careful anatomical and electrophysiological measurements should eventually provide an invaluable description of how neuronal networks function, it is unlikely to facilitate the treatment of neurological disorders on its own, therefore necessitating a molecular understanding. Determining the proteins and molecular mechanisms responsible for altered circuit function of specific neurons and discrete synapses in discovery-mode, represents an important opportunity that is only now becoming achievable [18].

Technological improvements in mass spectrometry (MS)-based proteomic analysis in combination with the recent development of several chemical strategies, have now made it possible to probe discrete neuronal and synaptic proteomes in vitro and in vivo. In this review article, we summarize recent success in this emerging area of neuroproteomic research and provide a preview of how we and others are combining these new tools to increase our understanding of underappreciated synaptic mechanisms. The hope is that by identifying small groups of altered proteins in specific neurons and synapses, we will be able to advance our understanding of synaptic dysfunction and discover new therapeutic targets.

2. The Synaptic Protein Averaging Problem

Historically, neurons have been classified based on the identity of the neurotransmitter (NT) that their axons secrete. For example, glutamate is released in synaptic vesicles by glutamatergic neurons and plays a key role in learning and memory as a major excitatory NT type in the brain [19]. Other major NTs, including adrenalin, noradrenaline, dopamine, gamma aminobutyric acid (GABA), acetylcholine, and serotonin, regulate discrete functional aspects of neuronal physiology and manage different physiological aspects such as mood, pleasure, and reward. Neurotransmission is a complex and intricate neurobiological process and we already know the identity of more than 100 NTs, with many more still likely to be discovered. It is expected that the machinery responsible for the release of synaptic vesicles containing different NTs require at least some of the same core proteins. However, this has yet to be deeply investigated and it is possible that the release of different NTs requires a unique set of protein factors [4]. Furthermore, it is well known that co-release or co-transmission in the same or distinct synaptic vesicles is also a common mechanism in the mammalian brain [20]. For NTs to function, they must each bind their cognate postsynaptic receptor and facilitate signaling as ligand-gated ion channels, transporters and G-protein coupled receptors (GPCRs). This high degree of fidelity guarantees the faithful incorporation of multiple signals and minimizes unwanted NT crosstalk.

Individual neurotransmitters bind to multiple receptor proteins and protein complexes. Take metabotropic glutamate receptors (mGluRs) as an example, there are eight different GPCRs that localize to both pre- and post-synaptic membranes. The expression of mGluR4 is very high in the cerebellum but mGluR6 is highly expressed only in the retina [21]. The spatially restricted expression pattern of these receptors suggests that synapses contain a unique array of proteins in both the cerebellum and retina. Let us consider the basal ganglia, situated at the base of the forebrain, as an illustration of regional synaptic heterogeneity. It is composed of the striatum, globus palladus, ventral pallidum, substantia nigra, and subthalamic nucleus and is associated with a variety of functions [22]. The basal ganglia possess two groups of spiny projection neurons (SPNs), direct pathway (dSPNs) and indirect pathway SPNs (iSPNs). Interestingly, dSPNs and iSPNs have nearly indistinguishable morphology and are both GABAergic, however they control contrasting aspects of motor control. Both types of SPNs receive glutamatergic input from the cortex and thalamus, and dopaminergic projections from substantia nigra. Dopamine receptor D1 (Drd1) is selectively expressed in dendritic spines of dSPNs, Drd2 is selectively expressed in iSPNs, and these receptors have distinct signaling activities [23]. Previous targeted biochemical studies and translation profiling experiments have confirmed significant difference in dSPNs and iSPNs expression profiles, however the synaptic proteomes of these neurons has yet to be effectively compared [24]. Taken altogether, the synaptic protein composition of dSPNs and iSPNs are likely to be different in at least a few key aspects. Interestingly, dSPN and iSPN have distinct roles in neurodegenerative diseases and disorders such as PD and schizophrenia. The most effective medicines for these diseases have targeted Drd2 SPNs, suggesting the malfunction of Drd2 synapses play a key role [25]. Moreover, a recent study showed that neuroligin-3 (Nlgn3) mutations selectively impaired dSPNs in nucleus accumbens (NAc) to boost autism spectrum disorder (ASD)-associated repetitive behaviors, suggesting the malfunction of Nlgn3 positive synapses in NAc dSPNs may be a driver of at least some behavioral symptoms in ASD [26].

The high degree of neuronal and synaptic heterogeneity in the brain poses a prohibitive obstacle for researchers trying to determine the protein composition of discrete synapses. Historically, the most widely used strategy has been to dissect individual brain regions, homogenize the tissue in sucrose buffer, perform differential centrifugation to purify synaptosomes, and use proteomics to identify the proteins present in the purified material. However, during the process of homogenization, the identity of all projection specific synaptic proteins will be lost and the resulting datasets will represent a composite 'average' measurement of synaptic protein content (Figure 1). Simply put, low abundance proteins present at all SPN synapses will be indistinguishable from those present at moderate levels only at iSPN or dSPN synapses. This is important since discrete neural circuits are affected in neurological disorders and by using the traditional synaptosome approach to evaluate altered synapses is suboptimal at best and misleading at worst. Thus, we need new approaches to facilitate in vivo proteomic characterization of discrete synapses in the context of rodent models of neurological disorders.

A

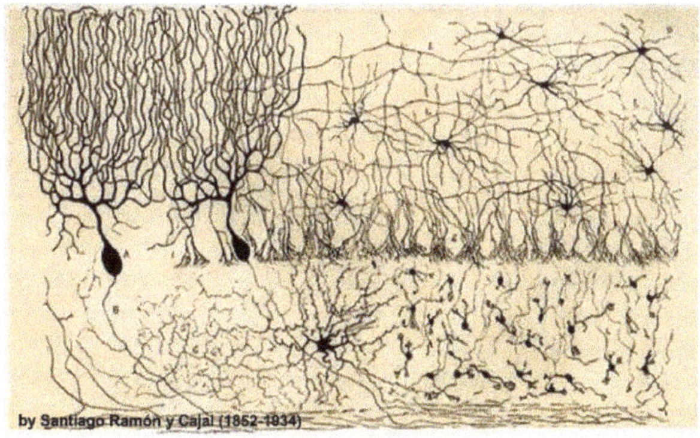

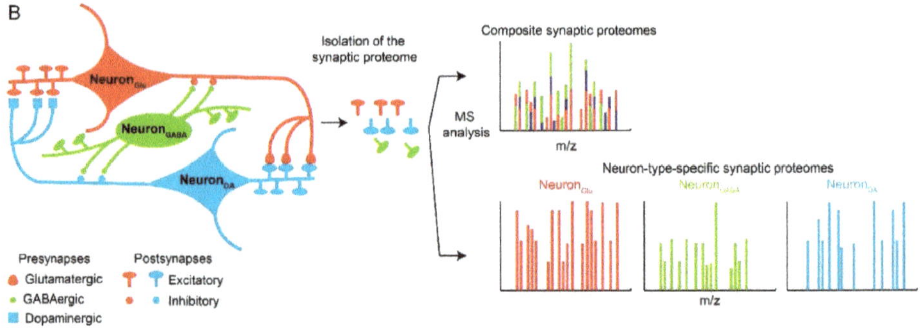

Figure 1. Neuronal and synaptic diversity complicates the interpretation of proteomic datasets. (**A**) Original drawing by Santiago Ramon Y Cajal showing a diverse neuronal population in chicken brain. Reproduced here without restriction since this work is in the public domain in its country of origin and other countries and areas where the copyright term is the author's life plus 70 years or less. (**B**) Traditional biochemical approaches with mass spectrometry (MS) analysis of pre- and postsynaptic proteomes are limited by the molecular averaging (top). Analysis of discrete synaptic proteomes represents a major advancement in our ability to understand synapse specific functions.

3. Strategies to Overcome Averaging at the Cellular Level

The accuracy and sensitivity of nearly all proteomic analysis workflows are limited by the degree of proteome complexity present in the extract. For example, the presence of even a few highly abundant synaptic proteins, such as CaMKII, tends to limit our ability to detect and quantify low abundant synaptic proteins in both targeted and untargeted MS-based analysis workflows [27]. The proteome complexity of synaptosomes prepared from rodent whole brain extracts poses an even greater challenge since they represent cumulative composite collections of proteins. Therefore, one commonly used strategy to surmount the averaging problem is to limit the heterogeneity of the protein extract [28]. With this goal in mind, multiple approaches have been developed and applied to the investigation of cell type and region-specific synaptic proteomes.

3.1. *Laser Capture Microdissection (LCM)*

LCM is a straightforward strategy to identify a tissue region of interest with a microscope and a laser to isolate it from intact brain tissue from any source for subsequent molecular characterization (Figure 2A). Typically, the brain tissue sections (5–50 micrometers thick) are mounted and examined with a light microscope. Regions of interest (i.e., certain group of neurons) are identified based on location or morphology, and targeted for isolation. Then the selected targets, which represent groups of cells or even a single cell, is precisely cut away from the brain section with an ultraviolet laser beam (usually 355 nm) and captured with an infrared laser [29]. LCM has high spatial resolution that can be precisely controlled within a few μms. Thus, the laser will only minimally damage the tissues adjacent to that of interest during the isolation and intact cellular structures such as synapses can be preserved. LCM has been widely used to study specific proteomes in discrete brain regions with laminar organization such as the hippocampus [30]. However, LCM has several significant limitations; it requires specific morphological characteristics and neuronal organization to identify synaptic regions of interest, time required to isolate the target tissue can compromise its integrity, and the depth of proteomic analysis is nearly always sample limited.

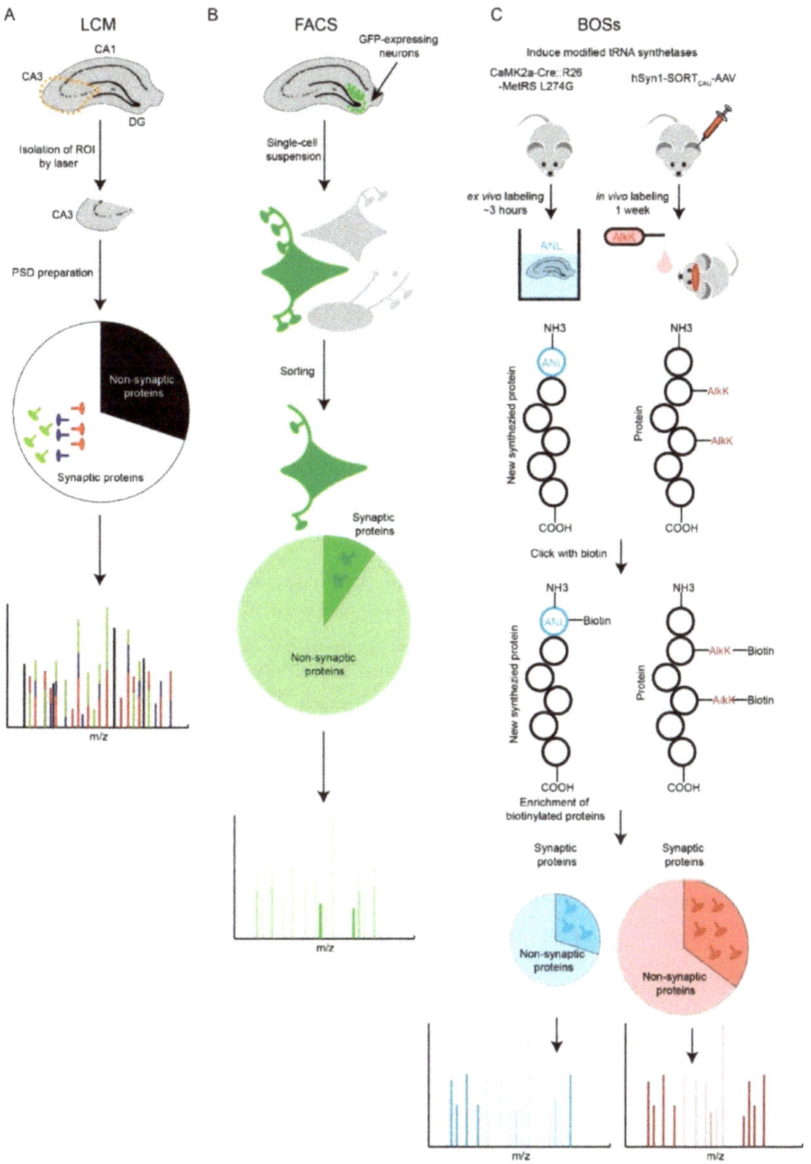

Figure 2. Comparison of Laser Capture Microdissection (LCM), Fluorescence-Activated Cell Sorting (FACS) and Bioorthogonal Strategies (BOSs) strategies. (**A**) LCM-based synaptic proteomic analysis is limited by potentially severe molecular averaging. (**B**) FACS-based proteomics is able to profile a specific type of neuron. However, loss of neurites and synapses during preparation of single-cell suspension is a shortfall. Most of the identified proteins are non-synaptic cytosolic proteins that localize to the soma. (**C**) Two BOSs-based proteomic strategies provide solutions for 'averaging problem' and potentially retain more synaptic information. AlkK is an economic alternative of CypK, which is affordable to feed rodents. Both BOSs can be used for ex vivo and in vivo labeling of cell-type specific proteins; due to space constraints, we illustrate one labeling strategy for each BOS.

3.2. Fluorescence-Activated Cell Sorting (FACS)

FACS is the most commonly used technology to isolate discrete populations of dissociated cells of interest in a stream of fluid with an electronic detector [31]. FACS requires fluorescently labeled cells via dyes, antibodies, or fluorescent proteins [32]. Several fluorescent labels can be used in parallel and the cells can be sorted and collected based on the combination of fluorescence signals. The advantage of FACS is that, with appropriate labeling, multiple cell types of interest can be confidently isolated at the same time. However, FACS may not be an optimal strategy to isolate intact neurons, since the cells need to be individually dissociated into a single-cell suspension (Figure 2B). During this process, synapses and neurites may become damaged or retract. To overcome this challenge, researchers have recently used FACS to sort fluorescence-labeled purified synaptosomes. This approach can be used to isolate synaptosomes from discrete synapse-types (e.g., VGluT1-GFP) and in combination with MS, begin to determine synapse type specific proteomes [33]. This strategy addresses the synapse-loss and specificity problems but may be somewhat hindered by the ability of FACS to sort small heterogeneous vesicles (i.e., synaptosomes) with weaker fluorescence signals and the potential for clumping of fluorescent and non-fluorescent particles. Thus, neuronal debris and mitochondria may co-purify with fluorescent synaptosomes, and potentially lead to co-purification of non-synaptic proteins and potentially limit the specificity of this strategy.

3.3. Bioorthogonal Strategies (BOSs)

Recently, genetically modified tRNA synthetases in combination with unnatural amino acids and BOS have facilitated the identification and measurement of cell type specific proteomes [34,35]. Alvarez-Castelao et al. developed a methionyl-tRNA synthetase (MetRS) L274G-based system; MetRS L274G primes methionine tRNAs with the methionine amino acid surrogate azidonorleucine (ANL). When MetRS L274G is selectively expressed in discrete cells with cell type-specific promoters or Cre recombinase, ANL is selectively incorporated into nascent polypeptides, which can be clicked with biotin and isolated with streptavidin. Krogager et al. used pyrrolysyl-tRNA synthetase—tRNAXXX with FLEx-adeno-associated virus (AAVs) and Cre recombinase strategy to measure neuron-type specific proteomes [34]. Pyrrolysyl-tRNA synthetase—tRNAXXX pair was only expressed in Cre-positive cells. In these cells, a non-canonical amino acid substrate with bioorthogonal cyclopropene group (CypK), leads to stochastic but low level labeling of the proteome with CypK. Proteins containing the non-canonical amino acids are clicked with biotin, and enriched with streptavidin. By combining BOS strategies with stable isotope-based quantitative proteomics, the measurement of discrete neuron-type specific proteomes in vivo can be achieved. BOSs are well suited to study cell type-specifically proteomes in vivo and represent an emerging field of chemical biology (Figure 2C). However, it is possible that ANL and CypK labeling impairs protein function. BOSs are dependent on new protein synthesis. It makes this strategy biased for those proteins with high translation rates and potentially slow degradation rates. Therefore, BOSs are suitable to probe neuron-type specific synaptic dynamic changes such as developmental maturation and activity-induced changes. Therefore, the measurement of long-live synaptic proteins will be in accessible [36].

4. Strategies to Overcome Averaging at the Molecular Level

Determining the proteome of a single type of neuron or synapse within the mammalian brain is a monumental challenge. Except for LCM, the strategies described above are not sufficiently sensitive or robust to identify and measure cell type-specific synaptic proteomes. To home in on synaptic proteins from the proteomic datasets generated from these approaches, it is common to filter protein lists based on publically available synaptic proteins databases, such as SynaptomeDB or G2C:Synapse Proteomics [37,38]. Overall, this strategy works well, however the confidence of one's findings will only be as specific as the databases used to filter the datasets are accurate. For example, this strategy may be misleading since these proteins may not be present at synapses in the specific neurons under investigation. FACS and BOS-based approaches are powerful but may be hindered by two limitations when used to measure cell type-specific synaptic proteomes. The first limitation is that these strategies will provide analyses with moderate sensitivity but are unlikely to provide the means to measure low abundance synaptic proteins since their signals will be quite low relative to the entire neuronal proteome. Second, these approaches may also suffer from the fact that many proteins localize to synapses in addition to other cellular locations. For example, many synaptic proteins are processed through the ER, Golgi apparatus, endosomal vesicles, and even lysosomes [39]. This situation is even more of a concern when investigating neurological disorders and diseases since non-synaptic pathogenic proteins accumulate or are removed from synaptic compartments in the context of disease [40,41]. For example, the loss of dopaminergic presynaptic terminals in PD brains is likely to have severe direct and compensatory effects on basal ganglia synaptic proteomes [42]. Collectively, the interpretation of "synaptic" proteomic datasets need to be interpreted with care and confirmed with additional experiments whenever possible. One interesting strategy that has provided an understanding of synapse type protein interaction networks has been the selective expression of tagged synaptic protein with affinity purification and MS analysis [43,44].

Recently, Ting and Roux have reported two breakthrough methods to identify and quantitate spatially restricted proteomes by promiscuously tagging proteins with biotin via ascorbate or horseradish peroxidase (i.e., APEX, HRP), and BioID respectively [45,46]. They are based on similar principles and workflow but use different biotin-tagging enzymes, modified peroxidases or *E. coli* biotin ligase (BirA-R118G, or BirA*). These enzymes have been fused to a wide-range of different subcellular targeting proteins and peptides in order to probe a panel of different subcellular specific localizations, including mitochondria and synapses [45,47,48]. These strategies are well suited to determining synapse-type specific proteomes in vitro and in vivo, by fusing HRP, APEX or BirA* to PSD95 (e.g., excitatory post-synapses) or gephyrin (e.g., inhibitory postsynapses). Then small molecules that trigger enzyme-catalyzed biotinylation of proximal endogenous proteins are administered to animals or living cells [49,50]. Subsequently, the biotinylated proteins are enriched with streptavidin or anti-biotin antibody conjugated beads, and their identities and levels are determined by MS (Figure 3). In this way, HRP-LRRTM2 was recently used to identify 199 glutamatergic and 42 GABAergic synaptic cleft proteins in cultured neurons, and BioID revealed an elaborate inhibitory postsynaptic protein network in the developing mouse cortex [48,51]. Compared to traditional affinity purifications in combination with MS-based protein identification and quantification, proximity tagging technologies represent complementary, and in some cases more sensitive, analysis workflows that allow the measurement of low abundant proteins [44,52]. One key advantage of HRP/APEX and BioID over traditional affinity purification-based strategies is that the biotin protein tagging occurs in situ, which eliminates the requirement that protein-protein interactions survive detergent based lysis and biochemical purification.

However, there are two major disadvantages of proximity tagging technologies worth noting. First, just like BOSs, expression of promiscuous biotin-tagging enzymes may alter synaptic function. Second, a small panel of high abundance proteins are robustly biotinylated (i.e., pyruvate carboxylase, propionyl-CoA carboxylase, acetyl-CoA carboxylase) and tend to dominate the MS spectra and thus limit the detection of low abundance tagged proteins of interest. We have found that extensive peptide fractionation, rather than depleting the endogenously biotinylated proteins, is an effective work around. Finally, both HRP/APEX and BioID require well-designed bioinformatic strategies to identify top protein candidates from potentially long lists [47].

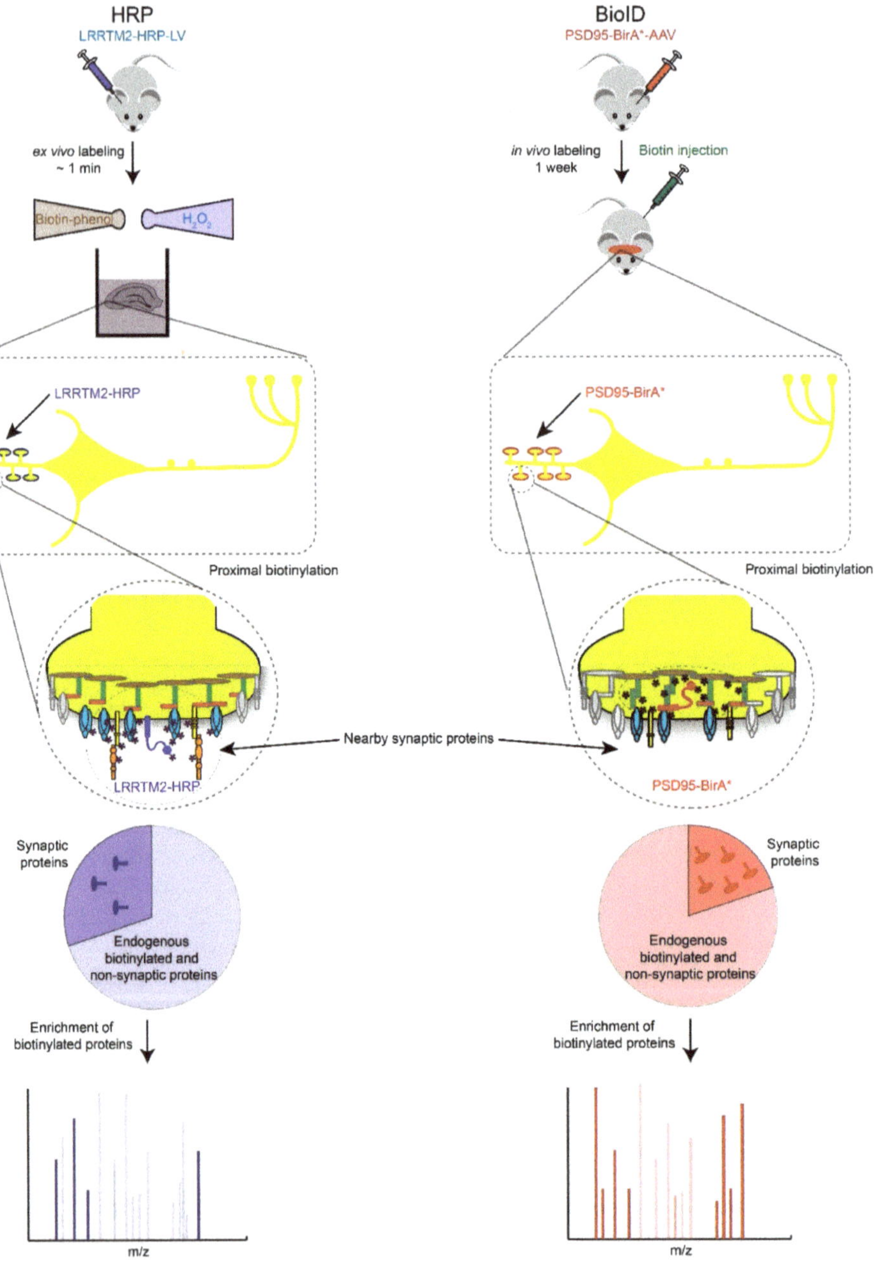

Figure 3. Proximity biotin-tagging strategies. Ascorbate or Horseradish Peroxidase (APEX or HRP)-based strategies are most suitable to profile synaptic proteomes in vitro or ex vivo because of biotin-phenol and H_2O_2 are toxic. Application of APEX or HRP based-strategies have been primarily used in cultured cells and neurons but we speculate here how these strategies could be used ex vivo. BioID-based proteomics works well in vivo but requires long incubation period (hours to days) to obtain high levels of biotinylated proteins, which raises the background biotinylation levels and increases the number of potential false positives.

5. Conclusions and Future Perspectives

The recent development of highly sensitive and robust RNA sequencing (RNA-Seq) technologies has led to a boom in single cell analyses and significantly increased our understanding of global gene expression patterns of discrete neuronal populations in healthy and stressed conditions [53–55]. However, single cell RNA-Seq does not provide the subcellular resolution or protein expression data needed to interrogate discrete synapses or provide insight into the actual synapse-type specific protein expression patterns. MS-based strategies are beginning to bridge this gap, and over the coming years, we will eventually reach a deep understanding of neuron-type and synapse-type specific proteomes. Besides the 'averaging problem', many other obstacles are hindering our understanding of synaptic proteomes. Post translation modification (PTM) is one such example. In many conditions, the initiation of long-term potentiation requires SUMOylation and phosphorylation of many pre- and postsynaptic proteins [56–58]. PTMs complicate peptide spectra and consequently reduce the depth of traditional proteomic analyses. Now researchers are developing new sample preparation approaches and MS search algorithms (e.g., open search) to better handle PTMs [59]. Altogether, the accumulating knowledge of neuron-type specific and synapse-type specific proteomes will accelerate our basic understanding of synapses and neuronal networks and may pave a path towards the effective treatment of neurological disorders and diseases.

Author Contributions: Both authors contributed equally to the preparation of the manuscript.

Funding: This research was funded by an Individual Biomedical Research Award from The Hartwell Foundation (J.N.S).

Acknowledgments: We would like to thank the members of the Savas laboratory for their comments and suggestions.

Conflicts of Interest: The authors declare no conflict of interest.

References

1. Laughlin, S.B.; Sejnowski, T.J. Communication in neuronal networks. *Science* **2003**, *301*, 1870–1874. [CrossRef] [PubMed]
2. Scheiffele, P. Cell-cell signaling during synapse formation in the CNS. *Annu. Rev. Neurosci.* **2003**, *26*, 485–508. [CrossRef] [PubMed]
3. Yogev, S.; Shen, K. Cellular and molecular mechanisms of synaptic specificity. *Annu. Rev. Cell Dev. Biol.* **2014**, *30*, 417–437. [CrossRef] [PubMed]
4. Shen, K.; Scheiffele, P. Genetics and cell biology of building specific synaptic connectivity. *Annu. Rev. Neurosci.* **2010**, *33*, 473–507. [CrossRef] [PubMed]
5. Mitchell, G.S.; Johnson, S.M. Neuroplasticity in respiratory motor control. *J. Appl. Physiol. (1985)* **2003**, *94*, 358–374. [CrossRef] [PubMed]
6. Mayford, M.; Siegelbaum, S.A.; Kandel, E.R. Synapses and memory storage. *Cold Spring Harb. Perspect. Biol.* **2012**, *4*, a005751. [CrossRef] [PubMed]
7. Albin, R.L.; Young, A.B.; Penney, J.B. The functional anatomy of basal ganglia disorders. *Trends Neurosci.* **1989**, *12*, 366–375. [CrossRef]
8. Lloyd, K.G.; Davidson, L.; Hornykiewicz, O. The neurochemistry of Parkinson's disease: Effect of L-dopa therapy. *J. Pharmacol. Exp. Ther.* **1975**, *195*, 453–464. [PubMed]
9. Andre, V.M.; Fisher, Y.E.; Levine, M.S. Altered Balance of Activity in the Striatal Direct and Indirect Pathways in Mouse Models of Huntington's Disease. *Front. Syst. Neurosci.* **2011**, *5*, 46. [CrossRef] [PubMed]
10. Maloney, S.E.; Rieger, M.A.; Dougherty, J.D. Identifying essential cell types and circuits in autism spectrum disorders. *Int. Rev. Neurobiol.* **2013**, *113*, 61–96. [PubMed]
11. Lewis, D.A.; Sweet, R.A. Schizophrenia from a neural circuitry perspective: Advancing toward rational pharmacological therapies. *J. Clin. Investig.* **2009**, *119*, 706–716. [CrossRef] [PubMed]
12. Mu, Y.; Gage, F.H. Adult hippocampal neurogenesis and its role in Alzheimer's disease. *Mol. Neurodegener.* **2011**, *6*, 85. [CrossRef] [PubMed]

13. Bereczki, E.; Branca, R.M.; Francis, P.T.; Pereira, J.B.; Baek, J.H.; Hortobagyi, T.; Winblad, B.; Ballard, C.; Lehtio, J.; Aarsland, D. Synaptic markers of cognitive decline in neurodegenerative diseases: A proteomic approach. *Brain* **2018**, *141*, 582–595. [CrossRef] [PubMed]
14. Savas, J.N.; Wang, Y.Z.; DeNardo, L.A.; Martinez-Bartolome, S.; McClatchy, D.B.; Hark, T.J.; Shanks, N.F.; Cozzolino, K.A.; Lavallee-Adam, M.; Smukowski, S.N.; et al. Amyloid Accumulation Drives Proteome-wide Alterations in Mouse Models of Alzheimer's Disease-like Pathology. *Cell Rep.* **2017**, *21*, 2614–2627. [CrossRef] [PubMed]
15. Herculano-Houzel, S.; Mota, B.; Lent, R. Cellular scaling rules for rodent brains. *Proc. Natl. Acad. Sci. USA* **2006**, *103*, 12138–12143. [CrossRef] [PubMed]
16. Herculano-Houzel, S. The remarkable, yet not extraordinary, human brain as a scaled-up primate brain and its associated cost. *Proc. Natl. Acad. Sci. USA* **2012**, *109*, 10661–10668. [CrossRef] [PubMed]
17. Zingg, B.; Hintiryan, H.; Gou, L.; Song, M.Y.; Bay, M.; Bienkowski, M.S.; Foster, N.N.; Yamashita, S.; Bowman, I.; Toga, A.W.; et al. Neural networks of the mouse neocortex. *Cell* **2014**, *156*, 1096–1111. [CrossRef] [PubMed]
18. Schreiner, D.; Savas, J.N.; Herzog, E.; Brose, N.; de Wit, J. Synapse biology in the 'circuit-age'-paths toward molecular connectomics. *Curr. Opin. Neurobiol.* **2017**, *42*, 102–110. [CrossRef] [PubMed]
19. Farber, N.B.; Newcomer, J.W.; Olney, J.W. The glutamate synapse in neuropsychiatric disorders. Focus on schizophrenia and Alzheimer's disease. *Prog. Brain Res.* **1998**, *116*, 421–437. [PubMed]
20. Vaaga, C.E.; Borisovska, M.; Westbrook, G.L. Dual-transmitter neurons: Functional implications of co-release and co-transmission. *Curr. Opin. Neurobiol.* **2014**, *29*, 25–32. [CrossRef] [PubMed]
21. Niswender, C.M.; Conn, P.J. Metabotropic glutamate receptors: Physiology, pharmacology, and disease. *Annu. Rev. Pharmacol. Toxicol.* **2010**, *50*, 295–322. [CrossRef] [PubMed]
22. Gerfen, C.R.; Surmeier, D.J. Modulation of striatal projection systems by dopamine. *Annu. Rev. Neurosci.* **2011**, *34*, 441–466. [CrossRef] [PubMed]
23. Zhai, S.; Tanimura, A.; Graves, S.M.; Shen, W.; Surmeier, D.J. Striatal synapses, circuits, and Parkinson's disease. *Curr. Opin. Neurobiol.* **2018**, *48*, 9–16. [CrossRef] [PubMed]
24. Heiman, M.; Schaefer, A.; Gong, S.; Peterson, J.D.; Day, M.; Ramsey, K.E.; Suarez-Farinas, M.; Schwarz, C.; Stephan, D.A.; Surmeier, D.J.; et al. A translational profiling approach for the molecular characterization of CNS cell types. *Cell* **2008**, *135*, 738–748. [CrossRef] [PubMed]
25. Beaulieu, J.M.; Gainetdinov, R.R. The physiology, signaling, and pharmacology of dopamine receptors. *Pharmacol. Rev.* **2011**, *63*, 182–217. [CrossRef] [PubMed]
26. Rothwell, P.E.; Fuccillo, M.V.; Maxeiner, S.; Hayton, S.J.; Gokce, O.; Lim, B.K.; Fowler, S.C.; Malenka, R.C.; Sudhof, T.C. Autism-associated neuroligin-3 mutations commonly impair striatal circuits to boost repetitive behaviors. *Cell* **2014**, *158*, 198–212. [CrossRef] [PubMed]
27. Peng, J.; Kim, M.J.; Cheng, D.; Duong, D.M.; Gygi, S.P.; Sheng, M. Semiquantitative proteomic analysis of rat forebrain postsynaptic density fractions by mass spectrometry. *J. Biol. Chem.* **2004**, *279*, 21003–21011. [CrossRef] [PubMed]
28. Butko, M.T.; Savas, J.N.; Friedman, B.; Delahunty, C.; Ebner, F.; Yates, J.R., 3rd; Tsien, R.Y. In vivo quantitative proteomics of somatosensory cortical synapses shows which protein levels are modulated by sensory deprivation. *Proc. Natl. Acad. Sci. USA* **2013**, *110*, E726–E735. [CrossRef] [PubMed]
29. Curran, S.; McKay, J.A.; McLeod, H.L.; Murray, G.I. Laser capture microscopy. *Mol. Pathol.* **2000**, *53*, 64–68. [CrossRef] [PubMed]
30. Kennard, J.T.; Guevremont, D.; Mason-Parker, S.E.; Abraham, W.C.; Williams, J.M. Redistribution of ionotropic glutamate receptors detected by laser microdissection of the rat dentate gyrus 48 h following LTP induction in vivo. *PLoS ONE* **2014**, *9*, e92972. [CrossRef] [PubMed]
31. Fulwyler, M.J. Electronic separation of biological cells by volume. *Science* **1965**, *150*, 910–911. [CrossRef] [PubMed]
32. Hickox, A.E.; Wong, A.C.; Pak, K.; Strojny, C.; Ramirez, M.; Yates, J.R., 3rd; Ryan, A.F.; Savas, J.N. Global Analysis of Protein Expression of Inner Ear Hair Cells. *J. Neurosci.* **2017**, *37*, 1320–1339. [CrossRef] [PubMed]
33. Biesemann, C.; Gronborg, M.; Luquet, E.; Wichert, S.P.; Bernard, V.; Bungers, S.R.; Cooper, B.; Varoqueaux, F.; Li, L.; Byrne, J.A.; et al. Proteomic screening of glutamatergic mouse brain synaptosomes isolated by fluorescence activated sorting. *EMBO J.* **2014**, *33*, 157–170. [CrossRef] [PubMed]

34. Krogager, T.P.; Ernst, R.J.; Elliott, T.S.; Calo, L.; Beranek, V.; Ciabatti, E.; Spillantini, M.G.; Tripodi, M.; Hastings, M.H.; Chin, J.W. Labeling and identifying cell-specific proteomes in the mouse brain. *Nat. Biotechnol.* **2018**, *36*, 156–159. [CrossRef] [PubMed]
35. Alvarez-Castelao, B.; Schanzenbacher, C.T.; Hanus, C.; Glock, C.; Tom Dieck, S.; Dorrbaum, A.R.; Bartnik, I.; Nassim-Assir, B.; Ciirdaeva, E.; Mueller, A.; et al. Cell-type-specific metabolic labeling of nascent proteomes in vivo. *Nat. Biotechnol.* **2017**, *35*, 1196–1201. [CrossRef] [PubMed]
36. Heo, S.; Diering, G.H.; Na, C.H.; Nirujogi, R.S.; Bachman, J.L.; Pandey, A.; Huganir, R.L. Identification of long-lived synaptic proteins by proteomic analysis of synaptosome protein turnover. *Proc. Natl. Acad. Sci. USA* **2018**, *115*, E3827–E3836. [CrossRef] [PubMed]
37. Pirooznia, M.; Wang, T.; Avramopoulos, D.; Valle, D.; Thomas, G.; Huganir, R.L.; Goes, F.S.; Potash, J.B.; Zandi, P.P. SynaptomeDB: An ontology-based knowledgebase for synaptic genes. *Bioinformatics* **2012**, *28*, 897–899. [CrossRef] [PubMed]
38. Bayes, A.; van de Lagemaat, L.N.; Collins, M.O.; Croning, M.D.; Whittle, I.R.; Choudhary, J.S.; Grant, S.G. Characterization of the proteome, diseases and evolution of the human postsynaptic density. *Nat. Neurosci.* **2011**, *14*, 19–21. [CrossRef] [PubMed]
39. Savas, J.N.; Ribeiro, L.F.; Wierda, K.D.; Wright, R.; DeNardo-Wilke, L.A.; Rice, H.C.; Chamma, I.; Wang, Y.Z.; Zemla, R.; Lavallee-Adam, M.; et al. The Sorting Receptor SorCS1 Regulates Trafficking of Neurexin and AMPA Receptors. *Neuron* **2015**, *87*, 764–780. [CrossRef] [PubMed]
40. McKinstry, S.U.; Karadeniz, Y.B.; Worthington, A.K.; Hayrapetyan, V.Y.; Ozlu, M.I.; Serafin-Molina, K.; Risher, W.C.; Ustunkaya, T.; Dragatsis, I.; Zeitlin, S.; et al. Huntingtin is required for normal excitatory synapse development in cortical and striatal circuits. *J. Neurosci.* **2014**, *34*, 9455–9472. [CrossRef] [PubMed]
41. Spires-Jones, T.L.; Hyman, B.T. The intersection of amyloid beta and tau at synapses in Alzheimer's disease. *Neuron* **2014**, *82*, 756–771. [CrossRef] [PubMed]
42. Picconi, B.; Piccoli, G.; Calabresi, P. Synaptic dysfunction in Parkinson's disease. *Adv. Exp. Med. Biol.* **2012**, *970*, 553–572. [PubMed]
43. Heller, E.A.; Zhang, W.; Selimi, F.; Earnheart, J.C.; Slimak, M.A.; Santos-Torres, J.; Ibanez-Tallon, I.; Aoki, C.; Chait, B.T.; Heintz, N. The biochemical anatomy of cortical inhibitory synapses. *PLoS ONE* **2012**, *7*, e39572. [CrossRef] [PubMed]
44. Selimi, F.; Cristea, I.M.; Heller, E.; Chait, B.T.; Heintz, N. Proteomic studies of a single CNS synapse type: The parallel fiber/purkinje cell synapse. *PLoS Biol.* **2009**, *7*, e83. [CrossRef] [PubMed]
45. Rhee, H.W.; Zou, P.; Udeshi, N.D.; Martell, J.D.; Mootha, V.K.; Carr, S.A.; Ting, A.Y. Proteomic mapping of mitochondria in living cells via spatially restricted enzymatic tagging. *Science* **2013**, *339*, 1328–1331. [CrossRef] [PubMed]
46. Roux, K.J.; Kim, D.I.; Raida, M.; Burke, B. A promiscuous biotin ligase fusion protein identifies proximal and interacting proteins in mammalian cells. *J. Cell Biol.* **2012**, *196*, 801–810. [CrossRef] [PubMed]
47. Loh, K.H.; Stawski, P.S.; Draycott, A.S.; Udeshi, N.D.; Lehrman, E.K.; Wilton, D.K.; Svinkina, T.; Deerinck, T.J.; Ellisman, M.H.; Stevens, B.; et al. Proteomic Analysis of Unbounded Cellular Compartments: Synaptic Clefts. *Cell* **2016**, *166*, 1295–1307. [CrossRef] [PubMed]
48. Uezu, A.; Kanak, D.J.; Bradshaw, T.W.; Soderblom, E.J.; Catavero, C.M.; Burette, A.C.; Weinberg, R.J.; Soderling, S.H. Identification of an elaborate complex mediating postsynaptic inhibition. *Science* **2016**, *353*, 1123–1129. [CrossRef] [PubMed]
49. Reinke, A.W.; Mak, R.; Troemel, E.R.; Bennett, E.J. In vivo mapping of tissue- and subcellular-specific proteomes in Caenorhabditis elegans. *Sci. Adv.* **2017**, *3*, e1602426. [CrossRef] [PubMed]
50. Chen, C.L.; Hu, Y.; Udeshi, N.D.; Lau, T.Y.; Wirtz-Peitz, F.; He, L.; Ting, A.Y.; Carr, S.A.; Perrimon, N. Proteomic mapping in live Drosophila tissues using an engineered ascorbate peroxidase. *Proc. Natl. Acad. Sci. USA* **2015**, *112*, 12093–12098. [CrossRef] [PubMed]
51. De Wit, J.; Sylwestrak, E.; O'Sullivan, M.L.; Otto, S.; Tiglio, K.; Savas, J.N.; Yates, J.R., 3rd; Comoletti, D.; Taylor, P.; Ghosh, A. LRRTM2 interacts with Neurexin1 and regulates excitatory synapse formation. *Neuron* **2009**, *64*, 799–806. [CrossRef] [PubMed]
52. Lambert, J.P.; Tucholska, M.; Go, C.; Knight, J.D.; Gingras, A.C. Proximity biotinylation and affinity purification are complementary approaches for the interactome mapping of chromatin-associated protein complexes. *J. Proteom.* **2015**, *118*, 81–94. [CrossRef] [PubMed]

53. Wu, Y.E.; Pan, L.; Zuo, Y.; Li, X.; Hong, W. Detecting Activated Cell Populations Using Single-Cell RNA-Seq. *Neuron* **2017**, *96*, 313–329. [CrossRef] [PubMed]
54. Zeisel, A.; Munoz-Manchado, A.B.; Codeluppi, S.; Lonnerberg, P.; La Manno, G.; Jureus, A.; Marques, S.; Munguba, H.; He, L.; Betsholtz, C.; et al. Brain structure. Cell types in the mouse cortex and hippocampus revealed by single-cell RNA-seq. *Science* **2015**, *347*, 1138–1142. [CrossRef] [PubMed]
55. Usoskin, D.; Furlan, A.; Islam, S.; Abdo, H.; Lonnerberg, P.; Lou, D.; Hjerling-Leffler, J.; Haeggstrom, J.; Kharchenko, O.; Kharchenko, P.V.; et al. Unbiased classification of sensory neuron types by large-scale single-cell RNA sequencing. *Nat. Neurosci.* **2015**, *18*, 145–153. [CrossRef] [PubMed]
56. Daniel, J.A.; Cooper, B.H.; Palvimo, J.J.; Zhang, F.P.; Brose, N.; Tirard, M. Analysis of SUMO1-conjugation at synapses. *eLife* **2017**, *6*, e26338. [CrossRef] [PubMed]
57. Martin, S.; Wilkinson, K.A.; Nishimune, A.; Henley, J.M. Emerging extranuclear roles of protein SUMOylation in neuronal function and dysfunction. *Nat. Rev. Neurosci.* **2007**, *8*, 948–959. [CrossRef] [PubMed]
58. Wilkinson, K.A.; Martin, S.; Tyagarajan, S.K.; Arancio, O.; Craig, T.J.; Guo, C.; Fraser, P.E.; Goldstein, S.A.N.; Henley, J.M. Commentary: Analysis of SUMO1-conjugation at synapses. *Front. Cell. Neurosci.* **2017**, *11*, 345. [CrossRef] [PubMed]
59. Chick, J.M.; Kolippakkam, D.; Nusinow, D.P.; Zhai, B.; Rad, R.; Huttlin, E.L.; Gygi, S.P. A mass-tolerant database search identifies a large proportion of unassigned spectra in shotgun proteomics as modified peptides. *Nat. Biotechnol.* **2015**, *33*, 743–749. [CrossRef] [PubMed]

proteomes

Article

Regional Diversity in the Postsynaptic Proteome of the Mouse Brain

Marcia Roy [1,†], Oksana Sorokina [2,†], Colin McLean [2], Silvia Tapia-González [3], Javier DeFelipe [3], J. Douglas Armstrong [2] and Seth G. N. Grant [1,*]

1 Centre for Clinical Brain Sciences, University of Edinburgh, Edinburgh EH16 4SB, UK; marciamroy@googlemail.com
2 School of Informatics, University of Edinburgh, Edinburgh EH8 9AB, UK; Oksana.Sorokina@ed.ac.uk (O.S.); cmclean5@staffmail.ed.ac.uk (C.M.); douglas.armstrong@ed.ac.uk (J.D.A.)
3 Departamento de Neurobiología Funcional y de Sistemas, Instituto Cajal (CSIC), Ave. Doctor Arce 37, 28002 Madrid and Laboratorio Cajal de Circuitos Corticales, Centro de Tecnología Biomédica (UPM), 28223 Pozuelo de Alarcón, Madrid, Spain; silvia.tapia@cajal.csic.es (S.T.-G.); defelipe@cajal.csic.es (J.D.)
* Correspondence: Seth.Grant@ed.ac.uk; Tel.: +44-131-650-1000
† These authors contributed equally to this work.

Received: 3 July 2018; Accepted: 27 July 2018; Published: 1 August 2018

Abstract: The proteome of the postsynaptic terminal of excitatory synapses comprises over one thousand proteins in vertebrate species and plays a central role in behavior and brain disease. The brain is organized into anatomically distinct regions and whether the synapse proteome differs across these regions is poorly understood. Postsynaptic proteomes were isolated from seven forebrain and hindbrain regions in mice and their composition determined using proteomic mass spectrometry. Seventy-four percent of proteins showed differential expression and each region displayed a unique compositional signature. These signatures correlated with the anatomical divisions of the brain and their embryological origins. Biochemical pathways controlling plasticity and disease, protein interaction networks and individual proteins involved with cognition all showed differential regional expression. Combining proteomic and connectomic data shows that interconnected regions have specific proteome signatures. Diversity in synapse proteome composition is key feature of mouse and human brain structure.

Keywords: synapse; postsynaptic; proteome; mass spectrometry; protein interaction networks; connectome

1. Introduction

Synapses are the specialized junctions between nerve cells and are present in vast numbers in the mammalian nervous system. During the 1990s, synapses were thought to be relatively simple connectors, but the application of proteomic mass spectrometry in 2000 revealed an unanticipated complexity in their protein composition [1]. Both the presynaptic and postsynaptic proteomes have since been systematically characterized in several vertebrate species and thousands of proteins have been identified [2–13]. Phosphoproteomic studies have shown that neural activity causes changes in large numbers of proteins [14,15]. These findings have transitioned the view of synapses to one where they are highly sophisticated and complex signaling machines that process information. The importance of understanding this complexity is underscored by the finding that over 130 human brain diseases are caused by mutations disrupting postsynaptic proteins [16,17].

It is of fundamental importance to understand how the high number of postsynaptic proteins are organized physically (within synapses) and spatially (between synapses). Biochemical studies have shown that postsynaptic proteins are typically assembled into a hierarchy of complexes and supercomplexes (complexes of complexes) [18–20]. The prototype of postsynaptic supercomplexes are those formed by the scaffolding protein PSD95 (also known as Dlg4). Dimers of PSD95 assemble with complexes of neurotransmitter receptors, ion channels, signaling and structural proteins into a family of high molecular weight (1–3 MDa) structures in excitatory synapses [18–20]. PSD93 (also known as Dlg2), which is a paralog of PSD95, co-assembles with PSD95 to bind *N*-methyl-D-aspartate (NMDA) receptors and these are an important functional subset of the PSD95 supercomplex family. Other combinations of proteins form other subtypes of PSD95 supercomplexes, such as those containing potassium channels and serotonin receptors [20–23]. Together, these members of the PSD95 supercomplex family confer diverse signal processing functions to the synapse.

The principles underlying the spatial organization of synapse proteomes in the brain is less well understood. To date, most studies of the synapse proteome have focused on defining composition from limited regions of the brain (or the whole brain). However, at the macroscopic level, brain architecture is characterized by regions with distinct functions [24]. It is therefore of importance to ask if synapse proteomes differ between brain regions and whether any differences might be relevant to their function or to the connectivity between these regions. In a recent study, we reported that regions of human neocortex differ in the composition of their postsynaptic proteomes and that these compositional differences correlate with functional properties [25]. The present study employs a similar analysis applied to the mouse brain, which allows us to ask if conserved principles may apply across these two species that evolved from a common ancestor ~90 million years ago.

Using a method suitable for the isolation and direct quantification of mouse synapse proteomes from small amounts of brain tissue, we compared and contrasted the synapse proteomes isolated from seven integral regions of the adult mouse brain. The postsynaptic proteome was analyzed to a depth of 1173 proteins and differential expression signatures were identified and characterized in each brain region. We used these datasets to analyze the spatial organization of the postsynaptic proteome in the mouse brain and identify organizational principles shared with humans [25,26]. This large-scale dataset is a useful resource for the field of neuroscience and future studies using mouse models of human disease.

2. Materials and Methods

2.1. Dissections of Mouse Brain Regions

This study was performed using 8-week-old male C57BL/6J mice. All experimental protocols involving the use of animals were performed in accordance with recommendations for the proper care and use of laboratory animals and under the authorization of the regulations and policies governing the care and use of laboratory animals (EU directive No. 86/609 and Council of Europe Convention ETS123, EU decree 2001-486 and Statement of Compliance with Standards for Use of Laboratory Animals by Foreign Institutions No. A5388-01, National Institutes of Health, Bethesda, MD, USA).

The mice (n = 6) were anesthetized with a pentobarbital dose of 40 mg/kg body weight and sacrificed by decapitation. The brains were rapidly removed and kept on ice while the areas of interest were dissected from the right hemisphere using the microdissection method of Palkovits [27]. Large regions were collected from the frontal, medial and caudal cortex, as well as the right caudate putamen, right hippocampus, whole hypothalamus, and cerebellum (right half), which was cut previously through the vermis (Figure S1). The samples were frozen on liquid nitrogen and stored at −80 °C until processed.

2.2. PSD Isolation and Protein Preparations for Mass Spectrometry

Dissected mouse brain regions were homogenized by performing 12 strokes with a Dounce homogenizer containing 2 mL ice-cold homogenization buffer (320 mM sucrose, 1 mM HEPES, pH 7.4) containing 1× Complete EDTA-free protease inhibitor (Roche) and 1× Phosphatase inhibitor cocktail set II (Calbiochem). Synaptosomes were isolated from homogenized mouse brain tissue as described [2]. Briefly, insoluble material was pelleted by centrifugation at 1000× g for 10 min at 4 °C. The supernatant (S1) was removed and the pellet resuspended in 1 mL homogenization buffer and an additional six strokes were performed. Following a second centrifugation at 1000× g for 10 min at 4 °C, the supernatant (S2) was removed and pooled with S1. The combined supernatants were then centrifuged at 18,500× g for 15 min at 4 °C. The pellet was resuspended in 0.25 mL homogenization buffer and 0.25 mL extraction buffer (50 mM NaCl, 1% DOC, 25 mM Tris-HCl, pH 8.0) containing 1× Complete EDTA-free protease inhibitor (Roche) and 1× Phosphatase inhibitor cocktail set II (Calbiochem) and incubated on ice for 1 h. The resulting PSD extracts were centrifuged at 10,000× g for 20 min at 4 °C and the resulting supernatant filtered through a 0.2 μm syringe filter (Millipore).

2.3. Sample Preparation and LC-MS/MS Analysis

All chemicals were purchased from Sigma-Aldrich unless otherwise stated. Acetonitrile and water for HPLC-MS/MS and sample preparation were HPLC quality and were purchased from Thermo Fisher Scientific (Loughborough, UK). Formic acid was supra-pure (90–100%) purchased from Merck KGaA (Darmstadt, Germany) while trypsin sequencing grade was purchased from Promega (Southampton, UK). All HPLC-MS connector fittings were either purchased from Upchurch Scientific (Hichrom) or Valco (RESTEK). Fifty micrograms of PSD proteins were acetone precipitated, protein pellets reconstituted in SDS-PAGE loading buffer, and briefly run on a 4–12% Bis-Tris gradient gel (Invitrogen) for ~10 min. Proteins were in-gel digested using a method similar to that of Shevchenko et al. (2006) [28]. Resulting peptide extracts were then acidified with 7 μL 0.05% TFA and were filtered with a Millex filter (Millipore) before HPLC-MS analysis. Nano-HPLC-MS/MS analysis was performed using an on-line system consisting of a nano-pump (Dionex Ultimate 3000, Thermo Fisher) coupled to a QExactive instrument (Thermo Fisher) with a pre-column of 300 μm × 5 mm (Acclaim Pepmap, 5 μm particle size) connected to a column of 75 μm × 50 cm (Acclaim Pepmap, 3 μm particle size). Samples were analyzed on a 90-min gradient in data-dependent analysis (one survey scan at 70 k resolution followed by the top ten MS/MS).

2.4. Mass Spectrometry and Data Analysis

Data from MS/MS spectra were searched using MASCOT version 2.4 (Matrix Science Ltd., London, UK) against the *Mus musculus* subset of the National Center for Biotechnology Information (NCBI) protein database (382,487 protein sequences) with maximum missed-cut value set to 2. The following features were used in all searches: (i) variable methionine oxidation; (ii) fixed cysteine carbamidomethylation; (iii) precursor mass tolerance of 10 ppm; (iv) MS/MS tolerance of 0.05 amu; (v) significance threshold (p) below 0.05 (MudPIT scoring); and (vi) final peptide score of 20.

Progenesis version 4 (Nonlinear Dynamics, Newcastle upon Tyne, UK) was used for HPLC-MS label-free quantitation. Only MS/MS peaks with a charge of 2+, 3+ or 4+ were considered for the total number of "Feature" (signal at one particular retention time and m/z) and only the five most intense spectra per "Feature" were included. Each LC-MS run was normalized by multiplying a scalar factor. The scalar factor is a ratio in log space of the median intensity of the selected features against the median intensity of the selected feature of a reference spectrum. The associated unique peptide ion intensities for a specific protein were then summed to generate an abundance value and transformed using an ArcSinH function. Based on the abundance values, within group means were calculated and from there the fold changes (in comparison to control) were evaluated. One-way analysis of variance (ANOVA) was used to calculate the p-value based on the transformed abundance values. p-values were adjusted for multiple comparisons and were calculated either from Progenesis version 4 (Nonlinear Dynamics) or using R (R Core Team, 2013) [29] based on Benjamini and Hochberg (1995) [30]. Further analysis was performed by extracting a Z-score calculated on ArcSinH average group.

Differentially expressed proteins were only considered significant in the current study if the following conditions were fulfilled: (i) adjusted p-values (pairwise) less than 0.05; (ii) number of unique peptides detected and used in quantification per protein was at least 2 for the 1173 dataset;

and (iii) absolute fold change was at least 1.3 for differentially abundant proteins and ≤ 0.667 for downregulated proteins.

2.5. Bioinformatic Analyses

The majority of the analysis was performed in the R software environment for statistical computing and graphics. Principal component analysis (PCA) and Tukey test was performed with R package *FactoMineR* and correlation analysis with the package *corrplot*. Differential stability (DS) analysis was performed as described [31]; briefly, for each protein from the list of 1173, the average Pearson correlation coefficient was estimated from 14 pairwise Pearson coefficients for six brain samples. Based on DS, Tukey and PCA analyses, we determined that the data from all six individuals could be combined and the mean protein abundances were then used for all downstream analyses. Heatmaps were generated with use of the *heatmap.2* function from *gplot* R library: parameters were set to default values with the exception of label and dendrogram visualization control. Hierarchical clustering validation and comparison of dendrograms were performed with package *dendextend* [32]. The number of stable clusters was independently assessed with *nbclust* package [33], which provides the list of indices to determine the optimal number of clusters. We selected a set of six clusters [the postsynaptic proteome modules (PPMs)] based on the best combination of indices provided by *nbclust* R package. Individual proteins in each of the six PPMs and their abundances across all seven integral regions are listed in Table S5 and proteins in each module were ranked by their abundance in each of the seven regions of the mouse brain examined.

We used Bioconductor package ClusterProfiler for Gene Ontology (GO) and KEGG enrichment analysis [34] and Bioconductor ReactomePA package for pathway over-representation analysis (http://bioconductor.org/packages/release/bioc/html/ReactomePA.html).

GO function and KEGG pathway enrichment for all proteins was performed using DAVID (https://david.ncifcrf.gov/). Disease enrichment for each brain region and each protein module was performed using DAVID (https://david.ncifcrf.gov/) and KEGG pathway enrichment was then performed by searching ranked protein lists obtained using GSea version 2.1.0 (http://software.broadinstitute.org/gsea/index.jsp) as previously described [35].

Circular hierarchical clustering of protein modules for the visualization of inter-region molecular interactions was performed using Circos (http://circos.ca) [36]. A Circos configuration file was created representing brain regions as "karyotypes". All proteins were grouped into "modules" according to their abundance similarity. Proteins that have positive abundance in more than one region were shown as links between regions. The width of each link is proportional to the fraction of the regional proteins that contributed to the link, while its color corresponds to that of the respective "module". All preprocessing of the relative abundance information and generation of appropriate Circos files were performed in R. Scripts are available on request.

For DS analysis, we used the described approach [31] on the MS intensity values obtained for all 1173 proteins identified with a minimum of two unique peptides in order to identify synaptic proteins with highly reproducible expression patterns across all six independent mouse brains. The average pairwise Pearson correlation ρ over the six individual mouse brains was quantified and obtained DS values ranged from 0.96 to -0.22 (avgCor, Table S3). As DS reflects the tendency of a gene to exhibit reproducible differential expression relationships across brain structures, the higher DS value represents a more reproducible relationship.

For correlation with mesoscale mouse connectome data [37], the mean voxel sum for each region was calculated with respect to all other regions and itself. The correlation of this matrix was then estimated with the matrix of protein abundances. The results are listed in Table S3.

2.6. Protein Interaction Identification and Mapping

The full postsynaptic proteome network was built from the list of 1173 proteins obtained in this study and protein-protein interactions (PPIs) obtained by mining publicly available databases:

BioGRID [38], IntAct [39] and DIP [40] both for mouse and human. The total network consists of 1016 proteins and 8105 PPIs. We applied weights to each interaction based on abundance values for specific brain regions as follows: mean (ExpA, ExpB), so that for each of the regions the specific weight for each of the interactions could be determined. Having varying abundances for interacting proteins in different brain regions, we estimated the region-specific edge that resulted in region-specific PPI. Each brain region network was clustered making use of the spectral properties of the network; the network being expressed in terms of its eigenvectors and eigenvalues, and partitioned recursively (using a fine-tuning step) into communities based on maximizing the clustering measure modularity [41–43]; the modularity of the networks was found to be 0.28–0.42. Modularity (Q) measures the quality of a network division into communities from the number of edges found relative to the number expected if placed at random. The modularity value lies in the range 0, which indicates clustering no better than random, to 1, with typical values for real networks ranging from 0.3 to 0.7 [42].

Enrichment for biological process and cellular component was performed using the topGO package (https://bioconductor.org/packages/release/bioc/html/topGO.html), while functional enrichment of synaptic proteins/gene groups that are known risk factors for schizophrenia was performed using the published schizophrenia risk factor dataset [44].

The stability of all interactions across the region were assessed by comparison of clustering results for each of seven region-specific networks and assigning each interaction a score of 0 if both proteins appeared in the same cluster and a score of 1 if it appeared in a different cluster. Scores were summed over all seven regions, resulting in sums ranging from 0 (proteins remain in the same cluster in all regions) to 7 (proteins never appear in the same cluster). For the "stable" network, we selected interactions with scores ≤ 2, which means that they persist in the same cluster in 5/7 (70%) regional networks.

PPI networks were visualized with Gephi (https://gephi.org).

For disease enrichment analysis, the community and protein robustness values within the range 0–1 were taken as edge weights. Each region network was then clustered and cluster enrichment was assessed using the TopOntop package (https://github.com/hxin/topOnto) and OMIM/Ensemble Var/genetic annotation data. For disease enrichment the annotation data were standardized using MetaMap [45–47] and NCBO Annotator (https://www.bioontology.org/annotator-service) to recognize terms found in the Human Disease Ontology (HDO) [48]. Recognized enriched disease ontology terms were then associated with gene identifiers and stored locally. Disease term enrichment could then be calculated using the topology-based elimination Fisher method [49] found in the topGO package (http://topgo.bioinf-mpi-inf.mpg.de/), together with the standardized OMIM/GeneRIF/Ensembl variation gene-disease annotation data (17,731 gene-disease associations), and the full HDO tree (3140 terms). Each region was then examined individually by performing the clustering analysis and enrichment for each of the clusters identified in each of the seven brain regions.

3. Results

3.1. Quantification of Postsynaptic Proteins from Brain Regions

Seven integral brain regions within the forebrain (prosencephalon) and hindbrain (rhombencephalon) were dissected from six eight-week-old C57BL/6J mice (Figure 1A and Figure S1). Forebrain regions included telencephalic structures: frontal cortex (CxF), medial cortex (CxM), caudal cortex (CxCA), hippocampus (Hip), and striatum (ST); the hypothalamus (Hyp) represented a diencephalic structure; and the hindbrain was represented by cerebellum (CB). These represent major brain regions with different structural and functional attributes and which can be relatively easy dissected from the brain.

PSD fractions were prepared from six mice and all 42 samples were analyzed using LC-MS/MS. Label-free quantitation of peptide intensity identified 1173 proteins across all seven brain regions (Table S1). We found a significant overlap between our dataset and those obtained in other mouse studies [4–6,10–12,22] (Figure S2A,B); the 61 proteins unique to this study are summarized in Table S2.

To determine the validity of pooling data from six mice, we performed several analyses. We first used the differential stability (DS) approach, which has been previously applied to transcriptomic and proteomic analyses of adult human brain regions [25,50] (Table S3). For this, we estimated the average pairwise Pearson correlation to identify the proteins that demonstrate similar patterns across all six brains. From 1173 synaptic proteins, roughly half (572) displayed a high DS correlation, or similar expression patterns across all brain regions (Table S3A,B). We found that these were functionally enriched in synaptic transmission proteins ($q = 8.25 \times 10^{-19}$), ATP metabolic processes ($q = 1.33 \times 10^{-12}$) and calcium ion transporting proteins ($q = 2.02 \times 10^{-9}$). Proteins involved in pathways associated with learning ($q = 8.08 \times 10^{-6}$), memory ($q = 6.23 \times 10^{-3}$) and behavior ($q = 1.39 \times 10^{-7}$) were also over-represented in this high DS subset. Components of several KEGG pathways were also highly correlated between individuals, including long-term potentiation ($q = 7.95 \times 10^{-7}$), calcium signaling pathways ($q = 1.13 \times 10^{-5}$), Huntington's ($q = 2.66 \times 10^{-7}$), Alzheimer's ($q = 2.14 \times 10^{-5}$) and Parkinson's disease ($q = 8.95 \times 10^{-6}$) (Table S3C). The most highly conserved proteins among the six individual mice with the highest DS values were STX1A ($p = 0.96$), STUM ($p = 0.96$), CDH13 ($p = 0.95$) and ATP1B2 ($p = 0.95$) (Table S3B). We also compared the distribution of synaptic protein abundances across individual mice by principal component analysis (PCA). This analysis (Figure S3A) indicates that the synaptic proteome of the mice largely overlap, with brains A–C corresponding to the central region of the distribution. As Tukey's HSD test shows no significant difference in the mean values of six individuals at a confidence level of 95% (Figure S3B), we determined that the data from all six individuals could be combined and the mean protein abundances were then used for all downstream analyses.

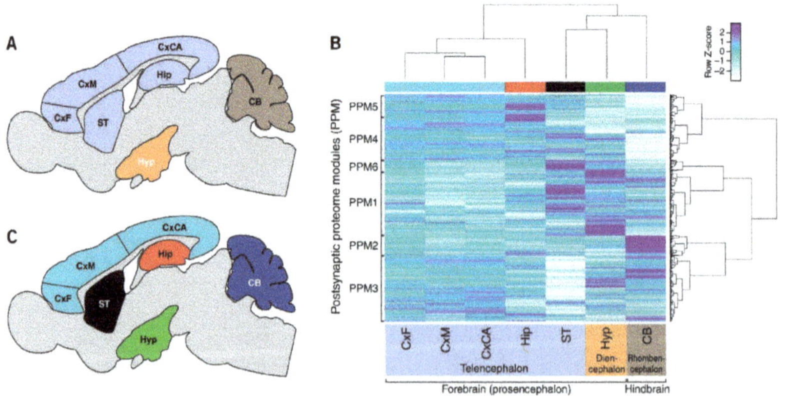

Figure 1. Signatures of postsynaptic proteome composition in mouse brain regions. (**A**) Seven integral brain regions in mouse: frontal cortex (CxF), medial cortex (CxM), caudal cortex (CxCA), hippocampus (Hip), striatum (ST), hypothalamus (Hyp), and cerebellum (CB). Color coded according to vertebrate embryological regions (as in (**B**)). (**B**) Hierarchical clustering by region (x-axis) and protein abundance (y-axis) shows that each region has a unique signature of postsynaptic proteome composition. (**C**) Neuroanatomical map of clusters in (**B**) showing proteome organization into forebrain and hindbrain structures: telencephalon, diencephalon and rhombencephalon.

3.2. *Regional Differences in Postsynaptic Proteome Composition*

To identify postsynaptic proteins with differential expression between brain regions, proteins having a mean peptide intensity of 1.5-fold or greater in one brain region compared with any other and determined to be significant with $p < 0.05$ were identified (Table S4). Eight hundred sixty-eight (74%) proteins were found to be differentially expressed in at least one region compared with all others (Table S4). The regions with the largest number of differentially expressed proteins were the cerebellum (251), hypothalamus (243) and striatum (161). By contrast, the frontal (14), medial (70) and caudal cortex (34) were found to contain the lowest number of differentially expressed proteins compared with all other regions (Figure 1C).

Hierarchical clustering of all proteins revealed that each region has a unique signature of expression. Moreover, these signatures are organized in line with the classical anatomical architecture of the brain: the three cortical regions showed greatest similarity, and the next most similar region was the hippocampus, then striatum, hypothalamus and cerebellum (Figure 1B). This clustering reflects the embryological divisions of the vertebrate brain into telencephalon, diencephalon and rhombencephalon (Figure 1C). Moreover, these results complement findings in the human neocortex, where unique signatures were also found for each region. Together, these findings indicate that compositional differences in the postsynaptic proteome reflect, at least in part, the embryological patterning mechanisms that define brain regions.

This clustering approach also allowed us to examine region-specific functions. We identified six sets of proteins, which we call postsynaptic proteome modules (PPM 1–6) (Figure 1B and Table S5). As indicated by the clustering and Circos plots (Figure S4), these PPMs were differentially distributed in brain regions. To understand the functional significance of differential protein expression in modules and regions, we analyzed the KEGG biochemical and disease pathways in PPMs (Figure 2A). The PPMs showed differential composition of pathways. For example, neurodegenerative diseases were found in PPM1, whereas synaptic plasticity (long-term potentiation and long-term depression) and relevant signaling pathways were in PPM2.

Examination of KEGG pathway enrichment in brain regions (Figure 2B) revealed three major groups. It is striking that very similar groupings were observed in the analysis of human neocortical regions. Group 1 contained terms including MAPK, chemokine, neurotrophin pathways; Group 2 included synaptic plasticity mechanisms and calcium signaling; and Group 3 included neurodegenerative diseases (Alzheimer's, Huntington's, and Parkinson's) and metabolic mechanisms (glycolysis/gluconeogenesis and oxidative phosphorylation). These findings suggest that biochemical pathways in the postsynaptic proteome are differentially distributed across brain regions and that the mechanisms controlling this distribution are species conserved.

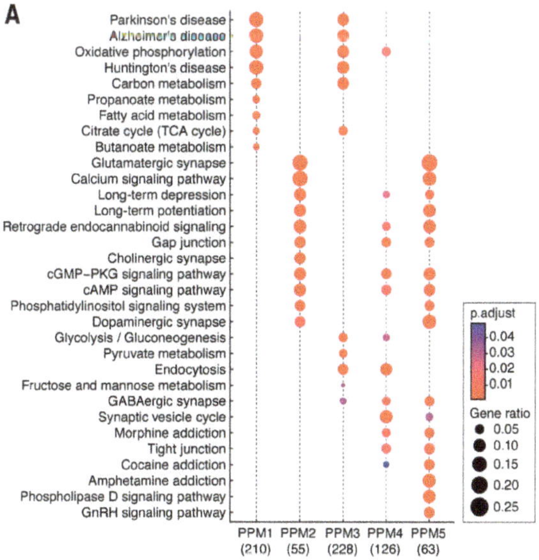

Figure 2. *Cont.*

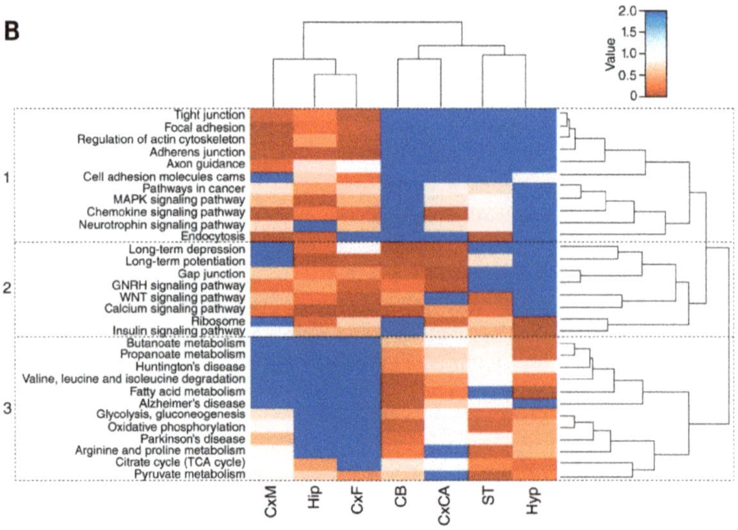

Figure 2. Biochemical pathways and functions in brain regions and postsynaptic proteome modules (PPMs). (**A**) KEGG pathway term (*y*-axis) enrichment in PPMs (*x*-axis) with the number of proteins contributing to KEGG enrichment indicated in brackets. Size of the dots represents the number of genes associated with that pathway (GeneRatio) and the significance indicated by the p-adjust color bar. (**B**) Heatmap of the KEGG biochemical pathway and disease enrichment terms (*y*-axis) based on the ranked abundance of postsynaptic proteins in each region (*x*-axis). Three clusters of KEGG terms are boxed: (1) many signal transduction mechanisms; (2) synaptic plasticity and other signaling processes; and (3) neurodegenerative diseases and metabolic mechanisms.

3.3. Distribution of Mechanism of Cognition and Protein Complexes

The seven regions of the brain examined in this study are thought to play distinct but interdependent roles in cognitive function. Therefore, we examined the distribution of 33 selected proteins that are known to play roles in cognition (Figure 3A). Hierarchical clustering shows that the three cortical regions examined (CxF, CxM, and CxCA) cluster together by similarity, while the Hip region clusters separately from all others. The Hyp and ST regions cluster together by similarity in their abundances of proteins involved in memory and cognition, while the CB clusters separately from all of the other six regions. The abundances of these proteins clustered into two main branches (Figure 3A).

To assess the heterogeneity of synaptic protein complexes throughout the brain, the abundances of the four MAGUK scaffold protein paralogs Dlg1 (also known as Sap97), PSD93 (Dlg2), Dlg3 (also known as Sap102) and PSD95 (Dlg4) were mapped across the various brain regions. We found that these four molecules, which play fundamental roles in synaptic transmission, were differentially distributed throughout the brain, with Dlg1 being most abundant in the synapses of the CxM and PSD93 most abundant in Hip, CxCA and CxF. By contrast, both Dlg3 and PSD95 showed similar protein abundance profiles across the various brain regions (Figure 3B).

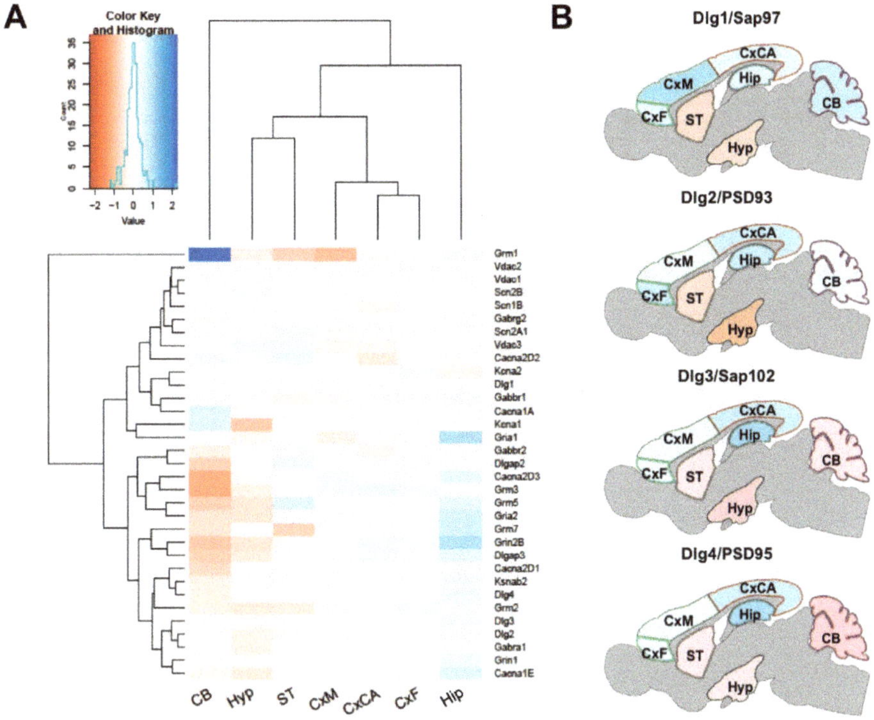

Figure 3. Distribution of 33 selected proteins known to play roles in cognition. (**A**) Hierarchical clustering indicates that the three cortical regions (CxF, CxM and CxCA) and Hyp and ST regions cluster together by similarity, but separately from Hip, while CB clusters separately from all six other regions. (**B**) The abundances of the four MAGUK scaffold protein paralogs Dlg1 (Sap97), PSD93 (Dlg2), Dlg3 (Sap102) and PSD95 (Dlg4) mapped across the various brain regions.

3.4. Correlations of Regional Synapse Proteomes with the Connectome

There are large-scale efforts to map the mouse brain connectome by identifying the projections of neurons between brain regions [37,51]. Because these connections are made at synapses, it follows that there may be a relationship between the molecular composition of synapses in one region and their interconnections. To address this, we asked if the synaptic proteins quantified in this study correlated with connectivity data from the Allen Brain Institute's Mouse Brain Connectivity Atlas (mesoscale connectome) [37] (Figure 4, and Table S6). Hierarchical clustering of postsynaptic proteome abundance and connection strength approximated from projection volume shows that regional connections are associated with distinct signatures of proteins. Moreover, two major branches separated cortex, striatum and hippocampus from cerebellum and hypothalamus, suggesting that hindbrain and basal forebrain connections have broadly distinct molecular properties compared with connections of other forebrain structures.

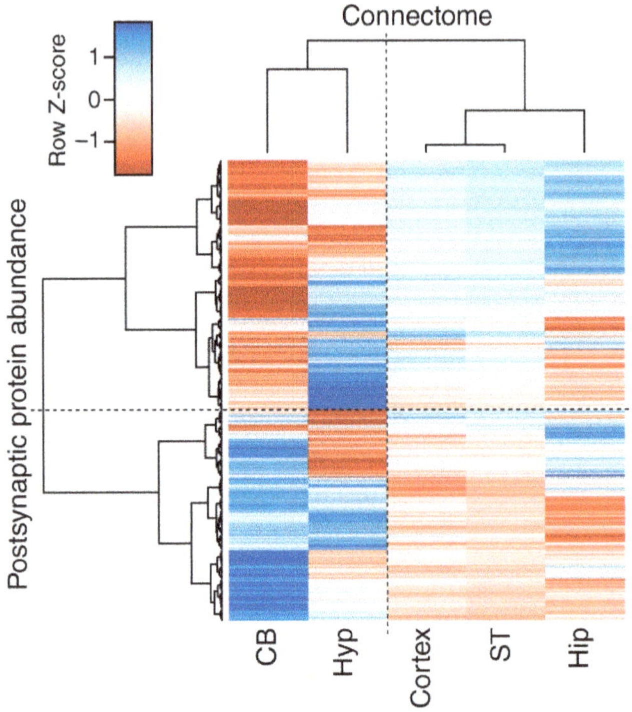

Figure 4. Correlation between brain region-specific postsynaptic protein abundance and mesoscale connectome. Clustering heatmap of the correlation between protein abundance and neuron projection volume in each brain region. Color key shows Z-transformed correlation values; red corresponds to negative correlation and blue to positive correlation.

We found that, in the hippocampus, Dlg3, PSD93 and PSD95 but not Dlg1 were highly correlated (R^2 = 0.7–0.8) with projection volume (Table S6). We then asked if the biochemical pathways that underlie brain connectivity were brain region specific, and performed functional enrichment on the synaptic proteins that were highly correlated ($R^2 \geq 0.6$) with neuron projection volume for each region. Pathways associated with glutamatergic synapses, calcium signaling, long-term potentiation (LTP), long-term depression (LTD) and insulin signaling were over-represented in the hippocampus, striatum and cortical regions, whereas pathways involved in Parkinson's, Huntington's and oxidative phosphorylation and mitochondrial components were enriched in the cerebellum and hypothalamus. Additionally, the molecular correlates of connectivity in the hippocampus are uniquely enriched in endocytosis ($q = 4.09 \times 10^{-9}$) and GABAergic synapses ($q = 7.53 \times 10^{-3}$), the cortical regions in components of the TCA cycle ($q = 1.72 \times 10^{-3}$) and the cerebellum in valine, leucine and isoleucine degradation pathways ($q = 1.11 \times 10^{-6}$) and fatty acids metabolism ($q = 4.6 \times 10^{-4}$) (Table S7A–E). Together, these findings indicate that synapse proteome composition may reflect functional differences between interconnected brain regions.

3.5. Regional Differences in Postsynaptic Protein Interaction Networks

The organization and function of synapse proteomes have been studied using protein-protein interaction (PPI) networks [18,22,52,53] and we used this approach to explore the organization of protein interaction networks in different brain regions. First, the total postsynaptic proteome network was built from the list of proteins obtained in this study and PPIs obtained by mining publicly available databases: BioGRID [38], IntAct [39] and DIP [40] for both mouse and human. The total network consists of 1016 proteins and 8105 PPIs. Using the differential abundance of proteins as an edge weight, we constructed individual networks for each region, identified clusters and their corresponding enrichments in biological and disease functions (Table S8). We found that each region-specific PPI network was split by the same method into a different number of clusters (cl. N), ranging from 82 to 112 clusters (results for a spectral clustering algorithm are shown in Table S8). We assessed the clustering structure of each region's PPIs for robustness and the resulting (consensus) clusters were examined for disease enrichment (Table S9).

Examination of disease enrichment showed that some diseases impact clusters across all brain regions whereas others had more discrete regional effects. For example, all brain regions (except CxF) contained one highly significantly enriched cluster for autism spectrum disorder (adjusted p-values as follows): CB (cl. 19, $p = 9.77 \times 10^{-8}$), CxCA (cl. 9, $p = 9.77 \times 10^{-8}$), CxM (cl. 10, $p = 2.76 \times 10^{-6}$), Hip (cl. 53, $p = 4.84 \times 10^{-7}$), Hyp (cl. 48, $p = 3.77 \times 10^{-6}$) and ST (cl. 24, $p = 8.17 \times 10^{-3}$). We also found clusters associated with intellectual disability (ID) to be enriched in the CB (cl. 19, $p = 4.36 \times 10^{-5}$) and the ST (cl. 25, $p = 6.64 \times 10^{-5}$). We found that PPI clusters associated with bipolar disorder were moderately enriched in the CxF (cl. 50, $p = 1.44 \times 10^{-2}$) and Hyp (cl. 66, $p = 2.05 \times 10^{-2}$), while those associated with schizophrenia were highly enriched in the CB, CxCA and CxM (cl. 10, $p = 1.29 \times 10^{-6}$; cl. 9, $p = 4.41 \times 10^{-6}$; and cl. 10, $p = 2.90 \times 10^{-5}$, respectively) (see Table S9 for all disease enrichment results).

We found regional variability in enrichment levels for neurodegenerative diseases. For Alzheimer's disease, the most enriched cluster was in the Hip (cl. 53, $p = 7.97 \times 10^{-6}$), while for Parkinson's disease a moderate enrichment occurred in the CxCA (cl. 32, $p = 6.64 \times 10^{-3}$). Clusters associated with dementia were significantly enriched only in the Hyp (cl. 6, $p = 4.63 \times 10^{-5}$).

3.6. Identifying a Stable Core Network

To define the network structures that are conserved across all brain regions, we identified 4205 binary interactions (52% of total) that were found in the same cluster in the majority of regional networks. We refer to this as the "stable" postsynaptic density (PSD) (1016 proteins, Figure 5 and Table S10). Spectral clustering generated 73 clusters in total, where the nine largest represent crucial synaptic proteins and neural housekeeping functions: cl. 37 corresponds to the postsynaptic signaling complexes composed of MAGUK scaffold proteins, AMPA and NMDA receptors [23], and other clusters contain ribosomal, metabolic enzymes and actin/myosin-associated proteins (Figure 5, Table S10). We compared the composition of these stable communities with previously detected PPMs and found significant overlaps. For example, cl. 1 containing ATPase and cytochrome-related proteins and cl. 8 containing mitochondrial complex I proteins were over-represented in protein module PPM1 associated with related terms ($p = 4.5 \times 10^{-4}$ and $p = 9.3 \times 10^{-5}$, respectively); cl 4 is composed mainly of metabolic enzymes and falls almost entirely in to PPM1 associated with Parkinson's disease, Alzheimer's disease and metabolic pathways by KEGG pathway enrichment (Figure 5; $p = 7.5 \times 10^{-5}$). Molecules involved in memory and cognition (e.g., MAGUKs) that populated cl. 6 are distributed between PPM5 ($p = 8.0 \times 10^{-3}$) and PPM4 ($p = 1.9 \times 10^{-2}$) associated with the terms "neurotransmitter receptor binding and downstream transmission in the postsynaptic cell", "long-term potentiation" and "long-term depression".

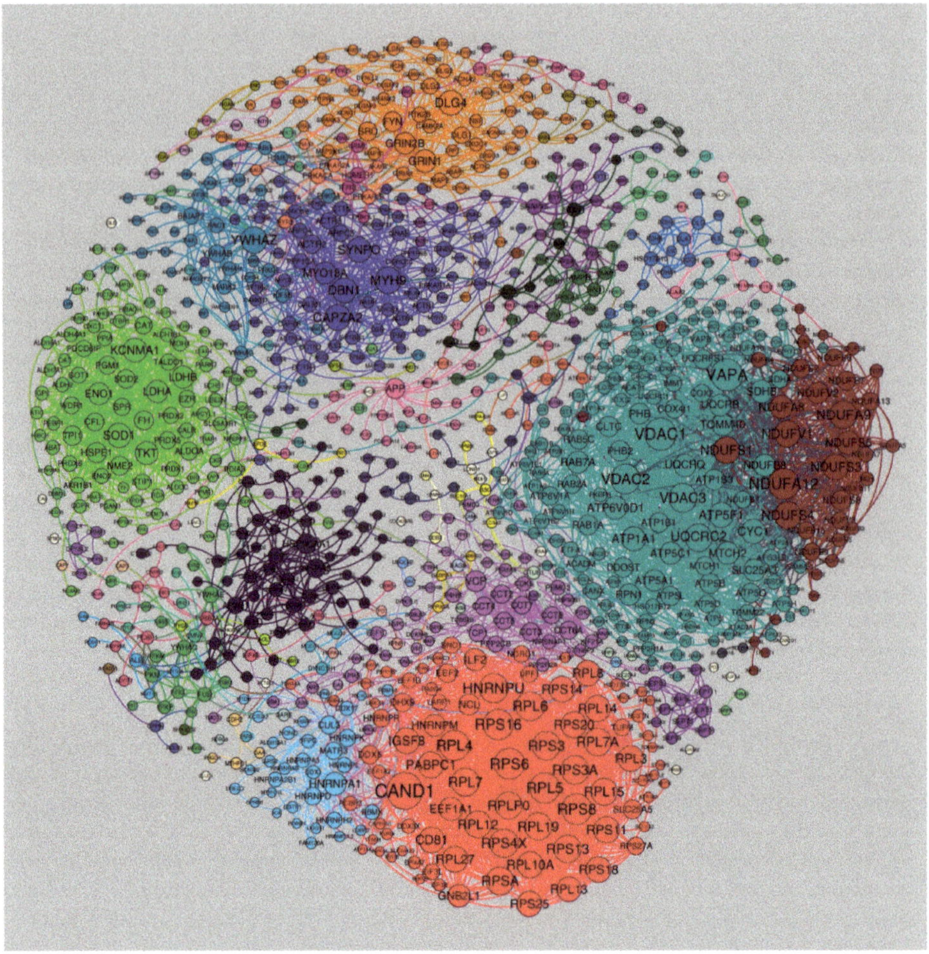

Figure 5. Postsynaptic proteome interaction network showing the cluster structure for the "stable network". A few large clusters with specific functionally related proteins could be detected: cl. 3 contains ribosomal proteins (red cluster at the bottom), cl. 4 contains metabolic enzymes (light green on the left), cl. 7 is enriched with actin-, myosin- and cytoskeleton remodeling-associated proteins (dark blue near the top), cl. 8 contains NADH-oxidoreductases (dark-red cluster on the right) and cl. 1 contains ATPases and voltage-dependent anion channels (light blue on the right). cl. 6 corresponds to key proteins involved in synaptic transmission and plasticity, including AMPA, NMDA receptors and MAGUK proteins (orange cluster at the top). Networks were visualized with Gephi.

4. Discussion

Using mass spectrometry we have examined the protein composition of the postsynaptic proteome of excitatory synapses from regions of the mouse brain and generated a freely available data resource (Edinburgh DataShare, http://dx.doi.org/10.7488/ds/2399). We found that a high percentage of proteins show abundant differences between brain regions. The postsynaptic proteome composition for each brain region forms a distinctive molecular signature. Because these proteomic data were obtained from tissue samples composed of many individual synapses, the proteomic signatures indicate that there might be synapse diversity at the single-synapse level. Consistent with this, we have recently examined the differential distribution of PSD95 and SAP102 in individual synapses across the whole mouse brain and found that these two proteins are differentially distributed into synapse subtypes [26]. Moreover, each brain region was composed of varying proportions of synapse subtypes, which results in each region having a "signature of subtypes". Together, these findings indicate that postsynaptic proteome diversity seen at the level of brain regions arises at the individual synapse level.

The advantage of proteomic mass spectrometry is that it examines the expression of large numbers of proteins and can therefore shed light on how sets of proteins are expressed. We found regional diversity in sets of proteins known to be associated with biochemical pathways controlling physiological processes (such as forms of synaptic plasticity and cognition) and diseases. We also found evidence that scaffold proteins involved with the supramolecular assembly of complexes and supercomplexes [20,23] were differentially distributed across the brain.

The analysis of weighted PPI networks supports previous findings that the postsynaptic proteome has a modular structure [52]. We now find that the regional variability of protein complex composition strongly depends on the relative protein abundance, thereby providing the heterogeneity and unique biochemical signaling potential of each region. In each brain region, we find that ~60% of the complexes/clusters are conserved in a stable network and that ~40% underpin regional specifications. Disease enrichment analysis, which was performed at the regional level, tends to identify the same clusters. Regional specificity results in effectively the same cluster being more or less enriched for each disease across the different brain regions.

The coordinate expression of sets and modules of proteins suggests the possibility that there is an underlying genetic mechanism coordinating the spatial expression of synapse proteins. Evidence in support of a coordinating genetic mechanism acting in the temporal domain was provided from transcriptome analyses of developing cultured neurons [54] and lifespan expression data in the brain of mouse and human [2], which show concerted regulation of postsynaptic proteins. Further evidence for an underlying genetic program regulating the differential spatial expression comes from the hierarchical clustering of protein expression that reveals a correlation with the early development of the nervous system. A similar result was obtained using the single-synapse resolution mapping: the regional signatures of synapse composition in the adult mouse brain are organized into three major groups corresponding to the earliest division of the neural tube [55]. The clustering of the hippocampus, a subcortical region, with cortical regions likely reflects their common developmental origins [55]. These multiple lines of evidence suggest that there is temporal and spatial regulation of postsynaptic proteome expression and that it produces diversity of synapse types [26].

The purpose of regional diversity is most likely to subserve region-specific physiological and behavioral functions. It might also be to provide regional specializations within the greater systems-level organization or global circuitry of the brain. Our analysis of connectome data indicates that the anatomical circuitry of the brain links areas with distinct synapse proteome compositions. Given the differences in biochemical pathways, this indicates that functional specializations in regions are integrated by the connectome. This is consistent with our recent findings showing that the regional composition of the neocortex is linked to behavioral functions observed using functional magnetic resonance imaging (fMRI) [25].

It is interesting that hypothalamus and striatal regions cluster together in protein abundance, whereas the cortex, striatum and hippocampus were separate from the cerebellum and hypothalamus when abundance and connection strength was analyzed. The protein abundance and connection strength correlation between the hypothalamus and cerebellum might relate to the involvement of these structures in several shared biological functions. For example, the cerebellum is known to be interconnected with the hypothalamus, an important center for feeding control, through direct bidirectional connections [56–58]. Most of the hypothalamic neurons receiving cerebellar projections are feeding related. The reciprocal connections between the cerebellum and hypothalamus might therefore play an important role in feeding motivation and in the regulation of feeding behavior [59].

Numerous forms of dementia show pathology in different regions of the brain. For example, dementia with Lewy bodies, which is the second most common form of neurodegenerative dementia after Alzheimer's disease [60], shows hypothalamic atrophy, potentially relevant to the enrichment of dementia pathways observed in this region, whereas this region is not affected in Alzheimer's disease patients [61]. Although the severe pathology of Huntington's disease is mostly related to the striatum, important changes in the hypothalamus and cerebellum have also been described [62], consistent with the observed enrichment of their respective disease pathways in these areas. Homeostatic control of emotions and of metabolism are disturbed early in Huntington's disease and there are alterations in the peptide expression of hypothalamic neurons known to be involved in the regulation of metabolism and emotion [63]. In cerebellum, a dysfunction of the Purkinje cells might contribute to motor impairment in a murine model of Huntington's disease [64], and the cerebellum appears to be commonly affected in juvenile Huntington's disease, as shown by a decrease of cerebellar volume [65–67]. The enrichment of Parkinson's disease pathways in hypothalamus is interesting since alterations in hypothalamus have been linked to non-motor symptoms such as sleep disturbances [68]. Furthermore, the cerebellum is known to be involved in Parkinsonian disorders with motor symptoms [69].

In conclusion, we provide a new data resource describing the composition of the postsynaptic proteome in excitatory synapses from regions of the mouse brain. These data indicate that molecular compositional differences in synapses in different brain regions are relevant to a broad range of physiological and disease processes.

Supplementary Materials: The following are available online at http://www.mdpi.com/2227-7382/6/3/31/s1, Figure S1. The seven regions dissected from the mouse brain. Figure S2. The overlap between proteomes obtained in this study and previous studies. Figure S3. The principal component analysis (PCA) of the abundances of the 1173 synaptic proteins quantified across the seven integral regions for all six mouse brains. Figure S4. Circos plots showing differential abundance of the six postsynaptic proteome modules (PPMs) across mouse brain regions. Table S1. LC-MS/MS quantitation and comparison of the mouse synaptic proteome across the seven integral regions of the mouse brain by one-way ANOVA. Table S2. New PSD proteins detected in this study compared to all other published mouse PSD studies. Table S3. Differential stability analysis and functional enrichment of synaptic proteins. Table S4. Summary of the 868 synaptic proteins found to be differentially abundant across the seven integral regions of the mouse brain. Table S5. Module gene list and abundances for all six modules. Table S6. Correlation of ABI functional connectivity with synaptic proteome abundances measured in this study. Table S7. Functional Enrichment of ABI and Roy Correlates by Brain region. Table S8. PSD protein interaction network analysis across the mouse brain. Table S9. Disease associated enrichment in the interactomes of the mouse PSD. Table S10. The "stable network": postsynaptic proteome protein-protein interaction networks found to be enriched in at least four different regions of the mouse brain.

Author Contributions: S.T.-G., J.D. brain samples; M.R. synapse biochemistry; M.R., O.S., C.M., J.D.A., bioinformatics; S.G.N.G. direction.

Funding: Support was obtained from the Medical Research Council (G0802238), European Union Seventh Framework Programme (FP7 grant agreement No. 604102) and Horizon 2020 (grant agreement No. 72027).

Acknowledgments: We thank T. Le Bihan and L. Imrie at SynthSys, University of Edinburgh, for mass spectrometry sample analysis. The LC-MS QExactive equipment was purchased by a Wellcome Trust Institutional Strategic Support Fund and a strategic award from the Wellcome Trust for the Centre for Immunity, Infection and Evolution (095831/Z/11/Z). Data were extracted from Neuroimaging Informatics Technology Initiative (NIFTI) files using a custom automated script written by J. Roy, MEMEX, Inc. (Burlington, Ontario, Canada). We thank K. Elsegood for laboratory management and D. Maizels for artwork; C.S. Davey, editing.

Conflicts of Interest: The authors declare no conflict of interest.

References

1. Husi, H.; Ward, M.A.; Choudhary, J.S.; Blackstock, W.P.; Grant, S.G. Proteomic analysis of NMDA receptor-adhesion protein signaling complexes. *Nat. Neurosci.* **2000**, 3, 661–669. [CrossRef] [PubMed]
2. Skene, N.G.; Roy, M.; Grant, S.G. A genomic lifespan program that reorganises the young adult brain is targeted in schizophrenia. *eLife* **2017**, *6*, e17915. [CrossRef] [PubMed]
3. Bayes, A.; Collins, M.O.; Reig-Viader, R.; Gou, G.; Goulding, D.; Izquierdo, A.; Choudhary, J.S.; Emes, R.D.; Grant, S.G. Evolution of complexity in the zebrafish synapse proteome. *Nat. Commun.* **2017**, *8*, 14613. [CrossRef] [PubMed]

4. Bayes, A.; Collins, M.O.; Croning, M.D.; van de Lagemaat, L.N.; Choudhary, J.S.; Grant, S.G. Comparative study of human and mouse postsynaptic proteomes finds high compositional conservation and abundance differences for key synaptic proteins. *PLoS ONE* **2012**, *7*, e46683. [CrossRef] [PubMed]
5. Collins, M.O.; Husi, H.; Yu, L.; Brandon, J.M.; Anderson, C.N.; Blackstock, W.P.; Choudhary, J.S.; Grant, S.G. Molecular characterization and comparison of the components and multiprotein complexes in the postsynaptic proteome. *J. Neurochem.* **2006**, *97* (Suppl. 1), 16–23. [CrossRef] [PubMed]
6. Distler, U.; Schmeisser, M.J.; Pelosi, A.; Reim, D.; Kuharev, J.; Weiczner, R.; Baumgart, J.; Boeckers, T.M.; Nitsch, R.; Vogt, J.; et al. In-depth protein profiling of the postsynaptic density from mouse hippocampus using data-independent acquisition proteomics. *Proteomics* **2014**, *14*, 2607–2613. [CrossRef] [PubMed]
7. Dosemeci, A.; Tao-Cheng, J.H.; Vinade, L.; Jaffe, H. Preparation of postsynaptic density fraction from hippocampal slices and proteomic analysis. *Biochem. Biophys. Res. Commun.* **2006**, *339*, 687–694. [CrossRef] [PubMed]
8. Jordan, B.A.; Fernholz, B.D.; Boussac, M.; Xu, C.; Grigorean, G.; Ziff, E.B.; Neubert, T.A. Identification and verification of novel rodent postsynaptic density proteins. *Mol. Cell. Proteom.* **2004**, *3*, 857–871. [CrossRef] [PubMed]
9. Peng, J.; Kim, M.J.; Cheng, D.; Duong, D.M.; Gygi, S.P.; Sheng, M. Semiquantitative proteomic analysis of rat forebrain postsynaptic density fractions by mass spectrometry. *J. Biol. Chem.* **2004**, *279*, 21003–21011. [CrossRef] [PubMed]
10. Trinidad, J.C.; Thalhammer, A.; Specht, C.G.; Lynn, A.J.; Baker, P.R.; Schoepfer, R.; Burlingame, A.L. Quantitative analysis of synaptic phosphorylation and protein expression. *Mol. Cell. Proteom.* **2008**, *7*, 684–696. [CrossRef] [PubMed]
11. Trinidad, J.C.; Thalhammer, A.; Specht, C.G.; Schoepfer, R.; Burlingame, A.L. Phosphorylation state of postsynaptic density proteins. *J. Neurochem.* **2005**, *92*, 1306–1316. [CrossRef] [PubMed]
12. Uezu, A.; Kanak, D.J.; Bradshaw, T.W.; Soderblom, E.J.; Catavero, C.M.; Burette, A.C.; Weinberg, R.J.; Soderling, S.H. Identification of an elaborate complex mediating postsynaptic inhibition. *Science* **2016**, *353*, 1123–1129. [CrossRef] [PubMed]
13. Yoshimura, Y.; Yamauchi, Y.; Shinkawa, T.; Taoka, M.; Donai, H.; Takahashi, N.; Isobe, T.; Yamauchi, T. Molecular constituents of the postsynaptic density fraction revealed by proteomic analysis using multidimensional liquid chromatography-tandem mass spectrometry. *J. Neurochem.* **2004**, *88*, 759–768. [CrossRef] [PubMed]
14. Coba, M.P.; Pocklington, A.J.; Collins, M.O.; Kopanitsa, M.V.; Uren, R.T.; Swamy, S.; Croning, M.D.; Choudhary, J.S.; Grant, S.G. Neurotransmitters drive combinatorial multistate postsynaptic density networks. *Sci. Signal.* **2009**, *2*, ra19. [CrossRef] [PubMed]
15. Li, J.; Zhang, W.; Yang, H.; Howrigan, D.P.; Wilkinson, B.; Souaiaia, T.; Evgrafov, O.V.; Genovese, G.; Clementel, V.A.; Tudor, J.C.; et al. Spatiotemporal profile of postsynaptic interactomes integrates components of complex brain disorders. *Nat. Neurosci.* **2017**, *20*, 1150–1161. [CrossRef] [PubMed]
16. Bayes, A.; van de Lagemaat, L.N.; Collins, M.O.; Croning, M.D.; Whittle, I.R.; Choudhary, J.S.; Grant, S.G. Characterization of the proteome, diseases and evolution of the human postsynaptic density. *Nat. Neurosci.* **2011**, *14*, 19–21. [CrossRef] [PubMed]
17. Kaizuka, T.; Takumi, T. Postsynaptic density proteins and their involvement in neurodevelopmental disorders. *J. Biochem.* **2018**, *163*, 447–455. [CrossRef] [PubMed]
18. Fernandez, E.; Collins, M.O.; Frank, R.A.W.; Zhu, F.; Kopanitsa, M.V.; Nithianantharajah, J.; Lempriere, S.A.; Fricker, D.; Elsegood, K.A.; McLaughlin, C.L.; et al. Arc Requires PSD95 for Assembly into Postsynaptic Complexes Involved with Neural Dysfunction and Intelligence. *Cell Rep.* **2017**, *21*, 679–691. [CrossRef] [PubMed]
19. Frank, R.A.; Grant, S.G. Supramolecular organization of NMDA receptors and the postsynaptic density. *Curr. Opin. Neurobiol.* **2017**, *45*, 139–147. [CrossRef] [PubMed]
20. Frank, R.A.W.; Zhu, F.; Komiyama, N.H.; Grant, S.G.N. Hierarchical organization and genetically separable subfamilies of PSD95 postsynaptic supercomplexes. *J. Neurochem.* **2017**, *142*, 504–511. [CrossRef] [PubMed]
21. Abbas, A.I.; Yadav, P.N.; Yao, W.D.; Arbuckle, M.I.; Grant, S.G.; Caron, M.G.; Roth, B.L. PSD-95 is essential for hallucinogen and atypical antipsychotic drug actions at serotonin receptors. *J. Neurosci.* **2009**, *29*, 7124–7136. [CrossRef] [PubMed]

22. Fernandez, E.; Collins, M.O.; Uren, R.T.; Kopanitsa, M.V.; Komiyama, N.H.; Croning, M.D.; Zografos, L.; Armstrong, J.D.; Choudhary, J.S.; Grant, S.G. Targeted tandem affinity purification of PSD-95 recovers core postsynaptic complexes and schizophrenia susceptibility proteins. *Mol. Syst. Biol.* **2009**, *5*, 269. [CrossRef] [PubMed]
23. Frank, R.A.; Komiyama, N.H.; Ryan, T.J.; Zhu, F.; O'Dell, T.J.; Grant, S.G. NMDA receptors are selectively partitioned into complexes and supercomplexes during synapse maturation. *Nat. Commun.* **2016**, *7*, 11264. [CrossRef] [PubMed]
24. Swanson, L.W. *Brain Maps III: Structure of the Rat Brain: An Atlas with Printed and Electronic Templates for Data, Models, and Schematics*, 3rd rev. ed.; Elsevier, Academic Press: Cambridge, MA, USA, 2004.
25. Roy, M.; Sorokina, O.; Skene, N.; Simonnet, C.; Mazzo, F.; Zwart, R.; Sher, E.; Smith, C.; Armstrong, J.D.; Grant, S.G.N. Proteomic analysis of postsynaptic proteins in regions of the human neocortex. *Nat. Neurosci.* **2018**, *21*, 130–138. [CrossRef] [PubMed]
26. Zhu, F.; Cizeron, M.; Qiu, Z.; Benavides-Piccione, R.; Kopanitsa, M.V.; Skene, N.; Koniaris, B.; DeFelipe, J.; Fransén, E.; Komiyama, N.H.; et al. Architecture of the mouse brain synaptome. *Neuron* **2018**, *172*, 143–150. [CrossRef]
27. Palkovits, M. *General Neurochemical Techniques*; Boulton, A.A., Baker, G.B., Eds.; Springer: Berlin, Germany, 1985; Volume 1, pp. 1–17.
28. Shevchenko, A.; Tomas, H.; Havlis, J.; Olsen, J.V.; Mann, M. In-gel digestion for mass spectrometric characterization of proteins and proteomes. *Nat. Protoc.* **2006**, *1*, 2856–2860. [CrossRef] [PubMed]
29. Team, R.C. *R: A Language and Environment for Statistical Computing*; R Foundation for Statistical Computing: Vienna, Austria, 2013.
30. Benjamini, Y.; Hochberg, Y. Controlling the false discovery rate: A practical and powerful approach to multiple testing. *J. R. Stat. Soc. Ser. B (Methodol.)* **1995**, *57*, 289–300.
31. Hawrylycz, M.; Miller, J.A.; Menon, V.; Feng, D.; Dolbeare, T.; Guillozet-Bongaarts, A.L.; Jegga, A.G.; Aronow, B.J.; Lee, C.K.; Bernard, A.; et al. Canonical genetic signatures of the adult human brain. *Nat. Neurosci.* **2015**, *18*, 1832–1844. [CrossRef] [PubMed]
32. Galili, T. Dendextend: An R package for visualizing, adjusting and comparing trees of hierarchical clustering. *Bioinformatics* **2015**, *31*, 3718–3720. [CrossRef] [PubMed]
33. Charrad, M.; Ghazzali, N.; Boiteau, V.; Niknafs, A. NbClust: An R Package for Determining the Relevant Number of Clusters in a Data Set. *J. Stat. Softw.* **2014**, *61*. [CrossRef]
34. Yu, G.; Wang, L.G.; Han, Y.; He, Q.Y. Clusterprofiler: An R package for comparing biological themes among gene clusters. *OMICS* **2012**, *16*, 284–287. [CrossRef] [PubMed]
35. Subramanian, A.; Tamayo, P.; Mootha, V.K.; Mukherjee, S.; Ebert, B.L.; Gillette, M.A.; Paulovich, A.; Pomeroy, S.L.; Golub, T.R.; Lander, E.S.; et al. Gene set enrichment analysis: A knowledge-based approach for interpreting genome-wide expression profiles. *Proc. Natl. Acad. Sci. USA* **2005**, *102*, 15545–15550. [CrossRef] [PubMed]
36. Krzywinski, M.; Schein, J.; Birol, I.; Connors, J.; Gascoyne, R.; Horsman, D.; Jones, S.J.; Marra, M.A. Circos: An information aesthetic for comparative genomics. *Genome Res.* **2009**, *19*, 1639–1645. [CrossRef] [PubMed]
37. Oh, S.W.; Harris, J.A.; Ng, L.; Winslow, B.; Cain, N.; Mihalas, S.; Wang, Q.; Lau, C.; Kuan, L.; Henry, A.M.; et al. A mesoscale connectome of the mouse brain. *Nature* **2014**, *508*, 207–214. [CrossRef] [PubMed]
38. Chatr-Aryamontri, A.; Oughtred, R.; Boucher, L.; Rust, J.; Chang, C.; Kolas, N.K.; O'Donnell, L.; Oster, S.; Theesfeld, C.; Sellam, A.; et al. The BioGRID interaction database: 2017 update. *Nucleic Acids Res.* **2017**, *45*, D369–D379. [CrossRef] [PubMed]
39. Kerrien, S.; Aranda, B.; Breuza, L.; Bridge, A.; Broackes-Carter, F.; Chen, C.; Duesbury, M.; Dumousseau, M.; Feuermann, M.; Hinz, U.; et al. The IntAct molecular interaction database in 2012. *Nucleic Acids Res.* **2012**, *40*, D841–D846. [CrossRef] [PubMed]
40. Xenarios, I.; Salwinski, L.; Duan, X.J.; Higney, P.; Kim, S.M.; Eisenberg, D. DIP, the Database of Interacting Proteins: A research tool for studying cellular networks of protein interactions. *Nucleic Acids Res.* **2002**, *30*, 303–305. [CrossRef] [PubMed]
41. Newman, M.E. Finding community structure in networks using the eigenvectors of matrices. *Phys. Rev. E Stat. Nonlin Soft Matter Phys.* **2006**, *74*, 036104. [CrossRef] [PubMed]
42. Newman, M.E.; Girvan, M. Finding and evaluating community structure in networks. *Phys. Rev. E Stat. Nonlin Soft Matter Phys.* **2004**, *69*, 026113. [CrossRef] [PubMed]

43. Simpson, T.I.; Armstrong, J.D.; Jarman, A.P. Merged consensus clustering to assess and improve class discovery with microarray data. *BMC Bioinform.* **2010**, *11*, 590. [CrossRef] [PubMed]
44. Lips, E.S.; Cornelisse, L.N.; Toonen, R.F.; Min, J.L.; Hultman, C.M.; International Schizophrenia; Holmans, P.A.; O'Donovan, M.C.; Purcell, S.M.; Smit, A.B.; et al. Functional gene group analysis identifies synaptic gene groups as risk factor for schizophrenia. *Mol. Psychiatry* **2012**, *17*, 996–1006. [CrossRef] [PubMed]
45. Aronson, A.R.; Lang, F.M. An overview of MetaMap: Historical perspective and recent advances. *J. Am. Med. Inform. Assoc.* **2010**, *17*, 229–236. [CrossRef] [PubMed]
46. Musen, M.A.; Noy, N.F.; Shah, N.H.; Whetzel, P.L.; Chute, C.G.; Story, M.A.; Smith, B.; Team, N. The National Center for Biomedical Ontology. *J. Am. Med. Inform. Assoc.* **2012**, *19*, 190–195. [CrossRef] [PubMed]
47. Whetzel, P.L.; Noy, N.F.; Shah, N.H.; Alexander, P.R.; Nyulas, C.; Tudorache, T.; Musen, M.A. BioPortal: Enhanced functionality via new Web services from the National Center for Biomedical Ontology to access and use ontologies in software applications. *Nucleic Acids Res.* **2011**, *39*, W541–W545. [CrossRef] [PubMed]
48. Schriml, L.M.; Arze, C.; Nadendla, S.; Chang, Y.W.; Mazaitis, M.; Felix, V.; Feng, G.; Kibbe, W.A. Disease Ontology: A backbone for disease semantic integration. *Nucleic Acids Res.* **2012**, *40*, D940–D946. [CrossRef] [PubMed]
49. Alexa, A.; Rahnenfuhrer, J.; Lengauer, T. Improved scoring of functional groups from gene expression data by decorrelating GO graph structure. *Bioinformatics* **2006**, *22*, 1600–1607. [CrossRef] [PubMed]
50. Kang, H.J.; Kawasawa, Y.I.; Cheng, F.; Zhu, Y.; Xu, X.; Li, M.; Sousa, A.M.; Pletikos, M.; Meyer, K.A.; Sedmak, G.; et al. Spatio-temporal transcriptome of the human brain. *Nature* **2011**, *478*, 483–489. [CrossRef] [PubMed]
51. Gamanut, R.; Kennedy, H.; Toroczkai, Z.; Ercsey-Ravasz, M.; Van Essen, D.C.; Knoblauch, K.; Burkhalter, A. The Mouse Cortical Connectome, Characterized by an Ultra-Dense Cortical Graph, Maintains Specificity by Distinct Connectivity Profiles. *Neuron* **2018**, *97*, 698–715. [CrossRef] [PubMed]
52. Pocklington, A.J.; Cumiskey, M.; Armstrong, J.D.; Grant, S.G. The proteomes of neurotransmitter receptor complexes form modular networks with distributed functionality underlying plasticity and behaviour. *Mol. Syst. Biol.* **2006**, *2*, 2006.0023. [CrossRef] [PubMed]
53. McLean, C.; He, X.; Simpson, T.I.; Armstrong, J.D. Improved Functional Enrichment Analysis of Biological Networks using Scalable Modularity Based Clustering. *J. Proteom. Bioinform.* **2016**, *9*, 9–18. [CrossRef]
54. Valor, L.M.; Charlesworth, P.; Humphreys, L.; Anderson, C.N.; Grant, S.G. Network activity-independent coordinated gene expression program for synapse assembly. *Proc. Natl. Acad. Sci. USA* **2007**, *104*, 4658–4663. [CrossRef] [PubMed]
55. Swanson, L.W. *Brain Architecture: Understanding the Basic Plan*, 2nd ed.; Oxford University Press: Oxford, UK, 2012.
56. Dietrichs, E.; Haines, D.E. Demonstration of hypothalamo-cerebellar and cerebello-hypothalamic fibres in a prosimian primate (Galago crassicaudatus). *Anat. Embryol. (Berl.)* **1984**, *170*, 313–318. [CrossRef] [PubMed]
57. Dietrichs, E.; Haines, D.E.; Roste, G.K.; Roste, L.S. Hypothalamocerebellar and cerebellohypothalamic projections–circuits for regulating nonsomatic cerebellar activity? *Histol. Histopathol.* **1994**, *9*, 603–614. [PubMed]
58. Haines, D.E.; Dietrichs, E.; Mihailoff, G.A.; McDonald, E.F. The cerebellar-hypothalamic axis: Basic circuits and clinical observations. *Int. Rev. Neurobiol.* **1997**, *41*, 83–107. [PubMed]
59. Zhu, J.N.; Wang, J.J. The cerebellum in feeding control: Possible function and mechanism. *Cell. Mol. Neurobiol.* **2008**, *28*, 469–478. [CrossRef] [PubMed]
60. Zaccai, J.; McCracken, C.; Brayne, C. A systematic review of prevalence and incidence studies of dementia with Lewy bodies. *Age Ageing* **2005**, *34*, 561–566. [CrossRef] [PubMed]
61. Whitwell, J.L.; Weigand, S.D.; Shiung, M.M.; Boeve, B.F.; Ferman, T.J.; Smith, G.E.; Knopman, D.S.; Petersen, R.C.; Benarroch, E.E.; Josephs, K.A.; et al. Focal atrophy in dementia with Lewy bodies on MRI: A distinct pattern from Alzheimer's disease. *Brain* **2007**, *130*, 708–719. [CrossRef] [PubMed]
62. Soneson, C.; Fontes, M.; Zhou, Y.; Denisov, V.; Paulsen, J.S.; Kirik, D.; Petersen, A.; Huntington Study Group. Early changes in the hypothalamic region in prodromal Huntington disease revealed by MRI analysis. *Neurobiol. Dis.* **2010**, *40*, 531–543. [CrossRef] [PubMed]
63. Gabery, S.; Murphy, K.; Schultz, K.; Loy, C.T.; McCusker, E.; Kirik, D.; Halliday, G.; Petersen, A. Changes in key hypothalamic neuropeptide populations in Huntington disease revealed by neuropathological analyses. *Acta Neuropathol.* **2010**, *120*, 777–788. [CrossRef] [PubMed]

64. Dougherty, S.E.; Reeves, J.L.; Lucas, E.K.; Gamble, K.L.; Lesort, M.; Cowell, R.M. Disruption of Purkinje cell function prior to huntingtin accumulation and cell loss in an animal model of Huntington disease. *Exp. Neurol.* **2012**, *236*, 171–178. [CrossRef] [PubMed]
65. Fennema-Notestine, C.; Archibald, S.L.; Jacobson, M.W.; Corey-Bloom, J.; Paulsen, J.S.; Peavy, G.M.; Gamst, A.C.; Hamilton, J.M.; Salmon, D.P.; Jernigan, T.L. In vivo evidence of cerebellar atrophy and cerebral white matter loss in Huntington disease. *Neurology* **2004**, *63*, 989–995. [CrossRef] [PubMed]
66. Nicolas, G.; Devys, D.; Goldenberg, A.; Maltete, D.; Herve, C.; Hannequin, D.; Guyant-Marechal, L. Juvenile Huntington disease in an 18-month-old boy revealed by global developmental delay and reduced cerebellar volume. *Am. J. Med. Genet. A* **2011**, *155A*, 815–818. [CrossRef] [PubMed]
67. Sakazume, S.; Yoshinari, S.; Oguma, E.; Utsuno, E.; Ishii, T.; Narumi, Y.; Shiihara, T.; Ohashi, H. A patient with early onset Huntington disease and severe cerebellar atrophy. *Am. J. Med. Genet. A* **2009**, *149A*, 598–601. [CrossRef] [PubMed]
68. Wilson, H.; Giordano, B.; Turkheimer, F.E.; Chaudhuri, K.R.; Politis, M. Serotonergic dysregulation is linked to sleep problems in Parkinson's disease. *Neuroimage Clin.* **2018**, *18*, 630–637. [CrossRef] [PubMed]
69. Joutsa, J.; Horn, A.; Hsu, J.; Fox, M.D. Localizing parkinsonism based on focal brain lesions. *Brain* **2018**, *141*, 2445–2456. [CrossRef] [PubMed]

proteomes

MDPI

Article

Mapping the Proteome of the Synaptic Cleft through Proximity Labeling Reveals New Cleft Proteins

Tony Cijsouw [1,*], Austin M. Ramsey [2], TuKiet T. Lam [3,4,5], Beatrice E. Carbone [1], Thomas A. Blanpied [2] and Thomas Biederer [1,*]

1 Department of Neuroscience, Tufts University School of Medicine, Boston, MA 02111, USA; Bea.Carbone@tufts.edu
2 Department of Physiology and Program in Neuroscience, University of Maryland School of Medicine, Baltimore, MD 21201, USA; austin.ramsey@umaryland.edu (A.M.R.); tblanpied@som.umaryland.edu (T.A.B.)
3 Yale/NIDA Neuroproteomics Center, New Haven, CT 06511, USA; tukiet.lam@yale.edu
4 W.M. Keck Biotechnology Resource Laboratory, Yale University School of Medicine, 300 George Street, New Haven, CT 06511, USA
5 Department of Molecular Biophysics and Biochemistry, Yale University School of Medicine, New Haven, CT 06520, USA
* Correspondence: tcijsouw@gmail.com (T.C.); thomas.biederer@tufts.edu (T.B.); Tel.: +1-617-636-2131 (T.B.)

Received: 16 October 2018; Accepted: 18 November 2018; Published: 28 November 2018

Abstract: Synapses are specialized neuronal cell-cell contacts that underlie network communication in the mammalian brain. Across neuronal populations and circuits, a diverse set of synapses is utilized, and they differ in their molecular composition to enable heterogenous connectivity patterns and functions. In addition to pre- and post-synaptic specializations, the synaptic cleft is now understood to be an integral compartment of synapses that contributes to their structural and functional organization. Aiming to map the cleft proteome, this study applied a peroxidase-mediated proximity labeling approach and used the excitatory synaptic cell adhesion protein SynCAM 1 fused to horseradish peroxidase (HRP) as a reporter in cultured cortical neurons. This reporter marked excitatory synapses as measured by confocal microcopy and was targeted to the edge zone of the synaptic cleft as determined using 3D dSTORM super-resolution imaging. Proximity labeling with a membrane-impermeant biotin-phenol compound restricted labeling to the cell surface, and Label-Free Quantitation (LFQ) mass spectrometry combined with ratiometric HRP tagging of membrane vs. synaptic surface proteins was used to identify the proteomic content of excitatory clefts. Novel cleft candidates were identified, and Receptor-type tyrosine-protein phosphatase zeta was selected and successfully validated. This study supports the robust applicability of peroxidase-mediated proximity labeling for synaptic cleft proteomics and its potential for understanding synapse heterogeneity in health and changes in diseases such as psychiatric disorders and addiction.

Keywords: synapse; synaptic cleft; trans-synaptic adhesion; proximity labeling; SynCAM; Cadm; Receptor-type tyrosine-protein phosphatase zeta; R-PTP-zeta; Ptprz1

1. Introduction

Synapses are the cellular units for information transfer in the central nervous system. The mammalian brain is comprised of functionally diverse synapse types connecting different neuronal populations into networks that enable complex behavior and responses to external and internal cues. Understanding this diversity will be fundamental to defining brain connectivity [1–3]. Synapse function is instructed by the diverse proteomic composition of these specialized neuronal contact sites. These molecular components guide synaptogenesis, maturation, and differentiation in development and in adulthood [4,5]. Function and composition are specific to neuron identity, local connectivity, and regional specificity and contribute to a diversified and specialized synapse proteome within the brain [3]. The molecular diversity of synapses is further controlled by processes such as neuronal activity and plasticity changes, as well as by secreted factors [2,6,7]. Different states of the diseased brain, e.g., neuro-degeneration or addiction, may further alter synapse composition and function [8–11].

Synapses are comprised of a presynaptic terminal and a postsynaptic compartment, each containing protein complexes unique to their function [12,13]. Importantly, the pre- and post-synapse are connected through the synaptic cleft, which is a protein-dense environment organized into molecularly distinct sub-domains [13,14]. Biochemical fractionation methods combined with mass spectrometry-based proteomics have been instrumental in determining the general composition of central nervous system synapses. These approaches have allowed for detailed studies of synaptosomes, which contain presynaptic terminals with postsynaptic sites remaining attached [15], presynaptic membranes [11], presynaptic vesicles [16,17], active zones [18,19], and postsynaptic specializations [20–22]. The combination of these methods with the use of genetic models to tag synaptic components has allowed for the specific isolation of excitatory synaptic compartments [15,23] or inhibitory complexes [24]. While these classical biochemical studies have mapped general synapse composition, the ability to parse out their heterogeneity has been limited, and the synaptic cleft as a separate compartment was intractable.

Recent proteomic advances have employed biotin-tagging of endogenous proteins within a specific cellular compartment using targeting of exogenous peroxidases that can create biotin-phenoxyl radicals or biotin ligases (BioID) without the need for cellular fractionation [25–27]. The enzyme horseradish peroxidase (HRP) and engineered peroxidases APEX and APEX2, derivatives from pea ascorbate peroxidase, have been used to map the proteome of mitochondria [28] and excitatory and inhibitory cleft proteomes [29]. These peroxidase-based labeling approaches are currently limited to cultured cells or small organisms that have been made permeable [30,31]. BioID involves the use of a promiscuous biotin ligase to biotinylate proximal proteins and has been used in mouse brain to map the proteome of inhibitory postsynaptic proteins [32]. However, this method requires hours to days of exogenous biotin application and the reactive biotin species has a relatively long lifetime. Recently, TurboID was developed using directed evolution and this engineered BioID mutant has enabled faster proximity labeling. Despite this advance, the temporal resolution is in the order of tens of minutes and background biotinylation is observed due to utilization of endogenous biotin by TurboID [33]. This limits the temporal and presumably the spatial resolution in contrast to peroxidase-mediated proximity labeling, which has a temporal resolution of tens of seconds and high ~20 nm spatial resolution due to the short lifetime of the enzyme-generated biotin radical [25].

Peroxidase-mediated proximity labeling has for the first time allowed to specifically dissect the proteome of the synaptic cleft of either excitatory or inhibitory synapses [29]. Loh, Ting and colleagues designed excitatory and inhibitory-specific cleft reporter proteins by fusion of HRP with excitatory synapse-specific LRRTM1 and LRRTM2 and inhibitory synapse-specific Neuroligin 2A and Slitrk3, and the authors made use of a ratiometric labeling technique for non-membrane enclosed cellular compartments. Biotinylated proteins of cleft-targeted HRP were identified and quantified using mass spectrometry and ratiometrically compared with biotinylated proteins of membrane-targeted HRP to detect cleft-enriched synaptic proteins. This resulted in proteomic lists representative of excitatory and inhibitory synapses, with a higher specificity and deeper coverage of trans-synaptic proteins than previous proteomes obtained after biochemical fractionation.

Among the prominently expressed synaptic cleft proteins are SynCAMs, a group of immunoglobulin molecules that engage in homo- and heterophilic interactions [34,35]. They are specific for excitatory synapses [29,34] and instruct the formation and guide maturation of these synapses [4,34,36,37]. The current study describes the use of peroxidase-mediated proximity labeling to map the proteome of excitatory synapses using SynCAM 1 as a reporter protein. We describe the use of SynCAM

1-HRP-mediated labeling and in silico filtering steps to select synaptic proteins over generic plasma membrane proteins and intracellular contaminants. This resulted in a list of proteins that were each enriched in multiple biological replicate experiments and spanned functional categories expected to be present in the synaptic cleft. Identified proteins included synaptic proteins that were reported earlier using classical biochemical and peroxidase-mediated proximity labeling approaches. In addition, several proteins on our list are novel synaptic cleft candidates that may add to the parts list of the excitatory synaptic cleft. As part of our validation approach, we show that the trans-membrane protein Receptor-type tyrosine-protein phosphatase zeta, or R-PTP-zeta, identified here as a synaptic cleft candidate is prominently expressed across forebrain regions of the mouse brain and localized at excitatory synaptic sites. This study corroborates proximity labeling as an approach to map the cleft proteome and underlines its applicability to analyze synapse diversity.

2. Materials and Methods

2.1. Animals

Pregnant Sprague Dawley timed-pregnancy rats were obtained from Charles River Laboratories (Wilmington, MA, USA). C57BL/6J background wild-type mice were obtained from Charles River Laboratories and maintained in the colony. All procedures were approved by the Institutional Animal Care and Use Committee (B2016-154) and in compliance with National Institutes of Health guidelines.

2.2. Neuronal Cell Culture

Dissociated cortical neuron cultures from embryonic day 18 (E18) rats were prepared as described previously [38]. In brief, pregnant rats were sacrificed and E18 embryos extracted. Embryo's brains were dissected, and cortices were isolated. Cortices were incubated in 0.05% trypsin (Invitrogen 25300054, Carlsbad, CA, USA) at 37 °C for 20 min, triturated to single cell suspension, and plated on poly-L-lysine (Sigma-Aldrich P1274, St. Louis, MO, USA)-coated surfaces (i.e., 12-mm glass coverslips or 10-cm culture dishes). Cytosine arabinoside (Sigma-Aldrich C1768) was added at a final concentration of 2 µM for 2–4 days-in-vitro (div). For mass-spectrometry experiments, per condition 3×10^6 cells/dish were plated on six times 10-cm cell culture dishes coated with poly-L-lysine. For immunocytochemistry, 100,000 cells were plated on 12-mm coverslips coated with poly-L-lysine in a well of a 24-well cell culture plate.

For dSTORM experiments, neuronal cell cultures were prepared as described before [14].

2.3. HEK293T Cell Culture

Human embryonic kidney (HEK) 293T cells (ATCC, Manassas, VA, USA) were seeded at 20% confluence in T75 flasks and cultured in Dulbecco's Modified Eagle Medium (DMEM) (Gibco via Fisher Scientific, Hampton, NH, USA, #11-965-118) supplemented with 10% fetal bovine serum, penicillin, and streptomycin at 37 °C under 5% CO_2. Cells were passaged at 80–90% confluence by trypsinization and reseeded. For biotinylation of HEK293T cells, 60,000 cells were reseeded to 12-mm uncoated coverslips.

2.4. Plasmids

Plasmid pCAGGS-SynCAM 1-APEX2 was cloned using general cloning procedures. Briefly, using restriction site *NheI*-flanking primers, FLAG-APEX2 (generating *NheI*-FLAG-APEX2-*NheI*) was amplified from plasmid pcDNA3-APEX2-NES (a gift from Alice Ting; Addgene plasmid #49386) and ligated into a pCR-BluntII-Topo vector. The resulting vector was then restriction digested using *NheI* and the excised fragment *NheI*-FLAG-APEX2-*NheI* was cloned into *NheI* restriction digested pCAGGS-SynCAM 1-(363-*NheI*) [34] resulting in pCAGGS-SynCAM 1-APEX2 with APEX2, N-terminal flanked with a FLAG-tag, at amino acid (AA) position 363 of mouse SynCAM 1.

Plasmid pAAV-CaMKIIa-HRP-TM (Membrane-TM) was cloned using general cloning procedures from pCAG-HRP-TM (a gift from Alice Ting; Addgene plasmid #44441). Briefly, pCAG-HRP-TM was restriction digested using *BamHI* and *HindIII*, and the excised HA-HRP-Myc-TM fragment was cloned into *BamHI* and *HindIII* restriction digested pAAV-CaMKIIa-EGFP (a gift from Bryan Roth; Addgene plasmid #50469), which removed EGFP but kept the CaMKIIa promoter.

Initially, a SynCAM 1-HRP version was cloned that had insufficient biotinylation activity in neurons (not shown), presumably due to a lack of flexible linkers adjacent the HRP. This initial plasmid pAAV-CaMKIIa-SynCAM 1-HRP was assembled using the NEBuilder High-Fidelity DNA Assembly Cloning kit (New England BioLabs, Ipswich, MA, USA, E5520S) according to the manufacturer's instructions and general cloning procedures. In brief, the fragments for the Gibson/Seamless cloning were: pAAV-CaMKIIa-EGFP (a gift from Bryan Roth; Addgene plasmid #50469) restriction digested using *BamHI* and *EcoRV* to remove EGFP, which served as vector backbone; 5′-fragment of SynCAM 1 containing amino acids 1–362 (of mouse SynCAM 1) amplified from pCR-BluntII-TOPO SynCAM 1(363-*NheI*) (see below); HRP (omitting start codon, HA-tag, and TM) amplified from pCAG-HRP-TM with 3′ of HRP a FLAG-tag (DYKDDDDKA) was introduced using additional sequences in the primers; and 3′ fragment of SynCAM 1 containing amino acids 363–445 (of mouse SynCAM 1) amplified from pCAGGS-SynCAM 1-(363-*NheI*) [34].

Then, a new SynCAM 1-HRP version was cloned that contained flanking linker sequences that sterically separated HRP-FLAG from SynCAM 1. This linker-containing plasmid pAAV-CaMKIIa-SynCAM 1-HRP was assembled using the NEBuilder High-Fidelity DNA Assembly Cloning kit (New England BioLabs, Ipswich, MA, USA, E5520S) according to the manufacturer's instructions and general cloning procedures. In brief, the fragments for the Gibson/Seamless cloning were: pAAV-CaMKIIa-EGFP (a gift from Bryan Roth; Addgene plasmid #50469) restriction digested using *BamHI* and *HindIII* to remove EGFP, which served as vector backbone; 5′-fragment of SynCAM 1 containing amino acids 1–362 amplified from pCR-BluntII-TOPO SynCAM 1(363-*NheI*) (see below); HRP-FLAG amplified from the initial pAAV- CaMKIIa-SynCAM 1-HRP plasmid with 5′ of HRP three repeats of a GGGGS-linker added and 3′ of the FLAG-tag three repeats of a GGGS-linker added using additional sequences in the primers; and 3′ fragment of SynCAM 1 containing amino acids 363–445 (of mouse SynCAM 1) amplified from the initial pAAV-CaMKIIa-SynCAM 1-HRP plasmid. The resulting plasmid pAAV-CaMKIIa-SynCAM 1-HRP, containing HRP-FLAG at AA position 363 of mouse SynCAM 1 and flanked with linkers, was used as SynCAM 1-HRP in all neuronal studies presented here.

The 5′-fragment of SynCAM 1 used in the Gibson assemblies above originated from a template where a *BamHI* restriction site at base pair position 25 (from start of coding sequence) in SynCAM 1 was mutated (synonymous) from GGATCC to GGTTCC using site-directed mutagenesis: pCR-BluntII-TOPO SynCAM 1(363-*NheI*).

Plasmids pCAGGS-SynCAM 1-APEX2 (Plasmid ID 119727), pAAV-CaMKIIa-SynCAM 1-HRP (Plasmid ID 119728), and pAAV-CaMKIIa-HRP-TM (Plasmid ID 119729) are available at Addgene.org.

2.5. Adeno-Associated Virus Production, Purification, and Titration

AAV was produced using the triple-transfection, helper-free method, with a modified version of a published protocol [39,40]. Briefly, AAV-293 cells (gift from Ralph DiLeone, Yale University, New Haven, CT, USA), a HEK-293-based cell line optimized for the packaging of AAV virions, were cultured in five 150-mm diameter cell culture dishes and transfected with pAAV-Reporter (i.e., HRP-fusion proteins), pHelper (gift from Ralph DiLeone, Yale University, New Haven, CT, USA), and pAAV-DJ-Rep-Cap (gift from Pascal Kaeser, Harvard University, Cambridge, MA, USA) plasmids using the acidified polyethylenimine (PEI; Polysciences, Inc., Warrington, PA, USA, 23966-2) method [41]. Cells were collected, pelleted, and resuspended in freezing buffer (0.15 M NaCl, 50 mM Tris, pH 8.0) 48–72 h after transfection. After four freeze-thaw cycles using liquid nitrogen and a 42 °C water bath, benzonase was added (Sigma-Aldrich, E1014; 50 U/mL, final) and incubated at 37 °C for

30 min. The lysate was spun at 3200× *g* for 30 min at 4 °C, and supernatant was added to an Optiseal centrifuge tube (Beckman Coulter, 361625, Brea, CA, USA) containing a 15%, 25%, 40%, and 60% iodixanol (Optiprep, 60%; Sigma-Aldrich, D1556) step gradient. The lysate on the step gradient was spun at 184,000× *g* (RCF average) for 3 h and 20 min at 10 °C (50,000 rpm, Beckman Optima LE-80K, Type 70 Ti Beckman rotor, Beckman Coulter) and the 40% fraction was collected. Iodixanol buffer solution was exchanged and AAV concentrated with 1× PBS containing 1 mM $MgCl_2$ and 2.5 mM KCl (PBS-MK) using Amicon Ultra centrifugal filters (100,000 NMWL;, UFC910024, Merck Millipore, Burlington, MA, USA). The purified virus was stored at −80 °C.

To titrate AAV, various quantities of purified virus were added to cultured neurons at 14–17 div. At 21–24 div, neurons were labeled and imaged as described below. Virus titer amount was selected based on the criteria that biotinylation was visible at distinct puncta for SynCAM 1-HRP at sites of Homer or was diffusely along the membrane for Membrane-HRP and overall transduction efficiency was >50%. At these expression levels, the FLAG and HA-antibodies to detect SynCAM 1-HRP or Membrane-HRP, respectively, were generally not sensitive enough to detect the reporters in immunocytochemistry and biotinylation served as marker of these reporters.

2.6. Transfection

For dSTORM imaging of SynCAM 1-HRP, cultured neurons on coverslips were transfected at 18 div using lipofectamine 2000 (1 µg/µL DNA) (Thermo Fisher Scientific, Waltham, MA, USA) and 1 µg/coverslip total pAAV SynCAM 1-HRP DNA in 50 µL opti-mem (Thermo Fisher Scientific) per coverslip. DNA was first added to half the total volume of opti-mem, subsequently pipetted into the other half the total volume of opti-mem containing lipofectamine and allowed to incubate for 5–20 min. 50 µL of the mixture was then pipetted drop-wise into each well containing a coverslip.

For HEK293T cell biotinylation and immunocytochemistry, APEX2 or HRP-fusion constructs were introduced by the acidified polyethylenimine (PEI; 23966-2, Polysciences, Inc., Warrington, PA, USA) method [41] one day after seeding.

2.7. Peroxidase-Mediated Biotinylation

For each specific neuron labeling experiment (i.e., neuronal cell biotinylation and immunocytochemistry, neuronal cell biotinylation and Western blot staining, and neuronal cell biotinylation and mass spectrometry), at indicated days cells were labeled live with 100 µM membrane-impermeant biotin-AEEA-phenol (Iris Biotech, Marktredwitz, Germany, LS-3490.0100) and 1 mM H_2O_2 (Sigma-Aldrich, 95321) in Tyrode's buffer (145 mM NaCl, 1.25 mM $CaCl_2$, 3 mM KCl, 1.25 mM $MgCl_2$, 0.5 mM NaH_2PO_4, 10 mM glucose, 10 mM HEPES (pH 7.4)) for 1 min at room temperature. After 1 min, the biotinylation reaction was quenched by washing the cells three times with Tyrode's buffer containing 10 mM sodium azide (#BDH7465-2, VWR, Radnor, PA, USA), 10 mM sodium ascorbate (Sigma-Aldrich, #PHR1279), and 2.5 mM Trolox (Acros Organics via VWR, #200008-026) as described [29]. Neurons for immunocytochemistry were fixed with 4% paraformaldehyde in "fixation buffer" (60 mM PIPES, 25 mM HEPES, 10 mM EGTA, 2 mM $MgCl_2$, 0.12 M sucrose [pH 7.4]) at room temperature for 10 min. Neurons for Western blot or mass spectrometry were immediately harvested, frozen in liquid nitrogen, and stored at −80 °C awaiting further processing.

For HEK293T cell biotinylation and immunocytochemistry, 1–2 days after transfection cells were labeled live with by adding a final concentration of 100 µM membrane-impermeant biotin-AEEA-phenol and 1 mM H_2O_2 to the cell culture for 1 min or by adding 500 µM membrane-permeant biotin-phenol (Iris Biotech, Marktredwitz, Germany, #LS-3500.0250) to the cell culture medium for 30 min at 37 °C and afterwards adding a final concentration of 1 mM H_2O_2 to the cell culture medium. Then, HEK293T cells were washed three times with PBS containing 10 mM sodium azide, 10 mM sodium ascorbate, and 2.5 mM Trolox to quench the biotinylation reaction, followed by fixation with 4% paraformaldehyde/4% sucrose in PBS for 10 min at room temperature.

2.8. Sample Preparation for Mass Spectrometry

Cell pellets were thawed on ice and each pellet was resuspended in 350 μL lysis buffer (1% SDS in 50 mM Tris-HCl (pH 8.0), including 10 mM sodium azide, 10 mM sodium ascorbate, and 2.5 mM Trolox, and the protease inhibitors at a final concentration of 1 mM PMSF, 2 μg/mL leupeptin, 1 μg/mL pepstatin, and 1 μg/mL aprotonin). Lysates were boiled at 95 °C for 5 min to dissociate the postsynaptic density (PSD) and diluted with 1400 μL 1.25× RIPA lysis buffer (50 mM Tris, 187.5 mM NaCl, 0.625% sodium deoxycholate, 1.25% Triton X-100) to a final 1× RIPA lysis buffer (50 mM Tris-HCl [pH 8.0], 150 mM NaCl, 0.2% SDS, 0.5% sodium deoxycholate, 1% Triton X-100). Lysates were cleared by centrifugation at 16,000× *g* for 10 min at 4 °C. Streptavidin-coated magnetic beads (Dynabeads M-270, Thermo Fisher Scientific, 65305) were equilibrated to room temperature and 50 μL resuspended bead slurry was washed twice with 1× RIPA buffer 50 mM Tris-HCl [pH 8.0], 150 mM NaCl, 0.2% SDS, 0.5% sodium deoxycholate, 1% Triton X-100). Beads were incubated with the lysate overnight at 4 °C with gentle rocking agitation to bind biotinylated proteins. Beads were then washed four times with 1 mL ice-cold 1× RIPA buffer followed by two washes with 1 mL 100 mM ammonium bicarbonate. In the second wash a final concentration of 0.01% RapiGest SF (#186001861, Waters Corporation, Milford, MA, USA) was added to avoid beads adhering to the tube.

For tryptic digestion of the bound biotinylated proteins, beads were incubated with 10 μL of 10 ng/μL trypsin (Sigma-Aldrich, 11418475001) in 100 mM ammonium bicarbonate containing 0.1% RapiGest SF overnight at 37 °C. For the first 15 min, samples were vortexed for 15 s every 2–3 min. After overnight digestion, another 10 μL of 10 ng/μL trypsin in 100 mM ammonium bicarbonate containing 0.1% RapiGest SF was added and incubated for 4 h at 37 °C. Afterwards, 100% formic acid (Fisher Chemical, Fisher Scientific, Hampton, NH, USA, A117-50) was added to a final concentration of 5% (*v/v*). Samples were frozen in liquid nitrogen and stored at −80 °C before shipment on dry ice to a mass spectrometry facility.

2.9. Mass Spectrometry and Label-Free Quantification

2.9.1. Mass Spectrometry Data Acquisition

The acidified samples were placed in an autosampler vial for analysis by reverse phase liquid chromatography-tandem mass spectrometry (RP)-LC-MS/MS. Reversed phase (RP)-LC-MS/MS was performed using nanoACQUITY UPLC system (Waters Corporation, Milford, MA, USA) connected to an Orbitrap Fusion Tribrid (Thermo Fisher Scientific, San Jose, CA, USA) mass spectrometer. After injection, samples were loaded into a trapping column (nanoACQUITY UPLC Symmetry C18 Trap column, 180 μm × 20 mm) at a flowrate of 5 μL/min and separated with a C18 column (nanoACQUITY column Peptide BEH C18, 75 μm × 250 mm). The compositions of mobile phases A and B were 0.1% formic acid in water and 0.1% formic acid in acetonitrile, respectively. Peptides were eluted with a gradient extending from 3% to 35% mobile phase B in 90 min at a flowrate of 300 nL/min and a column temperature of 37 °C. The data were acquired with the mass spectrometer operating in a top speed data-dependent mode. The full scan was performed in the range of 300–1500 *m/z* at an Orbitrap resolution of 120,000 at 200 *m/z* and automatic gain control (AGC) target value of 4×10^5. Full scan was followed by MS^2 event of the most intense ions above an intensity threshold of 5×10^4. The ions were iteratively isolated with a 1.6 Th window, injected with a maximum injection time of 110 msec, AGC target of 1×10^5, and fragmented with higher-energy collisional dissociation (HCD).

2.9.2. Data Processing for Identification

Raw data from the Orbitrap Fusion mass spectrometer (Thermo Fisher Scientific, Waltham, MA, USA) were processed using Proteome Discoverer software (version 2.1, Thermo Fisher Scientific). MS^2 spectra were searched using Mascot (Matrix Science, Boston, MA, USA) which was set up to search against the SwissProt rat database. The search criteria included 10 ppm precursor mass tolerance, 0.02 Da fragment mass tolerance, trypsin as proteolytic enzyme, and maximum missed cleavage sites of two. Potential dynamic modifications assigned were oxidation of methionine, de-amidation of asparagine, acetylation of N-terminus, and phosphorylation of serine, threonine and tyrosine.

Scaffold (version Scaffold_4.8.4, Proteome Software Inc., Portland, OR, USA) was used to validate MS/MS based peptide and protein identifications. Peptide identifications were accepted if they could be established at greater than 95.0% probability by the Peptide Prophet algorithm [42] with Scaffold delta-mass correction. Protein identifications were accepted if they could be established at greater than 99% probability and contained at least two identified peptides. Protein probabilities were assigned by the Protein Prophet algorithm [43]. Proteins that contained similar peptides and could not be differentiated based on MS/MS analysis alone were grouped to satisfy the principles of parsimony. Proteins sharing significant peptide evidence were grouped into clusters.

The mass spectrometry proteomics data have been deposited to the ProteomeXchange Consortium via the PRIDE partner repository [44] with the dataset identifier PXD011312.

2.9.3. Post Hoc In-Silico Filtering/Data Analysis

Data of all four biological replicate experiments were loaded together into Scaffold 4 (version 4.8.4) and GO annotations were added (Table S4) (NCBI, downloaded 20 December 2017). Per each biological replicate experiment, five conditions were included that each represented one sample (Figure 3B): non-transduced rat cortical neurons treated with biotin-AEEA-phenol and H_2O_2 (condition 1, c1); neurons transduced with Membrane-HRP rAAV and treated with biotin-AEEA-phenol omitting H_2O_2 (condition 2, c2) or with H_2O_2 (condition 3, c3); neurons transduced with SynCAM 1-HRP rAAV and treated with biotin-AEEA-phenol omitting H_2O_2 (condition 4, c4) or with H_2O_2 (condition 5, c5). Quantitative values, i.e., iBAQ (intensity-Based Absolute Quantitation) values [45] (no normalization), of detected proteins were calculated and exported to Microsoft Excel in which all further analysis was proceeded. Proteins identified within sub-clusters, keratin (a known impurity), and proteins with a probability <95% were excluded and protein iBAQ values were normalized to summed iBAQ values within one sample representing the molar abundance [46] or relative iBAQ (riBAQ) of an identified protein within a sample. Filtering was based upon previous protocol [29] for Filter 1 and 2 and modified as follows.

Filter 1

For each detected protein, the $\log_2$ of c5/c1 riBAQ value within a biological replicate was calculated, which measured the extent of biotinylation by a reporter protein. If a protein was not detected in c1 and hence a ratio could not be calculated the protein was regarded to be detected with high specificity and received the label "Specific". If a protein was not detected in c5, it received the label "N.A.". For Filter 1, in each biological replicate a c5/c1 riBAQ ratio cut-off value was determined above which a protein was retained using Receiver Operating Characteristic (ROC) analysis. First, proteins were labeled according to the following four groups [29]: (1) true positive or known synaptic proteins, TP1; (2) false positives or known intracellular proteins, FP1; (3) known surface proteins, FP2; (4) all other proteins. Proteins that existed in multiple groups were re-sorted to group 4. Endogenously biotinylated proteins and proteins with a cell surface GO-term were excluded from group 2. The c5/c1 riBAQ ratios of group 1 and group 2 were plotted in histograms (Figure 4B). An ROC curve analysis was then performed (Figure 4C) [47]. The remaining proteins were ranked in descending order according to the c5/c1 riBAQ ratio and then the riBAQ value. Proteins with "Specific" values ware placed on top and then ranked according to riBAQ values, and "N.A." proteins were excluded. In the ranked list, the True Positive Rate (TPR) for each protein was calculated as the summed number of group 1 proteins up till (and including) that protein divided by the total number of proteins in group 1. The False Positive Rate (FPR1) for each protein was calculated as the summed number of group 2 proteins up till that protein divided by the total number of proteins in group 2. For each ranked protein, TPR-FPR1 was calculated and plotted against its rank (Figure 4C). At maximum TPR-FPR, the associated ranked protein was found and its $\log_2$ of c5/c1 riBAQ value determined.

Filter 2

For each detected protein, the $\log_2$ c5/c3 riBAQ ratio within a biological replicate was calculated, which measured the extent of biotinylation by excitatory cleft-localized SynCAM 1-HRP over dendritic membrane-localized Membrane-HRP.

If a protein was not detected in c3 and hence a ratio could not be calculated the protein was regarded to be detected with high specificity and received the label "Specific". If a protein was not detected in c5, it received the label "N.A.". For Filter 2, in each biological replicate a c5/c3 riBAQ ratio cut-off value was determined above which a protein was retained using Receiver Operating Characteristic (ROC) analysis. First, proteins were labeled according to the following four groups [29]: (1) true positive or known synaptic proteins, TP1; (2) false positives or known intracellular proteins, FP1; (3) known surface proteins, FP2; (4) all other proteins. Proteins that existed in multiple groups were re-sorted to group 4. Endogenously biotinylated proteins and proteins with a cell surface GO-term were excluded from group 2. The c5/c3 riBAQ ratios of group 1 and group 3 were plotted in histograms (Figure 4B). An ROC curve analysis was then performed (Figure 4C) [47]. The remaining proteins were ranked in descending order according to the c5/c3 riBAQ ratio and then the riBAQ value. Proteins with "Specific" values ware placed on top and then ranked according to riBAQ values, and "N.A." proteins were excluded. In the ranked list, in descending order the True Positive Rate (TPR) for each protein was calculated as the summed number of group 1 proteins found from the top up till (and including) that protein divided by the total number of proteins in group 1. The False Positive Rate (FPR2) for each protein was calculated as the summed number of group 3 proteins found up till that protein divided by the total number of proteins in group 3. For each ranked protein, TPR-FPR2 was calculated and plotted against its rank (Figure 4C). At maximum TPR-FPR2, the associated ranked protein was found and its $\log_2$ of c5/c3 riBAQ value determined.

Filter 3

For each biological replicate, cut-off values for Filter 1 and Filter 2 were applied. Identified proteins in a biological replicate were retained when above the cut-off value or when labeled "Specific" for that filter. All identified proteins were then ranked in descending order to how many biological replicates they remained after filtering, how often they were labeled "Specific", and the average c5/c3 riBAQ ratio of all biological replicates. A ROC curve analysis was then performed (Figure 4F). In the ranked list, in descending order the True Positive Rate (TPR) for each protein was calculated as the summed number of group 1 proteins found from the top up till (and including) that protein divided by the total number of proteins in group 1. The False Positive Rate (FPR2) for each protein was calculated as the summed number of group 3 proteins found up till that protein divided by the total number of proteins in group 3. For each ranked protein, TPR-FPR2 was calculated and plotted against its rank (Figure 4F).

In the final selection step, proteins were removed that had only GO-terms associated with intracellular locations and proteins on the False Positive 1 (FP1)-list [29]. Serum albumin was removed from Table S3 as it is a likely carry-over from cell culture medium.

2.9.4. Calculation of Depth of Coverage

Depth of coverage for the excitatory synaptic cleft proteome was calculated as in [29]. In total, 10 excitatory synaptic cleft candidate proteins identified here (Supplementary Table S1) were among the list of 62 literature identified excitatory synaptic cleft proteins (TP2, Table S3).

2.10. *Immunochemistry*

For visualization of biotinylation in HEK293T cells, after the biotinylation reaction and fixation (see Peroxidase-mediated biotinylation), cells were washed three times with PBS and permeabilized with 0.1% Triton X-100 for 10 min at room temperature followed by blocking with 5% FBS in PBS 1 h at room temperature. Then, cells were incubated for 1 h at room temperature with Streptavidin-Alexa488

(Molecular Probes, S11223; 1:500) in 5% FBS in PBS, followed by three washes with PBS. SynCAM 1-APEX2 or APEX2-NES were detected by immunostaining using anti-FLAG M2 IgG1 antibodies raised in mouse (Sigma-Aldrich, F1804; 1:500) or anti-HA rabbit antibody (Cell Signaling Technologies, #3724; 1:500), resp., for 1 h at room temperature followed by three washes with PBS and secondary antibody staining with anti-IgG1 Alexa568-conjugated secondary antibodies (Invitrogen via Fisher, #A21124; 1:500) or anti-rabbit Alexa568-conjugated secondary antibodies (Invitrogen via Fisher, #A11036; 1:500), resp., for 1 h at room temperature followed by three washes with PBS. Coverslips were mounted onto glass slides (Aqua-Poly/Mount, Polysciences) and imaged with a Leica SP2 confocal microscope (Leica Camera Co., Wetzlar, Germany).

For biotinylation in neurons, after the biotinylation reaction and fixation (see Peroxidase-mediated biotinylation), non-permeabilized cells were blocked with 5% BSA (RMBIO, BSA-BAF) for 30 min at 4 °C followed by staining for biotin using NeutrAvidin protein, Dylight 488 (NA-488; Thermo Fisher Scientific, 22832; 1:250) for 10 min at 4 °C, immediately followed by one wash with 1 mM free biotin (Sigma-Aldrich, #B4501) in PBS and two washes in PBS. Staining with more concentrated NA-488 at 4 °C to reduce membrane protein mobility, followed with a biotin wash to block free biotin-binding sites in bound NA-488 was necessary to avoid that biotin-labeled surface proteins clustered in the membrane [48]. Cells were permeabilized with 0.1% Triton X-100 for 10 min at room temperature followed by blocking with 5% FBS in PBS 1 h at room temperature. From here on all permeabilization steps were with 0.1% Triton X-100 for 10 min, blocking was with 5% FBS in PBS 1 h at room temperature, and all primary and secondary stainings were done at room temperature for 1 h or overnight at 4 °C in 5% FBS in PBS and followed by three times 5-min washes with PBS at room temperature. Cells were stained with antibodies raised in rabbit against Homer (Synaptic Systems GmbH, Goettingen, Germany, 160 003; 1:500) and anti-rabbit Alexa568 secondary antibodies (1:500). Coverslips were mounted onto glass slides (Aqua-Poly/Mount, Polysciences) and imaged with a Leica SP8 confocal microscope (Leica Camera Co., Wetzlar, Germany).

For visualization of R-PTP-zeta in neurons, dissociated cortical neuron cultures from rats were fixed with fixation buffer (60 mM PIPES, 25 mM HEPES, 10 mM EGTA, 2 mM $MgCl_2$, 0.12 M sucrose [pH 7.4]) at 4 °C for 15 min and washed three times with PBS before blocking. Under non-permeabilizing conditions, R-PTP-zeta and SynCAM 1 were stained with anti-PTPζ IgM antibodies raised in mouse (Santa Cruz Biotechnology, Dallas, TX, USA, sc-33664; 1:500) and anti-SynCAM 1 antibodies raised in chicken (MBL Laboratories, Woods Hole, MA, USA), CM004-3; 1:500), respectively. Followed by staining with anti-IgM Alexa488 secondary antibodies (Life Technologies, Carlsbad, CA, USA, A21042; 1:500) and anti-chicken Alexa647 secondary antibodies (Life Technologies, A21449; 1:500). Cells were permeabilized followed by blocking. Cells were then either stained for Bassoon or Homer with anti-Bassoon mouse IgG2A antibodies (Enzo Life Sciences, Enzo Biochem., Inc., Farmingdale, NY, USA, ADI-VAM-PS003; 1:750) or anti-Homer rabbit (Synaptic Systems, 160 003; 1:500), respectively, and then anti-IgG2a Alexa568 (1:500) or anti-rabbit Alexa568 (1:500) secondary antibodies. Coverslips were mounted (Aqua-Poly/Mount, Polysciences) and imaged with a Leica SP8 confocal microscope (Leica Camera Co., Wetzlar, Germany).

For visualization of R-PTP-zeta in brain sections, adult mice were anaesthetized using ketamine/xylazine and perfused transcardially with 4% paraformaldehyde (PFA) in PBS. Brains were extracted and post-fixed overnight at 4 °C in 4% PFA in PBS. Brains were sectioned on a vibratome (Vibratome 1500, Leica Camera Co., Wetzlar, Germany) into 60 μm coronal sections. For detection of R-PTP-zeta, Homer, and Bassoon, individual sections were simultaneously permeabilized and blocked with 0.3% Triton X-100 and 3% horse serum for 1 h at room temperature, followed by a combined incubation with anti-PTPζ mouse IgM (Santa Cruz Biotechnology, sc-33664; 1:500), IgG2A antibodies raised against Bassoon in mouse (1:750) and anti-Homer antibodies raised in rabbit (1:500) (all in 0.3% Triton X-100 and 3% horse serum for 3 days at 4 °C). Followed by three washes with PBS and incubation with anti-IgM Alexa488 (1:500), anti-IgG2a Alexa647 (1:500) or anti-rabbit Alexa568 (1:500) antibodies in 0.3% Triton X-100 and 3% horse serum. Sections were mounted onto glass slides

(Aqua-Poly/Mount, Polysciences, Inc., Warrington, PA, USA) and imaged with Keyence microscope (BZ-X700, Keyence, Itasca, IL, USA). Secondary antibodies were purchased from Life Technologies.

2.11. Characterization of Biotinylation by SynCAM 1-HRP Using Western Blot

Cell pellets were thawed on ice and each cell pellet was resuspended in 100 μL lysis buffer (1% SDS in 50 mM Tris-HCl (pH 8.0), including 10 mM sodium azide, 10 mM sodium ascorbate, and 2.5 mM Trolox, and the protease inhibitors at final concentration of 1 mM PMSF, 2 μg/mL leupeptin, 1 μg/mL pepstatin, and 1 μg/mL aprotonin). Lysates were boiled at 95 °C for 5 min to dissociate the postsynaptic density (PSD) and diluted with 400 μL 1.25× RIPA lysis buffer (50 mM Tris, 187.5 mM NaCl, 0.625% sodium deoxycholate, 1.25% Triton X-100) to a final 1× RIPA lysis buffer (50 mM Tris-HCl [pH 8.0], 150 mM NaCl, 0.2% SDS, 0.5% sodium deoxycholate, 1% Triton X-100). Lysates were cleared by centrifugation at 16,000× *g* for 10 min at 4 °C. Cleared lysates were then boiled in sample buffer (containing final 50 mM DTT and 2% SDS in 60 mM Tris-HCl, pH 6.8) for 3 min and separated on a 10% polyacrylamide gel. Proteins were then blotted onto nitrocellulose blotting membrane and stained with Ponceau S for equal loading confirmation. Membrane was blocked in 5% BSA in TBST and incubated with NeutrAvidin-DyLight488 (Thermo Fisher Scientific, #22832; 1:500) in 5% BSA in TBST for 1 h at room temperature, followed by three washes with TBST and imaged (FluorChem M, ProteinSimple, San Jose, CA, USA).

2.12. Staining for dSTORM, and dSTORM Imaging and Analysis of SynCAM 1-HRP

For dSTORM experiments, neurons were fixed in 4% PFA with sucrose for 15 min at 21 div then washed for 3 times for 5 min each in PBS containing glycine (PBS/Gly). Samples were blocked in donkey serum (10%) for 20 min. Surface labeling was done by inverting coverslips on 50 μL droplets containing the anti-FLAG M2 antibody (Mouse IgG1) at 1:500 on parafilm. Coverslips were washed two times in PBS/Gly. Permeabilization was achieved with 0.1% TX-100 for 20 min. This was exchanged with PBS/Gly containing 10% donkey serum and 0.1% TX-100 and incubated for an additional 20 min. Coverslips were then transferred and inverted onto 50 μL droplets on parafilm containing anti-Homer1 primary antibody (Synaptic Systems; 1:500) at 4 °C overnight (<15 h). Primary contains 0.03% TX-100 and 5% donkey serum. Coverslips were washed 3 times for 5 min each in PBS/Gly the next day. Coverslips were then transferred and inverted onto parafilm with 50 μL droplets containing secondary antibody for 1 h at room temperature in the dark (Donkey anti-Rabbit CF555, 1:500; Donkey α-Mouse Alexa-647, 1:200). Coverslips were then washed 3 times for 5 min each in PBS/Gly and were then imaged according to [14].

Images were denoised using Thunderstorm's Wavelet (basis spline) filter ([49]; http://imagej.net/Fiji). Approximate localization of molecules was done using the local maximum approach using the Thunderstorm plug-in in Fiji/ImageJ (dev-2016-01-01-b1, github.com/zitmen/thunderstorm/). Sub-pixel localization of stochastic blinking was achieved by fitting peaks with an integrated Gaussian function using the Thunderstorm plug-in. For visualization, pixels were magnified 10× and detected localizations were binned into pixels. Poorly fit localizations were filtered out using the Thunderstorm software. Drift correction was performed in Thunderstorm using the cross-correlation method. The visualization method used is the Average Shifted Histograms method in Thunderstorm.

All synaptic data was analyzed by identifying synaptic clusters containing both Homer1 and SynCAM 1-HRP, in order to identify synapses from transfected cells. For line scan analyses, synaptic pairs were selected, then a line (20 pixels wide) was drawn perpendicular to the synaptic axis, as to pass through both Homer1 and SynCAM 1-HRP. SynCAM 1-HRP puncta were counted by eye in ImageJ. For the inter-cluster distance analysis, center was determined by eye and a line was drawn between the two clusters and the distance was estimated by rounding to the nearest pixel.

3. Results

3.1. Peroxidase-Mediated Proximity Labeling Using a SynCAM 1-Horseradish Peroxidase (HRP)-Fusion Protein

Peroxidase-mediated proximity labeling allows for the biotinylation of endogenous proteins proximal to a recombinant peroxidase-reporter fusion protein that is exogenously expressed at a cellular compartment. Peroxidases that have been used for proximity labeling are HRP and APEX/APEX2, which are metalloenzymes that catalyze the oxidation of organic substrates by hydrogen peroxide (H_2O_2). These peroxidases can oxidize in presence of H_2O_2 biotin-phenol compounds to generate short-lived biotin-phenoxyl radicals. These radicals can form covalent bonds with tyrosine and other electron-rich amino acids at proximal proteins [25,50]. APEX2, a more active variant of the engineered peroxidase APEX [51], is suitable for intracellular and extracellular applications, while HRP is only active in the oxidizing environments of the secretory pathway and the cell surface due to its structurally essential disulfide-bonds.

To map the excitatory synaptic cleft proteome, a fusion protein needs to target a peroxidase to this specific compartment. We here used SynCAM 1 as it is a synaptic cell adhesion protein that localizes exclusively to excitatory synapses during development and in mature synapses [25,34]. HRP or APEX2 was inserted at amino acid position 363 of SynCAM 1, located between the last immunoglobulin (Ig) domain and the trans-membrane (TM) domain. This placed HRP at the base of the extracellular region of SynCAM 1, creating a SynCAM 1-HRP fusion protein (Figure 1A) or SynCAM 1-APEX2 fusion protein. Recombinant proteins or tags inserted at this position do not alter the synaptic localization of SynCAM 1 [52]. SynCAM 1 is a single-pass type 1 membrane protein that after biogenesis requires trafficking through the secretory pathway to reach the cell surface. To restrict biotin labeling mediated by peroxidase-fusion proteins that traffic through the secretory pathway to the cell surface, including the synaptic surface, a membrane-impermeant biotin-phenol compound containing a polar linker is required [29]. We therefore selected the commercially available compound biotin-AEEA-phenol, which is membrane-impermeant.

To verify that it allows for biotinylation only at the cell surface, biotin-AEEA-phenol was added to HEK293T cells that expressed a membrane-bound SynCAM 1-APEX2 proximity reporter. In presence, but not in absence, of H_2O_2, exogenous biotin-AEEA-phenol induced biotinylation at the surface of HEK293T cells expressing SynCAM 1-APEX2 (Figure 1B). As control whether this biotin compound is membrane impermeable, biotin-AEEA-phenol was added to HEK293T cells expressing soluble APEX2 fused to a nuclear export sequence (APEX2-NES) [51]. This did not result in detectable intracellular biotinylation (Figure 1C). As expected from a previous study [28], membrane-permeable biotin-phenol allowed for intracellular biotinylation in HEK293T cells expressing APEX2-NES (not shown). Biotin-AEEA-phenol hence restricts proximity labeling to the cell surface. To efficiently use the proximity labeling approach in cultured neurons, HRP fusion proteins were used in all proximity labeling experiments in neurons as HRP is more active than APEX2 in the extracellular space [29].

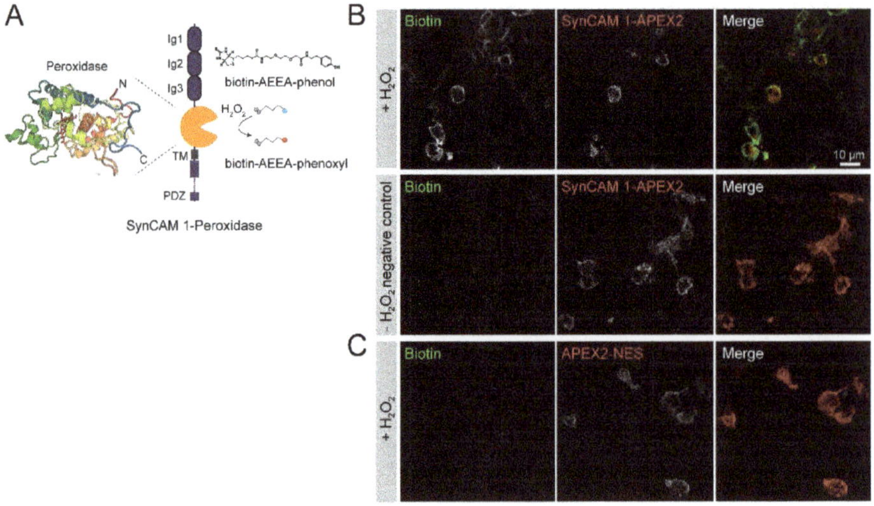

Figure 1. SynCAM 1-peroxidase fusion protein peroxidase-mediated proximity labeling in the synaptic cleft. (**A**) APEX2 or HRP (image RCSB PDB [53,54] (www.rcsb.org) of PDB ID 1HCH [54]) peroxidase was inserted at the base of the SynCAM 1 extracellular domain, with immunoglobulin (Ig) domains, trans-membrane (TM) region, and intracellular PDZ domain interaction sequence indicated. APEX2 or HRP catalyzes the formation of a short-lived biotin-AEEA-phenoxyl radical (red dot) after exogenous addition of H_2O_2 and membrane-impermeable biotin-AEEA-phenol (blue dot). (**B**) Exogenous biotin-AEEA-phenol induced biotinylation only at the cell surface. Staining for biotin (visualized by StreptAvidin-Alexa488) in HEK293T cells expressing SynCAM 1-APEX2 in presence (+) but not in absence (−) of H_2O_2. (**C**) Exogenous biotin-AEEA-phenol did not induce biotinylation in HEK293T cells expressing cytosolic APEX2-NES.

3.2. Subsynaptic Distribution of the SynCAM 1-HRP Reporter

Previously, the synaptic expression of endogenous SynCAM 1 has been analyzed using STED and 3D dSTORM super-resolution imaging in cultured neurons and by immuno-EM in brain sections. These approaches determined that SynCAM 1 localizes to the synaptic cleft of excitatory, asymmetric synapses and is predominantly present in the postsynaptic membrane, where it was detected around the edge of the postsynaptic density [14]. To analyze the localization of the SynCAM 1-HRP reporter, a FLAG-tag was inserted C-terminal of the HRP. Cultured rat hippocampal neurons were transfected with this construct and analyzed by two-color 3D dSTORM super-resolution imaging. Immunostaining for FLAG to detect SynCAM 1-HRP and for the postsynaptic excitatory scaffolding protein Homer followed by 3D dSTORM showed that super-resolved SynCAM 1-HRP localized adjacent to clusters of Homer (Figure 2A and Figure S1). Distribution of SynCAM 1-HRP ensembles were within ~100 nm of the edge of the PSD as outlined by super-resolved Homer clusters (Figure 2B and Figure S1), in agreement with the sub-synaptic distribution of endogenous SynCAM 1 in cultured rat hippocampal neurons [14]. This demonstrated that the SynCAM 1-based reporter targets HRP to the synaptic cleft and is enriched at the cleft border zone.

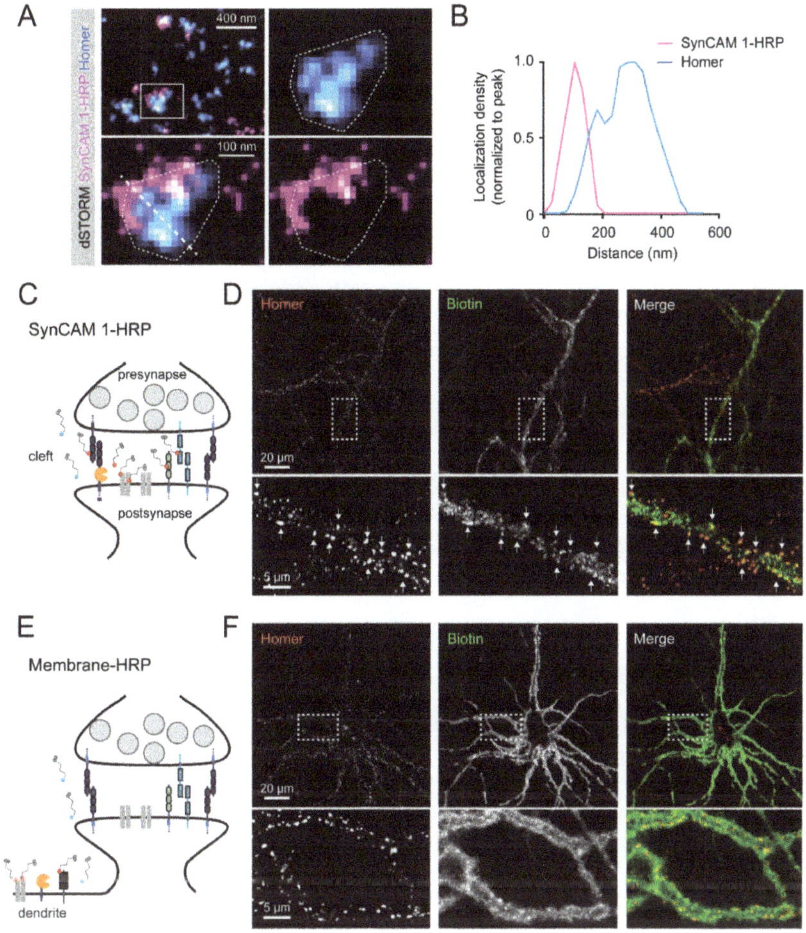

Figure 2. Synaptic SynCAM 1-HRP expression and biotinylation. (**A**) Two-color dSTORM reconstruction of synapses from 21 days-in-vitro (div) rat hippocampal neurons surface-labeled by immunostaining with anti-FLAG antibodies against exogenous SynCAM 1-HRP containing the FLAG epitope (magenta) and the endogenous excitatory postsynaptic marker Homer (cyan). Top left, overview. Enlarged panels show one synapse with SynCAM 1-HRP and Homer localizations fit by a convex hull to demarcate the PSD border (dotted outline). SynCAM 1-HRP localizations are at the periphery of the PSD. Diagonal line, line scan used for (**B**). See Supplemental Figure S1 for additional examples and quantitative analyses. (**B**) Protein localization distribution perpendicular to the trans-synaptic axis. Densities were determined by dSTORM and normalized to the peak of each channel and measured over the distance shown in A. (**C**) Model of HRP targeting by the reporter SynCAM 1-HRP to excitatory synaptic clefts for biotinylation of proximal surface proteins. (**D**) Following the proximity labeling reaction with membrane-impermeant biotin-AEEA-phenol, biotin staining was detected in SynCAM 1-HRP transduced rat cortical neurons along dendrites at excitatory synaptic sites visualized by immunostaining of Homer (arrows). (**E**) Targeting of HRP by the reporter Membrane-HRP to the plasma membrane of dendrites. (**F**) Staining for biotin was detected in Membrane-HRP transduced rat cortical neuronal cultures along dendrites after proximity labeling with biotin-AEEA-phenol.

3.3. SynCAM 1-HRP Validation for Synaptic Proximity Labeling

Synaptic clefts are open cellular compartments that are not enclosed by membranes. This requires to further improve the specificity of protein identification in the synaptic cleft. A ratiometric peroxidase-mediated labeling approach was previously applied to synaptic cleft proteins [29] and mitochondrial inter-membrane space proteins [55]. We aimed to implement this ratiometric approach and compared the proteins biotinylated by the SynCAM 1-HRP cleft reporter (Figure 2C) versus a broadly surface-expressed HRP fused extracellularly to a trans-membrane domain (Membrane-HRP; Figure 2E) [28]. To verify proper generation of biotinylation reaction products by these two reporters in neurons, the reporters were transduced into cultures of dissociated rat cortical neurons using recombinant AAV (rAAV) at 14 days-in-vitro (div). rAAV allowed the large-scale transduction of cultured neurons, and virus particles were titrated to balance high transduction efficiency with moderate protein overexpression. Neurons were subjected to brief peroxidase-mediated proximity labeling at 21 div upon addition of H_2O_2 and biotin-AEEA-phenol. Staining of biotinylation products using NeutrAvidin-DyLight488 visualized biotin-protein conjugates along dendrites at excitatory synaptic locations positive for Homer in SynCAM 1-HRP expressing neurons (Figure 2D), with some extra-synaptic labeling. In contrast, Membrane-HRP transduced into cultured rat cortical neurons that underwent proximity labeling showed biotin labeling along dendrites that was not enriched at synaptic locations (Figure 2F). These results supported that the SynCAM 1-HRP and the Membrane-HRP reporters can be used to obtain proximity-labeled protein samples from excitatory synaptic clefts and the neuronal cell surface, respectively, for comparative ratiometric analysis in silico.

3.4. Robust Identification of Synaptic Cleft Candidate Proteins Using Proximity Labeling

To map the proteome of excitatory synaptic clefts, large-scale rat cortical cultures were prepared and treated following the work flow shown in Figure 3A. rAAV encoding SynCAM 1-HRP or Membrane-HRP was added to separate neuronal cultures at 14 div. Cultures underwent at 21 div peroxidase-mediated labeling upon addition of H_2O_2 and biotin-AEEA-phenol for 60 s. Biotinylated proteins were purified and peptides were generated by on-bead digestion with trypsin. Peptides were analyzed by Label-Free Quantitation (LFQ) proteomics, which compares peptide intensities across samples [56–58]. Per each biological replicate experiment, five conditions were included that each represented one sample (Figure 3B): non-transduced rat cortical neurons treated with biotin-AEEA-phenol and H_2O_2 (condition 1); neurons transduced with Membrane-HRP rAAV and treated with biotin-AEEA-phenol omitting H_2O_2 (condition 2) or with H_2O_2 (condition 3); neurons transduced with SynCAM 1-HRP rAAV and treated with biotin-AEEA-phenol omitting H_2O_2 (condition 4) or with H_2O_2 (condition 5).

For biochemical analysis, protein samples from treated neurons were separated and biotinylated proteins were visualized by Western blotting (Figure 3C). This showed that proteins were biotinylated in SynCAM 1-HRP or Membrane-HRP expressing neurons in presence, but not in absence, of H_2O_2 and conjugates spanned a large molecular weight range (Figure 3C, lanes 3, 5). Endogenously biotinylated proteins (e.g., histones and mitochondrial carboxylases) [29,59–62] were detected independent of the biotinylation reaction, as expected (Figure 3C).

For proteomic analysis, four independent biological replicate experiments were performed and six pair-wise comparisons of protein levels of identified proteins across all four biological replicate experiments examined the robustness of this approach. As reliable measure for protein levels, intensity-based absolute quantification (iBAQ) levels were calculated [45] as iBAQ levels are proportional to molar abundance [46]. Pair-wise comparisons across biological replicates of protein iBAQ values or molar abundance of condition 5 showed strong correlation between biological replicates over minimally 5 orders of magnitude (Figure 3D). Several known synaptic proteins expected to be present at the synaptic cleft, including Neurexin-1, Neuroligin-3, Latrophilin-3, Contactin-1, Kilon, Hapln1, and Noelin-1, were found across the molar abundance spectrum. Retention of proteins due to interaction with other specifically bound proteins was unlikely as proteins were denatured before bead-binding. Further, endogenously biotinylated mitochondrial carboxylases and nuclear histones were among the most abundant proteins detected. The raw data also included intracellular proteins, e.g., GAPDH and Erlin-2, which likely unspecifically adsorbed to the beads. These non-synaptic hits identified in SynCAM 1-HRP samples further highlighted the need for a data filtering approach that analyzes a protein's extent of biotinylation by a cleft reporter relative to a control to determine actual synaptic cleft abundance.

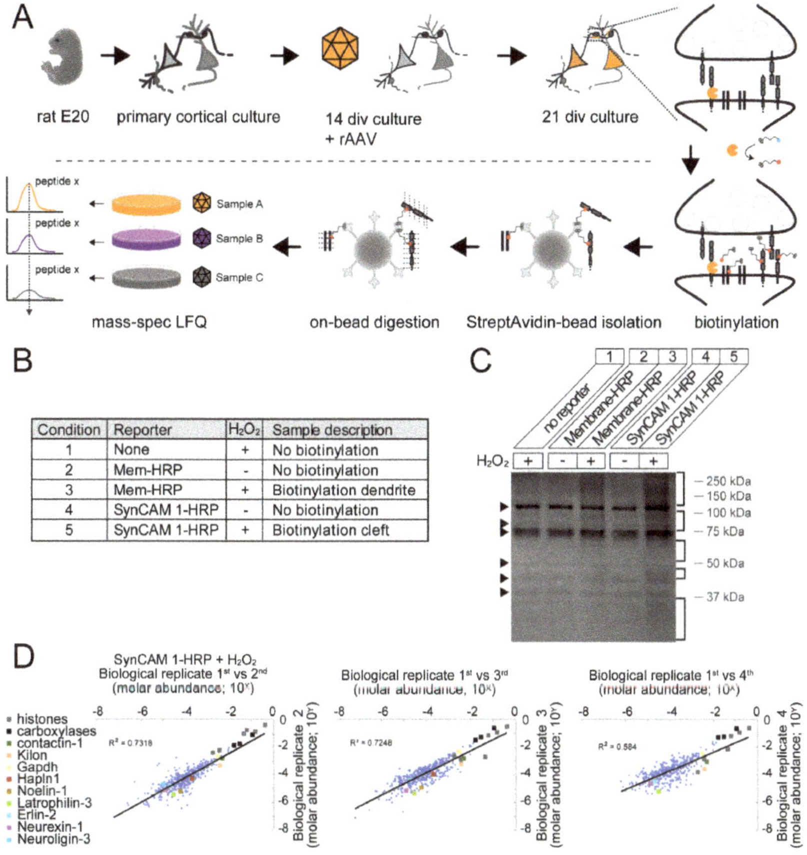

Figure 3. Scaled proximity labeling in cultured neurons. (**A**) Rat cortical neurons were transduced with rAAV encoding HRP reporters (orange) at 14 div and underwent proximity labeling at 21 div. SynCAM 1-HRP (orange) biotinylates synaptic proteins that are purified and digested for LFQ. (**B**) Each biological replicate included samples from the 5 conditions shown. (**C**) Western Blot of samples visualized by NeutrAvidin-DyLight488. Endogenously biotinylated proteins (arrow heads) are marked and weight ranges (brackets) wherein exogenously biotinylated proteins were detected are indicated. (**D**) Example comparisons of molar abundance ranges. Four biological replicates were performed and molar abundance (relative iBAQ or riBAQ) of all detected proteins in SynCAM 1-HRP samples were compared. Graphs show pair-wise comparisons of the 1st biological replicate with the 2nd, 3rd, and 4th replicate, respectively. Endogenously biotinylated proteins (histones, mitochondrial carboxylases), cytosolic proteins (GAPDH, Erlin-2) and synaptic cleft proteins (Contactin-1, Kilon, Hapln1, Noelin-1, Latrophilin-3, Neurexin-1, Neuroligin-1) are indicated.

3.5. Ratiometric Analysis of Proximity Labeled Protein Hits

In silico post hoc filtering was implemented to identify biotinylated, synaptic cleft-enriched proteins, following a previously-described ratiometric approach [29]. Specifically, proteins identified in any of the conditions were sorted into four groups: (i) true positives, i.e., known synaptic proteins, TP1; (ii) false positives, i.e., known intracellular proteins, FP1; (iii) false positives, i.e., known surface proteins that are not synapse-enriched, FP2; (iv) all other proteins (Figure 4A). In filter step 1, the $\log_2$ ratio of normalized iBAQ values (i.e., relative to total iBAQ per sample; riBAQ) for condition 5 (c5) over condition 1 (c1) assessed how likely a protein was to be biotinylated by SynCAM 1-HRP. In filter step 2, the ratios of riBAQ for condition 5 (c5) over condition 3 (c3) was calculated to indicate how likely a protein was biotinylated by SynCAM 1-HRP compared to Membrane-HRP. As first internal quality control of the data filtering, the set of known synaptic surface proteins TP1 was used [29], for which c5/c1 riBAQ ratios are expected to be higher than for known intracellular proteins (FP1). As second internal quality control, the c5/c3 riBAQ ratios were assessed, which are expected to be higher for the list of synaptic membrane proteins (TP1) over the set of known surface proteins (FP2). Indeed, the distribution of c5/c1 riBAQ values of known synaptic proteins was bimodal with a first peak at the distribution of known intracellular proteins and a second peak at higher c5/c1 values (Figure 4B). Similarly, the distribution of known synaptic proteins was shifted towards higher c5/c3 riBAQ values compared with known surface proteins. As expected, the ratio of SynCAM 1-HRP plus H_2O_2 over SynCAM 1-HRP minus H_2O_2, i.e., c5/c4 riBAQ values, had comparable distributions as c5/c1 riBAQ values (not shown).

This analysis was expanded to select a cutoff for each filter step and obtain from the filtered data an enriched selection of proteins with maximal synaptic cleft proteins and minimal intracellular or dendritic surface proteins. Specifically, a Receiver Operating Characteristic (ROC) curve analysis was performed [47] to determine the optimal c5/c1 riBAQ and c5/c3 riBAQ cut-off values (Figure 4C). In brief, proteins in a biological replicate were ranked according to c5/c1 riBAQ values (filter 1) or c5/c3 riBAQ values (filter 2) (see Methods for ranking rules). For each protein, the rate of finding a false positive (False Positive Rate, FPR1 for intracellular protein and FPR2 for surface proteins) was subtracted from the rate of finding a true positive (True Positive Rate, TPR). TPR-FPR was plotted against the riBAQ values for each ranked protein (see Methods). At the maximum TPR-FPR, the matching ranked protein and corresponding c5/c1 riBAQ or c5/c3 riBAQ value was determined and used as minimum cut-off to retain proteins for filter 1 or 2, respectively (Figure 4C).

In all four biological replicate experiments, a total of 706 proteins were identified (Figure 4D). After filter 1, total protein number was reduced to 50% (or 353 proteins), the group of known intracellular proteins was reduced by 50%, and the group of known surface proteins not enriched at synapses was reduced to 55%. All known synaptic proteins were retained. After filter 2, total protein number was reduced to 28% (or 200 proteins), known intracellular proteins to 27%, and known surface proteins to 20%, while 97% of known synaptic proteins were retained (Figure 4E). A fraction of proteins that passed filter 1 and 2 (33.5%) showed the same selection in multiple rounds of biological replicate experiments. A third filtering step was introduced that used the information for how frequently a protein passed filter 1 and 2 across biological replicates. Proteins were ranked according to c5/c3 riBAQ values and plotted against TPR-FPR2. The proteins that passed filter 1 and 2 in at least two of the four biological replicates showed a strong increase in TPR-FPR2, indicating that there was a strong enrichment for true positives (Figure 4F).

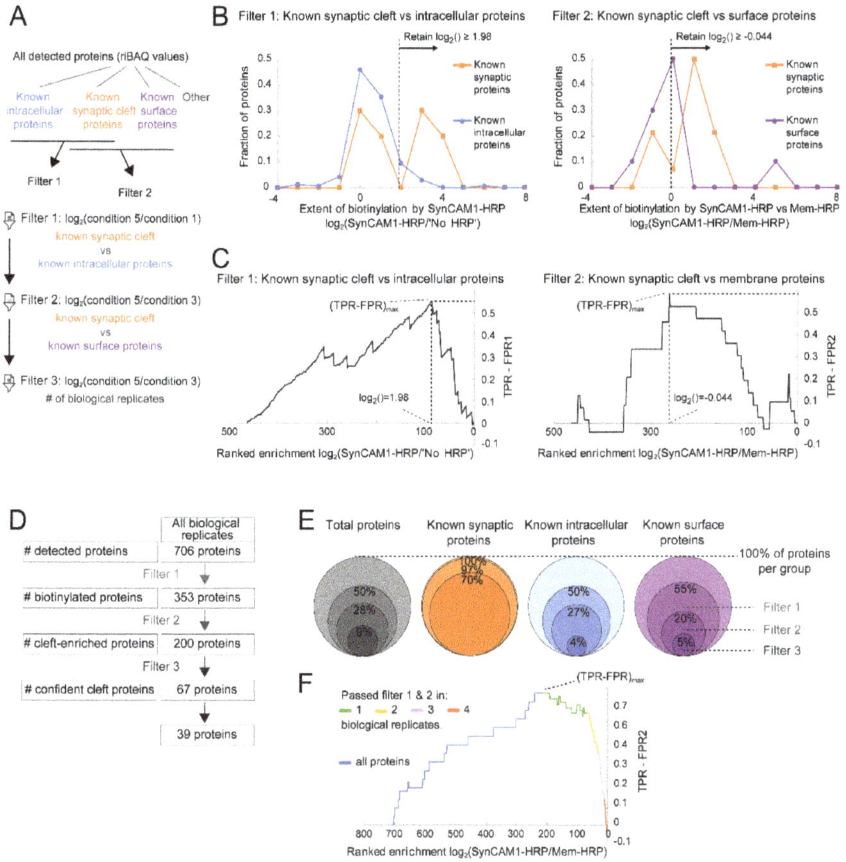

Figure 4. Data filtering based on proteins of known sub-cellular localizations. (**A**) Detected proteins were sorted into 4 categories based upon literature and these categories were used in Filter 1 and 2. Filter 3 used the number of biological replicates a protein passed Filter 1 and 2. (**B**) Histograms of the extent of biotinylation by SynCAM 1-HRP vs. no reporter control for Filter 1 (left plot) and vs. Mem-HRP for Filter 2 (right plot). Cut-off values determined in (**C**) for Filter 1 and Filter 2 are indicated. (**C**) Receiver Operating Characteristic (ROC) curve analysis to determine optimal enrichment for Filter 1 (left plot) or Filter 2 (right plot) by plotting the True Positive Rate (TPR: known synaptic proteins) minus the False Positive Rate (FPR). FPR1: known intracellular proteins; FPR2: known membrane proteins. At maximum (TPR-FPR), the $\log_2$ value is determined at the corresponding protein in the ranked protein list, which served as a cut-off value in (**B**). (**D**) Enrichment for excitatory synaptic cleft proteins through filtering. All proteins combined in four biological replicate experiments were subjected to Filter 1, Filter 2, and Filter 3. A final manual curation step removed remaining false positives. (**E**) Relative enrichment per group of proteins of known sub-cellular localization. Per group, all proteins (100% per group) are depicted as a circle of unitary size. Each filter step reduces protein number and circle area proportionally. Note that after Filter 1, the total proteins in group of known synaptic proteins is 100% and area of circle is unitary. (**F**) ROC analysis of TPR-FPR2 for Filter 3. Indicated are proteins that passed Filters 1 and 2 four times (red), three times (purple), two times (yellow), and one time (green).

Filter 3 encompassed this strict inclusion criterion to include only proteins that passed Filter 1 and 2 in minimally 2 biological replicates, and when applied, 67 proteins were retained (Figure 4E). After applying Filter 1 (selecting for biotinylated proteins), Filter 2 (selecting for synaptic cleft-biotinylated proteins), and Filter 3, 9% of total protein, 4% of known intracellular proteins, and 5%

of known surface proteins were retained. Importantly, 70% of known synaptic proteins were still retained after all filters had been applied. It was expected that the resulting list still contained false positives, specifically intracellular proteins due to the designed filtering steps that minimized but not necessarily eliminated false positives. In a final selection step, proteins were removed that had only GO-terms associated with intracellular locations and proteins on the False Positive 1 (FP1)-list [29]. The resulting list contained 39 synaptic cleft candidate proteins (Supplementary Table S1). Twenty-six of the 123 proteins that were identified in 1 biological replicate neither had a GO-terms associated with intracellular locations nor were on the False Positive 1-list and were added to the list synaptic cleft-enriched candidates (Supplementary Table S1). Together, this resulted in a set of proteins with high confidence of synaptic cleft localization that were cleft-enriched in multiple biological replicates and a set of proteins that were enriched in one biological replicate.

3.6. Molecular Class and Gene Ontology Analysis of Synaptic Cleft Hits

The list of 39 proteins (Figure 5: black, underlined gene names) found in multiple biological replicates and the 26 additional proteins (Figure 5, grey gene names) found in one biological replicate were characterized for Molecular Class according to the Human Protein Reference Database (HPRD) [63–65]. The HPRD classifies each protein into one category, which simplifies functional protein characterization. Proteins were classified across major categories expected in the synaptic cleft: 24% of proteins were associated with either adhesion molecule (18%), immunoglobulin (3%) or adhesion molecule activity (3%); 20% of proteins were associated with cell surface receptor (14%), G protein-coupled receptor (3%), or receptor tyrosine phosphatase (3%); 20% of proteins were associated with either Membrane transport protein (11%), Extracellular ligand-gated channel (5%), Ion channel (2%), or Voltage-gated channel (2%); 14% were associated with membrane transport protein (11%) or Transport/cargo protein (3%).

Previous proteomic studies of synapses have targeted different synaptic compartments or protein complexes [11,15–22,66–68]. A comparison with these studies found that almost all of 65 cleft-enriched proteins identified here, except for Contactin-5, were reported earlier to be synaptic. However, none of these studies specifically characterized the synaptic cleft. The only study so far targeting the synaptic cleft by Loh and colleagues had an estimated 69% excitatory synaptic cleft proteome coverage and reported several novel synaptic cleft candidates [29]. Compared with this previous work, 30 proteins were solely enriched in this study using SynCAM 1-HRP-mediated proximity labeling of the synaptic cleft (Supplementary Table S2). Eleven synaptic cleft candidates identified here were neither enriched in the previous proteomic study targeting the (excitatory and inhibitory) synaptic cleft [29], nor did they contain a GO-term associated with synapses. Several of these proteins, such as CD166 antigen and Leucine-rich glioma-inactivated protein 1, were identified in previous proteomic studies of the active zone and other synaptic compartments [11,15–22,66–68]. However, these studies did not demonstrate that these 17 proteins were synaptic cleft-enriched, and a literature and GO-term analysis did not find that these were synaptic cleft-enriched proteins. Hence, these proteins were termed candidate excitatory synaptic cleft orphans (Supplementary Table S3).

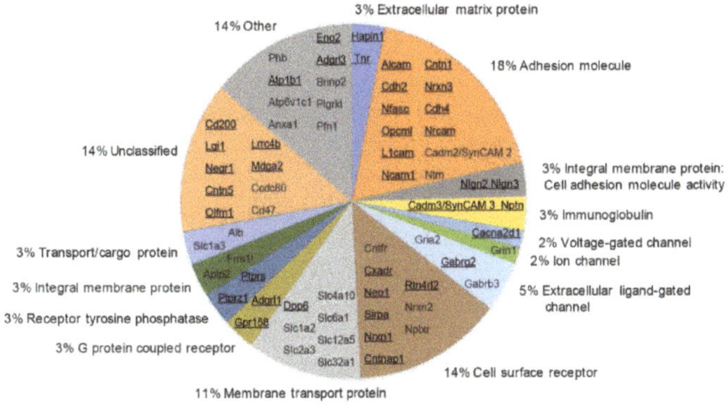

Figure 5. Pie chart showing relative abundance of protein functional groups (categories according to the Human Protein Reference Database) and synaptic cleft candidates in each group. Synaptic cleft candidates found in minimally two biological replicates are underlined. Candidates found only in one biological replicate are shown in grey font.

3.7. Validation of the Synaptic Cleft Candidate Receptor-Type Tyrosine-Protein Phosphatase Zeta

To validate the peroxidase-mediated proximity labeling approach and filtering process for identification of novel synaptic cleft proteins, a candidate protein not previously identified in a synaptic cleft proteomics study (Supplementary Table S2) was selected for further study. Receptor-type tyrosine-protein phosphatase zeta, or R-PTP-zeta (gene name: *Ptprz1*) was previously detected in proteomic studies of postsynaptic density and synaptosomes of rodent and human brain [15,20–22]. Here, coronal sections of mouse brain stained for R-PTP-zeta showed immunofluorescence in hippocampus and cortex (Figure 6). Immunostaining for R-PTP-zeta was particularly strong in areas with strong immunostaining for postsynaptic marker Homer and presynaptic marker Bassoon, suggesting that R-PTP-zeta is associated with excitatory synapses.

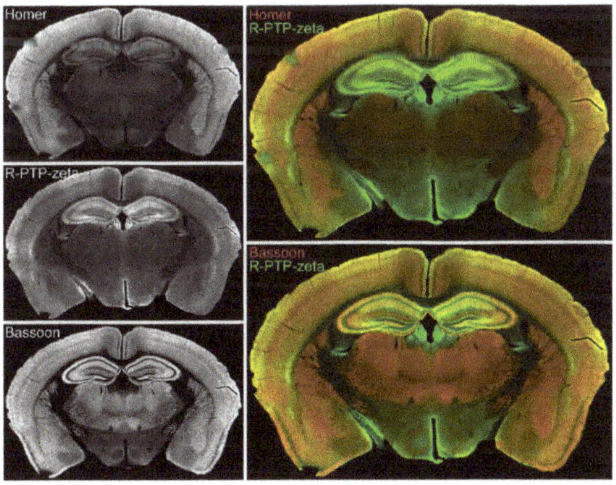

Figure 6. Characterization of R-PTP-zeta expression in vivo. Left, Immuno-histochemical staining of rat brain at P66 for the presynaptic marker Bassoon, excitatory postsynaptic marker Homer and R-PTP-zeta (grey in single color images). Right, yellow in the merged composite (top) image indicates co-expression of Homer (red) and R-PTP-zeta (green). Yellow in the bottom merged image indicates co-expression of Bassoon (red) and R-PTP-zeta (green).

Indeed, immunostaining of cultured rat cortical neurons for surface-expressed R-PTP-zeta under non-permeabilized conditions showed its punctate localization along dendrites. R-PTP-zeta puncta colocalized with puncta positive for the presynaptic marker Bassoon. Moreover, these puncta were also immunopositive for surface-expressed SynCAM 1 (Figure 7A), as expected for a protein identified by proximity-reporter SynCAM 1-HRP. Similarly, extracellular R-PTP-zeta colocalized with puncta positive for the excitatory postsynaptic marker Homer that were also positive for extracellular SynCAM 1 (Figure 7B). These results agree with a previous study that used antibodies recognizing R-PTP-zeta to detect immunoreactivity at PSD-95-positive spines of pyramidal neurons in cerebral cortex and hippocampus of rats, specifically the postsynaptic membrane of dendritic spines and shafts [69]. Our results support that endogenous SynCAM 1 and R-PTP-zeta are co-expressed at the cleft of the same excitatory synapses and may be in close proximity, in agreement with the proteomic identification of R-PTP-zeta by SynCAM 1-HRP.

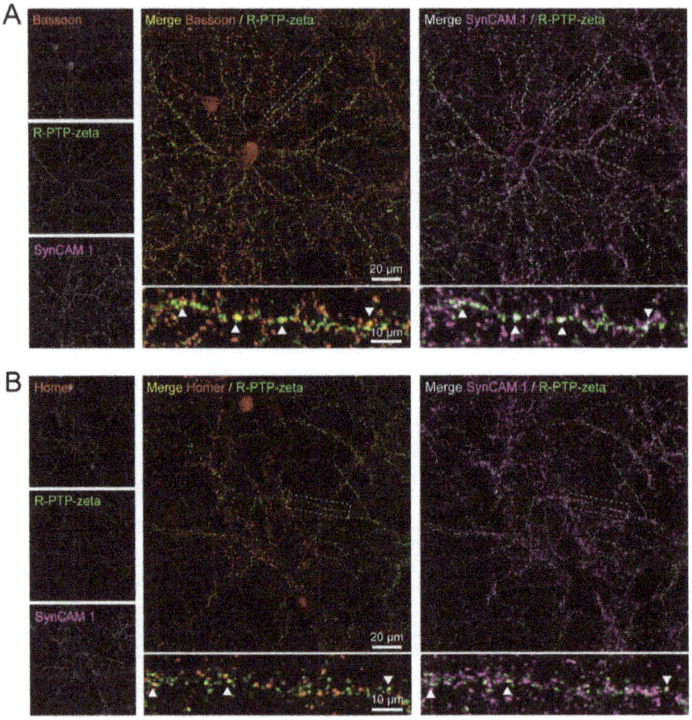

Figure 7. Characterization of R-PTP-zeta expression and synaptic markers in vitro. (**A**) Immunostaining for Bassoon, R-PTP-zeta, and SynCAM 1 in cultured neurons. R-PTP-zeta (grey in single color image) and SynCAM 1 (grey in single color image) were stained under non-permeabilizing conditions using antibodies detecting extracellular epitopes and Bassoon (grey in single color image) was stained under permeabilizing conditions. Yellow in larger composite (left) image indicates colocalization of Bassoon (red) and extracellular R-PTP-zeta (green). White in larger composite (right) image indicates colocalization of extracellular SynCAM 1 (magenta) and extracellular R-PTP-zeta (green). Panels below show enlarged dendritic segments from the composite images to visualize colocalization (arrowheads). (**B**) Immunostaining for Homer and R-PTP-zeta and SynCAM 1 in cultured neurons. Extracellular R-PTP-zeta (grey in single color image) and SynCAM 1 (grey in single color image) were immunostained under non-permeabilizing conditions as in (**A**) and Homer (grey in single color image) was immunodetected under permeabilizing conditions. Yellow in larger composite image (left) indicates colocalization of Homer (red) and extracellular R-PTP-zeta (green). White in larger composite image (right) indicates colocalization of extracellular SynCAM 1 (magenta) and extracellular R-PTP-zeta (green). Panels below show enlarged dendrites from the composite images. Arrowheads mark sites of colocalization.

4. Discussion

Brain circuits are anatomically and functionally highly diverse, which corresponds with the different neuronal cell types found across regions [70]. In agreement with the distinct expression profiles of these neuron types, the synapses they form are highly heterogenous in function and composition [2]. Specifically, the postsynaptic proteome of the brain exhibits unique compositional signatures, which correlate with anatomical divisions of the brain both in mice [71,72] and humans [73,74]. Within a brain region, inputs originating from different neuronal populations may synapse onto one particular neuron and these inputs on the same target have specific functional characteristics [75–78]. This involves synapse-organizing mechanisms to which synaptic cell-surface proteins such as adhesion proteins of the immunoglobulin and leucine-rich repeat protein super-families contribute in the hippocampus [79]. Instructive roles of adhesion molecules in synapse specification are underlined by the roles of immunoglobulin proteins and cadherins in shaping connectivity in the retina [80,81]. Hence, synaptic cleft proteins are positioned to play an essential role in establishing synapse connectivity and function and specify the identity of synapses. It is therefore important to determine to what extent synapses differ based on the cleft proteins they contain. Moreover, the ability to answer this question allows to assess how disease states or substance abuse remodel the cleft and change its synapse-organizing properties.

This study applied a peroxidase-mediated proximity labeling approach to examine the protein composition of excitatory synaptic clefts in cultured cortical neuron [25]. The excitatory synaptic cell adhesion protein SynCAM 1 of the immunoglobulin superfamily was utilized as a new reporter to target HRP to synaptic clefts of glutamatergic synapses and label and identify its proteomic content. This generated a list of synaptic cleft candidates of which several are novel. The proximity labeling approach is robust as shown by its reproducibility across biological replicates and stringent in silico data filtering. The list of proteins this approach identified in our study was enriched for synaptic membrane proteins and depleted for general surface proteins and intracellular contaminants. 39 proteins were detected as synaptic cleft candidates in multiple biological replicates, and 26 additional proteins were found in only one of the four biological replicates. Of these 65 proteins, 30 proteins are novel compared with a previous study that used this approach to map the proteome of excitatory synaptic clefts with different HRP reporters [29] (Supplementary Table S2). These novel cleft candidates included adhesion proteins, receptors, and secreted proteins/extracellular matrix proteins. The fact that a differential set of proteins was enriched, despite this study having a lower estimated coverage (16%) of the excitatory cleft proteome compared with Loh et al. may be explained by several factors. First, SynCAM 1 (this study) and LRRTM1/2 (used to design reporters by Loh et al.), may have differential expression across synapse types. SynCAM 1 is expressed in the cortex of rodent brain [35,36], similarly, LRRTM1-2 are expressed in the cortex [82,83]. Yet, it is unclear to what extent these proteins may be expressed at the same synapse. Second, the possibility exists that SynCAM 1 and LRRTMs localize to different sub-cleft regions and that reporters based on these molecules probe different sub-synaptic cleft proteomes. Endogenous SynCAM 1 was primarily found at the edges of the area marked by the postsynaptic density [14] consistent with the results obtained in this study for SynCAM 1-HRP, while LRRTM2 resides more closer towards the center of the postsynaptic density [84]. The radius of biotinylated proteins proximal to peroxidase-fusion proteins after proximity labeling is as previously reported to be <20 nm [28,55], which is a fraction of the synaptic cleft length [12,85,86]. Hence, HRP-fusion proteins of these reporters may probe differential environments within the synaptic cleft, opening the possibility to map sub-cleft proteomes and increase the molecular definition of the cleft.

The identification of synaptic cleft candidates that were previously not described calls for their validation to attest to proximity labeling as a tool to map cleft proteomes. The protein selected in this study for validation, Receptor-type tyrosine-protein phosphatase zeta, or R-PTP-zeta (gene product of *Ptprz1*) was enriched in three out of four biological replicate experiments providing confidence in a possible synaptic cleft localization. R-PTP-zeta is predominantly expressed in the central nervous system and plays a role during development and adulthood in myelination and learning and memory processes [87,88], and *PTPRZ1* may be a potential schizophrenia susceptibility gene [89–91]. R-PTP-zeta (RPTPζ/β) was previously detected at some PSD-95-positive spines of pyramidal neurons in cerebral cortex and hippocampus of rats, specifically at the postsynaptic membrane of dendritic spines and at shafts [69]. R-PTP-zeta is expressed as three structurally distinct isoforms: a long, membrane-integral isoform; a short, membrane-integral isoform; and a soluble isoform [92–94]. While it is unclear which specific variant was found enriched in this proteomics study, in situ hybridizations of Ptprz1 mRNA by the Allen Institute for Brain Science suggest that the long isoform of R-PTP-zeta is mainly expressed in the olfactory areas and the cerebellum. A closer examination finds that these probes match a region of exon 12 that is excluded from the short membrane isoform. Results here show that R-PTP-zeta is expressed in cortical neurons in dissociated cultures derived from cortices and cortical and hippocampal regions of the mouse brain (Figures 6 and 7). Hence, the short isoform is likely the variant detected by our immunohistochemical staining in the cortex and hippocampus. Notably, we observed in dissociated cortical neurons a strong co-localization of R-PTP-zeta with synaptic markers, including the excitatory marker Homer 1, and with endogenous SynCAM 1. These results validate the identification of R-PTP-zeta in this screen.

Together, our data support that proximity labeling using synaptic HRP reporters is a robust approach to identify the molecular composition of the cleft, a compartment not readily accessible to previous biochemical studies. Future applications can include testing changes in the makeup of the synaptic cleft under disease-linked conditions that alter synapse structure, e.g., mouse models relevant for developmental disorders. Moreover, synapse structure is altered by repeat administration of psycho-stimulants, and the synapse organizer SynCAM 1 acts in medium spiny neurons to control the number of their dendritic spines and the remodeling of spine morphology in these neurons upon cocaine exposure [95]. This provides evidence that cleft components can contribute to the synaptic changes upon exposure to drugs of abuse and warrants future proteomic studies to map these changes. Moreover, peroxidase-mediated proximity labeling offers the opportunity for performing acute manipulations to measure acute synaptic cleft remodeling. Recently, an improved BioID method has been introduced, TurboID, which has a temporal resolution that is in the order of tens of minutes and may be used to measure long-term remodeling of the synaptic cleft proteome in vivo. These approaches can therefore be utilized in future studies to analyze the activity-dependent re-organization of the synaptic cleft that is supported by the redistribution of SynCAM 1 and Neuroligin-1 after induction of long-term depression [14,84]. Proximity labeling of cleft proteins may hence provide valuable insights into roles of the dynamic cleft in shaping synapses [13].

Supplementary Materials: The following are available online at http://www.mdpi.com/2227-7382/6/4/48/s1. Table S1. Synaptic cleft candidates identified in proximity labeling with SynCAM 1-HRP. Table S2. Comparison with excitatory synaptic cleft parts list [29]. Table S3. Candidate excitatory synaptic cleft orphans. Table S4. GO-terms used in post hoc in-silico filtering/data analysis. Figure S1. Characterization of super-resolved SynCAM 1-HRP and postsynaptic Homer.

Author Contributions: Conceptualization, T.C. and T.B.; Data curation, T.C. and T.T.L.; Formal analysis, T.C. and A.M.R.; Funding acquisition, T.C., T.T.L., T.A.B. and T.B.; Investigation, T.C., A.M.R. and B.E.C.; Methodology, T.C.; Project administration, T.C. and T.B.; Resources, T.T.L., T.A.B. and T.B.; Supervision, T.A.B. and T.B.; Visualization, T.C. and A.M.R.; Writing—original draft, T.C.; Writing—review & editing, T.C., T.T.L. and T.B.

Funding: This research was funded by NIH R01 DA018928 (to T.B.), MH080046 (to T.A.B.), MH116583 (to A.M.R.) and NIH P30 DA018343 (to T.T.L, and to T.C. for a pilot award). The Orbitrap Fusion used for mass spectral data collection was funded by NIH SIG from the Office of The Director, National Institutes of Health of the National Institutes of Health under Award Number S10OD018034. The content is solely the responsibility of the authors and does not necessarily represent the official views of the National Institutes of Health.

Acknowledgments: We would like to acknowledge Dr. Aihui Tang (University of Maryland, and University of Science and Technology of China) for initial Super-resolution imaging analysis, Dr. Navin Rauniyar for support with proteomics sample collection and data analysis, Dr. Pascal Kaeser (Harvard University) for the gift of an AAV cloning vector, and Dr. Ralph DiLeone (Yale University) for the gift of AAV cloning vectors and a cell line.

Conflicts of Interest: The authors declare no conflict of interest.

References

1. Sudhof, T.C. Synaptic Neurexin Complexes: A Molecular Code for the Logic of Neural Circuits. *Cell* **2017**, *171*, 745–769. [CrossRef] [PubMed]
2. O'Rourke, N.A.; Weiler, N.C.; Micheva, K.D.; Smith, S.J. Deep molecular diversity of mammalian synapses: why it matters and how to measure it. *Nat. Rev. Neurosci.* **2012**, *13*, 365–379. [CrossRef] [PubMed]
3. Emes, R.D.; Grant, S.G. Evolution of synapse complexity and diversity. *Annu. Rev. Neurosci.* **2012**, *35*, 111–131. [CrossRef] [PubMed]
4. Missler, M.; Sudhof, T.C.; Biederer, T. Synaptic cell adhesion. *Cold Spring Harb. Perspect. Biol.* **2012**, *4*, a005694. [CrossRef] [PubMed]
5. Shen, K.; Scheiffele, P. Genetics and cell biology of building specific synaptic connectivity. *Annu. Rev. Neurosci.* **2010**, *33*, 473–507. [CrossRef] [PubMed]
6. Maeder, C.I.; Shen, K. Genetic dissection of synaptic specificity. *Curr. Opin. Neurobiol.* **2011**, *21*, 93–99. [CrossRef] [PubMed]
7. Goda, Y.; Davis, G.W. Mechanisms of synapse assembly and disassembly. *Neuron* **2003**, *40*, 243–264. [CrossRef]
8. Andreev, V.P.; Petyuk, V.A.; Brewer, H.M.; Karpievitch, Y.V.; Xie, F.; Clarke, J.; Camp, D.; Smith, R.D.; Lieberman, A.P.; Albin, R.L.; et al. Label-free quantitative LC-MS proteomics of Alzheimer's disease and normally aged human brains. *J. Proteome Res.* **2012**, *11*, 3053–3067. [CrossRef] [PubMed]
9. Moron, J.A.; Devi, L.A. Use of proteomics for the identification of novel drug targets in brain diseases. *J. Neurochem.* **2007**, *102*, 306–315. [CrossRef] [PubMed]
10. Abul-Husn, N.S.; Devi, L.A. Neuroproteomics of the synapse and drug addiction. *J. Pharmacol. Exp. Ther.* **2006**, *318*, 461–468. [CrossRef] [PubMed]
11. Abul-Husn, N.S.; Bushlin, I.; Moron, J.A.; Jenkins, S.L.; Dolios, G.; Wang, R.; Iyengar, R.; Ma'ayan, A.; Devi, L.A. Systems approach to explore components and interactions in the presynapse. *Proteomics* **2009**, *9*, 3303–3315. [CrossRef] [PubMed]
12. Harris, K.M.; Weinberg, R.J. Ultrastructure of synapses in the mammalian brain. *Cold Spring Harb. Perspect. Biol.* **2012**, *4*, a005587. [CrossRef] [PubMed]
13. Biederer, T.; Kaeser, P.S.; Blanpied, T.A. Transcellular Nanoalignment of Synaptic Function. *Neuron* **2017**, *96*, 680–696. [CrossRef] [PubMed]
14. Perez de Arce, K.; Schrod, N.; Metzbower, S.W.R.; Allgeyer, E.; Kong, G.K.; Tang, A.H.; Krupp, A.J.; Stein, V.; Liu, X.; Bewersdorf, J.; et al. Topographic Mapping of the Synaptic Cleft into Adhesive Nanodomains. *Neuron* **2015**, *88*, 1165–1172. [CrossRef] [PubMed]
15. Biesemann, C.; Gronborg, M.; Luquet, E.; Wichert, S.P.; Bernard, V.; Bungers, S.R.; Cooper, B.; Varoqueaux, F.; Li, L.; Byrne, J.A.; et al. Proteomic screening of glutamatergic mouse brain synaptosomes isolated by fluorescence activated sorting. *EMBO J.* **2014**, *33*, 157–170. [CrossRef] [PubMed]
16. Morciano, M.; Burre, J.; Corvey, C.; Karas, M.; Zimmermann, H.; Volknandt, W. Immunoisolation of two synaptic vesicle pools from synaptosomes: a proteomics analysis. *J. Neurochem.* **2005**, *95*, 1732–1745. [CrossRef] [PubMed]
17. Takamori, S.; Holt, M.; Stenius, K.; Lemke, E.A.; Gronborg, M.; Riedel, D.; Urlaub, H.; Schenck, S.; Brugger, B.; Ringler, P.; et al. Molecular anatomy of a trafficking organelle. *Cell* **2006**, *127*, 831–846. [CrossRef] [PubMed]
18. Boyken, J.; Gronborg, M.; Riedel, D.; Urlaub, H.; Jahn, R.; Chua, J.J. Molecular profiling of synaptic vesicle docking sites reveals novel proteins but few differences between glutamatergic and GABAergic synapses. *Neuron* **2013**, *78*, 285–297. [CrossRef] [PubMed]
19. Morciano, M.; Beckhaus, T.; Karas, M.; Zimmermann, H.; Volknandt, W. The proteome of the presynaptic active zone: from docked synaptic vesicles to adhesion molecules and maxi-channels. *J. Neurochem.* **2009**, *108*, 662–675. [CrossRef] [PubMed]
20. Bayes, A.; Collins, M.O.; Croning, M.D.; van de Lagemaat, L.N.; Choudhary, J.S.; Grant, S.G. Comparative study of human and mouse postsynaptic proteomes finds high compositional conservation and abundance differences for key synaptic proteins. *PLoS One* **2012**, *7*, e46683. [CrossRef] [PubMed]
21. Distler, U.; Schmeisser, M.J.; Pelosi, A.; Reim, D.; Kuharev, J.; Weiczner, R.; Baumgart, J.; Boeckers, T.M.; Nitsch, R.; Vogt, J.; et al. In-depth protein profiling of the postsynaptic density from mouse hippocampus using data-independent acquisition proteomics. *Proteomics* **2014**, *14*, 2607–2613. [CrossRef] [PubMed]

22. Collins, M.O.; Husi, H.; Yu, L.; Brandon, J.M.; Anderson, C.N.; Blackstock, W.P.; Choudhary, J.S.; Grant, S.G. Molecular characterization and comparison of the components and multiprotein complexes in the postsynaptic proteome. *J. Neurochem.* **2006**, *97*, 16–23. [CrossRef] [PubMed]
23. Selimi, F.; Cristea, I.M.; Heller, E.; Chait, B.T.; Heintz, N. Proteomic studies of a single CNS synapse type: the parallel fiber/purkinje cell synapse. *PLoS Biol.* **2009**, *7*, e83. [CrossRef] [PubMed]
24. Heller, E.A.; Zhang, W.; Selimi, F.; Earnheart, J.C.; Slimak, M.A.; Santos-Torres, J.; Ibanez-Tallon, I.; Aoki, C.; Chait, B.T.; Heintz, N. The biochemical anatomy of cortical inhibitory synapses. *PLoS One* **2012**, *7*, e39572. [CrossRef] [PubMed]
25. Han, S.; Li, J.; Ting, A.Y. Proximity labeling: spatially resolved proteomic mapping for neurobiology. *Curr. Opin. Neurobiol.* **2018**, *50*, 17–23. [CrossRef] [PubMed]
26. Chen, C.L.; Perrimon, N. Proximity-dependent labeling methods for proteomic profiling in living cells. *Wiley Interdiscip. Rev. Dev. Biol.* **2017**, *6*. [CrossRef] [PubMed]
27. Varnaite, R.; MacNeill, S.A. Meet the neighbors: Mapping local protein interactomes by proximity-dependent labeling with BioID. *Proteomics* **2016**, *16*, 2503–2518. [CrossRef] [PubMed]
28. Rhee, H.W.; Zou, P.; Udeshi, N.D.; Martell, J.D.; Mootha, V.K.; Carr, S.A.; Ting, A.Y. Proteomic mapping of mitochondria in living cells via spatially restricted enzymatic tagging. *Science* **2013**, *339*, 1328–1331. [CrossRef] [PubMed]
29. Loh, K.H.; Stawski, P.S.; Draycott, A.S.; Udeshi, N.D.; Lehrman, E.K.; Wilton, D.K.; Svinkina, T.; Deerinck, T.J.; Ellisman, M.H.; Stevens, B.; et al. Proteomic Analysis of Unbounded Cellular Compartments: Synaptic Clefts. *Cell* **2016**, *166*, 1295–1307.e21. [CrossRef] [PubMed]
30. Reinke, A.W.; Balla, K.M.; Bennett, E.J.; Troemel, E.R. Identification of microsporidia host-exposed proteins reveals a repertoire of rapidly evolving proteins. *Nat. Commun.* **2017**, *8*, 14023. [CrossRef] [PubMed]
31. Reinke, A.W.; Mak, R.; Troemel, E.R.; Bennett, E.J. In vivo mapping of tissue- and subcellular-specific proteomes in Caenorhabditis elegans. *Sci. Adv.* **2017**, *3*, e1602426. [CrossRef] [PubMed]
32. Uezu, A.; Kanak, D.J.; Bradshaw, T.W.; Soderblom, E.J.; Catavero, C.M.; Burette, A.C.; Weinberg, R.J.; Soderling, S.H. Identification of an elaborate complex mediating postsynaptic inhibition. *Science* **2016**, *353*, 1123–1129. [CrossRef] [PubMed]
33. Branon, T.C.; Bosch, J.A.; Sanchez, A.D.; Udeshi, N.D.; Svinkina, T.; Carr, S.A.; Feldman, J.L.; Perrimon, N.; Ting, A.Y. Efficient proximity labeling in living cells and organisms with TurboID. *Nat. Biotechnol.* **2018**, *36*, 880–887. [CrossRef] [PubMed]
34. Fogel, A.I.; Akins, M.R.; Krupp, A.J.; Stagi, M.; Stein, V.; Biederer, T. SynCAMs organize synapses through heterophilic adhesion. *J. Neurosci.* **2007**, *27*, 12516–12530. [CrossRef] [PubMed]
35. Thomas, L.A.; Akins, M.R.; Biederer, T. Expression and adhesion profiles of SynCAM molecules indicate distinct neuronal functions. *J. Comp. Neurol.* **2008**, *510*, 47–67. [CrossRef] [PubMed]
36. Biederer, T.; Sara, Y.; Mozhayeva, M.; Atasoy, D.; Liu, X.; Kavalali, E.T.; Sudhof, T.C. SynCAM, a synaptic adhesion molecule that drives synapse assembly. *Science* **2002**, *297*, 1525–1531. [CrossRef] [PubMed]
37. Robbins, E.M.; Krupp, A.J.; Perez de Arce, K.; Ghosh, A.K.; Fogel, A.I.; Boucard, A.; Sudhof, T.C.; Stein, V.; Biederer, T. SynCAM 1 adhesion dynamically regulates synapse number and impacts plasticity and learning. *Neuron* **2010**, *68*, 894–906. [CrossRef] [PubMed]
38. Biederer, T.; Scheiffele, P. Mixed-culture assays for analyzing neuronal synapse formation. *Nat. Protoc.* **2007**, *2*, 670–676. [CrossRef] [PubMed]
39. Zolotukhin, S.; Byrne, B.J.; Mason, E.; Zolotukhin, I.; Potter, M.; Chesnut, K.; Summerford, C.; Samulski, R.J.; Muzyczka, N. Recombinant adeno-associated virus purification using novel methods improves infectious titer and yield. *Gene Ther.* **1999**, *6*, 973–985. [CrossRef] [PubMed]
40. Hommel, J.D.; Sears, R.M.; Georgescu, D.; Simmons, D.L.; DiLeone, R.J. Local gene knockdown in the brain using viral-mediated RNA interference. *Nat. Med.* **2003**, *9*, 1539–1544. [CrossRef] [PubMed]
41. Fukumoto, Y.; Obata, Y.; Ishibashi, K.; Tamura, N.; Kikuchi, I.; Aoyama, K.; Hattori, Y.; Tsuda, K.; Nakayama, Y.; Yamaguchi, N. Cost-effective gene transfection by DNA compaction at pH 4.0 using acidified, long shelf-life polyethylenimine. *Cytotechnology* **2010**, *62*, 73–82. [CrossRef] [PubMed]
42. Keller, A.; Nesvizhskii, A.I.; Kolker, E.; Aebersold, R. Empirical statistical model to estimate the accuracy of peptide identifications made by MS/MS and database search. *Anal. Chem.* **2002**, *74*, 5383–5392. [CrossRef] [PubMed]

43. Nesvizhskii, A.I.; Keller, A.; Kolker, E.; Aebersold, R. A statistical model for identifying proteins by tandem mass spectrometry. *Anal. Chem.* **2003**, *75*, 4646–4658. [CrossRef] [PubMed]
44. Vizcaino, J.A.; Csordas, A.; del-Toro, N.; Dianes, J.A.; Griss, J.; Lavidas, I.; Mayer, G.; Perez-Riverol, Y.; Reisinger, F.; Ternent, T.; et al. 2016 update of the PRIDE database and its related tools. *Nucleic Acids Res.* **2016**, *44*, D447–D456. [CrossRef] [PubMed]
45. Schwanhausser, B.; Busse, D.; Li, N.; Dittmar, G.; Schuchhardt, J.; Wolf, J.; Chen, W.; Selbach, M. Global quantification of mammalian gene expression control. *Nature* **2011**, *473*, 337–342. [CrossRef] [PubMed]
46. Krey, J.F.; Wilmarth, P.A.; Shin, J.B.; Klimek, J.; Sherman, N.E.; Jeffery, E.D.; Choi, D.; David, L.L.; Barr-Gillespie, P.G. Accurate label-free protein quantitation with high- and low-resolution mass spectrometers. *J. Proteome Res.* **2014**, *13*, 1034–1044. [CrossRef] [PubMed]
47. Hung, V.; Udeshi, N.D.; Lam, S.S.; Loh, K.H.; Cox, K.J.; Pedram, K.; Carr, S.A.; Ting, A.Y. Spatially resolved proteomic mapping in living cells with the engineered peroxidase APEX2. *Nat. Protoc.* **2016**, *11*, 456–475. [CrossRef] [PubMed]
48. Stanly, T.A.; Fritzsche, M.; Banerji, S.; Garcia, E.; Bernardino de la Serna, J.; Jackson, D.G.; Eggeling, C. Critical importance of appropriate fixation conditions for faithful imaging of receptor microclusters. *Biol. Open* **2016**, *5*, 1343–1350. [CrossRef] [PubMed]
49. Ovesny, M.; Krizek, P.; Borkovec, J.; Svindrych, Z.; Hagen, G.M. ThunderSTORM: a comprehensive ImageJ plug-in for PALM and STORM data analysis and super-resolution imaging. *Bioinformatics* **2014**, *30*, 2389–2390. [CrossRef] [PubMed]
50. Udeshi, N.D.; Pedram, K.; Svinkina, T.; Fereshetian, S.; Myers, S.A.; Aygun, O.; Krug, K.; Clauser, K.; Ryan, D.; Ast, T.; et al. Antibodies to biotin enable large-scale detection of biotinylation sites on proteins. *Nat. Methods* **2017**, *14*, 1167–1170. [CrossRef] [PubMed]
51. Lam, S.S.; Martell, J.D.; Kamer, K.J.; Deerinck, T.J.; Ellisman, M.H.; Mootha, V.K.; Ting, A.Y. Directed evolution of APEX2 for electron microscopy and proximity labeling. *Nat. Methods* **2015**, *12*, 51–54. [CrossRef] [PubMed]
52. Stagi, M.; Fogel, A.I.; Biederer, T. SynCAM 1 participates in axo-dendritic contact assembly and shapes neuronal growth cones. *Proc. Natl. Acad. Sci. USA* **2010**, *107*, 7568–7573. [CrossRef] [PubMed]
53. Berman, H.M.; Westbrook, J.; Feng, Z.; Gilliland, G.; Bhat, T.N.; Weissig, H.; Shindyalov, I.N.; Bourne, P.E. The Protein Data Bank. *Nucleic Acids Res.* **2000**, *28*, 235–242. [CrossRef] [PubMed]
54. Berglund, G.I.; Carlsson, G.H.; Smith, A.T.; Szoke, H.; Henriksen, A.; Hajdu, J. The catalytic pathway of horseradish peroxidase at high resolution. *Nature* **2002**, *417*, 463–468. [CrossRef] [PubMed]
55. Hung, V.; Zou, P.; Rhee, H.W.; Udeshi, N.D.; Cracan, V.; Svinkina, T.; Carr, S.A.; Mootha, V.K.; Ting, A.Y. Proteomic mapping of the human mitochondrial intermembrane space in live cells via ratiometric APEX tagging. *Mol. Cell* **2014**, *55*, 332–341. [CrossRef] [PubMed]
56. Bantscheff, M.; Lemeer, S.; Savitski, M.M.; Kuster, B. Quantitative mass spectrometry in proteomics: critical review update from 2007 to the present. *Analyt. Bioanal. Chem.* **2012**, *404*, 939–965. [CrossRef] [PubMed]
57. Schulze, W.X.; Usadel, B. Quantitation in mass-spectrometry-based proteomics. *Annu. Rev. Plant Biol.* **2010**, *61*, 491–516. [CrossRef] [PubMed]
58. Ankney, J.A.; Muneer, A.; Chen, X. Relative and Absolute Quantitation in Mass Spectrometry-Based Proteomics. *Annu. Rev. Analyt. Chem.* **2018**, *11*, 49–77. [CrossRef] [PubMed]
59. Hassan, Y.I.; Zempleni, J. A novel, enigmatic histone modification: biotinylation of histones by holocarboxylase synthetase. *Nutr. Rev.* **2008**, *66*, 721–725. [CrossRef] [PubMed]
60. Kuroishi, T.; Rios-Avila, L.; Pestinger, V.; Wijeratne, S.S.; Zempleni, J. Biotinylation is a natural, albeit rare, modification of human histones. *Mol. Genet. Metab.* **2011**, *104*, 537–545. [CrossRef] [PubMed]
61. Jitrapakdee, S.; Wallace, J.C. The biotin enzyme family: conserved structural motifs and domain rearrangements. *Curr. Protein Pept. Sci.* **2003**, *4*, 217–229. [CrossRef] [PubMed]
62. Tytgat, H.L.; Schoofs, G.; Driesen, M.; Proost, P.; Van Damme, E.J.; Vanderleyden, J.; Lebeer, S. Endogenous biotin-binding proteins: an overlooked factor causing false positives in streptavidin-based protein detection. *Microb. Biotechnol.* **2015**, *8*, 164–168. [CrossRef] [PubMed]
63. Peri, S.; Navarro, J.D.; Amanchy, R.; Kristiansen, T.Z.; Jonnalagadda, C.K.; Surendranath, V.; Niranjan, V.; Muthusamy, B.; Gandhi, T.K.; Gronborg, M.; et al. Development of human protein reference database as an initial platform for approaching systems biology in humans. *Genome Res.* **2003**, *13*, 2363–2371. [CrossRef] [PubMed]

64. Mishra, G.R.; Suresh, M.; Kumaran, K.; Kannabiran, N.; Suresh, S.; Bala, P.; Shivakumar, K.; Anuradha, N.; Reddy, R.; Raghavan, T.M.; et al. Human protein reference database—2006 update. *Nucleic Acids Res.* **2006**, *34*, D411–D414. [CrossRef] [PubMed]
65. Keshava Prasad, T.S.; Goel, R.; Kandasamy, K.; Keerthikumar, S.; Kumar, S.; Mathivanan, S.; Telikicherla, D.; Raju, R.; Shafreen, B.; Venugopal, A.; et al. Human Protein Reference Database–2009 update. *Nucleic Acids Res.* **2009**, *37*, D767–D772. [CrossRef] [PubMed]
66. Weingarten, J.; Lassek, M.; Mueller, B.F.; Rohmer, M.; Lunger, I.; Baeumlisberger, D.; Dudek, S.; Gogesch, P.; Karas, M.; Volknandt, W. The proteome of the presynaptic active zone from mouse brain. *Mol. Cell. Neurosci.* **2014**, *59*, 106–118. [CrossRef] [PubMed]
67. Pocklington, A.J.; Cumiskey, M.; Armstrong, J.D.; Grant, S.G. The proteomes of neurotransmitter receptor complexes form modular networks with distributed functionality underlying plasticity and behaviour. *Mol. Syst. Biol.* **2006**, *2*, 0023. [CrossRef] [PubMed]
68. Schwenk, J.; Baehrens, D.; Haupt, A.; Bildl, W.; Boudkkazi, S.; Roeper, J.; Fakler, B.; Schulte, U. Regional diversity and developmental dynamics of the AMPA-receptor proteome in the mammalian brain. *Neuron* **2014**, *84*, 41–54. [CrossRef] [PubMed]
69. Hayashi, N.; Oohira, A.; Miyata, S. Synaptic localization of receptor-type protein tyrosine phosphatase ζ/β in the cerebral and hippocampal neurons of adult rats. *Brain Res.* **2005**, *1050*, 163–169. [CrossRef] [PubMed]
70. Lein, E.S.; Hawrylycz, M.J.; Ao, N.; Ayres, M.; Bensinger, A.; Bernard, A.; Boe, A.F.; Boguski, M.S.; Brockway, K.S.; Byrnes, E.J.; et al. Genome-wide atlas of gene expression in the adult mouse brain. *Nature* **2007**, *445*, 168–176. [CrossRef] [PubMed]
71. Roy, M.; Sorokina, O.; McLean, C.; Tapia-Gonzalez, S.; DeFelipe, J.; Armstrong, J.D.; Grant, S.G.N. Regional Diversity in the Postsynaptic Proteome of the Mouse Brain. *Proteomes* **2018**, *6*, 31. [CrossRef] [PubMed]
72. Zhu, F.; Cizeron, M.; Qiu, Z.; Benavides-Piccione, R.; Kopanitsa, M.V.; Skene, N.G.; Koniaris, B.; DeFelipe, J.; Fransen, E.; Komiyama, N.H.; et al. Architecture of the Mouse Brain Synaptome. *Neuron* **2018**, *99*, 781–799.e10. [CrossRef] [PubMed]
73. Roy, M.; Sorokina, O.; Skene, N.; Simonnet, C.; Mazzo, F.; Zwart, R.; Sher, E.; Smith, C.; Armstrong, J.D.; Grant, S.G.N. Proteomic analysis of postsynaptic proteins in regions of the human neocortex. *Nat. Neurosci.* **2018**, *21*, 130–138. [CrossRef] [PubMed]
74. Carlyle, B.C.; Kitchen, R.R.; Kanyo, J.E.; Voss, E.Z.; Pletikos, M.; Sousa, A.M.M.; Lam, T.T.; Gerstein, M.B.; Sestan, N.; Nairn, A.C. A multiregional proteomic survey of the postnatal human brain. *Nat. Neurosci.* **2017**, *20*, 1787–1795. [CrossRef] [PubMed]
75. Martin, E.A.; Muralidhar, S.; Wang, Z.; Cervantes, D.C.; Basu, R.; Taylor, M.R.; Hunter, J.; Cutforth, T.; Wilke, S.A.; Ghosh, A.; et al. The intellectual disability gene Kirrel3 regulates target-specific mossy fiber synapse development in the hippocampus. *Elife* **2015**, *4*, e09395. [CrossRef] [PubMed]
76. Sylwestrak, E.L.; Ghosh, A. Elfn1 regulates target-specific release probability at CA1-interneuron synapses. *Science* **2012**, *338*, 536–540. [CrossRef] [PubMed]
77. Williams, M.E.; de Wit, J.; Ghosh, A. Molecular mechanisms of synaptic specificity in developing neural circuits. *Neuron* **2010**, *68*, 9–18. [CrossRef] [PubMed]
78. Williams, M.E.; Wilke, S.A.; Daggett, A.; Davis, E.; Otto, S.; Ravi, D.; Ripley, B.; Bushong, E.A.; Ellisman, M.H.; Klein, G.; et al. Cadherin-9 regulates synapse-specific differentiation in the developing hippocampus. *Neuron* **2011**, *71*, 640–655. [CrossRef] [PubMed]
79. de Wit, J.; Ghosh, A. Specification of synaptic connectivity by cell surface interactions. *Nat. Rev. Neurosci.* **2016**, *17*, 22–35. [CrossRef] [PubMed]
80. Yamagata, M.; Sanes, J.R. Dscam and Sidekick proteins direct lamina-specific synaptic connections in vertebrate retina. *Nature* **2008**, *451*, 465–469. [CrossRef] [PubMed]
81. Duan, X.; Krishnaswamy, A.; De la Huerta, I.; Sanes, J.R. Type II cadherins guide assembly of a direction-selective retinal circuit. *Cell* **2014**, *158*, 793–807. [CrossRef] [PubMed]
82. Linhoff, M.W.; Lauren, J.; Cassidy, R.M.; Dobie, F.A.; Takahashi, H.; Nygaard, H.B.; Airaksinen, M.S.; Strittmatter, S.M.; Craig, A.M. An unbiased expression screen for synaptogenic proteins identifies the LRRTM protein family as synaptic organizers. *Neuron* **2009**, *61*, 734–749. [CrossRef] [PubMed]
83. De Wit, J.; Sylwestrak, E.; O'Sullivan, M.L.; Otto, S.; Tiglio, K.; Savas, J.N.; Yates, J.R.; Comoletti, D.; Taylor, P.; Ghosh, A. LRRTM2 interacts with Neurexin1 and regulates excitatory synapse formation. *Neuron* **2009**, *64*, 799–806. [CrossRef] [PubMed]

84. Chamma, I.; Letellier, M.; Butler, C.; Tessier, B.; Lim, K.H.; Gauthereau, I.; Choquet, D.; Sibarita, J.B.; Park, S.; Sainlos, M.; et al. Mapping the dynamics and nanoscale organization of synaptic adhesion proteins using monomeric streptavidin. *Nat. Commun.* **2016**, *7*, 10773. [CrossRef] [PubMed]
85. Santuy, A.; Rodriguez, J.R.; DeFelipe, J.; Merchan-Perez, A. Study of the Size and Shape of Synapses in the Juvenile Rat Somatosensory Cortex with 3D Electron Microscopy. *eNeuro* **2018**, *5*, 17. [CrossRef] [PubMed]
86. Schikorski, T.; Stevens, C.F. Quantitative ultrastructural analysis of hippocampal excitatory synapses. *J. Neurosci.* **1997**, *17*, 5858–5867. [CrossRef] [PubMed]
87. Tamura, H.; Fukada, M.; Fujikawa, A.; Noda, M. Protein tyrosine phosphatase receptor type Z is involved in hippocampus-dependent memory formation through dephosphorylation at Y1105 on p190 RhoGAP. *Neurosci. Lett.* **2006**, *399*, 33–38. [CrossRef] [PubMed]
88. Niisato, K.; Fujikawa, A.; Komai, S.; Shintani, T.; Watanabe, E.; Sakaguchi, G.; Katsuura, G.; Manabe, T.; Noda, M. Age-dependent enhancement of hippocampal long-term potentiation and impairment of spatial learning through the Rho-associated kinase pathway in protein tyrosine phosphatase receptor type Z-deficient mice. *J. Neurosci.* **2005**, *25*, 1081–1088. [CrossRef] [PubMed]
89. Buxbaum, J.D.; Georgieva, L.; Young, J.J.; Plescia, C.; Kajiwara, Y.; Jiang, Y.; Moskvina, V.; Norton, N.; Peirce, T.; Williams, H.; et al. Molecular dissection of NRG1-ERBB4 signaling implicates PTPRZ1 as a potential schizophrenia susceptibility gene. *Mol. Psychiatry* **2008**, *13*, 162–172. [CrossRef] [PubMed]
90. Takahashi, N.; Sakurai, T.; Bozdagi-Gunal, O.; Dorr, N.P.; Moy, J.; Krug, L.; Gama-Sosa, M.; Elder, G.A.; Koch, R.J.; Walker, R.H.; et al. Increased expression of receptor phosphotyrosine phosphatase-β/ζ is associated with molecular, cellular, behavioral and cognitive schizophrenia phenotypes. *Transl. Psychiatry* **2011**, *1*, e8. [CrossRef] [PubMed]
91. Cressant, A.; Dubreuil, V.; Kong, J.; Kranz, T.M.; Lazarini, F.; Launay, J.M.; Callebert, J.; Sap, J.; Malaspina, D.; Granon, S.; et al. Loss-of-function of PTPR γ and ζ, observed in sporadic schizophrenia, causes brain region-specific deregulation of monoamine levels and altered behavior in mice. *Psychopharmacology* **2017**, *234*, 575–587. [CrossRef] [PubMed]
92. Nishiwaki, T.; Maeda, N.; Noda, M. Characterization and Developmental Regulation of Proteoglycan-Type Protein Tyrosine Phosphatase ζ/RPTP β Isoforms. *J. Biochem.* **1998**, *123*, 458–467. [CrossRef] [PubMed]
93. Chow, J.P.; Fujikawa, A.; Shimizu, H.; Suzuki, R.; Noda, M. Metalloproteinase- and γ-secretase-mediated cleavage of protein-tyrosine phosphatase receptor type Z. *J. Biol. Chem.* **2008**, *283*, 30879–30889. [CrossRef] [PubMed]
94. Canoll, P.D.; Petanceska, S.; Schlessinger, J.; Musacchio, J.M. Three forms of RPTP-β are differentially expressed during gliogenesis in the developing rat brain and during glial cell differentiation in culture. *J. Neurosci. Res.* **1996**, *44*, 199–215. [CrossRef]
95. Giza, J.I.; Jung, Y.; Jeffrey, R.A.; Neugebauer, N.M.; Picciotto, M.R.; Biederer, T. The synaptic adhesion molecule SynCAM 1 contributes to cocaine effects on synapse structure and psychostimulant behavior. *Neuropsychopharmacology* **2013**, *38*, 628–638. [CrossRef] [PubMed]

proteomes

MDPI

Article

Development of Targeted Mass Spectrometry-Based Approaches for Quantitation of Proteins Enriched in the Postsynaptic Density (PSD)

Rashaun S. Wilson [1,2,3], Navin Rauniyar [4], Fumika Sakaue [5,6], TuKiet T. Lam [1,2,3], Kenneth R. Williams [1,3] and Angus C. Nairn [1,6,*]

1 Yale/NIDA Neuroproteomics Center, New Haven, CT 06511, USA; rashaun.wilson@yale.edu (R.S.W.); tukiet.lam@yale.edu (T.T.L.); kenneth.williams@yale.edu (K.R.W.)
2 W.M Keck Biotechnology Resource Laboratory, Yale University School of Medicine, New Haven, CT 06511, USA
3 Molecular Biophysics and Biochemistry, Yale University School of Medicine, New Haven, CT 06511, USA
4 Tanvex BioPharma Inc., San Diego, CA 92121, USA; navin.rauniyar@tanvex.com
5 Department of Neurology and Neurological Science, Tokyo Medical and Dental University, Tokyo 113-8519, Japan; sakaue.nuro@tmd.ac.jp
6 Department of Psychiatry, Yale School of Medicine, Connecticut Mental Health Center, New Haven, CT 06511, USA
* Correspondence: angus.nairn@yale.edu

Received: 11 March 2019; Accepted: 28 March 2019; Published: 2 April 2019

Abstract: The postsynaptic density (PSD) is a structural, electron-dense region of excitatory glutamatergic synapses, which is involved in a variety of cellular and signaling processes in neurons. The PSD is comprised of a large network of proteins, many of which have been implicated in a wide variety of neuropsychiatric disorders. Biochemical fractionation combined with mass spectrometry analyses have enabled an in-depth understanding of the protein composition of the PSD. However, the PSD composition may change rapidly in response to stimuli, and robust and reproducible methods to thoroughly quantify changes in protein abundance are warranted. Here, we report on the development of two types of targeted mass spectrometry-based assays for quantitation of PSD-enriched proteins. In total, we quantified 50 PSD proteins in a targeted, parallel reaction monitoring (PRM) assay using heavy-labeled, synthetic internal peptide standards and identified and quantified over 2100 proteins through a pre-determined spectral library using a data-independent acquisition (DIA) approach in PSD fractions isolated from mouse cortical brain tissue.

Keywords: postsynaptic density; PSD; parallel reaction monitoring; PRM; targeted proteomics; data-independent acquisition; DIA; quantitative mass spectrometry

1. Introduction

The postsynaptic density (PSD) is an electron-dense region of excitatory glutamatergic synapses located just beneath the postsynaptic membrane. The PSD was first discovered by electron microscopy in 1956 [1] and was later found to consist of 30–50 nm-thick, disc-shaped protein structures [2,3]. Within these protein structures are several classes of protein families, many of which are involved in processes such as scaffolding and signal transduction. Each of these families are organized in two different structural layers of the PSD: the core and the pallium [4]. The core is the structural layer located near the postsynaptic membrane, while the pallium is positioned beneath the core and is thought to be more labile.

One group of proteins that has previously been identified in the PSD core is the membrane-associated guanylate kinases (MAGUKs) [5–8] (Figure 1A). These proteins are comprised of three main domains including the PDZ, SH3, and guanylate kinase (GK) domains [5,9,10]. One of the most abundant proteins within the MAGUK family is PSD-95 (also known as DLG4 or SAP90) [7,8], which is involved in structural maintenance and signaling through interactions with integral membrane proteins and receptors, protein complexes, and other structural proteins within the PSD [10–12]. In addition to PSD-95, the MAGUK family includes PSD-93 (DLG2), SAP-102 (DLG3), and SAP-97 (DLG1).

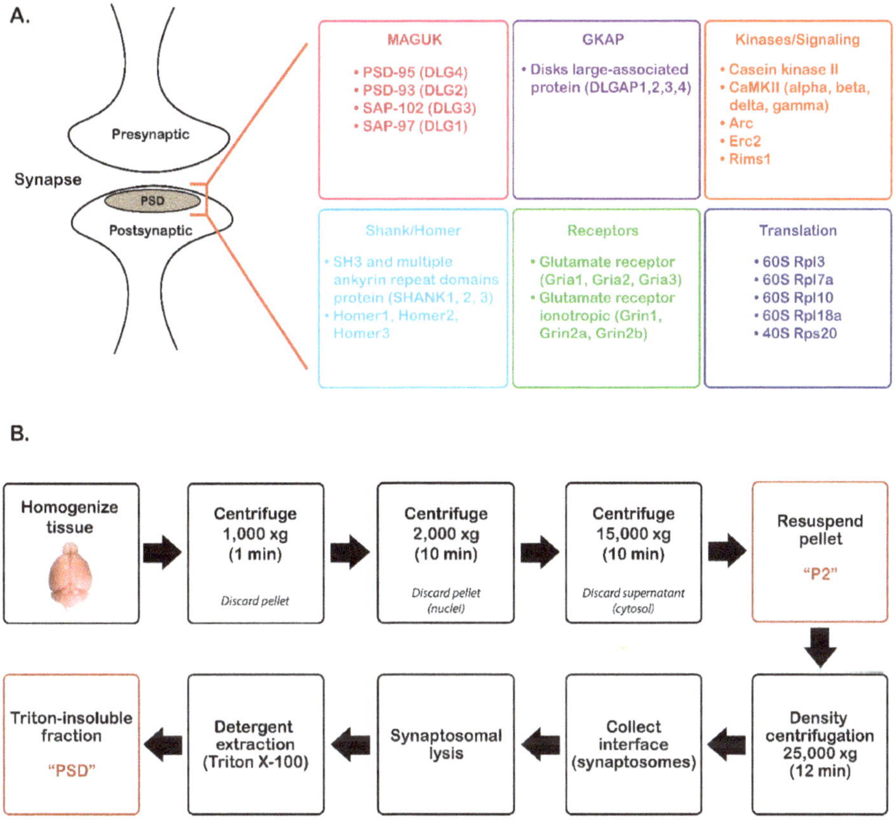

Figure 1. Overview of postsynaptic density (PSD) protein enrichment from mouse cortical tissue. (**A**) List of groups of commonly identified proteins in the PSD. (**B**) Steps for PSD enrichment starting from tissue homogenization to Triton X-100 precipitation. MAGUK, membrane-associated guanylate kinases, GKAP, guanylate kinase-associated proteins, DLGAP, disks large-associated proteins.

Guanylate kinase-associated proteins (GKAPs) are another class of proteins found in the PSD core. This family was first isolated by Kim et al. [13] and found to directly bind to the GK domains of MAGUKs through co-immunoprecipitation assays and immunohistochemistry. GKAPs are often referred to as disks large-associated proteins (DLGAPs, also referred to as SAPAPs), which include four different isoforms designated DLGAP1,2,3, and 4 (Figure 1A). These isoforms enable the formation of protein complexes with MAGUKs and proteins found in the pallial layer of the PSD.

An additional protein family present in the PSD pallium and associated with these complexes is the SH3 and multiple ankyrin repeat domain protein (Shank) family. As their name suggests, these proteins contain an SH3 domain, as well as ankyrin repeats, a PDZ domain, a proline-rich domain, and a SAM domain. These proteins were first identified by Naisbitt et al. [14], who demonstrated that the C-terminal region of GKAP binds to the PDZ domain of Shank. There are three Shank isoforms (Shank1,2,3) that are capable of binding to both MAGUKs and GKAPs and have been shown to form a PSD-95–SAPAP–SHANK complex [9,13,15]. This complex has been implicated in scaffolding and organization of signaling complexes at glutamatergic synapses [15,16].

The Homer family of proteins is also found in the PSD pallium. The two characteristic structural regions of Homer proteins include an Enabled/vasodilator-stimulated phosphoprotein homology 1 (EVH1) domain [17–20] and a carboxyl-terminal coiled-coil domain [17,19,21–25]. Homer proteins self-polymerize and interact with Shank proteins, creating a matrix-like structure [4,24,26]. This scaffolding structure is involved in excitatory signal transduction as well as in receptor plasticity [27]. There are three different *Homer* genes (1–3) that are differentially expressed throughout the brain [27].

In addition to structural proteins, protein kinases are an important component of the signaling pathways within the PSD pallium. Ca^{2+}/calmodulin-dependent protein kinase II (CaMKII) is a serine-threonine kinase that comprises approximately 1–2% of the total proteome in the cerebral cortex and hippocampus [28]. Studies have shown a marked accumulation of CaMKII at the PSD with increasing levels of neuronal excitation [4,29–32]. CaMKII has also been implicated in NMDA-dependent long-term potentiation (LTP) [4,33,34] through regulation of its activity by Ca^{2+}/calmodulin and autophosphorylation.

There is increasing interest in understanding the functions of proteins in the intricate PSD network because of their potential involvement in a wide variety of neuropsychiatric disorders. For instance, several reports have linked Shank3 to autism spectrum disorder (ASD) [15,35,36]. Specifically, deletion of the Shank3B isoform in mice resulted in an ASD-like, compulsive grooming phenotype, which was more prominent than the other Shank3-associated phenotypes investigated in this report [15]. Another study observed a similar excessive grooming phenotype of a different genetic Shank3B knockout mouse, further implicating Shank3B in ASD-like behavioral disorders [35]. Similarly, mutations in either PSD-95 or SynGAP have been shown to be associated with intellectual disorders and autism [37–39]. Furthermore, PSD proteins such as DLG isoforms, DLGAP1, Gria2/3, Grin2a/b, CaMKII, and Homer isoforms have all been implicated in schizophrenia, among many other disorders [40–42]. It is apparent from this growing list that studying the organization and function of proteins within the PSD has become an important focus in neuroscience research. Fractionation methods for enriching PSD proteins were developed decades ago [2,43–45]; however, because the PSD is not enclosed in a bilayer, it can be challenging to minimize contamination of the PSD fraction with other subcellular proteins [46]. Apart from the enrichment method, the structure of synapses themselves can make analyses difficult. Synapses differ significantly from one another and can change their composition rapidly, making reproducibility and accuracy of the analysis important [47–50]. Despite these challenges, researchers have made significant efforts to study the proteome of the PSD, particularly through the use of mass spectrometry [27,42,51–58]. Note, however, that many of the PSD fractionation methods use Triton-X100, a detergent that is not compatible with mass spectrometry analysis. Therefore, care must be taken when preparing a PSD fraction to minimize detergent interference.

Mass spectrometry analysis of PSD fractions from mouse and human cortical tissue identified 1556 and 1461 proteins, respectively [53,54]. Interestingly, there was a 70% overlap of proteins in the mouse and human PSDs. A later study identified 2876 PSD-associated proteins from mouse brain tissue using immunopurification prior to mass spectrometry analysis [57]. Recently, label-free quantitation was performed on 48 PSD samples from 12 human neocortical brain regions, identifying 1213 proteins in total [51]. While these discovery studies have made significant progress identifying PSD proteins, targeted mass spectrometry-based assays are needed to provide the highest possible sensitivity, quantification precision, and accuracy [59]. Our group previously used multiple reaction monitoring (MRM) coupled with stable-isotope peptide standards (SIS) to quantify 112 rat synaptic proteins [58]. Though this assay has made significant improvements in the quantitation of PSD proteins, it lacks high mass accuracy.

Currently, there are two major approaches for targeted, high-mass-accuracy quantitative mass spectrometry. The first method is data-independent analysis (DIA), which was first proposed by Venable et al. [60]. DIA uses sequential window acquisition to fragment and quantify all precursor and product ions within a sample [61–64]. Unlike data-dependent methods, DIA offers high reproducibility and quantitation, while maintaining sensitivity at higher levels of multiplexing [65]. One study has already demonstrated the use of DIA analysis on fractionated PSD samples from mouse hippocampal tissue, which resulted in the identification of 2102 protein groups in the PSD fractions [66]. The second approach is a more targeted method called parallel reaction monitoring (PRM). Like MRM/selected reaction monitoring (SRM) methods, PRM offers similar accuracy and reproducibility; however, it provides a wider dynamic range and improved selectivity [67–69].

Given the advantages of these targeted methods, we developed new DIA and PRM assays to quantify PSD proteins. For the PRM assay, heavy labeled peptides were synthesized and used as internal standards for accurate protein quantitation. Two different mouse datasets were used to evaluate the performance of these methods: PSD-enriched fractions versus pre-fractionation, and wild-type (WT) versus Shank3B knockout (KO) PSD fractions. These assays enabled accurate quantitation of PSD proteins and provide promising tools for future PSD proteomics studies.

2. Materials and Methods

2.1. Tissue Collection

Wild-type and Shank3B mouse cortical tissue was isolated and frozen on dry ice prior to protein extraction.

2.2. PSD Enrichment

PSD isolation was adapted from previously described methods [2]. In brief, mouse cortical brain tissue (~100 mg/sample) was homogenized on ice in 1 mL Buffer A (5 mM HEPES, 10% sucrose (w/v), 1X cOmplete, Mini, EDTA-free protease inhibitor cocktail (Roche Diagnostics GmbH, Mannheim, Germany)) using a rotary homogenizer (Glas-Col, LLC, Terre Haute, IN, motor size: 4.38"w × 4.38"d × 5.50"h) for 10 strokes at a speed of 40. The lysate was spun in a tabletop centrifuge at 1000× *g* for 1 min at 4 °C (Figure 1) to remove cellular debris. The supernatant was transferred to a new Eppendorf tube and centrifuged at 2000× g for 10 min at 4 °C to remove the nuclei. The supernatant was transferred to a new tube and centrifuged at 15,000× g for 10 min at 4 °C. The supernatant (cytosolic fraction) was discarded, and the pellet, which contains synaptosome/synaptoneurosomes, was resuspended in three volumes of Buffer A (P2 fraction). The sample was applied to the top of a Percoll gradient (3-23% in Buffer A; GE Healthcare, Chicago, IL, USA) and centrifuged in an Optima MAX-XP Ultracentrifuge (Beckman Coulter, Brea, CA, USA) at 25,000× g (MLA-55 rotor) for 12 min at 4 °C. The interface containing synaptosomes was collected between 15–23% Percoll. The synaptosomal fraction was subjected to hypotonic lysis by suspending in three volumes of Buffer B (5 mM HEPES, 1 mM DTT, 1X cOmplete, Mini, EDTA-free protease inhibitor cocktail) for 30 min on ice. The lysate was centrifuged at 25,000× g (MLA-55 rotor) for 30 min at 4 °C. The pellet was resuspended in 2 mL Buffer C (0.75% Triton X-100 in Buffer A) and incubated on ice for 15 min (detergent extraction). The sample was centrifuged at 63,000× g (MLA-55 rotor) for 30 min at 4 °C. The supernatant (detergent-soluble fraction) was removed, and the pellet (detergent-insoluble PSD fraction) was washed three times with 1 mL phosphate-buffered saline (PBS). The pellet was resuspended in 8 M urea, 400 mM ammonium bicarbonate, and stored at −20 °C.

2.3. Immunoblot Analysis

Proteins (10 µg) were resolved using 4–20% gradient gels (Invitrogen, Carlsbad, CA, USA) and then transferred to PVDF membranes that then were blocked for 1 h with blocking buffer (LI-COR Biosciences, Lincoln, NE). Primary antibodies were diluted 1:5000 in blocking buffer prior to membrane incubation overnight at 4 °C. Blots were washed four times with phosphate buffered saline with Tween 20 (PBST) (0.05% v/v) and incubated with IRDye secondary antibody (LI-COR Biosciences) (1:10,000 dilution in PBST (0.5% v/v)). Blots were imaged using a LI-COR Odyssey Imaging System (LI-COR Biosciences). Immunoblot quantitation was performed using Image Studio Software v. 5.2.5 (LI-COR Biosciences).

2.4. Sample Preparation for LC–MS/MS

PSD protein fractions were quantified using the Bradford method [70]. Proteins (50 µg) were placed into an Eppendorf tube, and the volume was brought to 100 µL with 8 M Urea, 400 mM ammonium bicarbonate. Proteins were reduced with 10 µL of 45 mM DTT and incubated at 37 °C for 30 min. They were then alkylated with 10 µL of 100 mM iodoacetamide (IAM) and incubated in the dark at room temperature for 30 min. After diluting with water to bring urea concentration to 2 M, sequencing-grade trypsin (Promega, Madison, WI, USA) was added at a weight ratio of 1:20 (trypsin/protein), and the fractions were incubated at 37 °C for 16 h. The samples were desalted using C18 spin columns (The Nest Group, Inc., Southborough, MA, USA) and dried in a rotary evaporator. The samples were resuspended in 0.2% trifluoroacetic acid (TFA) and 2% acetonitrile (ACN) in water prior to LC–MS/MS analysis.

2.5. Parallel Reaction Monitoring (PRM) Method Development

2.5.1. Peptide Design and Synthesis

A list of peptides was generated from previous DDA and DIA analyses of PSD fractions isolated from rat brain tissue. Candidate PRM peptides were selected from this list on the basis of the following criteria: (1) the peptide must be 8–30 amino acids in length and have the same sequence in both mice and rats, (2) the peptide must contain a minimal number of modifiable residues (Met, Cys, Ser, Thr, Tyr), and (3) the peptide must have a minimal number of flanking Arg and Lys residues to avoid miscleavage events. Stable-isotope-labeled (SIL) peptides were synthesized as SpikeTides TQL PLUS peptides and then robotically pooled by JPT Peptide Technologies, GmbH (Berlin, Germany).

2.5.2. SIL Peptide Dilution Series (Neat)

A six-point, two-fold dilution series was performed from 75–3000 fmol per peptide. The peptides were reduced with 45 mM DTT and incubated at 37 °C for 30 min. The peptides were alkylated with 100 mM IAM and incubated at room temperature for 30 min in the dark. Sequencing-grade trypsin (Promega, Madison, WI, USA) was added at a weight ratio of 1:20 (trypsin/protein), and the samples were incubated at 37 °C for 16 h to remove the C-terminal QTag that can be cleaved by tryptic digestion. The samples were desalted using C18 spin columns (The Nest Group, Inc., Southborough, MA, USA) and dried in a rotary evaporator. The samples were resuspended in 0.2% TFA and 2% ACN in water. Each dilution was injected in technical triplicates, resulting in 25, 50, 100, 250, 500, and 1,000 fmol of each peptide being injected on the column. Results (peak area intensities, dot products (dotp), mass error, and retention times) from this analysis are displayed in Table S1.

2.5.3. SIL Peptide Dilution Series in Fixed Biological Peptide Matrix

A six-point, two-fold dilution series was performed from 75–3000 fmol per peptide in triplicate. Each dilution was added to a fixed amount (10 µg) of three independent biological protein extracts from mouse brain tissue. Each dilution was reduced with 45 mM DTT and incubated at 37 °C for

30 min. The peptides were alkylated with 100 mM IAM and incubated at room temperature for 30 min in the dark. Sequencing-grade trypsin (Promega, Madison, WI, USA) was added at a ratio of 1:20 (trypsin:protein), and the samples were incubated at 37 °C for 16 h. The samples were desalted using C18 spin columns (The Nest Group, Inc., Southborough, MA, USA) and dried in a rotary evaporator. The samples were resuspended in 0.2% TFA and 2% ACN in water. Each dilution was injected in technical triplicates, resulting in 25, 50, 100, 250, 500, and 1000 fmol per SIL peptide and 2–3 µg biological peptide matrix injected on the column. Results (peak area intensities, dotp, mass error, and retention times) from this analysis are displayed in Table S2. Response ratios (heavy vs light peak areas) from this analysis are listed for each peptide in Table S3 along with the corresponding linear performance in Table S4. A linear performance comparison of the SIL peptide in the neat versus fixed matrix analysis series can be found in Table S5.

2.6. LC–MS/MS

2.6.1. Data-Independent Acquisition (DIA)

DIA LC–MS/MS was performed using a nanoACQUITY UPLC system (Waters Corporation, Milford, MA, USA) connected to an Orbitrap Fusion Tribrid (ThermoFisher Scientific, San Jose, CA, USA) mass spectrometer. After injection, the samples were loaded into a trapping column (nanoACQUITY UPLC Symmetry C18 Trap column, 180 µm × 20 mm) at a flow rate of 5 µL/min and separated with a C18 column (nanoACQUITY column Peptide BEH C18, 75 µm × 250 mm). The compositions of mobile phases A and B were 0.1% formic acid in water and 0.1% formic acid in ACN, respectively. The peptides were eluted with a gradient extending from 6% to 35% mobile phase B in 90 min and then to 85% mobile phase B in additional 15 min at a flow rate of 300 nL/min and a column temperature of 37 °C. The data were acquired with the mass spectrometer operating in a data-independent mode with an isolation window width of 25 m/z. The full scan was performed in the range of 400–1,000 m/z with "Use Quadrupole Isolation" enabled at an Orbitrap resolution of 120,000 at 200 m/z and automatic gain control (AGC) target value of 4×10^5. Fragment ions from each peptide MS2 were generated in the C-trap with higher-energy collision dissociation (HCD) at a collision energy of 28% and detected in the Orbitrap at a resolution of 60,000.

2.6.2. Parallel Reaction Monitoring (PRM)

PRM LC–MS/MS was performed using a nanoACQUITY UPLC system (Waters Corporation, Milford, MA, USA) connected to an Orbitrap Fusion Tribrid (ThermoFisher Scientific, San Jose, CA, USA) mass spectrometer. After injection, the samples were loaded into a trapping column (nanoACQUITY UPLC Symmetry C18 Trap column, 180 µm × 20 mm) at a flow rate of 5 µL/min and separated with a C18 column (nanoACQUITY column Peptide BEH C18, 75 µm × 250 mm). The compositions of mobile phases A and B were 0.1% formic acid in water and 0.1% formic acid in ACN, respectively. The peptides were eluted with a gradient extending from 6% to 35% mobile phase B in 90 min and then to 85% mobile phase B in additional 15 min at a flow rate of 300 nL/min and a column temperature of 37○C. The data were acquired with the mass spectrometer operating in targeted mode with a MS^2 isolation window of 1.6 m/z. The full scan was performed in the range of 350–1,200 m/z with "Use Quadrupole Isolation" enabled at an Orbitrap resolution of 120,000 at 200 m/z and AGC target value of 4×10^5. The MS^2 scan range was set to 100–2,000 m/z. Fragment ions from each peptide MS^2 were generated in the C-trap with HCD at a collision energy of 28% and were detected in the Orbitrap at a resolution of 60,000.

2.7. Data Analysis

2.7.1. Data-Independent Acquisition (DIA)

DIA spectra were searched against a peptide library generated from DDA spectra using Scaffold DIA software v. 1.1.1 (Proteome Software, Portland, OR, USA). Within Scaffold DIA, raw files were first converted to the mzML format using ProteoWizard v. 3.0.11748. The samples were then aligned by retention time and individually searched against a *Mus musculus* proteome database exported from UniProt with a peptide mass tolerance of 10 ppm and a fragment mass tolerance of 10 ppm. The data acquisition type was set to "Non-Overlapping DIA", and the maximum missed cleavages was set to 1. Fixed modifications included carbamidomethylation of cysteine residues (+57.02). Peptides with charge states between 2 and 3 and 6–30 amino acids in length were considered for quantitation, and the resulting peptides were filtered by Percolator v. 3.01 at a threshold FDR of 0.01. Peptide quantification was performed by EncyclopeDIA v. 0.6.12 [71], and six of the highest quality fragment ions were selected for quantitation. Proteins containing redundant peptides were grouped to satisfy the principles of parsimony, and proteins were filtered at a threshold of two peptides per protein and an FDR of 1%. Significance was determined using a two-tailed student's *t*-test.

2.7.2. Parallel Reaction Monitoring (PRM)

PRM spectra were analyzed by Skyline software v. 4.2.0.19009 (MacCoss Lab, University of Washington) [72]. Three to six transition ion peak area intensities were integrated and summed for each peptide (heavy and light) (See mass list in Table S6). The ratio of light/heavy peak areas was calculated and mean-normalized to obtain a final quantification value for each peptide. Protein quantitation values were then calculated by summation of the peptide quantitative values. Significance was determined using a two-tailed student's t-test.

3. Results

3.1. Validation of PSD Enrichment

A previously optimized enrichment protocol, which requires density centrifugation with a Percoll gradient followed by Triton-X100 precipitation of the PSD fraction (Figure 1B), was used to enrich PSD proteins from four biological WT and Shank3B KO mouse brain tissue and three additional biological replicate WT mouse brain tissue samples. Immunoblot analysis compared protein expression of PSD-95 (PSD marker), GAPDH (cytosolic marker), and prohibitin (mitochondrial marker) in the P2 and PSD fractions isolated from each biological replicate of WT and Shank3B KO tissue (Figure S1A) and in pre-fractionation (PF) (supernatant from Step 2 of Figure 1B) and PSD-enriched (PSD) samples isolated from wild-type tissue (Figure S2A). Immunoblot quantitation revealed that the PSD-enriched fraction displayed a higher ratio of PSD-95/GAPDH expression when compared to the P2 fraction (Figure S1B) or the pre-fractionation samples (Figure S2B) in all biological replicates. From these results, it was apparent that the PSD fraction isolated from all biological replicates was enriched for a PSD marker while also being depleted of cytosolic and mitochondrial contaminants, indicating that these samples were suitable for mass spectrometry-based quantitation of PSD proteins.

3.2. *DIA Results Indicated Minor Differences Between WT and Shank3B KO PSD-Enriched Proteins*

DIA analysis was first performed on WT and Shank3B KO PSD-enriched fractions to demonstrate the utility of this assay by its ability to detect decreased expression of the Shank3B protein in the KO extracts. Data were analyzed using Scaffold DIA software. Across all samples, a total of 12,699 peptides were identified corresponding to 1862 proteins at two peptides per protein and a 1% protein FDR. The results from this analysis are displayed in Table S7. Between the two samples, the WT and KO fractions displayed similar median intensities of 4.53×10^6 and 4.33×10^6, respectively (Figure 2A) after quartile median normalization. The quantitative CV graph (Figure 2B) indicates that both the WT and KO CV values were below 5% over the entire range of intensities, suggesting low biological variability between samples within each group. In addition, both groups displayed a normal intensity distribution, which was calculated using a Gaussian kernel density estimate. Principal Component Analysis (PCA) was also performed using Scaffold DIA to observe differences between sample groups. These results showed PC1 and PC2 having a 52% and 17% explained variance, respectively, at a 95% confidence interval (Figure 2C). These results indicated a significant overlap of the WT and KO groups when plotting PC1 against PC2, suggesting minor differences between the samples in each group. A two-tailed t-test was then performed between WT and KO samples to determine which proteins had significant differences in expression ($p < 0.05$). A volcano plot was generated to display the $\log_{10}$ p-value as a function of the corresponding $\log_2$ fold change (WT/KO) in expression for all of the identified proteins (Figure 2D). In this plot, the points highlighted in green represent proteins whose expression significantly differed ($p < 0.05$) between WT and KO samples, while the proteins whose expression was not significantly changed are shown in black. In total, the 140 proteins that are listed in Table S8 were found to have significant differences in expression between these two groups. The 140 proteins that had statistically significant differences in expression levels were then displayed in a heatmap, which also shows hierarchical clustering between groups (Figure 2E).

3.3. *Expression Profiles from DIA Analysis of Wild-Type and Shank3B KO PSD Fractions Revealed Shank3-Associated Patterns*

Next, individual expression patterns were examined for some of the proteins identified in the analysis. Not surprisingly, Shank3 displayed a five-fold significant decrease in expression in the KO versus WT fractions, while no significant differences in expression were observed for Shank 1 or 2 (Figure 3A). Since the Shank3 protein has 10 expressed isoforms in mice, it can be expected that partial expression of Shank3 will be present even in the absence of the Shank3B isoform. In addition, three out of four of the CaMKII isoforms displayed a significant increase in expression in KO fractions compared to WT fractions (Figure 3B). This result was particularly interesting, as several CaMKII isoforms have previously been shown to interact with Shank3 [41,73]. In addition, several other known Shank3-interacting proteins were found to have significantly different expression in KO compared to WT fractions (Figure 3C) [56,57].

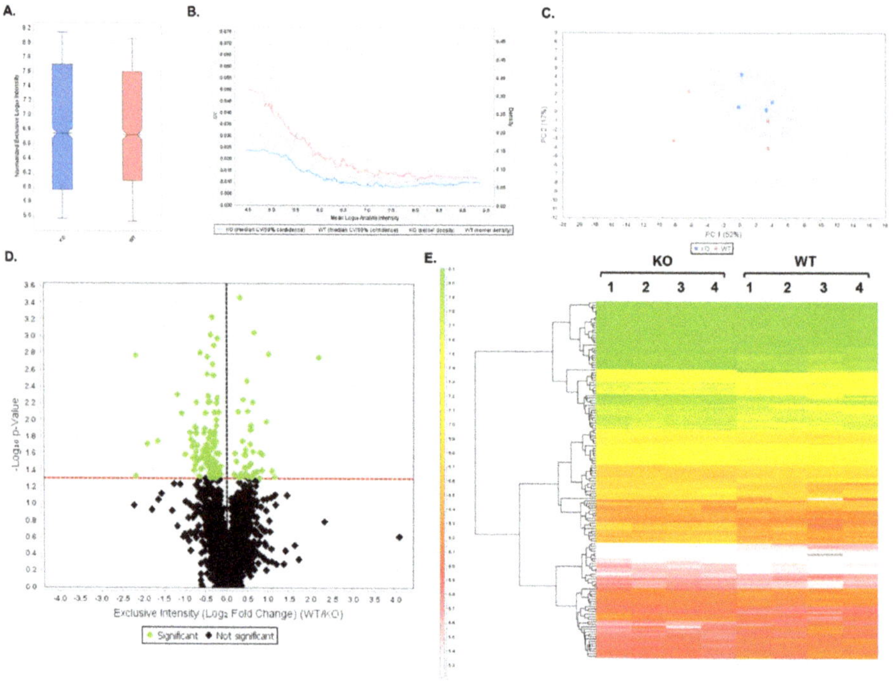

Figure 2. Data-independent analysis (DIA) results comparing wild-type (WT) and Shank3B knockout (KO) samples. (**A**) Box plot displaying quartile, median-normalized $\log_{10}$ intensities for each sample group. (**B**) Quantitative CVs chart. The bold lines show the relationship between the mean $\log_{10}$ protein intensity and the CV values for WT (pink) and KO (blue) samples. The shaded areas around the plotted lines represent the 50% confidence interval for the CV values. The faint lines indicate the intensity distribution for all proteins within WT (pink) and KO (blue) samples, which were calculated using a Gaussian kernel density estimate. (**C**) Principal Component Analysis (PCA). PCA plot displays the distribution of PC1 and PC2 in WT (pink) and KO (blue) samples. The percentages (%) in each axis represent the explained variance for each Principal Component. (**D**) Volcano plot displaying the $\log_{10}$ p-values for each protein as a function of $\log_2$ fold change (WT/KO) values after performing a t-test. Proteins that are significantly ($p < 0.05$, uncorrected values) changing in expression between the two groups are highlighted in green, while non-significant proteins are shown in black. (**E**) Heatmap of differentially expressed proteins ($p < 0.05$) after t-test statistical analysis. In total, 140 proteins were differentially expressed between WT and KO replicate samples. Hierarchical clustering tree is displayed on the left of the heatmap. The heatmap scale units are in $\log_{10}$ intensity.

3.4. *DIA Analyses Indicated Significant Differences in Protein Expression between Pre-Fractionation and PSD-Enriched Samples*

To quantify changes in abundance between proteins present prior to fractionation compared to those in the PSD-enriched fractions, DIA analysis was first performed on three biological samples per group, and the resulting data were analyzed using Scaffold DIA. This experiment demonstrated the utility of the DIA assay to analyze the same set of proteins in both PSD-enriched and unfractionated mouse brain samples. Across all samples, a total of 14,273 peptides were identified corresponding to 2134 proteins at 2 peptides per protein and a 1% protein FDR. Results from this analysis are displayed in Table S9. Between the two samples, the PSD-enriched and pre-fractionation samples displayed median intensities of 4.17×10^6 and 5.50×10^6, respectively (Figure 4A) after quartile median normalization. The quantitative CV graph (Figure 4B) indicated that both pre-fractionation and PSD-enriched CV values were below 5% over the entire range of intensities, suggesting low biological variability between samples within each group. In addition, both groups displayed a normal intensity distribution, which was calculated using a Gaussian kernel density estimate. PCA analysis performed using Scaffold DIA showed PC1 and PC2 having a 92% and 3.9% explained variance, respectively, at a 95% confidence interval (Figure 4C). These results indicated a significant divergence between the PSD-enriched and the pre-fractionation groups when plotting PC1 against PC2. A two-tailed t-test was then performed between PSD-enriched and pre-fractionation samples to determine which proteins had significant differences in expression ($p < 0.05$). A volcano plot was generated to display the $\log_{10}$ p-value as a function of the corresponding $\log_2$ fold change (PSD-enriched/pre-fractionation) for all of the identified proteins (Figure 4D). In this plot, the points highlighted in green represent proteins whose expression significantly differed ($p < 0.05$) between samples, while the proteins whose expression did not significantly differ are shown in black. In total, 1721 proteins, listed in Table S10, were found to have significantly different expression between groups. These proteins with significantly different expression levels were then displayed in a heatmap, which also shows hierarchical clustering between groups (Figure 4E).

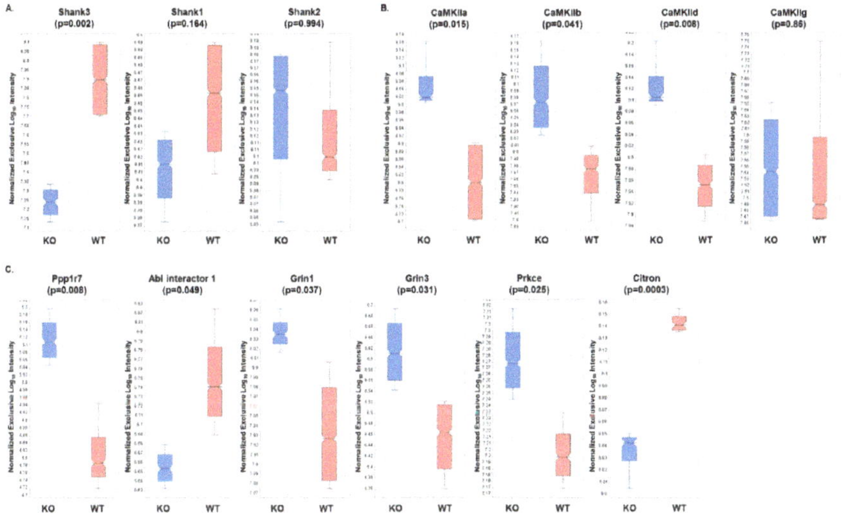

Figure 3. Relative expression levels of several mouse brain proteins based on DIA analyses of WT and KO samples. Expression levels and associated p-values (t-test) are displayed for (**A**) Shank isoforms, (**B**) CaMKII subunits, and (**C**) known Shank3-interacting proteins.

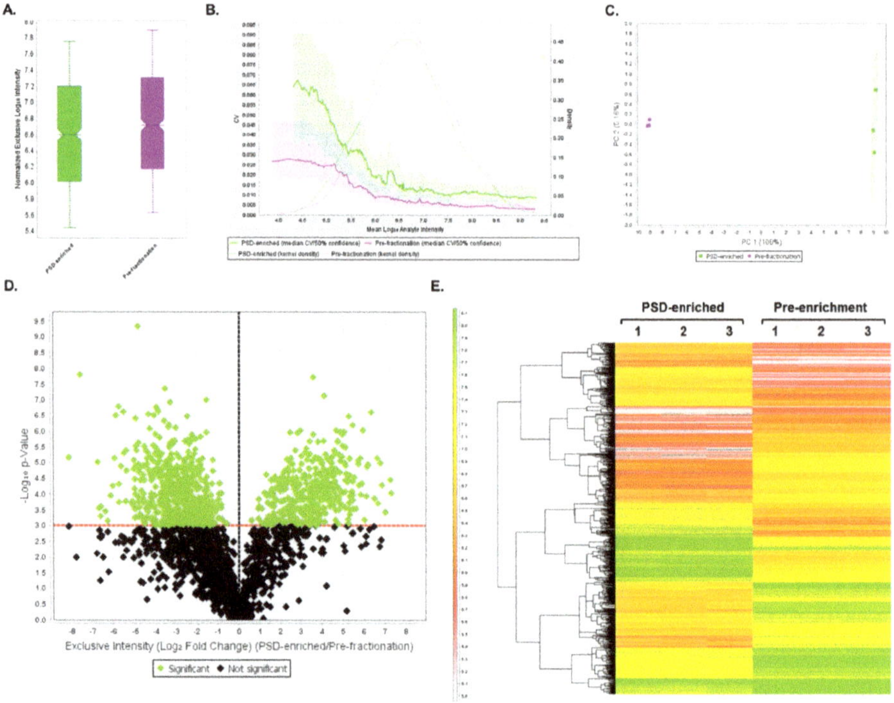

Figure 4. DIA results comparing PSD-enriched and pre-fractionation samples. (**A**) Box plot displaying quartile, median-normalized $\log_{10}$ intensities for each sample group. (**B**) Quantitative CVs chart. The bold lines show the relationship between the mean $\log_{10}$ protein intensity and the CV values for PSD-enriched (green) and pre-fractionation (purple) samples. The shaded areas around the plotted lines represent the 50% confidence interval for the CV values. The faint lines indicate the intensity distribution for all proteins within PSD-enriched (green) and pre-fractionation (purple) samples, which were calculated using a Gaussian kernel density estimate. (**C**) PCA. PCA plot displays the distribution of PSD-enriched (green) and pre-fractionation (purple) samples between PC1 and PC2. The percentages (%) in each axis represent the explained variance for each Principal Component. (**D**) Volcano plot displaying the $\log_{10}$ p-values for each protein as a function of $\log_2$ fold change (PSD-enriched/Pre-fractionation) values after performing a t-test. Proteins that are significantly ($p < 0.05$, uncorrected values) changing in expression between the two groups are highlighted in green, while proteins whose expression does not significantly differ are shown in black. (**E**) Heatmap of significantly differentially expressed proteins ($p < 0.05$) after t-test statistical analysis. In total, 1721 proteins were differentially expressed between PSD-enriched and pre-fractionation replicate samples. A hierarchical clustering tree is displayed on the left of the heatmap. The heatmap scale units are in $\log_{10}$ intensity.

3.5. DIA Expression Profiles Displayed Enrichment of PSD Proteins and Depletion of Contaminants in PSD-Enriched Fractions Comparerd to Pre-Fractionation Samples

To quantify the degree of PSD enrichment, expression profiles of PSD protein families were analyzed (Figure 5). Significant increases ($p < 0.05$) in protein expression in PSD fractions compared to pre-fractionation samples were observed for the Shank family (Figure 5A), CaMKII subunits (Figure 5B), ionotropic glutamate receptors (Figure 5C), Disks-large family (Figure 5D), and Homer family (Figure 5E). Conversely, expression patterns of PSD contaminating proteins such as histones (nuclear), GAPDH (cytoplasmic), and alpha spectrin (cytoskeletal) were all significantly decreased in the PSD fractions compared to the pre-fractionation samples (Figure 5E). These results confirmed that the PSD fractions were significantly enriched for known PSD proteins and depleted of other cellular contaminants. Furthermore, this suggests that the DIA assay can be utilized for quantitation of both fractionated and unfractionated brain samples.

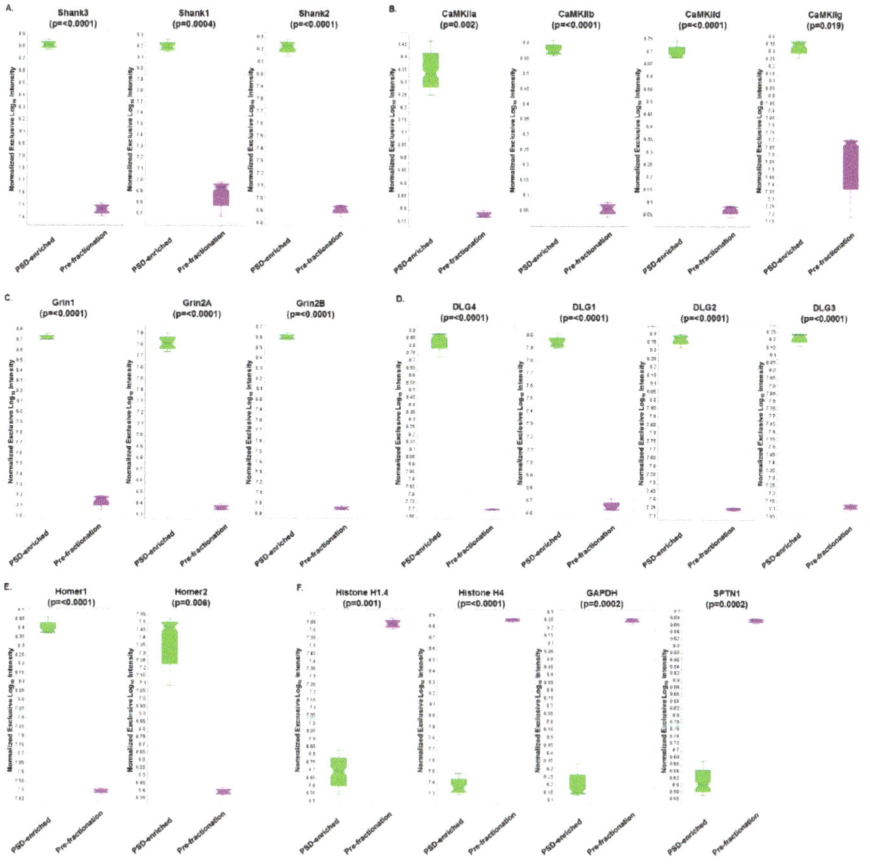

Figure 5. Expression profile results from DIA analysis comparing PSD-enriched and pre-fractionation samples. Expression profiles and associated p-values (t-test) are displayed for (**A**) Shank isoforms, (**B**) CaMKII subunits, (**C**) Glutamate receptors (NMDA), (**D**) Disks-large isoforms, (**E**) Homer isoforms, and (**F**) PSD contaminants.

3.6. Peptide Design for PRM Analysis

The PSD/PRM assay contains 47 proteins that were shown to be from 1.2 to 3.6-fold enriched in the PSD compared to the P2 fraction. In addition, this assay also includes another PSD protein, Csnk2a1 (Casein Kinase 2), and two other synaptic proteins, NEDD4 and Synpo, that were included to support another research project (Table 1). A list of candidate peptides corresponding to the 50 proteins was generated, and these peptides were then filtered through a set of criteria to select the optimal peptides for quantitative analysis. These criteria included minimizing the number of modifiable residues (e.g., Met, Cys, Tyr, Ser, Thr) as well as the number of flanking lysine and arginine residues to avoid potential miscleavage events. In addition, only nonredundant peptides were selected to ensure quantitation specificity. After performing this filtering, a list of 138 peptides (1–3 peptides per protein) was generated for synthesis of stable-isotope-labeled peptides. Notably, of the proteins selected for targeted PRM analysis, several contaminants were included to monitor the quality of PSD enrichment, such as GFAP, MBP, piccolo, bassoon, alpha spectrin, and various ribosomal proteins.

Table 1. List of target proteins and peptides for parallel reaction monitoring (PRM) analysis.

Protein#.	Gene Name	Protein Description	Peptide #	Peptide Sequence
1	Anks1b	Ankyrin repeat & sterile alpha motif domain-containing protein 1B	1 2 3	TLANLPWIVEPGQEAK LIFQSCDYK ILQAIQLLPK
2	Arc	Activity-regulated cytoskeleton-associated protein	4 5	GGPAAKPNVILQIGK TLEQLIQR
3	Baiap2	Brain-specific angiogenesis inhibitor 1-associated protein 2	6 7 8	EGDLITLLVPEAR AFHNELLTQLEQK AIFSHAAGDNSTLLSFK
4	Bsn	Protein bassoon	9 10 11	ATAEFSTQTPSLTPSSDIPR HGGGSGGPDLVPYQPQHGPGLNAPQGLASLR ATSVPGPTQATAPPEVGR
5	Camk2a	Calcium/calmodulin-dependent protein kinase type II subunit alpha	12 13 14	FTEEYQLFEELGK VLAGQEYAAK ITQYLDAGGIPR
6	Camk2b	Calcium/calmodulin-dependent protein kinase type II subunit beta	15 16 17	TTEQLIEAVNNGDFEAYAK GSLPPAALEPQTTVIHNPVDGIK ESSDSTNTTIEDEDAK
7	Camk2d	Calcium/calmodulin-dependent protein kinase type II subunit delta	18 19	FTDEYQLFEELGK IPTGQEYAAK
8	Camk2g	Calcium/calmodulin-dependent protein kinase type II subunit gamma	20 21 22	FYFENLLSK ITEQLIEAINNGDFEAYTK FTDDYQLFEELGK
9	Cldn11	Claudin-11	23	FYYSSGSSSPTHAK
10	Csnk2a1	CK2	24 25 26	GGPNIITLADIVKDPVSR TPALVFEHVNNTDFK LIDWGLAEFYHPGQEYNVR
11	Dlg2	Disks large homolog 2	27 28 29	DSGLPSQGLSFK GQEDLILSYEPVTR FIEAGQYNDNLYGTSVQSVR
12	Dlg3	Disks large homolog 3	30 31 32	VNEVDVSEVVHSR ILSVNGVNLR LLAVNNTNLQDVR
13	Dlg4	PSD-95	33 34 35	NAGQTVTIIAQYKPEEYSR EVTHSAAVEALK IIPGGAAAQDGR
14	Dlgap1	Disks large-associated protein 1	36 37 38	AVSEVSINR FQSVGVQVEEEK SLDSLDPAGLLTSPK
15	Dlgap2	Disks large-associated protein 2	39 40 41	TQGLFSYR CSSIGVQDSEFPDHQPYPR TSPTVALRPEPLLK
16	Dlgap3	Disks large-associated protein 3	42 43 44	EAEDYELPEEILEK FLELQQLK GPAGPGPGPGSGAAPEAR
17	Erc2	ERC protein 2	45 46 47	DLNHLLQQESGNR VNALQAELTEK IAELESLTLR
18	Gfap	Glial fibrillary acidic protein	48 49 50	ALAAELNQLR ITIPVQTFSNLQIR LADVYQAELR
19	Gja1	Gap junction alpha-1 protein	51	SDPYHATTGPLSPSK
20	Gria2	Glutamate receptor 2	52 53 54	LTIVGDGK ADIAIAPLTITLVR GADQEYSAFR
21	Gria3	Glutamate receptor 3	55 56 57	GSALGNAVNLAVLK NTQNFKPAPATNTQNYATYR ADIAVAPLTITLVR

Table 1. *Cont.*

Protein#.	Gene Name	Protein Description	Peptide #	Peptide Sequence
22	Grin1	Glutamate receptor ionotropic, NMDA 1	58 59 60	VIILSASEDDAATVYR HNYESAAEAIQAVR IPVLGLTTR
23	Grin2a	Glutamate receptor ionotropic, NMDA 2A	61 62 63	FSYIPEAK GVEDALVSLK YLPEEVAHSDISETSSR
24	Grin2b	Glutamate receptor ionotropic, NMDA 2B	64 65	FQRPNDFSPPFR SDVSDISTHTVTYGNIEGNAAK
25	Homer1	Homer1	66 67 68	LTAALLESTANVK HAVTVSYFYDSTR ANTVYGLGFSSEHHLSK
26	Ina	Alpha-internexin	69 70 71	ALEAELAALR FANLNEQAAR HSAEVAGYQDSIGQLESDLR
27	Kcnj4	Inward rectifier potassium channel 4	72 73 74	FEPVVFEEK SSYLASEILWGHR TYEVAGTPCCSAR
28	Lrrc7	Leucine-rich repeat-containing protein 7	75 76 77	VLNLSDNR ALIPLQTEAHPETK IVGVPLELEQSTHR
29	Mbp	Myelin basic protein	78 79 80	DTGILDSIGR TPPPSQGK TQDENPVVHFFK
30	Mog	Myelin-oligodendrocyte glycoprotein	81 82 83	ALVGDEAELPCR DQDAEQAPEYR FSDEGGYTCFFR
31	Myo1d	Unconventional myosin-1d	84 85 86	VVSVIAELLSTK HQVEYLGLLENVR IGELVGVLVNHFK
32	Nedd4	E3 ubiquitin-protein ligase NEDD4	87 88 89	EWFFLISK LLDGFFIRPFYK LLQFVTGTSR
33	Nrn1	Neuritin	90 91	FSTFSGSITGPLYTHR GFSDCLLK
34	Pclo	Protein piccolo	92 93 94	NYVLIDDIGDITK AQEAEALDVSFGHSSSSAR AAAGPLPPISADTR
35	Plec	Plectin	95 96 97	DSQDAGGFGPEDR IISLETYNLFR LGFHLPLEVAYQR
36	Rims1	Regulating synaptic membrane exocytosis protein 1	98 99 100	ATTLTVPEQQR ESGALLGLK ETSPISSHPVTWQPSK
37	Rpl3	60S ribosomal protein L3	101 102 103	VACIGAWHPAR IGQGYLIKDGK NNASTDYDLSDK
38	Rpl7a	60S ribosomal protein L7a	104 105 106	NFGIGQDIQPK LKVPPAINQFTQALDR AGVNTVTTLVENK
39	Rpl10	60S ribosomal protein L10	107	VHIGQVIMSIR
40	Rpl18a	60S ribosomal protein L18a	108 109 110	IFAPNHVVAK VKNFGIWLR DLTTAGAVTQCYR
41	Rps20	40S ribosomal protein S20	111 112 113	DTGKTPVEPEVAIHR VCADLIR LIDLHSPSEIVK
42	Shank1	SH3 and multiple ankyrin repeat domains protein 1	114 115 116	ALTASPPAAR LESGGSSGGYGAYAAGSR GSSTEDGPGVPPPSPR
43	Shank2	SH3 and multiple ankyrin repeat domains protein 2	117 118 119	AASVPALADLVK LLDPSSPLALALSAR IFLSGITEEER

Table 1. *Cont.*

Protein#.	Gene Name	Protein Description	Peptide #	Peptide Sequence
44	Shank3	SH3 and multiple ankyrin repeat domains protein 3	120 121 122	AALAVGSPGPVGGSFAR LDPTAPVWAAK VLSIGEGGFWEGTVK
45	Sptan1	Spectrin alpha chain, non-erythrocytic 1	123 124 125	ELPTAFDYVEFTR SSLSSAQADFNQLAELDR HQAFEAELSANQSR
46	Srcin1	SRC kinase signaling inhibitor 1	126 127 128	GEGLYADPYGLLHEGR AGAGGPLYGDGYGFR LLEETQAELLK
47	Syngap1	Ras GTPase-activating protein SynGAP	129 130 131	AGYVGLVTVPVATLAGR GGEPPGDTFAPFHGYSK SASGDTVFWGEHFEFNNLPAVR
48	Synpo	Synaptopodin	132 133 134	YVIESSGHAELAR AASPAKPSSLDLVPNLPR VASEEEEVPLVVYLK
49	Tomm20	Mitochondrial import receptor subunit TOM20	135	LPDLKDAEAVQK
50	Vdac2	Voltage-dependent anion-selective channel protein 2	136 137 138	GFGFGLVK YQLDPTASISAK WCEYGLTFTEK

[1] List of proteins and corresponding tryptic peptides targeted in the PSD PRM assay. Stable-isotope-labeled (SIL) peptides were synthesized with the label incorporated in the C-terminal arginine (R) or lysine (K) residue of each peptide.

3.7. PRM Analysis of PSD Target Proteins Revealed Quantitative Differences in Protein Expression in WT Versus Shank3B KO Mouse Brain Samples

To absolutely quantify PSD proteins in a more targeted approach, a PRM assay was developed for 50 known PSD and selected contaminating proteins (Table 1). Stable-isotope-labeled peptides were synthesized for 138 peptides corresponding to the 50 proteins and used as internal standards for absolute quantitation. The same sample sets that were used in the DIA assay were also used for PRM analysis. However, Sample 3 of the pre-fractionation group was injected in technical duplicate, and both were included in the quantitation. The resulting data were analyzed using Skyline software, which quantified the peak area intensities for each heavy and corresponding light peptide. The response ratios were then summed and mean-normalized for each protein (Figure S3). A protein expression heatmap was generated for each analysis (Figure 6), and a two-tailed *t*-test was performed between the two groups to determine statistical significance.

In total, there were 31 proteins that were significantly differentially expressed (as indicated by the asterisks preceding the accession names of these proteins) in the pre-fractionation versus PSD-enriched analysis (Figure 6A). These results are displayed in Table S11. Like the DIA assay, the PSD-enriched fractions displayed significantly increased expression levels of PSD proteins, including those in the MAGUK, Shank, and GKAP families. Three out of four of the CaMKII subunits had significantly increased abundance in the PSD fractions, with CaMKIIb trending in a similar direction ($p=0.059$). Interestingly, AMPA receptor Gria2 displayed significantly decreased expression in the PSD-enriched fractions compared to pre-fractionation samples, which was the inverse of the results observed in the DIA analysis. However, after assessment of the peptides identified for Gria2 in the DIA analysis (26 total), it seemed that this discrepancy was largely driven by the 24 Gria2 peptides that were unique to the DIA assay. That is, the two peptides ADIAIAPLTITLVR and LTIVGDGK, which were common to both the DIA and PRM assays showed similar trends in expression in both assays. Furthermore, expression profiles of PSD-contaminating proteins including alpha spectrin, myelin-oligodendrocyte glycoprotein (Mog), GFAP, and plectin indicated significant decreases in protein expression in PSD fractions compared to pre-fractionation samples.

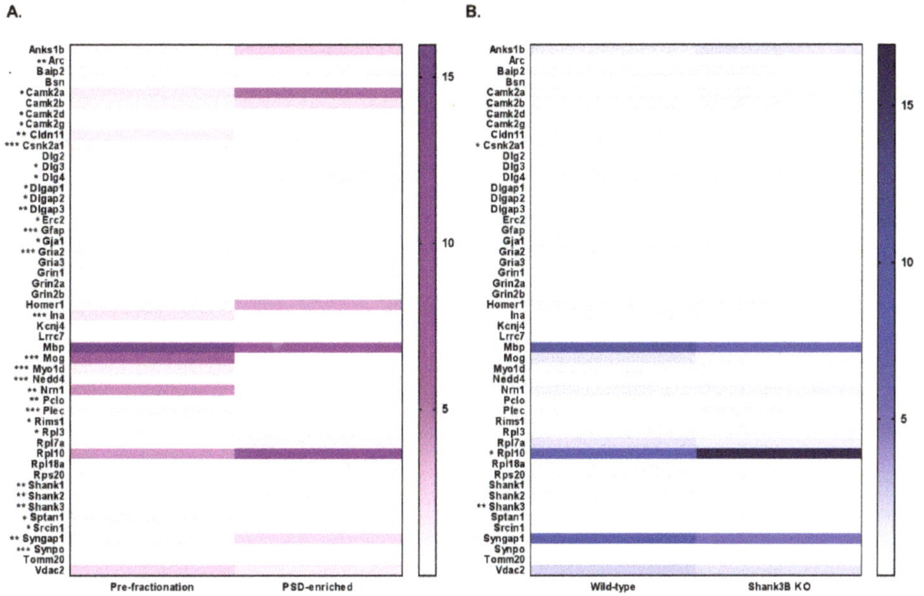

Figure 6. Mean-normalized protein abundance heatmap results from PRM LC–MS/MS analysis. Heatmap of analysis comparing (**A**) Pre-fractionation vs PSD-enriched samples (purple) and (**B**) WT vs Shank3B KO samples (blue). Protein accessions are listed to the left of the heatmap, and the degree of statistical significance between the two groups is designated for each protein (* = $p < 0.05$, ** = $p < 0.01$, *** = $p < 0.005$). Protein abundance is plotted as mean-normalized intensity response ratios (light/heavy), which are directly correlated with color intensity within the gradient displayed on the right of the heatmap.

The second PRM analysis compared WT and KO Shank3B fractions and revealed three significantly, differentially expressed proteins, including a 12-fold decrease (p=0.005) in Shank3 protein in KO fractions (Figure 6B), a decrease that was also observed in the DIA assay. These results are displayed in Table S12. Again, a low level of Shank3 expression was still present in the KO fractions, since the three selected Shank3 target peptides were not exclusive to the Shank3B isoform. For instance, while peptides AALAVGSPGPVGGSFAR and LDPTAPVWAAK were not present in the Shank3B sequence, they were found in eight and one other Shank3 isoforms, respectively. Conversely, both Shank3B and three other isoforms contained the third Shank3 peptide, VLSIGEGGFWEGTVK, in the PRM assay (Figure S4). In addition to Shank3, Csnk2a1 (CK2) and ribosomal protein L10 (Rpl10) were found to be significantly differentially expressed in WT versus KO samples. A significant increase in Csnk2a1 ($p = 0.017$) expression in WT compared to KO fractions was observed, while the inverse was true for Rpl10 ($p = 0.048$). Although the DIA expression profiles for these proteins were trending in similar directions as in the PRM assays, the levels of differential expression seen in the DIA assays were not statistically significant. A complete list of experimental results for both PRM analyses can be found in Table S13. Collectively, these results indicated that the PRM assay can be utilized for accurate quantitation of PSD proteins in both fractionated and unfractionated samples for biological characterization.

4. Discussion

Collectively, these assays demonstrated the power and selectivity of targeted mass spectrometry for quantitation of PSD proteins. Performing PRM and DIA assays in parallel enabled the identification and quantitation of over 2000 proteins before and after enrichment of the PSD from mouse cortical tissue. Many of these proteins displayed similar trends in both assays, including the scaffolding protein Shank3, which had significantly decreased expression in Shank3B knockout PSD samples compared to wild-type samples. Furthermore, proteins that have routinely been identified in PSD fractions in other proteomics studies, such as PSD-95, DLGAPs, and glutamate receptors (Gria), displayed significantly increased expression in PSD fractions compared to pre-enrichment samples in both PRM and DIA assays.

Though many of the proteins displayed similar expression profiles in both assays, there were also some discrepancies which can be attributed to differences in the number and specific peptides identified and quantified in each protein. For instance, CaMKIIa was significantly increased ($p = 0.015$) in Shank3B KO PSD samples compared to WT after DIA analysis, which identified 14 total peptides for CaMKIIa. However, PRM analysis of three peptides corresponding to CaMKIIa in the same samples resulted in a quantitative profile that was trending in a similar direction, but the expression difference was not significant with a t-test. Conversely, Csnk2a1 displayed a significant decrease ($p = 0.017$) in expression in KO versus WT samples after PRM analysis, while there was no significant difference in expression after DIA analysis of the same samples. However, the DIA analysis used five total peptides to quantify Csnk2a1, and only two out of the three PRM target peptides were identified and included in the DIA quantitation. These differences illustrate the importance of careful design, optimization, and validation of targeted assays for quantitative proteomics. In addition to mass spectrometry method development, sample selection also becomes important to determine the utility of the assays. This is one reason why two different sample sets were used for initial validation of the PSD targeted assays.

The quantification of proteins from pre-fractionated samples and PSD-enriched samples of mouse cortical tissues was initially performed to demonstrate the selectivity and utility of these assays for different sample types. Determining the limit of detection and quantitation of these proteins allows one to assess the degree of PSD enrichment and the level of contaminating proteins, which is commonly performed using methods such as immunoblot analysis. The second comparison of PSD proteins from Shank3B KO and WT mice was selected on the basis of prior Shank3-related proteomic analyses [15,74]. The Shank3B knockout line used in our study was originally generated by homologous recombination that resulted in the disruption of the PDZ domain of Shank3B (exon 13-16) [15]. Initial proteomic characterization of this knockout line was performed in striatal synapses using immunoblot analysis, which revealed a significant decrease in protein expression of many characteristic PSD proteins in KO versus WT, including PSD-95, glutamate receptors, and CaMKIIa [15]. Interestingly, DIA analysis of mouse cortical tissue revealed inverse results to those seen in the Peça et al. study; however, this difference may be attributed to the brain region analyzed, as Shank3, but not Shank1 or Shank 2, is highly expressed in the striatum of mouse brain [15]. Another study used ion-mobility-enhanced DIA analysis to assess changes in the striatal and hippocampal proteomes of Shank3Δ11$^{-/-}$ knockout mice, revealing significant decreases in the expression of glutamate receptors, including Grin1, Grin2B, Gria1, and Gria2, compared to wild-type [74]. Both PRM and DIA analysis of mouse cortical tissues did not show significant differences in glutamate receptors between WT and Shank3B KO animals, which again could be due to differences in the brain regions analyzed and to differences in the knockout mouse lines used.

In conclusion, we report on the validation and utilization of both PRM and DIA assays for quantitation of PSD proteins, which have now been demonstrated on two different sample sets. These assays provide a high-mass-accuracy, reproducible method for quantitation of PSD proteins that can be used as tools for a variety of applications in mouse or rat brain tissue. Together, the results from these analyses show promise for future studies of PSD proteomics and neurological disorders.

Supplementary Materials: The following are available online at http://www.mdpi.com/2227-7382/7/2/12/s1, Figures S1–S4; Tables S1–S13.

Author Contributions: R.S.W. performed sample preparation, analysis, and manuscript preparation. N.R. performed mass spectrometry data collection and analysis. F.S. contributed to original DDA data collection. Other authors (T.T.L., K.R.W., A.C.N.) contributed to manuscript writing.

Funding: We acknowledge primary support from the NIH (Yale/NIDA Neuroproteomics Center DA018343). Support was also obtained from the NIH (AG04270), the State of Connecticut, Department of Mental Health and Addiction Services. The Q-Exactive Plus mass spectrometer was funded in part by NIH SIG from the Office of The Director, National Institutes of Health, under Award Number (S10OD018034). The content is solely the responsibility of the authors and does not necessarily represent the official views of the National Institutes of Health.

Acknowledgments: We would like to thank Guoping Feng at MIT for his generous donation of Shank3B mutant mouse brain tissue. We also appreciate the assistance with SIL peptide synthesis of JPT Peptide Technologies.

Conflicts of Interest: The authors declare no conflict of interest.

References

1. Palay, S.L. Synapses in the Central Nervous System. *J. Cell Biol.* **1956**, *2*, 193–201. [CrossRef]
2. Cohen, R.S.; Blomberg, F.; Berzins, K.; Siekevitz, P. The structure of postsynaptic densities isolated from dog cerebral cortex. *J. Cell Biol.* **1977**, *74*, 181–203. [CrossRef]
3. Harris, K.M.; Weinberg, R.J. Ultrastructure of Synapses in the Mammalian Brain. *Cold Spring Harb. Perspect. Biol.* **2012**, *4*, 1–30. [CrossRef] [PubMed]
4. Dosemeci, A.; Weinberg, R.J.; Reese, T.S.; Tao-Cheng, J.H. The postsynaptic density: There is more than meets the eye. *Front. Synaptic Neurosci.* **2016**, *8*. [CrossRef]
5. Cho, K.-O.; Hunt, C.A.; Kennedy, M.B. The Rat Brain Postsynaptic Density Fraction Contains a Homolog of the Drosophila Discs-Large Tumor Suppressor Protein. *Neuron* **1992**, *9*, 929–942. [CrossRef]
6. Kistner, U.; Wenzel, B.M.; Vehs, R.W.; Cases-langhoff, C.; Garner, A.M.; Appeltauer, U.; Voss, B.; Gundelfinger, E.D.; Garner, C.C. SAP90, a Rat Presynaptic Protein Related to the Product of the Drosophila Tumor Suppressor Gene dlg-A. *J. Biol. Chem.* **1993**, *268*, 4580–4583. [PubMed]
7. Chen, X.; Vinade, L.; Leapman, R.D.; Petersen, J.D.; Nakagawa, T.; Phillips, T.M.; Sheng, M.; Reese, T.S. Mass of the postsynaptic density and enumeration of three key molecules. *Proc. Natl. Acad. Sci. USA* **2005**, *102*, 11551–11556. [CrossRef]
8. Sugiyama, Y.; Kawabata, I.; Sobue, K.; Okabe, S. Determination of absolute protein numbers in single synapses by a GFP-based calibration technique. *Nat. Methods* **2005**, *2*, 677–684. [CrossRef]
9. Takeuchi, M.; Hata, Y.; Hirao, K.; Toyoda, A.; Irie, M.; Takai, Y. SAPAPs: A family of PSD-95/SAP90-associated proteins localized at postsynaptic density. *J. Biol. Chem.* **1997**, *272*, 11943–11951. [CrossRef] [PubMed]
10. Ehrlich, I.; Malinow, R. Postsynaptic Density 95 controls AMPA Receptor Incorporation during Long-Term Potentiation and Experience-Driven Synaptic Plasticity. *J. Neurosci.* **2004**, *24*, 916–927. [CrossRef]
11. Garner, C.C.; Nash, J.; Huganir, R.L. PDZ domains in synapse assembly and signalling. *Trends Cell Biol.* **2000**, *10*, 274–280. [CrossRef]
12. Sheng, M.; Sala, C. PDZ Domains and the Organization of Supramolecular Complexes. *Annu. Rev. Neurosci.* **2001**, *24*, 1–29. [CrossRef] [PubMed]
13. Kim, E.; Naisbitt, S.; Hsueh, Y.-P.; Rao, A.; Rothschild, A.; Craig, A.M.; Sheng, M. GKAP, a Novel Synaptic Protein That Interacts with the Guanylate Kinase-like Domain of the PSD-95. *J. Cell Biol.* **1997**, *136*, 669–678. [CrossRef] [PubMed]
14. Naisbitt, S.; Kim, E.; Tu, J.C.; Xiao, B.; Sala, C.; Valtschanoff, J.; Weinberg, R.J.; Worley, P.F.; Sheng, M. Shank, a Novel Family of Postsynaptic Density Proteins that Binds to the NMDA Receptor/PSD-95/GKAP Complex and Cortactin. *Neuron* **1999**, *23*, 569–582. [CrossRef]
15. Peça, J.; Feliciano, C.; Ting, J.T.; Wang, W.; Wells, M.F.; Venkatraman, T.N.; Lascola, C.D.; Fu, Z.; Feng, G. Shank3 mutant mice display autistic-like behaviours and striatal dysfunction. *Nature* **2011**, *472*, 437–442. [CrossRef]

16. Hung, A.Y.; Futai, K.; Sala, C.; Valtschanoff, J.G.; Ryu, J.; Woodworth, M.A.; Kidd, F.L.; Sung, C.C.; Miyakawa, T.; Bear, M.F.; et al. Smaller Dendritic Spines, Weaker Synaptic Transmission, but Enhanced Spatial Learning in Mice Lacking Shank1. *Neurosci. Res.* **2009**, *28*, 1697–1708. [CrossRef] [PubMed]
17. Kato, A.; Inokuchi, K.; Ozawa, F.; Fukazawa, Y.; Saitoh, Y.; Sugiyama, H. Novel Members of the Vesl/Homer Family of PDZ Proteins That Bind Metabotropic Glutamate Receptors. *J. Biol. Chem.* **1998**, *273*, 23969–23975. [CrossRef] [PubMed]
18. Brakeman, P.R.; Lanahan, A.A.; O'Brien, R.; Roche, K.; Barnes, C.A.; Huganir, R.L.; Worley, P.F. Homer: A protein that selectively binds metabotropic glutamate receptors. *Nature* **1997**, *386*, 284–288. [CrossRef]
19. Xiao, B.; Tu, J.C.; Petralia, R.S.; Yuan, J.P.; Doan, A.; Breder, C.D.; Ruggiero, A.; Lanahan, A.A.; Wenthold, R.J.; Worley, P.F. Homer regulates the association of group 1 metabotropic glutamate receptors with multivalent complexes of Homer-related, synaptic proteins. *Neuron* **1998**, *21*, 707–716. [CrossRef]
20. Barzik, M.; Carl, U.D.; Schubert, W.D.; Frank, R.; Wehland, J.; Heinz, D.W. The N-terminal domain of Homer/Vesl is a new class II EVH1 domain. *J. Mol. Biol.* **2001**, *309*, 155–169. [CrossRef]
21. Sun, J.; Tadokoro, S.; Imanaka, T.; Murakami, S.D.; Nakamura, M.; Kashiwada, K.; Ko, J.; Nishida, W.; Sobue, K. Isolation of PSD-Zip45, a novel Homer/vesl family protein containing leucine zipper motifs, from rat brain 1. *FEBS Lett.* **1998**, *437*, 304–308. [CrossRef]
22. Tadokoro, S.; Tachibana, T.; Imanaka, T.; Nishida, W.; Sobue, K. Involvement of unique leucine-zipper motif of PSD-Zip45 (Homer 1c/vesl-1L) in group 1 metabotropic glutamate receptor clustering. *Proc. Natl. Acad. Sci. USA* **1999**, *96*, 13801–13806. [CrossRef]
23. Hayashi, M.K.; Ames, H.M.; Hayashi, Y. Tetrameric Hub Structure of Postsynaptic Scaffolding Protein Homer. *J. Neurosci.* **2006**, *26*, 8492–8501. [CrossRef]
24. Hayashi, M.K.; Tang, C.; Verpelli, C.; Narayanan, R.; Stearns, M.H.; Xu, R.-M.; Li, H.; Sala, C.; Hayashi, Y. The Postsynaptic Density Proteins Homer and Shank Form a Polymeric Network Structure. *Cell* **2009**, *137*, 159–171. [CrossRef]
25. Shiraishi-Yamaguchi, Y.; Sato, Y.; Sakai, R.; Mizutani, A.; Knöpfel, T.; Mori, N.; Mikoshiba, K.; Furuichi, T. Interaction of Cupidin/Homer2 with two actin cytoskeletal regulators, Cdc42 small GTPase and Drebrin, in dendritic spines. *BMC Neurosci.* **2009**, *10*, 1–14. [CrossRef]
26. Baron, M.K.; Boeckers, T.M.; Vaida, B.; Faham, S.; Gingery, M.; Sawaya, D.; Gundelfinger, E.D.; Bowie, J.U. An architectural framework that may lie at the core of the postsynaptic density. *Science* **2006**, *311*, 531–535. [CrossRef]
27. Goulding, S.P.; Szumlinski, K.K.; Contet, C.; MacCoss, M.J.; Wu, C.C. A mass spectrometry-based proteomic analysis of Homer2-interacting proteins in the mouse brain. *J. Proteomics* **2017**, *166*, 127–137. [CrossRef]
28. Erondu, N.E.; Kennedy, M.B. Regional distribution of type II Ca2+/calmodulin-dependent protein kinase in rat brain. *J. Neurosci.* **1985**, *5*, 3270–3277. [CrossRef]
29. Dosemeci, A.; Tao-Cheng, J.H.; Vinade, L.; Winters, C.A.; Pozzo-Miller, L.; Reese, T.S. Glutamate-induced transient modification of the postsynaptic density. *Proc. Natl. Acad. Sci. USA* **2001**, *98*, 10428–10432. [CrossRef] [PubMed]
30. Hu, B.R.; Park, M.; Martone, M.E.; Fischer, W.H.; Ellisman, M.H.; Zivin, J.A. Assembly of proteins to postsynaptic densities after transient cerebral ischemia. *J. Neurosci.* **1998**, *18*, 625–633. [CrossRef] [PubMed]
31. Suzuki, T.; Okumura-Noji, K.; Tanaka, R.; Tada, T. Rapid Translocation of Cytosolic Ca2+/Calmodulin-Dependent Protein Kinase II into Postsynaptic Density After Decapitation. *J. Neurochem.* **1994**, *63*, 1529–1537. [CrossRef] [PubMed]
32. Martone, M.E.; Jones, Y.Z.; Young, S.J.; Ellisman, M.H.; Zivin, J.A.; Hu, B.-R. Modification of Postsynaptic Densities after Transient Cerebral Ischemia: A Quantitative and Three-Dimensional Ultrastructural Study. *J. Neurosci.* **1999**, *19*, 1988–1997. [CrossRef] [PubMed]
33. Lisman, J.; Yasuda, R.; Raghavachari, S. Mechanisms of CaMKII action in long-term potentiation. *Nat. Rev. Neurosci.* **2012**, *13*, 169–182. [CrossRef]
34. Shonesy, B.C.; Jalan-Sakrikar, N.; Cavener, V.S.; Colbran, R.J. CaMKII: A molecular substrate for synaptic plasticity and memory. *Prog. Mol. Biol. Transl. Sci.* **2014**, *122*, 61–87. [CrossRef] [PubMed]
35. Dhamne, S.C.; Silverman, J.L.; Super, C.E.; Lammers, S.H.T.; Hameed, M.Q.; Modi, M.E.; Copping, N.A.; Pride, M.C.; Smith, D.G.; Rotenberg, A.; et al. Replicable in vivo physiological and behavioral phenotypes of the Shank3B null mutant mouse model of autism. *Mol. Autism* **2017**, *8*, 1–19. [CrossRef] [PubMed]

36. Peixoto, R.T.; Wang, W.; Croney, D.M.; Kozorovitskiy, Y.; Sabatini, B.L. Early hyperactivity and precocious maturation of corticostriatal circuits in Shank3B-/- mice. *Nat. Neurosci.* **2016**, *19*, 716–724. [CrossRef]
37. Berryer, M.H.; Hamdan, F.F.; Klitten, L.L.; Møller, R.S.; Carmant, L.; Schwartzentruber, J.; Patry, L.; Dobrzeniecka, S.; Rochefort, D.; Neugnot-Cerioli, M.; et al. Mutations in SYNGAP1 Cause Intellectual Disability, Autism, and a Specific Form of Epilepsy by Inducing Haploinsufficiency. *Hum. Mutat.* **2013**, *34*, 385–394. [CrossRef]
38. Hamdan, F.F.; Gauthier, J.; Spiegelman, D.; Noreau, A.; Yang, Y.; Pellerin, S.; Dobrzeniecka, S.; Cote, M.; Perreau-Linck, E.; Carmant, L.; et al. Mutations in SYNGAP1 in Autosomal Nonsyndromic Mental Retardation. *N. Engl. J. Med.* **2009**, *360*, 599–605. [CrossRef]
39. Parker, M.J.; Fryer, A.E.; Shears, D.J.; Lachlan, K.L.; Mckee, S.A.; Magee, A.C.; Mohammed, S.; Vasudevan, P.C.; Park, S.M.; Benoit, V.; et al. De novo, heterozygous, loss-of-function mutations in SYNGAP1 cause a syndromic form of intellectual disability. *Am. J. Med. Genet. Part A* **2015**, *167*, 2231–2237. [CrossRef]
40. Bertram, L.; Lange, C.; Mullin, K.; Parkinson, M.; Hsiao, M.; Hogan, M.F.; Schjeide, B.M.M.; Hooli, B.; DiVito, J.; Ionita, I.; et al. Genome-wide Association Analysis Reveals Putative Alzheimer's Disease Susceptibility Loci in Addition to APOE. *Am. J. Hum. Genet.* **2008**, *83*, 623–632. [CrossRef]
41. Stephenson, J.R.; Wang, X.; Perfitt, T.L.; Parrish, W.P.; Shonesy, B.C.; Marks, C.R.; Mortlock, D.P.; Nakagawa, T.; Sutcliffe, J.S.; Colbran, R.J. A Novel Human CAMK2A Mutation Disrupts Dendritic Morphology and Synaptic Transmission, and Causes ASD-Related Behaviors. *J. Neurosci.* **2017**, *37*, 2216–2233. [CrossRef]
42. Fernández, E.; Collins, M.O.; Uren, R.T.; Kopanitsa, M.V.; Komiyama, N.H.; Croning, M.D.R.; Zografos, L.; Armstrong, J.D.; Choudhary, J.S.; Grant, S.G.N. Targeted tandem affinity purification of PSD-95 recovers core postsynaptic complexes and schizophrenia susceptibility proteins. *Mol. Syst. Biol.* **2009**, *5*, 1–17. [CrossRef]
43. Carlin, R.K.; Grab, D.J.; Cohen, R.S.; Siekevitz, P. Isolation and characterization of postsynaptic densities from various brain regions: Enrichment of different types of postsynaptic densities. *J. Cell Biol.* **1980**, *86*. [CrossRef]
44. De Robertis, E.; Azcurra, J.M.; Fiszer, S. Ultrastructure and cholinergic binding capacity of junctional complexes isolated from rat brain. *Brain Res.* **1967**, *5*, 45–56. [CrossRef]
45. Davis, G.A.; Bloom, F.E. Isolation of synaptic junctional complexes from rat brain. *Brain Res.* **1973**, *62*, 135–153. [CrossRef]
46. Zeng, M.; Shang, Y.; Araki, Y.; Guo, T.; Huganir, R.L.; Zhang, M. Phase Transition in Postsynaptic Densities Underlies Formation of Synaptic Complexes and Synaptic Plasticity. *Cell* **2016**, *166*, 1163–1175.e12. [CrossRef] [PubMed]
47. Zeng, M.; Chen, X.; Guan, D.; Xu, J.; Wu, H.; Tong, P.; Zhang, M. Reconstituted Postsynaptic Density as a Molecular Platform for Understanding Synapse Formation and Plasticity. *Cell* **2018**, *174*, 1172–1187.e16. [CrossRef] [PubMed]
48. Bartol, T.M.; Bromer, C.; Kinney, J.; Chirillo, M.A.; Bourne, J.N.; Harris, K.M.; Sejnowski, T.J. Nanoconnectomic upper bound on the variability of synaptic plasticity. *Elife* **2015**, *4*, 1–18. [CrossRef] [PubMed]
49. Matsuzaki, M.; Honkura, N.; Ellis-Davies, G.C.R.; Kasai, H. Structural basis of long-term potentiation in single dendritic spines. *Nature* **2004**, *429*, 761–765. [CrossRef] [PubMed]
50. Nishiyama, J.; Yasuda, R. Biochemical Computation for Spine Structural Plasticity. *Neuron* **2015**, *87*, 63–75. [CrossRef]
51. Roy, M.; Sorokina, O.; Skene, N.; Simonnet, C.; Mazzo, F.; Zwart, R.; Sher, E.; Smith, C.; Armstrong, J.D.; Grant, S.G.N. Proteomic analysis of postsynaptic proteins in regions of the human neocortex. *Nat. Neurosci.* **2018**, *21*, 130–141. [CrossRef]
52. Bayés, Á.; Van De Lagemaat, L.N.; Collins, M.O.; Croning, M.D.R.; Whittle, I.R.; Choudhary, J.S.; Grant, S.G.N. Characterization of the proteome, diseases and evolution of the human postsynaptic density. *Nat. Neurosci.* **2011**, *14*, 19–21. [CrossRef]
53. Bayés, À; Collins, M.O.; Croning, M.D.R.; van de Lagemaat, L.N.; Choudhary, J.S.; Grant, S.G.N. Comparative Study of Human and Mouse Postsynaptic Proteomes Finds High Compositional Conservation and Abundance Differences for Key Synaptic Proteins. *PLoS ONE* **2012**, *7*, 1–13. [CrossRef]

54. Bayés, À.; Collins, M.O.; Galtrey, C.M.; Simonnet, C.; Roy, M.; Croning, M.D.R.; Gou, G.; Van De Lagemaat, L.N.; Milward, D.; Whittle, I.R.; Smith, C.; et al. Human post-mortem synapse proteome integrity screening for proteomic studies of postsynaptic complexes. *Mol. Brain* **2014**, *7*, 1–11. [CrossRef] [PubMed]
55. Peng, J.; Kim, M.J.; Cheng, D.; Duong, D.M.; Gygi, S.P.; Sheng, M. Semi-quantitative Proteomic Analysis of Rat Forebrain Postsynaptic Density Fractions by Mass Spectrometry. *J. Biol. Chem.* **2004**, *179*, 21003–21011. [CrossRef]
56. Li, J.; Wilkinson, B.; Clementel, V.A.; Hou, J.; O'Dell, T.J.; Coba, M.P. Long-term potentiation modulates synaptic phosphorylation networks and reshapes the structure of the postsynaptic interactome. *Sci. Signal.* **2016**, *9*, rs8. [CrossRef]
57. Li, J.; Zhang, W.; Yang, H.; Howrigan, D.P.; Wilkinson, B.; Souaiaia, T.; Evgrafov, O.V.; Genovese, G.; Clementel, V.A.; Tudor, J.C.; et al. Spatiotemporal profile of postsynaptic interactomes integrates components of complex brain disorders. *Nat. Neurosci.* **2017**, *20*, 1150–1161. [CrossRef] [PubMed]
58. Colangelo, C.M.; Ivosev, G.; Chung, L.; Abbott, T.; Shifman, M.; Sakaue, F.; Cox, D.; Kitchen, R.R.; Burton, L.; Tate, S.A.; et al. Development of a highly automated and multiplexed targeted proteome pipeline and assay for 112 rat brain synaptic proteins. *Proteomics* **2015**, *15*, 1202–1214. [CrossRef]
59. Gillette, M.A.; Carr, S.A. Quantitative analysis of peptides and proteins in biomedicine by targeted mass spectrometry. *Nat. Methods* **2013**, *10*, 28–34. [CrossRef] [PubMed]
60. Venable, J.D.; Dong, M.Q.; Wohlschlegel, J.; Dillin, A.; Yates, J.R. Automated approach for quantitative analysis of complex peptide mixtures from tandem mass spectra. *Nat. Methods* **2004**, *1*, 39–45. [CrossRef] [PubMed]
61. Gillet, L.C.; Navarro, P.; Tate, S.; Röst, H.; Selevsek, N.; Reiter, L.; Bonner, R.; Aebersold, R. Targeted Data Extraction of the MS/MS Spectra Generated by Data-independent Acquisition: A New Concept for Consistent and Accurate Proteome Analysis. *Mol. Cell. Proteomics* **2012**, *11*, O111.016717. [CrossRef]
62. Bruderer, R.; Bernhardt, O.M.; Gandhi, T.; Miladinović, S.M.; Cheng, L.-Y.; Messner, S.; Ehrenberger, T.; Zanotelli, V.; Butscheid, Y.; Escher, C.; et al. Extending the Limits of Quantitative Proteome Profiling with Data-Independent Acquisition and Application to Acetaminophen-Treated Three-Dimensional Liver Microtissues. *Mol. Cell. Proteomics* **2015**, *14*. [CrossRef]
63. Geiger, T.; Cox, J.; Mann, M. Proteomics on an Orbitrap Benchtop Mass Spectrometer Using All-ion Fragmentation. *Mol. Cell. Proteomics* **2010**, *9*, 2252–2261. [CrossRef]
64. Egertson, J.D.; Kuehn, A.; Merrihew, G.E.; Bateman, N.W.; MacLean, B.X.; Ting, Y.S.; Canterbury, J.D.; Marsh, D.M.; Kellmann, M.; Zabrouskov, V.; et al. Multiplexed MS/MS for improved data-independent acquisition. *Nat. Methods* **2013**, *10*, 744–746. [CrossRef] [PubMed]
65. Selevsek, N.; Chang, C.-Y.; Gillet, L.C.; Navarro, P.; Bernhardt, O.M.; Reiter, L.; Cheng, L.-Y.; Vitek, O.; Aebersold, R. Reproducible and Consistent Quantification of the Saccharomyces cerevisiae Proteome by SWATH-mass spectrometry. *Mol. Cell. Proteomics* **2015**, *14*, 739–749. [CrossRef]
66. Distler, U.; Schmeisser, M.J.; Pelosi, A.; Reim, D.; Kuharev, J.; Weiczner, R.; Baumgart, J.; Boeckers, T.M.; Nitsch, R.; Vogt, J.; et al. In-depth protein profiling of the postsynaptic density from mouse hippocampus using data-independent acquisition proteomics. *Proteomics* **2014**, *14*, 2607–2613. [CrossRef]
67. Peterson, A.C.; Russell, J.D.; Bailey, D.J.; Westphall, M.S.; Coon, J.J. Parallel Reaction Monitoring for High Resolution and High Mass Accuracy Quantitative, Targeted Proteomics. *Mol. Cell. Proteomics* **2012**, *11*, 1475–1488. [CrossRef] [PubMed]
68. Gallien, S.; Duriez, E.; Crone, C.; Kellmann, M.; Moehring, T.; Domon, B. Targeted Proteomic Quantification on Quadrupole-Orbitrap Mass Spectrometer. *Mol. Cell. Proteomics* **2012**, *11*, 1709–1723. [CrossRef] [PubMed]
69. Ankney, J.A.; Muneer, A.; Chen, X. Relative and Absolute Quantitation in Mass Spectrometry–Based Proteomics. *Annu. Rev. Anal. Chem.* **2018**, *11*, 49–77. [CrossRef] [PubMed]
70. Bradford, M.M. Sistema séptico domiciliario | Rotomoldeo en Colombia Tanques Plasticos En Colombia Rotoplast. *Anal. Biochem.* **1976**, *72*, 248–254. [CrossRef]
71. Searle, B.C.; Pino, L.K.; Egertson, J.D.; Ting, Y.S.; Lawrence, R.T.; Villen, J.; MacCoss, M.J. Comprehensive peptide quantification for data independent acquisition mass spectrometry using chromatogram libraries. *bioRxiv* **2018**, 277822. [CrossRef]

72. MacLean, B.; Tomazela, D.M.; Shulman, N.; Chambers, M.; Finney, G.L.; Frewen, B.; Kern, R.; Tabb, D.L.; Liebler, D.C.; MacCoss, M.J. Skyline: An open source document editor for creating and analyzing targeted proteomics experiments. *Bioinformatics* **2010**, *26*, 966–968. [CrossRef] [PubMed]
73. Baucum, A.J.; Shonesy, B.C.; Rose, K.L.; Colbran, R.J. Quantitative Proteomics Analysis of CaMKII Phosphorylation and the CaMKII Interactome in the Mouse Forebrain. *ACS Chem. Neurosci.* **2015**, *6*, 615–631. [CrossRef] [PubMed]
74. Reim, D.; Distler, U.; Halbedl, S.; Verpelli, C.; Sala, C.; Bockmann, J.; Tenzer, S.; Boeckers, T.M.; Schmeisser, M.J. Proteomic Analysis of Post-synaptic Density Fractions from Shank3 Mutant Mice Reveals Brain Region Specific Changes Relevant to Autism Spectrum Disorder. *Front. Mol. Neurosci.* **2017**, *10*, 26. [CrossRef] [PubMed]

proteomes

Review

Proteomic Approaches for the Discovery of Biofluid Biomarkers of Neurodegenerative Dementias

Becky C. Carlyle, Bianca A. Trombetta and Steven E. Arnold *

Massachusetts General Hospital Department of Neurology, Charlestown, MA 02129, USA; bcarlyle@mgh.harvard.edu (B.C.C.); btrombetta@partners.org (B.A.T.)

* Correspondence: searnold@mgh.harvard.edu

Received: 20 July 2018; Accepted: 29 August 2018; Published: 31 August 2018

Abstract: Neurodegenerative dementias are highly complex disorders driven by vicious cycles of intersecting pathophysiologies. While most can be definitively diagnosed by the presence of disease-specific pathology in the brain at postmortem examination, clinical disease presentations often involve substantially overlapping cognitive, behavioral, and functional impairment profiles that hamper accurate diagnosis of the specific disease. As global demographics shift towards an aging population in developed countries, clinicians need more sensitive and specific diagnostic tools to appropriately diagnose, monitor, and treat neurodegenerative conditions. This review is intended as an overview of how modern proteomic techniques (liquid chromatography mass spectrometry (LC-MS/MS) and advanced capture-based technologies) may contribute to the discovery and establishment of better biofluid biomarkers for neurodegenerative disease, and the limitations of these techniques. The review highlights some of the more interesting technical innovations and common themes in the field but is not intended to be an exhaustive systematic review of studies to date. Finally, we discuss clear reporting principles that should be integrated into all studies going forward to ensure data is presented in sufficient detail to allow meaningful comparisons across studies.

Keywords: neurodegeneration; Alzheimer's disease; cerebrospinal fluid; plasma; serum; proteomics; biomarkers; LC-MS/MS

1. Introduction

Clinical neuroscientists and practitioners have gained access to an increasing array of tools to assist in the diagnosis of neurodegenerative disease dementias. Various neuroimaging techniques and a number of cerebrospinal fluid (CSF) biomarkers can now complement diagnosis that was once based solely on careful clinical and neuropsychological assessments of symptoms and only positively confirmed at autopsy [1]. These additional biomarkers can be extremely informative, as many neurological diseases present with similar sets of cognitive, behavioral, and/or movement symptoms, particularly in early disease stages. While neuroimaging-based techniques, including structural and functional Magnetic Resonance Imaging (MRI) and Positron Emission Tomography (PET), are currently the most commonly used diagnostic measures, these require sophisticated on-site technologies and expertise in specialized centers and they are expensive [2]. The field could benefit from increasing availability of biomarkers in blood, CSF, or other biofluids, which are more widely attainable through minimally invasive means, simpler to interpret, and performed on more routine diagnostic equipment [3].

A series of National Institute on Aging and Alzheimer Association consensus conferences suggested a number of criteria that a biomarker of neurodegenerative disease should fulfill [4]. A putative marker should be linked to the fundamental neuropathology of the disease and validated in neuropathologically confirmed cases. Ideally, a marker would be able to detect the disease before the onset of symptoms, distinguish between neurodegenerative disorders, and not be affected by treatment with symptom-relieving drugs. Practically, a marker should be non- or minimally invasive, simple to execute, and relatively inexpensive. Based on these principles, a new research framework, "AT(N)", was proposed for clear delineation of Alzheimer's disease (AD) from other disorders. In this framework [1], an indication of amyloid pathology (A+) by amyloid PET or in CSF is necessary for assigning a subject to an AD diagnosis. The disease can be further classified by the presence or absence of tau fibrillation (T), measured by PET or phosphorylated-tau (pTau) in CSF, and the extent of neurodegeneration (N) as measured by structural MRI or total tau in CSF. Despite this improvement in defining AD in biological terms, these markers alone do not allow for clear staging and AD prognosis. For example, the definition of a case as A+T+ may predict progression of a subject from mild cognitive impairment (MCI) to dementia but with a highly variable timeframe. As a result of this variability, the AT(N) framework was designed to flexibly accommodate the addition of further biomarker groups such as vascular and synuclein markers that may aid in the overall characterization of neurodegenerative disorders as distinct clinical entities and likely treatment groups.

Biofluids fulfill the practicality recommendations for a biomarker, being relatively easily and economically attainable. CSF is the primary fluid of choice, being in intimate contact with the interstitial fluid of the brain and carrying molecules secreted by neurons and glia, excreted metabolic waste, and material from dying synapses, axons, and cells that indicate neurodegeneration [5–7]. However, although the lumbar puncture procedure to obtain CSF is generally considered straightforward, safe, and tolerable, it is not routinely performed in many neurology clinics due to patient and clinician disinclination [8,9]. The procedure is also not particularly well suited to multiple short-term repeat measures, such as those used to assess target engagement, pharmacokinetics, or acute pharmacodynamic response of a novel drug. This had led to a widespread belief that the "holy grail" of neurodegenerative disease research lies in a blood-based biomarker [10].

In blood-derived fluids (plasma and serum), central nervous system (CNS)-specific proteins are diluted by proteins from all other peripheral tissue sources, leading to potentially low concentrations that require ultrasensitive quantification [6,7]. Proteins may be regulated and modified by different processes in the CNS versus the periphery, resulting in a lack of correlation between abundance in CSF and blood [11,12]. Blood may also be presumed to be more labile, being in contact with many more secretory and excretory tissues than CSF. Finally, blood, and to a lesser extent CSF, is a complex mixture of proteins and metabolites that span a large range of abundances. In plasma, protein concentrations range from the most abundant protein, human serum albumin at 50 mg/mL, to signaling proteins in the low pg/mL range, such as IL-6 [13–15]. These large differences in protein abundance mean there is currently no perfect technique for quantifying a large number of analytes that span this dynamic range.

Proteomic approaches are an excellent companion in the search for novel neurodegenerative disease biomarkers. Recent improvements in reproducibility and sensitivity of liquid chromatography tandem mass spectrometry (LC-MS/MS) instrumentation [16], coupled with the development of immunoassay-based single molecule quantification and multiplexing [17–22], offer a wide range of tools to allow for hypothesis-free target discovery through to the ability to accurately, sensitively, and simultaneously quantify a specific small number of targets. While proteomic techniques are available that together span most of the range of protein abundances in a complex biofluid, from ultrasensitive (~0.05 pg/mL) through to extremely abundant (~50 mg/mL), careful experimental selection and design is important to maximize the likelihood of accurately quantifying a target of interest (Figure 1). In this review, we introduce a toolbox of techniques available to the biomarker researcher, the advantages and disadvantages of the major technologies, and finally, some of the key discoveries to date in the field of protein biomarkers for neurodegeneration.

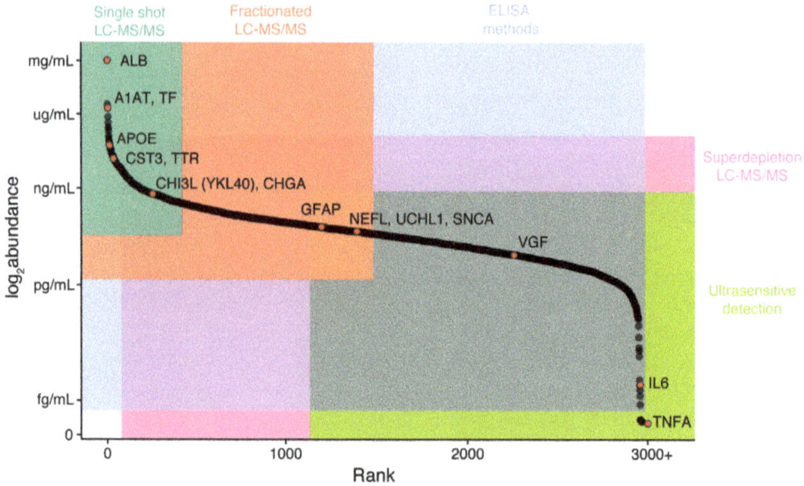

Figure 1. Different proteomic techniques are more suited to different concentration ranges of biofluid analytes. In this plot, cerebrospinal fluid (CSF) proteins are ranked according to their abundance, with the location of specific proteins placed according to their concentrations in enzyme-linked immunoassays (ELISAs), Multiple-Reaction-Monitoring (MRM), and in-house (unpublished) label-free experiments [23–26]. It is of note that there is a large amount of disagreement between experiments on the exact concentrations of these analytes, and so their place on this plot should be considered illustrative. Of particular note is VGF, an analyte that exists as multiple processed peptides, which is easily detected by single-shot LC-MS/MS but detected in the low pg/mL ranges by ELISA. Single-shot LC-MS/MS will generally quantify 300–500 abundant proteins in CSF (turquoise), and protein identifications can be increased by offline fractionation of samples (orange). While ELISA-based methods measure analytes across the widest concentration range, these techniques require a strong hypothesis for target selection and rely on the availability of an appropriate antibody pair for the analyte. At low analyte concentrations, super depletion can be combined with LC-MS/MS to reveal low-abundance proteins, but there are concerns over nonspecific depletion of some target analytes. Finally, ultrasensitive platforms can be used to measure proteins such as cytokines in CSF, which are present in the low pg/mL to fg/mL range.

2. LC-MS/MS Strategies

Most basic LC-MS/MS proteomic workflows derive from the same underlying tandem mass-spectrometry method [27]. A protease-digested peptide mixture is injected onto a liquid chromatography column, then eluted from the column with a solvent gradient over a period of time. Peptides enter the tandem mass spectrometer, where they are ionized ("precursor ion"), separated by mass charge ratio, and detected. In data-dependent methods, the first "MS1" detection is generally used to quantify the peptides. In most workflows, a subset of precursor ions is isolated and fragmented ("fragment ions") for a second round of mass spectrometry (MS2). MS2 fragments can be used for both confident identification of a peptide and for peptide quantification [28]. Almost every step of this simple workflow, including sample preparation, can be tweaked to optimize the parameters of the experiment, providing an extremely flexible basic platform for biomarker discovery across a range of analyte concentrations [29–31].

2.1. Data-Dependent LC-MS/MS

Label-free methods are the simplest LC-MS/MS workflows. In these experiments, an unlabeled peptide sample is injected directly onto the instrument-coupled LC column and quantified by MS1 intensity or spectral counting [32–35]. Peptides are identified by matching of the MS2 fragmentation products to the spectral properties of known peptides in a database. As only a single "snapshot" MS2 measurement is taken, accurate MS2 level quantification is not possible. Each sample is injected independently, and experimental reproducibility is highest if these injections are performed consecutively with careful monitoring of LC performance [36,37]. For this reason, it may be difficult to directly compare quantification from two label-free experiments carried out at different times in different labs or with a different LC setup.

While this method enables truly hypothesis-free biomarker discovery without the need for antibodies, there are a number of disadvantages to using label-free techniques that are of particular importance in biofluids. The greatest disadvantage is that peptides from high-abundance proteins such as albumin can mask or interfere with peptides from lower-abundance proteins, decreasing the sensitivity of the experiment [38,39]. While it is possible to simplify the peptide mixture entering the instrument by increasing the length of the elution from the LC, the number of protein identifications in brain tissue currently tends to plateau at between 3000 and 5000 proteins [40]. In biofluids such as blood, where albumin and the immunoglobulins make up more than 75% of total protein weight, and a further 20 proteins account for more than 24% of the total weight, this masking is profound. A standard long-gradient (>2.5 h) label-free experiment in blood yields identification of approximately 300 of the most abundant proteins [13], which may not be sufficiently sensitive (Figure 1).

Two main approaches have been used to increase the sensitivity of data-dependent approaches. In the first, samples are prefractionated offline, simplifying the injection mixture and spreading out spectra to decrease the impact of peptide masking from abundant peptides [30]. In unlabeled experiments, this can lead to quantification difficulties, as normalizing across multiple injections is complex. To get around this issue, individual samples can be labeled using a sample-specific isobaric tag (TMT or iTRAQ) [41–43]. Tagging results in coelution of isobaric precursors from all multiplexed samples that can then be assigned to individual samples at the MS2 fragment stage. Peptides are quantified at the MS2 level, and a relative abundance is obtained for each peptide in each sample, removing the need to normalize across injections. While the sensitivity of this technique to small fold changes is high, large fold changes may be compressed [44–46]. This approach improves the overall depth of the experiment to an extent determined by the number of offline fractions run [42,47] but is not always sensitive enough to detect proteins only found in a small number of the multiplexed samples. In their proof-of-principle paper, Russell et al. [48] leveraged this potential weakness by combining CSF samples with microglial cell line (BV2) lysate samples to improve detection of immune related proteins, which are low abundance and generally difficult to detect by LC-MS/MS in CSF. Presence of strong MS1 spectra driven by the BV2 cell calibrator drives data-dependent MS2 level acquisition, allowing for quantification of peptides that would not normally be acquired in CSF samples alone. Forty-one proteins that had not previously been identified in CSF were found to differ in abundance between AD and control subjects. The utility of this approach to drive acquisition of data from low-abundance CNS-derived proteins in plasma should be tested.

The second approach to increasing the sensitivity of data-dependent experiments is to deplete samples of the most abundant proteins to decrease interference from these proteins. The standard approach is immunodepletion, using immobilized antibodies to remove abundant proteins from the biofluid sample. While this technique does increase the sensitivity to a subset of lower abundance proteins, nonspecific interactions between the immunodepletion matrix and specific protein–protein interactions between the depletion targets and other proteins can lead to off-target depletion of proteins [49,50]. Therefore, it is important to run pilot experiments or search publicly available data to assess the effect that immunodepletion may have on particular proteins of interest. In plasma, where the dominance by abundant proteins is more extreme, Keshishian et al. [51] reported using a super depletion technique (of approximately 60 of the most abundant proteins) that was combined with isobaric labeling and offline fractionation to confidently identify over 5000 proteins in plasma samples, highlighting several novel candidates for detecting early myocardial infarction. While these approaches may prove useful in discovery experiments, it is likely that such a procedure would introduce much variation and be too costly for routine clinical or large-scale research use.

2.2. Targeted LC-MS/MS Acquisition

If an investigator already has an analytes(s) of interest, then a targeted approach such as selected reaction monitoring (SRM) [52,53] or parallel/multiple reaction monitoring (PRM/MRM) [54–56] may be the preferred approach. These methods quantify at the MS2 level, allowing for better precision and more accurate peptide quantification than data-dependent methods [57]. From a user perspective, the main difference between SRM and PRM is the number of peptides that can be quantified [58]. In SRM, each precursor-fragment pair ("transition") must be independently scanned for quantification, whereas in PRM, all fragments from the same precursor are simultaneously scanned, allowing quantification of a greater number of targets. Work-up time is also therefore shorter for PRM, as individual transitions do not need to be manually selected [59,60]. Scheduling (looking for a precursor only at a specific retention time range) can increase the number of targets included in either method but may lead to missing data in cases where there is significant LC drift. In both methods, it is best to use data-dependent acquired libraries generated on the same LC setup and instrument that the targeted methods will be performed on to begin the precursor and fragment selection process. Due to the lower number of targets quantified, targeted experiments, particularly those using SRM, are often performed using heavy labeled standards, and as a result, are currently seen as the gold standard in LC/MS-MS quantification of proteins, lipids, and metabolites [54].

2.3. Data-Independent Acquisition

Data-independent acquisition (DIA/"Sequential Window Acquisition of All Theoretical Spectra" (SWATH)) sits at the intersection between data-dependent acquisition (DDA) and targeted approaches [61,62]. In a DIA method, acquisition is untargeted, with data acquired from tiled fragment scans that together span the whole mass/charge range. Each tile is repeated every instrument cycle, which allows for repeat measures and quantification for each MS2 fragment. Tiling of fragment scans results in a greater sensitivity than DDA approaches, allowing for higher throughput and shorter LC elution gradients. DIA is intermediate in accuracy between DDA and targeted methods and requires no advance work up [63,64]. Instead, data can be manually curated postacquisition, and removing poor quality fragments and peptides (such as those that exhibit interference from other ions) can vastly improve the precision of DIA, bringing it close to targeted methods. The sensitivity of DIA to lower abundance peptides was initially mostly dependent on the quality and depth of the libraries used to deconvolute MS2 data. These libraries can be generated on the instrument by preliminary DDA runs [65,66], but recently, there has been a proliferation in a number of tools that allow high-depth DIA analysis without the need for a comprehensive, user-generated peptide libraries (Spectronaut Pulsar, DIA Umpire [67], PeCan [68], EncyclopeDIA [69]). Scanning with variable size windows and overlapping tiles can also attain smaller but significant improvements in specificity and sensitivity [70,71]. A recent publication from Meier et al. [72] used DIA-like tiling approaches to replace the full m/z scan at the MS1 level, reducing suppression from abundant peptides and increasing ion injection time. Early data suggests this approach may greatly increase the depth of single-shot label-free techniques, allowing quantification of up to 10,000 proteins in an hour-long scan, with sensitivity down to attomolar levels. Fold change sensitivity and performance of this technique across large, multiday experiments is still to be established.

2.4. Candidate Disease Markers from LC-MS/MS Studies

Despite significant improvements in LC-MS/MS technology and an increasing adoption of these techniques, their utility thus far has been limited by low-powered studies, often utilizing pooling strategies that limit the assessment of individual heterogeneity of potential markers. The neurodegenerative disease biofluid biomarker field is currently dominated by studies of AD, with only a handful of studies on other conditions. In a review of LC-MS/MS studies performed in the last five years (see references [3,73] for comprehensive reviews of work prior to this), only a handful of potential targets were highlighted as significant between clinical groups by three or more studies in CSF, and there was no consensus from studies of blood. In plasma, there have been a number of hits in the complement factor cascade pathway but little agreement over which exact components may be dysregulated [74–82]. In CSF, potential targets fell into two main functional categories: neuropeptides (Chromogranin-A, Secretogranin-2, Secretogranin-3, Neurosecretory Protein VGF) and proteins that interact with amyloid precursor protein (APP) or its resulting peptides (Figure 2A). For all these proteins, there were studies that disagreed on the direction of change or that showed no abundance differences between AD and control (Table 1). There are also currently no markers that appear specific to a single neurodegenerative disease. The relatively low power of all of these studies (n per group ranging from 3 to 134 with a substantial right skew; it is also worth noting that the best powered study [83] found only one between-group difference that survived multiple testing correction) combined with differences in approach may account for a large amount of disagreement between studies. Targeted studies with fewer multiple tests are more likely to find significant outcomes, and correction is not always performed appropriately. Because original data is very rarely presented in these studies, it is difficult to re-examine data distributions, the effect of normalization, and assess whether a peptide was borderline significant or highly variable. In the Considerations for Accurate and Reproducible Findings section of this review, we discuss the adoption of minimum reporting standards to ensure improved reproducibility and comparability of future studies.

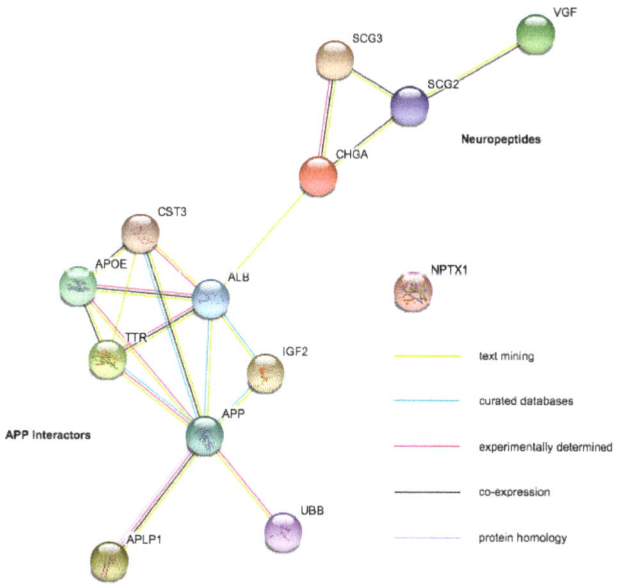

Figure 2. String [84] diagram shows functional protein relationships of proteins highlighted as potential CSF biomarkers of Alzheimer's disease. These proteins currently fall into two main groups: neuropeptides and proteins that interact with amyloid precursor protein (APP, the precursor to beta-amyloid). The type of interaction can be determined from the key in the bottom right. Where peptides from the same protein differ in their significance, the reference is shown in more than one group.

A final reason for the discrepancies in this data may be that many proteins in biofluids exist not as intact peptides but as multiple processed peptides with differing functions, abundance, and stability [85–87]. The existence of these different proteoforms means that protein-level abundance values may vary wildly depending on which peptides are selected or detected in an assay. While targeted methods can be designed towards individual processed peptides to explicitly address this question, untargeted experiments quantified at the protein level only may produce confusing or conflicting results (Table 1). As understanding of the relationship of proteoforms to disease susceptibility increases, it is likely that there will be an expanding need for top-down proteomic methods, where intact peptides can be identified and quantified [88].

Table 1. Summary table showing cross-study results from the proteins illustrated in Figure 2. The arrow shows the direction of change in the neurodegenerative disease compared to controls. PD: Parkinson's Disease, LBD: Lewy Body Dementia, APS: Atypical Parkinsonism, FTD: Frontotemporal Dementia.

Protein	Gene Symbol	Mild Cognitive Impairment	Alzheimer's Disease	Amyotrophic Lateral Sclerosis	Other Diseases
Serum albumin	ALB	↔ [89,90]	↓ [48,91,92] ↑ [48,92] ↔ [89,90]	↔ [93–95]	
Amyloid Beta Precursor Like Protein	APLP1	↑ [96] ↔ [89,90]	↔ [89,90,96–98] ↓ [91] ↑ [98]	↔ [94,95]	↓ PD [98]
Apolipoprotein E	APOE	↓ [89] ↔ [90]	↑ [48,92,99,100] ↔ [83,90,91,97] ↓ [89]	↔ [93–95]	↔ PD [98,99] ↑ LBD [99]
Amyloid Precursor Protein	APP	↔ [90]	↔ [83,89,90,96] ↓ [97]	↔ [93–95]	↔ PD [98,99] ↑ LBD [99] ↓ APS [101]
Chromogranin A	CHGA	↔ [89,90]	↓ [91,97,102] ↔ [89,90,92]	↔ [93–95]	
Chitinase 3 Like 1 (YKL-40)	CHI3L	↔ [89,90]	↑ [90,99,100] ↔ [83,89]	↔ [93,94] ↑ [95]	↔ PD [99] ↑ LBD [99] ↑ FTD [103] ↑ APS [101]
Cystatin-C	CST3	↔ [89,90]	↓ [102] ↑ [92,99,100] ↔ [89–91,97]	↔ [93,95] ↓ [94]	↔ PD [98,99] ↑ LBD [99]
Insulin Like Growth Factor-2	IGF2	↔ [89]	↑ [99,100] ↔ [89]	↓ [93] ↔ [95]	↔ PD [99] ↑ LBD [99]
Neuronal Pentraxin 1	NPTX1	↓ [89] ↔ [96]	↓ [89,102] ↔ [83,96]	↔ [93–95]	↔ PD [98] ↓ APS [101]
Secretogranin-2	SCG2	↔ [96]	↓ [91,102] ↔ [83,96]	↔ [93,95] ↓ [94]	↓ APS [101]
Secretogranin-3	SCG3	↔ [89,96]	↔ [83,89,96] ↓ [91,97] ↑ [48]	↔ [93–95]	↓ APS [101]
Transthyretin	TTR	↑ [89,90]	↑ [90,92,99] ↔ [83,91,97,100]	↔ [93,94]	↔ PD [99] ↔ LBD [99]
Ubiquitin (mono/poly)	UBB		↑ [48,99,104,105] ↔ [83]	↔ [94,95,104]	↔ FTD [104] ↔ APS [105] ↑ LBD [99] ↔ PD [99,104,105]
Neurosecretory Protein VGF	VGF	↔ [89,96]	↓ [91,97,102] ↔ [83,89,96]	↔ [93–95]	↓ APS [101]

3. Capture-Based Strategies

Antibodies have long been the bedrock of protein quantification strategies, particularly in biofluid biomarker development. Antibodies are specific and flexible protein tools that can be easily conjugated to a number of different reporters and immobilized on a variety of matrices, allowing for their use in enzyme-linked immunoassays (ELISAs), Western blotting, and immunohistochemistry. Here, we focus on recent technological developments that allow for multiplexing of targets on ELISA-like platforms and ultrasensitive protein quantification, which may prove exceptionally useful in the detection of very low levels of CNS specific proteins in blood-derived biofluids. The reliance on antibodies for these techniques may result in problems, however [106]. The process of antibody production, particularly for polyclonal antibodies, can be subjected to large batch variation in antibody specificity. Antibody specificity can be difficult to test in human biofluids, where knockdown of a protein is not possible. Antibodies are commonly tested for cross-reactivity with spiked-in proteins that are structurally similar to the target, but nonspecificity can be difficult to predict and this approach is not exhaustive. It is therefore of critical importance to keep comprehensive documentation of lot numbers and batch numbers when performing antibody-based proteomic experiments to monitor potential unexpected causes of variation.

3.1. Multiplexed Immunoassays

Although conventional colorimetric ELISA methods have remained the primary workhorse for measuring biomarker levels in biofluids, the emergence of electrochemiluminescent (ECL) immunoassay technology has allowed for the simultaneous measurement of multiple analytes across a broad dynamic range, leading ECL immunoassays to quickly become the new standard in the field [20,21]. ECL immunoassays are similar in workflow to traditional ELISAs. With plate-based immunoassays, such as those developed by MesoScale Discovery (MSD), carbon electrodes are coated with capture antibodies coated onto discrete spots in each plate well to allow multiplexing of up to 10 targets per sample. Secondary detection antibodies are conjugated to ECL labels that emit light when electricity is applied to the electrodes [21]. In contrast to ELISAs, which depend on developing colorimetric substrates over time, ECL immunoassays have heightened sensitivity with the application of multiple excitation cycles, which amplifies light intensity at lower levels and improves the signal-to-background ratio, enabling accurate measurements in the low pg/mL range [107]. Elimination of the chemical substrate also allows for more consistent and replicable detection, as ECL signal intensity does not vary over time. The increased sensitivity coupled with multiplexing capabilities allows for reduced sample volumes, lower per sample cost, and decreased processing time [108], which are critical considerations when working with valuable and limited biospecimens such as CSF.

Luminex Multi-Analyte Profiling (xMAP) technology uses color-coded beads bound to capture antibodies in order to multiplex up to 500 targets in a single assay [109,110]. Analytes are quantified by the binding of a biotinylated target-specific detector antibody to a streptavidin-coated fluorescent dye, which then passes through two lasers. The first laser decodes the color-coded bead, while the second quantifies the fluorescence intensity of the associated detector dye. The detection system can be flow based or magnetic based; in the latter, beads are anchored to a specific location by a magnet for imaging. The flow system has a higher multiplexing capability, as immunocomplexes are analyzed individually and sequentially [111].

Both the MSD and Luminex immunoassays run into similar pitfalls as other antibody-based techniques, namely, antibody specificity and cross-reactivity, which restrict the number of multiplexable targets. While Luminex boasts the simultaneous measurement of up to 500 analytes, realistically, it is limited to a panel of approximately 30 targets due to antibody cross-reactivity [112]. Although immunoassays are considered high-throughput for sample quantification, the number of multiplexable targets available through these techniques requires the development of a strong hypothesis in order to be used efficiently. Initial biomarker discovery may be more suited to LC-MS/MS strategies, which can then be extended into an ECL immunoassay approach once a select set of proteins of interest has been identified.

3.2. Adaptations of Standard Capture Methods

The shortcomings of antibody-based detection techniques have driven the development of new technologies to detect proteins in biofluids. In an attempt to decrease the influence of nonspecific cross-reactivity, OLink proteomics developed the Proximity Extension Assay (PEA) [113,114]. Instead of using one capture and one labeled detection antibody, complementary DNA oligonucleotides are conjugated to both antibodies. The probes only anneal if both antibodies are bound to the same protein. Quantification is performed by qPCR on annealed oligonucleotides, allowing for multiplexing of up to 92 targets with higher sensitivity than a standard ELISA.

In SOMAscan technology from Somalogic [115], antibodies are entirely replaced by short (20–60 nucleotide) fluorescently labeled DNA Slow Offrate Modified Aptamers (SOMAmers) that can specifically bind over 1100 protein targets. After biotinylation and multiple rounds of washing, aptamers that successfully bind protein targets are bound to a DNA array and quantified by fluorescence intensity. DNA SOMAmers are unlikely to suffer from batch effects as severely as antibodies given they can be easily synthesized, but design and testing of specific probes for thousands of targets requires multiple rounds of optimization and careful quality control procedures. The SOMAscan assay has been shown to have extremely reliable technical reproducibility, with intra- and interplate Coefficients of Variation (CVs) in the ~5% range [19]. As with traditional immunoassays, sources that can introduce variability and contribute to poor (>20%) CVs include dilution factors and proximity to detection limits.

The interpretation of both the Proximity Extension Assay and SOMAscan data is heavily dependent on post-data collection processing algorithms and normalization procedures [19,116]. There are several data treatment methods currently developed for transforming PEA and SOMAscan data, each designed to focus on minimizing a specific source of variability. Differences in data processing can also drastically affect intersite replicability and lead to inconsistencies between reported findings. Standardized data-treatment procedures are necessary in order to ensure concordant interpretation of the data and comparability between study centers.

3.3. Ultrasensitive Detection Methods

In a traditional ELISA, sensitivity to lower abundance analytes is reduced due to the dilution of capture-target-detector complexes (immunocomplexes) in a relatively large liquid volume. The limits of detection are therefore related to the optical sensitivity of the detection system. In novel ultrasensitive methods such as single molecule counting (SMC, EMD Millipore) [22] and single molecule array (Simoa, Quanterix) [17], microfluidic technologies spatially isolate immunocomplexes, allowing for significantly more sensitive detection of low-concentration analytes through counting single molecules. In SMC systems, detector antibodies from immunocomplexes are cleaved off to pass through a laser that excites fluorescent tags, allowing each individual detector to be counted as it passes through. Currently, this technology only allows for measurement of a single analyte. In Simoa, intact immunocomplexes are washed into a bead array, where each immunocomplex occupies a single well. This spatial localization allows for detection of a single immunocomplex on each bead, and coupling with different fluorophores allows for multiplexing of up to six analytes. Although SMC and Simoa technology are still antibody-based techniques and maintain similar matrix interference issues to ELISA immunoassays, increased spatial localization allows an algorithm to model the binding of low-abundance antigens, increasing the dynamic range of the system. Analyte concentrations as low as femtogram/mL can now be quantified, as higher dilution factors can be employed without causing analyte concentrations to fall below the detection limits of the assay.

3.4. Candidate Disease Markes from Capture-Based Studies

The improved sensitivity and the reduced impact of extreme abundance proteins in capture-based studies in comparison to LC-MS/MS techniques has led to their being used to great effect in blood-derived biofluids. Neurofilament light chain (NfL) may prove to be a useful biomarker of overall neurodegeneration ("N"). In both blood and CSF, NfL is elevated in the presence of neuronal damage, although it is not disease specific [18,117]. Although 50 times more concentrated in CSF than in blood, differences in NfL levels between controls and cognitively impaired individuals are still evident in blood. Although NfL data across various platforms tends to be consistent, the measurements do not always perfectly correlate, and in some cases, significant outcomes are only evident on particular platforms [18]. Such variability between platforms is not peculiar to NfL and has been observed for a number of analytes in multiple studies [111,118]. YKL-40 is another emerging biomarker in Alzheimer's disease that shows promise in linking neuroinflammation to neurodegeneration. Concentrations of YKL-40 were significantly elevated in CSF (and more modestly increased in plasma) in individuals across various states of dementia [119,120]. However, YKL-40, like NfL, may be reflective of general neuroinflammation and may not necessarily be disease specific. The lack of agreement between different immunoassay technologies can contribute to mixed findings and discrepancies in reported absolute concentrations, complicating the overall understanding of neurodegenerative diseases at a population level.

Many studies have also proposed panels of various combinations of plasma or serum biomarkers associated with cognitive decline or disease severity that have the potential to profile different aspects of neurodegeneration. Some of the most consistently investigated candidates include proinflammatory cytokine TNF-α, microvascular injury markers ICAM-1 and VCAM-1, and clusterin, an extracellular shuttling protein reported to be associated with Alzheimer's disease progression [7,121–124]. Within the literature, there have been discussions regarding conflicting reports of significant associations between proposed markers and disease staging or differential diagnoses [7,124,125], which are attributed to differences in platforms, methods, data processing, and a lack of standardization and reproducibility. Of particular concern is the general under-reporting of nonsignificant analytes in studies that use large-scale multiplexes such as SOMAscan and antibody array-like methods. By only including data of a small subset of analytes (commonly, those that are found to be the most significant) and not making data on the full range of analytes publicly available, it is impossible to tell which of the remaining analytes were confidently detected but not significantly altered with disease. This is an important distinction, as it can inform whether the analyte may still be of interest as opposed to not reliably quantifiable due to limitations of the technology used.

4. Considerations for Accurate and Reproducible Findings

If the field wishes to discover reliable, quantifiable biomarkers for neurodegenerative dementias, then data from multiple large studies across heterogeneous populations must be comparable. In the final section of this review, we will discuss some technical considerations important for the accuracy of these techniques and recommendations for reporting that will improve our ability to compare data and achieve sufficient sample sizes to draw population-level conclusions above the variability of human samples.

4.1. Preanalytical Effects

In addition to post-data collection processing and platform-specific variability, preanalytical factors can affect the accuracy and reproducibility of measured analytes. The effects of preanalytical factors have already been systematically reviewed [126–128]. Here, we aim to emphasize the importance of standardizing these factors to ensure reliable measurements across multiple centers. Preanalytical factors are divided into two subgroups: in vivo and in vitro factors. These factors include but are not limited to: collection methods and materials, hemolytic contamination of samples, sample handling, storage temperature, thaw conditions, sample stability prior to processing, and kit lot-to-lot variability [129,130]. Much has been written on the importance of collecting and storing CSF only with polypropylene plasticware, as polystyrene or other materials can bind very sticky proteins such as amyloid-β or prion proteins [131,132]. Freeze-thaw cycles (the number of times a stored sample is thawed and refrozen) are often investigated as a cause of protein degradation over repeated uses [133]. Protein integrity varies across analytes and biofluids and maximum acceptable freeze-thaw cycles are specific to each platform, depending on detection sensitivity. Ideally, sample collection methods and times should be strictly controlled to minimize diurnal effects, as well as accounting for possible differences in analyte concentrations between fasting and nonfasting biofluids, which can affect levels of hormones, triglycerides, and other metabolic-pathway-related markers. Levels of certain proteins may vary widely day to day, and thus it is also important to examine the biotemporal stability of an analyte before considering its use as a biomarker [23].

4.2. Matrix Effects

Biofluid composition is also an important consideration when using a multiplex immunoassay system. Matrix effects can negatively impact the ability of highly sensitive immunoassays to accurately quantify certain analytes [134]. As with label-free proteomics techniques, complex matrices with high abundance of albumin and immunoglobulins can affect antibody binding and increase background, masking low-abundance proteins. These low-abundance proteins often approach immunoassay limits of detection, increasing the difficulty of accurate quantification. In some cases, such as with CSF, increasing sample volume may allow for the detection of these low abundance proteins. However, for more complex biofluids, the sample matrix has been found to inhibit detection of certain analytes in spike-recovery experiments, and increasing sample volume would not improve quantification [135]. In a comparison between standards of known concentrations spiked in immunoassay buffer versus serum and plasma matrices, analyte quantification was significantly lower in the presence of either human sample matrix compared to the buffer. This inhibitory effect has been investigated by a number of other studies researching the quantification of low-abundance proteins in complex biofluids [136,137].

These sources of interference in immunoassay detection can lead to misinterpretation of assay results, which can affect clinical or research outcomes. Inhibitory effects may vary between immunoassay detection systems and contribute to inaccurate measurements, increasing the difficulty of comparing quantification across multiple platforms. Due to possible matrix effects, it is generally recommended that the interpretation of analyte quantification in undiluted samples be relative rather than absolute; that is, the measurement should be interpreted in relation to other sample concentrations measured using the same platform. Dilution of samples in immunoassay buffers often improves quantification accuracy by mitigating such matrix effects, resulting in more absolute quantification. When investigating a new immunoassay, it is important to take into consideration possible sources of interference and assess dilution linearity and spike-recovery performance to determine optimal sample conditions. Some assays may not be suited to analyte detection in all matrices, as each sample matrix requires individual optimization. For CSF, dilution factors may be necessary for absolute quantification but can cause analyte measurements to fall below the limit of detection.

4.3. Data Processing

The difficult challenge of how to standardize data comes from the technical aspects of the proteomic workflow. The adoption of different quantification techniques for proteins of variable abundance makes comparison across studies difficult. LC performance can vary substantially over time and can introduce significant variability to an experiment [36]. Simple measures can be taken to improve monitoring of day-to-day instrument variability and demonstrate instrument reliability, such as spiking with retention time calibrators and monitoring of abundant peptides in automatic QC systems like AutoQC in Panorama [138].

How to appropriately normalize data and compare across studies is a more difficult problem with very little consensus, and the field should consider a series of questions. The first regards whether input protein concentration should be normalized before proteomic quantification, as is standard in LC-MS/MS workflows, or whether the same volume of each fluid should be used per assay (as applies to ELISA workflows). The second is whether distribution-based normalization methods (e.g., median or quantile normalization) are appropriate in this context, given that they are based on the assumption that most analytes will not change in abundance between conditions, and that a roughly symmetric proportion of proteins will increase and decrease in abundance. If the integrity of the blood brain barrier is compromised by a neurodegenerative process, this may lead to proteome-wide increases in CSF protein concentration, invalidating the assumption that most proteins will not change in abundance between conditions [139]. Where panels of proteins have been selected on the basis that they are likely to vary between disease conditions, the same assumption is also invalidated and distribution-based normalization may be rendered inappropriate. The alternative approach, to select a subset of "housekeeping" proteins to which to normalize, is also problematic, as a number of studies have shown significant disease-related differences in the abundant biofluid proteins, which would be the most obvious candidates for selection. We would argue that there is currently insufficient high-quality data available to select a panel of normalizing peptides/proteins that may be stable across neurodegenerative conditions, and establishing whether such stable proteins exist should be an additional priority of hypothesis-free proteomic experiments. The current gold standard in quantification and reproducibility, therefore, may be smaller-scale targeted experiments, where ratiometric comparisons to a heavy-labeled standard with proven linearity or a standard curve allowing reporting of a concentration may be the most reliable means of quantification. As this approach does not allow for hypothesis-free discovery, these approaches should be used in replication cohorts for findings that arise from untargeted methods.

4.4. Multisite Variability

It is important to conduct replication studies to assess intersite and interuser variability using the same platform and data-processing methods. Seemingly trivial or unapparent differences in techniques, materials, or environmental conditions can affect results. It is not sufficient to assume that employing the same sample-processing procedures, the same multiplex assay kits or LC setup, and standardized data reporting will necessarily eliminate variability. In an extensive multisite study involving six different labs, Breen et al. [118] found that each analyte measured showed at least one significant lab or assay lot-to-lot effect despite following a consensus protocol across all sites. Care should be taken to establish systems of determining assay reproducibility, such as including standardized plate-to-plate controls to minimize plate effects across multiple sites and batch ordering assays to ensure lot consistency. Even so, controlling for every source of variability and assessing the performance of all available technologies and platforms is often unrealistic due to financial and resource limitations.

5. Future Directions

Proteomics is a relatively new and rapidly growing field and has yet to develop clear standards for reporting data and consistent methods to allow for confident comparison of datasets. The complexity of and similarity between neurodegenerative diseases means that studies of large, diverse populations are required to define biomarkers that are both sensitive and specific. It is therefore of critical importance that the field as a whole adopts stringent and detailed reporting criteria to build knowledge on a scale that will help delineate and stratify subjects across populations in a biologically informative manner. While proteomic-specific journals have begun to adopt set reporting criteria, clinical journals do not generally require this level of detail, and the field suffers as a result. At a bare minimum, a data table that includes every peptide and/or protein confidently detected in each proteomic experiment (including retention time and mz data for LC-MS/MS), abundance in each individual sample, and per group summary statistics should be provided for every study. A list of significantly changed proteins with a fold change and p/q value is not sufficient for thorough examination of the data. As a field, a decision should be made to use a standardized protein reference, as switching between Uniprot IDs [140], gene names, and other reference formats often leads to errors and data loss. We propose the use of both the Ensembl gene ID [141], which is clearly linked to genomic locus and reference version, and a more descriptive gene ID such as the gene symbol for ease of understanding results. Similarly, clinical and demographic data should be provided on an individual subject level to allow for modeling of age, sex, and other important demographic variables. The development and adoption of user-friendly resources such as the CSF Proteome Resource and Plasma Proteome Database [14,142] to allow for cross-study comparison is also critically important. Adoption of standards along these lines will likely lead to leaps forward in the biomarker discovery pipeline equivalent to the speed at which the discovery technology is improving.

Funding: B.C.C., B.A.T., and S.E.A. are funded by grants from the NIH (AG059856, AG005134, AG039478, AG059856) and a gift from the Challenger Foundation.

Conflicts of Interest: The authors declare no conflict of interest.

References

1. Jack, C.R.; Bennett, D.A.; Blennow, K.; Carrillo, M.C.; Dunn, B.; Haeberlein, S.B.; Holtzman, D.M.; Jagust, W.; Jessen, F.; Karlawish, J.; et al. NIA-AA Research Framework: Toward a biological definition of Alzheimer's disease. *Alzheimer's Dement.* **2018**, *14*, 535–562. [CrossRef] [PubMed]
2. Sheikh-Bahaei, N.; Sajjadi, S.A.; Manavaki, R.; Gillard, J.H. Imaging Biomarkers in Alzheimer's Disease: A Practical Guide for Clinicians. *J. Alzheimer's Dis. Rep.* **2017**, *1*, 71–88. [CrossRef]
3. Blennow, K.; Zetterberg, H.; Fagan, A.M. Fluid biomarkers in Alzheimer disease. *Cold Spring Harb. Perspect. Med.* **2012**, *2*, a006221. [CrossRef] [PubMed]
4. Lewczuk, P.; Riederer, P.; O'Bryant, S.E.; Verbeek, M.M.; Dubois, B.; Visser, P.J.; Jellinger, K.A.; Engelborghs, S.; Ramirez, A.; Parnetti, L.; et al. Cerebrospinal fluid and blood biomarkers for neurodegenerative dementias: An update of the Consensus of the Task Force on Biological Markers in Psychiatry of the World Federation of Societies of Biological Psychiatry. *World J. Biol. Psychiatry* **2018**, *19*, 244–328. [CrossRef] [PubMed]
5. Spector, R.; Robert Snodgrass, S.; Johanson, C.E. A balanced view of the cerebrospinal fluid composition and functions: Focus on adult humans. *Exp. Neurol.* **2015**, *273*, 57–68. [CrossRef] [PubMed]
6. Zetterberg, H. Applying fluid biomarkers to Alzheimer's disease. *Am. J. Physiol.-Cell Physiol.* **2017**, *313*, C3–C10. [CrossRef] [PubMed]
7. Snyder, H.M.; Carrillo, M.C.; Grodstein, F.; Henriksen, K.; Jeromin, A.; Lovestone, S.; Mielke, M.M.; O'Bryant, S.; Sarasa, M.; Sjøgren, M.; et al. Developing novel blood-based biomarkers for Alzheimer's disease. *Alzheimers Dement.* **2014**, *10*, 109–114. [CrossRef] [PubMed]
8. Engelborghs, S.; Niemantsverdriet, E.; Struyfs, H.; Blennow, K.; Brouns, R.; Comabella, M.; Dujmovic, I.; van der Flier, W.; Frölich, L.; Galimberti, D.; et al. Consensus guidelines for lumbar puncture in patients with neurological diseases. *Alzheimer's Dement.* **2017**, *8*, 111–126. [CrossRef] [PubMed]
9. Duits, F.H.; Martinez-Lage, P.; Paquet, C.; Engelborghs, S.; Lleó, A.; Hausner, L.; Molinuevo, J.L.; Stomrud, E.; Farotti, L.; Ramakers, I.H.; et al. Performance and complications of lumbar puncture in memory clinics: Results of the multicenter lumbar puncture feasibility study. *Alzheimer's Dement.* **2016**, *12*, 154–163. [CrossRef] [PubMed]
10. Shi, L.; Baird, A.L.; Westwood, S.; Hye, A.; Dobson, R.; Thambisetty, M.; Lovestone, S. A Decade of Blood Biomarkers for Alzheimer's Disease Research: An Evolving Field, Improving Study Designs, and the Challenge of Replication. *J. Alzheimer's Dis.* **2018**, *62*, 1181–1198. [CrossRef] [PubMed]
11. Kusminski, C.M.; McTernan, P.G.; Schraw, T.; Kos, K.; O'Hare, J.P.; Ahima, R.; Kumar, S.; Scherer, P.E. Adiponectin complexes in human cerebrospinal fluid: Distinct complex distribution from serum. *Diabetologia* **2007**, *50*, 634–642. [CrossRef] [PubMed]
12. Mattsson, N.; Zetterberg, H.; Janelidze, S.; Insel, P.S.; Andreasson, U.; Stomrud, E.; Palmqvist, S.; Baker, D.; Tan Hehir, C.A.; Jeromin, A.; et al. ADNI Investigators Plasma tau in Alzheimer disease. *Neurology* **2016**, *87*, 1827–1835. [CrossRef] [PubMed]
13. Geyer, P.E.; Holdt, L.M.; Teupser, D.; Mann, M. Revisiting biomarker discovery by plasma proteomics. *Mol. Syst. Biol.* **2017**, *13*, 942. [CrossRef] [PubMed]

14. Nanjappa, V.; Thomas, J.K.; Marimuthu, A.; Muthusamy, B.; Radhakrishnan, A.; Sharma, R.; Ahmad Khan, A.; Balakrishnan, L.; Sahasrabuddhe, N.A.; Kumar, S.; et al. Plasma Proteome Database as a resource for proteomics research: 2014 update. *Nucleic Acids Res.* **2014**, *42*, D959–D965. [CrossRef] [PubMed]
15. Anderson, N.L.; Anderson, N.G. The human plasma proteome: History, character, and diagnostic prospects. *Mol. Cell. Proteom.* **2002**, *1*, 845–867. [CrossRef]
16. Mann, M.; Kulak, N.A.; Nagaraj, N.; Cox, J. The coming age of complete, accurate, and ubiquitous proteomes. *Mol. Cell* **2013**, *49*, 583–590. [CrossRef] [PubMed]
17. Wilson, D.H.; Rissin, D.M.; Kan, C.W.; Fournier, D.R.; Piech, T.; Campbell, T.G.; Meyer, R.E.; Fishburn, M.W.; Cabrera, C.; Patel, P.P.; et al. The Simoa HD-1 Analyzer. *J. Lab. Autom.* **2016**, *21*, 533–547. [CrossRef] [PubMed]
18. Kuhle, J.; Barro, C.; Andreasson, U.; Derfuss, T.; Lindberg, R.; Sandelius, Å.; Liman, V.; Norgren, N.; Blennow, K.; Zetterberg, H. Comparison of three analytical platforms for quantification of the neurofilament light chain in blood samples: ELISA, electrochemiluminescence immunoassay and Simoa. *Clin. Chem. Lab. Med.* **2016**, *54*, 1655–1661. [CrossRef] [PubMed]
19. Candia, J.; Cheung, F.; Kotliarov, Y.; Fantoni, G.; Sellers, B.; Griesman, T.; Huang, J.; Stuccio, S.; Zingone, A.; Ryan, B.M.; et al. Assessment of Variability in the SOMAscan Assay. *Sci. Rep.* **2017**, *7*, 14248. [CrossRef] [PubMed]
20. Blackburn, G.F.; Shah, H.P.; Kenten, J.H.; Leland, J.; Kamin, R.A.; Link, J.; Peterman, J.; Powell, M.J.; Shah, A.; Talley, D.B. Electrochemiluminescence detection for development of immunoassays and DNA probe assays for clinical diagnostics. *Clin. Chem.* **1991**, *37*, 1534–1539. [PubMed]
21. Gross, E.M.; Maddipati, S.S.; Snyder, S.M. A review of electrogenerated chemiluminescent biosensors for assays in biological matrices. *Bioanalysis* **2016**, *8*, 2071–2089. [CrossRef] [PubMed]
22. Ledger, K.S.; Agee, S.J.; Kasaian, M.T.; Forlow, S.B.; Durn, B.L.; Minyard, J.; Lu, Q.A.; Todd, J.; Vesterqvist, O.; Burczynski, M.E. Analytical validation of a highly sensitive microparticle-based immunoassay for the quantitation of IL-13 in human serum using the Erenna® immunoassay system. *J. Immunol. Methods* **2009**, *350*, 161–170. [CrossRef] [PubMed]
23. Trombetta, B.A.; Carlyle, B.C.; Koenig, A.M.; Shaw, L.M.; Trojanowski, J.Q.; Wolk, D.A.; Locascio, J.J.; Arnold, S.E. The technical reliability and biotemporal stability of cerebrospinal fluid biomarkers for profiling multiple pathophysiologies in Alzheimer's disease. *PLoS ONE* **2018**, *13*, e0193707. [CrossRef] [PubMed]
24. Ren, Y.; Zhu, W.; Cui, F.; Yang, F.; Chen, Z.; Ling, L.; Huang, X. Measurement of cystatin C levels in the cerebrospinal fluid of patients with amyotrophic lateral sclerosis. *Int. J. Clin. Exp. Pathol.* **2015**, *8*, 5419–5426. [PubMed]
25. Blennow, K.; Davidsson, P.; Wallin, A.; Ekman, R. Chromogranin A in cerebrospinal fluid: A biochemical marker for synaptic degeneration in Alzheimer's disease? *Dementia* **1995**, *6*, 306–311. [CrossRef] [PubMed]
26. Percy, A.J.; Yang, J.; Chambers, A.G.; Simon, R.; Hardie, D.B.; Borchers, C.H. Multiplexed MRM with Internal Standards for Cerebrospinal Fluid Candidate Protein Biomarker Quantitation. *J. Proteome Res.* **2014**, *13*, 3733–3747. [CrossRef] [PubMed]
27. Xie, F.; Liu, T.; Qian, W.-J.; Petyuk, V.A.; Smith, R.D. Liquid chromatography-mass spectrometry-based quantitative proteomics. *J. Biol. Chem.* **2011**, *286*, 25443–25449. [CrossRef] [PubMed]
28. Karpievitch, Y.V.; Polpitiya, A.D.; Anderson, G.A.; Smith, R.D.; Dabney, A.R. Liquid Chromatography Mass Spectrometry-Based Proteomics: Biological and Technological Aspects. *Ann. Appl. Stat.* **2010**, *4*, 1797–1823. [CrossRef] [PubMed]
29. Drabik, A. Quantitative Measurements in Proteomics: Mass Spectrometry. *Proteom. Profiling Anal. Chem.* **2016**, 145–160. [CrossRef]
30. Mostovenko, E.; Hassan, C.; Rattke, J.; Deelder, A.M.; van Veelen, P.A.; Palmblad, M. Comparison of peptide and protein fractionation methods in proteomics. *EuPA Open Proteom.* **2013**, *1*, 30–37. [CrossRef]
31. Bauer, M.; Ahrné, E.; Baron, A.P.; Glatter, T.; Fava, L.L.; Santamaria, A.; Nigg, E.A.; Schmidt, A. Assessment of current mass spectrometric workflows for the quantification of low abundant proteins and phosphorylation sites. *Data Br.* **2015**, *5*, 297–304. [CrossRef] [PubMed]
32. Bubis, J.A.; Levitsky, L.I.; Ivanov, M.V.; Tarasova, I.A.; Gorshkov, M.V. Comparative evaluation of label-free quantification methods for shotgun proteomics. *Rapid Commun. Mass Spectrom.* **2017**, *31*, 606–612. [CrossRef] [PubMed]

33. Wilm, M.; Shevchenko, A.; Houthaeve, T.; Breit, S.; Schweigerer, L.; Fotsis, T.; Mann, M. Femtomole sequencing of proteins from polyacrylamide gels by nano-electrospray mass spectrometry. *Nature* **1996**, *379*, 466–469. [CrossRef] [PubMed]
34. Link, A.J.; Eng, J.; Schieltz, D.M.; Carmack, E.; Mize, G.J.; Morris, D.R.; Garvik, B.M.; Yates, J.R. Direct analysis of protein complexes using mass spectrometry. *Nat. Biotechnol.* **1999**, *17*, 676–682. [CrossRef] [PubMed]
35. Aebersold, R.; Mann, M. Mass spectrometry-based proteomics. *Nature* **2003**, *422*, 198–207. [CrossRef] [PubMed]
36. Rudnick, P.A.; Clauser, K.R.; Kilpatrick, L.E.; Tchekhovskoi, D.V.; Neta, P.; Blonder, N.; Billheimer, D.D.; Blackman, R.K.; Bunk, D.M.; Cardasis, H.L.; et al. Performance metrics for liquid chromatography-tandem mass spectrometry systems in proteomics analyses. *Mol. Cell. Proteom.* **2010**, *9*, 225–241. [CrossRef] [PubMed]
37. Tabb, D.L.; Vega-Montoto, L.; Rudnick, P.A.; Variyath, A.M.; Ham, A.-J.L.; Bunk, D.M.; Kilpatrick, L.E.; Billheimer, D.D.; Blackman, R.K.; Cardasis, H.L.; et al. Repeatability and Reproducibility in Proteomic Identifications by Liquid Chromatography-Tandem Mass Spectrometry. *J. Proteome Res.* **2010**, *9*, 761–776. [CrossRef] [PubMed]
38. Sandberg, A.; Branca, R.M.M.; Lehtiö, J.; Forshed, J. Quantitative accuracy in mass spectrometry based proteomics of complex samples: The impact of labeling and precursor interference. *J. Proteom.* **2014**, *96*, 133–144. [CrossRef] [PubMed]
39. Michalski, A.; Cox, J.; Mann, M. More than 100,000 Detectable Peptide Species Elute in Single Shotgun Proteomics Runs but the Majority is Inaccessible to Data-Dependent LC−MS/MS. *J. Proteome Res.* **2011**, *10*, 1785–1793. [CrossRef] [PubMed]
40. Carlyle, B.C.; Kitchen, R.R.; Kanyo, J.E.; Voss, E.Z.; Pletikos, M.; Sousa, A.M.M.; Lam, T.T.; Gerstein, M.B.; Sestan, N.; Nairn, A.C. A multiregional proteomic survey of the postnatal human brain. *Nat. Neurosci.* **2017**, *20*, 1787–1795. [CrossRef] [PubMed]
41. Westbrook, J.A.; Noirel, J.; Brown, J.E.; Wright, P.C.; Evans, C.A. Quantitation with chemical tagging reagents in biomarker studies. *PROTEOMICS—Clin. Appl.* **2015**, *9*, 295–300. [CrossRef] [PubMed]
42. Lapek, J.D.; Greninger, P.; Morris, R.; Amzallag, A.; Pruteanu-Malinici, I.; Benes, C.H.; Haas, W. Detection of dysregulated protein-association networks by high-throughput proteomics predicts cancer vulnerabilities. *Nat. Biotechnol.* **2017**, *35*, 983–989. [CrossRef] [PubMed]
43. Gygi, S.P.; Rist, B.; Gerber, S.A.; Turecek, F.; Gelb, M.H.; Aebersold, R. Quantitative analysis of complex protein mixtures using isotope-coded affinity tags. *Nat. Biotechnol.* **1999**, *17*, 994–999. [CrossRef] [PubMed]
44. Latosinska, A.; Vougas, K.; Makridakis, M.; Klein, J.; Mullen, W.; Abbas, M.; Stravodimos, K.; Katafigiotis, I.; Merseburger, A.S.; Zoidakis, J.; et al. Comparative Analysis of Label-Free and 8-Plex iTRAQ Approach for Quantitative Tissue Proteomic Analysis. *PLoS ONE* **2015**, *10*, e0137048. [CrossRef] [PubMed]
45. Li, Z.; Adams, R.M.; Chourey, K.; Hurst, G.B.; Hettich, R.L.; Pan, C. Systematic Comparison of Label-Free, Metabolic Labeling, and Isobaric Chemical Labeling for Quantitative Proteomics on LTQ Orbitrap Velos. *J. Proteome Res.* **2012**, *11*, 1582–1590. [CrossRef] [PubMed]
46. Wang, H.; Alvarez, S.; Hicks, L.M. Comprehensive Comparison of iTRAQ and Label-free LC-Based Quantitative Proteomics Approaches Using Two *Chlamydomonas reinhardtii* Strains of Interest for Biofuels Engineering. *J. Proteome Res.* **2012**, *11*, 487–501. [CrossRef] [PubMed]
47. Paulo, J.A.; Gygi, S.P. A comprehensive proteomic and phosphoproteomic analysis of yeast deletion mutants of 14-3-3 orthologs and associated effects of rapamycin. *Proteomics* **2015**, *15*, 474–486. [CrossRef] [PubMed]
48. Russell, C.L.; Heslegrave, A.; Mitra, V.; Zetterberg, H.; Pocock, J.M.; Ward, M.A.; Pike, I. Combined tissue and fluid proteomics with Tandem Mass Tags to identify low-abundance protein biomarkers of disease in peripheral body fluid: An Alzheimer's Disease case study. *Rapid Commun. Mass Spectrom.* **2017**, *31*, 153–159. [CrossRef] [PubMed]
49. Bellei, E.; Bergamini, S.; Monari, E.; Fantoni, L.I.; Cuoghi, A.; Ozben, T.; Tomasi, A. High-abundance proteins depletion for serum proteomic analysis: Concomitant removal of non-targeted proteins. *Amino Acids* **2011**, *40*, 145–156. [CrossRef] [PubMed]
50. Günther, R.; Krause, E.; Schümann, M.; Blasig, I.E.; Haseloff, R.F. Depletion of highly abundant proteins from human cerebrospinal fluid: A cautionary note. *Mol. Neurodegener.* **2015**, *10*, 53. [CrossRef] [PubMed]

51. Keshishian, H.; Burgess, M.W.; Gillette, M.A.; Mertins, P.; Clauser, K.R.; Mani, D.R.; Kuhn, E.W.; Farrell, L.A.; Gerszten, R.E.; Carr, S.A. Multiplexed, Quantitative Workflow for Sensitive Biomarker Discovery in Plasma Yields Novel Candidates for Early Myocardial Injury. *Mol. Cell. Proteom.* **2015**, *14*, 2375–2393. [CrossRef] [PubMed]
52. Kuhn, E.; Wu, J.; Karl, J.; Liao, H.; Zolg, W.; Guild, B. Quantification of C-reactive protein in the serum of patients with rheumatoid arthritis using multiple reaction monitoring mass spectrometry and ^{13}C-labeled peptide standards. *Proteomics* **2004**, *4*, 1175–1186. [CrossRef] [PubMed]
53. Picotti, P.; Bodenmiller, B.; Mueller, L.N.; Domon, B.; Aebersold, R. Full Dynamic Range Proteome Analysis of S. cerevisiae by Targeted Proteomics. *Cell* **2009**, *138*, 795–806. [CrossRef] [PubMed]
54. Gillette, M.A.; Carr, S.A. Quantitative analysis of peptides and proteins in biomedicine by targeted mass spectrometry. *Nat. Methods* **2013**, *10*, 28–34. [CrossRef] [PubMed]
55. Zhu, X.; Desiderio, D.M. Peptide quantification by tandem mass spectrometry. *Mass Spectrom. Rev.* **1996**, *15*, 213–240. [CrossRef]
56. Peterson, A.C.; Russell, J.D.; Bailey, D.J.; Westphall, M.S.; Coon, J.J. Parallel reaction monitoring for high resolution and high mass accuracy quantitative, targeted proteomics. *Mol. Cell. Proteom.* **2012**, *11*, 1475–1488. [CrossRef] [PubMed]
57. Liebler, D.C.; Zimmerman, L.J. Targeted quantitation of proteins by mass spectrometry. *Biochemistry* **2013**, *52*, 3797–3806. [CrossRef] [PubMed]
58. Rauniyar, N. Parallel Reaction Monitoring: A Targeted Experiment Performed Using High Resolution and High Mass Accuracy Mass Spectrometry. *Int. J. Mol. Sci.* **2015**, *16*, 28566–28581. [CrossRef] [PubMed]
59. Lange, V.; Picotti, P.; Domon, B.; Aebersold, R. Selected reaction monitoring for quantitative proteomics: A tutorial. *Mol. Syst. Biol.* **2008**, *4*, 222. [CrossRef] [PubMed]
60. Gallien, S.; Duriez, E.; Crone, C.; Kellmann, M.; Moehring, T.; Domon, B. Targeted proteomic quantification on quadrupole-orbitrap mass spectrometer. *Mol. Cell. Proteom.* **2012**, *11*, 1709–1723. [CrossRef] [PubMed]
61. Frederick, K. SWATH-MS: Data Acquisition and Analysis. *Proteom. Profiling Anal. Chem.* **2016**, 161–173. [CrossRef]
62. Gillet, L.C.; Navarro, P.; Tate, S.; Röst, H.; Selevsek, N.; Reiter, L.; Bonner, R.; Aebersold, R. Targeted Data Extraction of the MS/MS Spectra Generated by Data-independent Acquisition: A New Concept for Consistent and Accurate Proteome Analysis. *Mol. Cell. Proteom.* **2012**, *11*, O111–016717. [CrossRef] [PubMed]
63. Hu, A.; Noble, W.S.; Wolf-Yadlin, A. Technical advances in proteomics: New developments in data-independent acquisition. *F1000Research* **2016**, *5*. [CrossRef] [PubMed]
64. Shi, T.; Song, E.; Nie, S.; Rodland, K.D.; Liu, T.; Qian, W.-J.; Smith, R.D. Advances in targeted proteomics and applications to biomedical research. *Proteomics* **2016**, *16*, 2160–2182. [CrossRef] [PubMed]
65. Krasny, L.; Bland, P.; Kogata, N.; Wai, P.; Howard, B.A.; Natrajan, R.C.; Huang, P.H. SWATH mass spectrometry as a tool for quantitative profiling of the matrisome. *J. Proteome* **2018**. [CrossRef] [PubMed]
66. Bruderer, R.; Bernhardt, O.M.; Gandhi, T.; Miladinović, S.M.; Cheng, L.-Y.; Messner, S.; Ehrenberger, T.; Zanotelli, V.; Butscheid, Y.; Escher, C.; et al. Extending the limits of quantitative proteome profiling with data-independent acquisition and application to acetaminophen-treated three-dimensional liver microtissues. *Mol. Cell. Proteom.* **2015**, *14*, 1400–1410. [CrossRef] [PubMed]
67. Tsou, C.-C.; Avtonomov, D.; Larsen, B.; Tucholska, M.; Choi, H.; Gingras, A.-C.; Nesvizhskii, A.I. DIA-Umpire: Comprehensive computational framework for data-independent acquisition proteomics. *Nat. Methods* **2015**, *12*, 258–264. [CrossRef] [PubMed]
68. Ting, Y.S.; Egertson, J.D.; Bollinger, J.G.; Searle, B.C.; Payne, S.H.; Noble, W.S.; MacCoss, M.J. PECAN: Library-free peptide detection for data-independent acquisition tandem mass spectrometry data. *Nat. Methods* **2017**, *14*, 903–908. [CrossRef] [PubMed]
69. Searle, B.C.; Pino, L.K.; Egertson, J.D.; Ting, Y.S.; Lawrence, R.T.; Villen, J.; MacCoss, M.J. Comprehensive peptide quantification for data independent acquisition mass spectrometry using chromatogram libraries. *bioRxiv* **2018**, 277822. [CrossRef]
70. Schilling, B.; Gibson, B.W.; Hunter, C.L. Generation of High-Quality SWATH® Acquisition Data for Label-free Quantitative Proteomics Studies Using TripleTOF® Mass Spectrometers. *Methods Mol. Biol.* **2017**, *1550*, 223–233. [CrossRef] [PubMed]

71. Zhang, Y.; Bilbao, A.; Bruderer, T.; Luban, J.; Strambio-De-Castillia, C.; Lisacek, F.; Hopfgartner, G.; Varesio, E. The Use of Variable Q1 Isolation Windows Improves Selectivity in LC–SWATH–MS Acquisition. *J. Proteome Res.* **2015**, *14*, 4359–4371. [CrossRef] [PubMed]
72. Meier, F.; Geyer, P.E.; Virreira Winter, S.; Cox, J.; Mann, M. BoxCar acquisition method enables single-shot proteomics at a depth of 10,000 proteins in 100 minutes. *Nat. Methods* **2018**. [CrossRef] [PubMed]
73. Olsson, B.; Lautner, R.; Andreasson, U.; Öhrfelt, A.; Portelius, E.; Bjerke, M.; Hölttä, M.; Rosén, C.; Olsson, C.; Strobel, G.; et al. CSF and blood biomarkers for the diagnosis of Alzheimer's disease: A systematic review and meta-analysis. *Lancet Neurol.* **2016**, *15*, 673–684. [CrossRef]
74. Muenchhoff, J.; Poljak, A.; Song, F.; Raftery, M.; Brodaty, H.; Duncan, M.; McEvoy, M.; Attia, J.; Schofield, P.W.; Sachdev, P.S. Plasma protein profiling of mild cognitive impairment and Alzheimer's disease across two independent cohorts. *J. Alzheimer's Dis.* **2015**, *43*, 1355–1373. [CrossRef] [PubMed]
75. Ashton, N.J.; Kiddle, S.J.; Graf, J.; Ward, M.; Baird, A.L.; Hye, A.; Westwood, S.; Wong, K.V.; Dobson, R.J.; Rabinovici, G.D.; et al. Blood protein predictors of brain amyloid for enrichment in clinical trials? *Alzheimer's Dement. Diagn. Assess. Dis. Monit.* **2015**, *1*, 48–60. [CrossRef] [PubMed]
76. Westwood, S.; Leoni, E.; Hye, A.; Lynham, S.; Khondoker, M.R.; Ashton, N.J.; Kiddle, S.J.; Baird, A.L.; Sainz-Fuertes, R.; Leung, R.; et al. Blood-Based Biomarker Candidates of Cerebral Amyloid Using PiB PET in Non-Demented Elderly. *J. Alzheimer's Dis.* **2016**, *52*, 561–572. [CrossRef] [PubMed]
77. Song, F.; Poljak, A.; Kochan, N.A.; Raftery, M.; Brodaty, H.; Smythe, G.A.; Sachdev, P.S. Plasma protein profiling of Mild Cognitive Impairment and Alzheimer's disease using iTRAQ quantitative proteomics. *Proteome Sci.* **2014**, *12*, 5. [CrossRef] [PubMed]
78. Zabel, M.; Schrag, M.; Mueller, C.; Zhou, W.; Crofton, A.; Petersen, F.; Dickson, A.; Kirsch, W.M. Assessing Candidate Serum Biomarkers for Alzheimer's Disease: A Longitudinal Study. *J. Alzheimer's Dis.* **2012**, *30*, 311–321. [CrossRef] [PubMed]
79. Bennett, S.; Grant, M.; Creese, A.J.; Mangialasche, F.; Cecchetti, R.; Cooper, H.J.; Mecocci, P.; Aldred, S. Plasma Levels of Complement 4a Protein are Increased in Alzheimer's Disease. *Alzheimer Dis. Assoc. Disord.* **2012**, *26*, 329–334. [CrossRef] [PubMed]
80. Xu, Z.; Lee, A.; Nouwens, A.; Henderson, R.D.; McCombe, P.A. Mass spectrometry analysis of plasma from amyotrophic lateral sclerosis and control subjects. *Amyotroph. Lateral Scler. Frontotemporal Degener.* **2018**, 1–15. [CrossRef] [PubMed]
81. Suzuki, I.; Noguchi, M.; Arito, M.; Sato, T.; Omoteyama, K.; Maedomari, M.; Hasegawa, H.; Suematsu, N.; Okamoto, K.; Kato, T.; et al. Serum peptides as candidate biomarkers for dementia with Lewy bodies. *Int. J. Geriatr. Psychiatry* **2015**, *30*, 1195–1206. [CrossRef] [PubMed]
82. Dayon, L.; Wojcik, J.; Núñez Galindo, A.; Corthésy, J.; Cominetti, O.; Oikonomidi, A.; Henry, H.; Migliavacca, E.; Bowman, G.L.; Popp, J. Plasma Proteomic Profiles of Cerebrospinal Fluid-Defined Alzheimer's Disease Pathology in Older Adults. *J. Alzheimer's Dis.* **2017**, *60*, 1641–1652. [CrossRef] [PubMed]
83. Spellman, D.S.; Wildsmith, K.R.; Honigberg, L.A.; Tuefferd, M.; Baker, D.; Raghavan, N.; Nairn, A.C.; Croteau, P.; Schirm, M.; Allard, R.; et al. Development and evaluation of a multiplexed mass spectrometry based assay for measuring candidate peptide biomarkers in Alzheimer's Disease Neuroimaging Initiative (ADNI) CSF. *PROTEOMICS—Clin. Appl.* **2015**, *9*, 715–731. [CrossRef] [PubMed]
84. Szklarczyk, D.; Franceschini, A.; Wyder, S.; Forslund, K.; Heller, D.; Huerta-Cepas, J.; Simonovic, M.; Roth, A.; Santos, A.; Tsafou, K.P.; et al. STRING v10: Protein-protein interaction networks, integrated over the tree of life. *Nucleic Acids Res.* **2015**, *43*, D447–D452. [CrossRef] [PubMed]
85. Vialaret, J.; Schmit, P.-O.; Lehmann, S.; Gabelle, A.; Wood, J.; Bern, M.; Paape, R.; Suckau, D.; Kruppa, G.; Hirtz, C. Identification of multiple proteoforms biomarkers on clinical samples by routine Top-Down approaches. *Data Br.* **2018**, *18*, 1013–1021. [CrossRef] [PubMed]
86. Lehmann, S.; Gabelle, A.; Vialaret, J.; Schmit, P.-O.; Hirtz, C. Profiling of Intact Proteins in the CSF of Alzheimer's Disease Patients using Top Down Clinical Proteomics (TDCP): A New Approach Giving Access to Isoform Specific Information of Neurodegenerative Biomarkers. *Alzheimer's Dement.* **2016**, *12*, P183–P184. [CrossRef]
87. Fania, C.; Arosio, B.; Capitanio, D.; Torretta, E.; Gussago, C.; Ferri, E.; Mari, D.; Gelfi, C. Protein signature in cerebrospinal fluid and serum of Alzheimer's disease patients: The case of apolipoprotein A-1 proteoforms. *PLoS ONE* **2017**, *12*, e0179280. [CrossRef] [PubMed]

88. Schmit, P.-O.; Vialaret, J.; Wessels, H.J.C.T.; van Gool, A.J.; Lehmann, S.; Gabelle, A.; Wood, J.; Bern, M.; Paape, R.; Suckau, D.; et al. Towards a routine application of Top-Down approaches for label-free discovery workflows. *J. Proteom.* **2018**, *175*, 12–26. [CrossRef] [PubMed]
89. Wang, J.; Cunningham, R.; Zetterberg, H.; Asthana, S.; Carlsson, C.; Okonkwo, O.; Li, L. Label-free quantitative comparison of cerebrospinal fluid glycoproteins and endogenous peptides in subjects with Alzheimer's disease, mild cognitive impairment, and healthy individuals. *PROTEOMICS—Clin. Appl.* **2016**, *10*, 1225–1241. [CrossRef] [PubMed]
90. Wildsmith, K.R.; Schauer, S.P.; Smith, A.M.; Arnott, D.; Zhu, Y.; Haznedar, J.; Kaur, S.; Mathews, W.R.; Honigberg, L.A. Identification of longitudinally dynamic biomarkers in Alzheimer's disease cerebrospinal fluid by targeted proteomics. *Mol. Neurodegener.* **2014**, *9*, 22. [CrossRef] [PubMed]
91. Hölttä, M.; Minthon, L.; Hansson, O.; Holmén-Larsson, J.; Pike, I.; Ward, M.; Kuhn, K.; Rüetschi, U.; Zetterberg, H.; Blennow, K.; et al. An Integrated Workflow for Multiplex CSF Proteomics and Peptidomics—Identification of Candidate Cerebrospinal Fluid Biomarkers of Alzheimer's Disease. *J. Proteome Res.* **2015**, *14*, 654–663. [CrossRef] [PubMed]
92. Choi, Y.S.; Hou, S.; Choe, L.H.; Lee, K.H. Targeted human cerebrospinal fluid proteomics for the validation of multiple Alzheimer's disease biomarker candidates. *J. Chromatogr. B* **2013**, *930*, 129–135. [CrossRef] [PubMed]
93. Chen, Y.; Liu, X.-H.; Wu, J.-J.; Ren, H.-M.; Wang, J.; Ding, Z.-T.; Jiang, Y.-P. Proteomic analysis of cerebrospinal fluid in amyotrophic lateral sclerosis. *Exp. Ther. Med.* **2016**, *11*, 2095–2106. [CrossRef] [PubMed]
94. Collins, M.A.; An, J.; Hood, B.L.; Conrads, T.P.; Bowser, R.P. Label-Free LC–MS/MS Proteomic Analysis of Cerebrospinal Fluid Identifies Protein/Pathway Alterations and Candidate Biomarkers for Amyotrophic Lateral Sclerosis. *J. Proteome Res.* **2015**, *14*, 4486–4501. [CrossRef] [PubMed]
95. Thompson, A.G.; Gray, E.; Thézénas, M.-L.; Charles, P.D.; Evetts, S.; Hu, M.T.; Talbot, K.; Fischer, R.; Kessler, B.M.; Turner, M.R. Cerebrospinal fluid macrophage biomarkers in amyotrophic lateral sclerosis. *Ann. Neurol.* **2018**, *83*, 258–268. [CrossRef] [PubMed]
96. Begcevic, I.; Brinc, D.; Brown, M.; Martinez-Morillo, E.; Goldhardt, O.; Grimmer, T.; Magdolen, V.; Batruch, I.; Diamandis, E.P. Brain-related proteins as potential CSF biomarkers of Alzheimer's disease: A targeted mass spectrometry approach. *J. Proteom.* **2018**, *182*, 12–20. [CrossRef] [PubMed]
97. Hendrickson, R.C.; Lee, A.Y.H.; Song, Q.; Liaw, A.; Wiener, M.; Paweletz, C.P.; Seeburger, J.L.; Li, J.; Meng, F.; Deyanova, E.G.; et al. High Resolution Discovery Proteomics Reveals Candidate Disease Progression Markers of Alzheimer's Disease in Human Cerebrospinal Fluid. *PLoS ONE* **2015**, *10*, e0135365. [CrossRef] [PubMed]
98. Shi, M.; Movius, J.; Dator, R.; Aro, P.; Zhao, Y.; Pan, C.; Lin, X.; Bammler, T.K.; Stewart, T.; Zabetian, C.P.; et al. Cerebrospinal fluid peptides as potential Parkinson disease biomarkers: A staged pipeline for discovery and validation. *Mol. Cell. Proteom.* **2015**, *14*, 544–555. [CrossRef] [PubMed]
99. Heywood, W.E.; Galimberti, D.; Bliss, E.; Sirka, E.; Paterson, R.W.; Magdalinou, N.K.; Carecchio, M.; Reid, E.; Heslegrave, A.; Fenoglio, C.; et al. Identification of novel CSF biomarkers for neurodegeneration and their validation by a high-throughput multiplexed targeted proteomic assay. *Mol. Neurodegener.* **2015**, *10*, 64. [CrossRef] [PubMed]
100. Paterson, R.W.; Heywood, W.E.; Heslegrave, A.J.; Magdalinou, N.K.; Andreasson, U.; Sirka, E.; Bliss, E.; Slattery, C.F.; Toombs, J.; Svensson, J.; et al. A targeted proteomic multiplex CSF assay identifies increased malate dehydrogenase and other neurodegenerative biomarkers in individuals with Alzheimer's disease pathology. *Transl. Psychiatry* **2016**, *6*, e952. [CrossRef] [PubMed]
101. Magdalinou, N.K.; Noyce, A.J.; Pinto, R.; Lindstrom, E.; Holmén-Larsson, J.; Holtta, M.; Blennow, K.; Morris, H.R.; Skillbäck, T.; Warner, T.T.; et al. Identification of candidate cerebrospinal fluid biomarkers in parkinsonism using quantitative proteomics. *Parkinsonism Relat. Disord.* **2017**, *37*, 65–71. [CrossRef] [PubMed]
102. Brinkmalm, G.; Sjödin, S.; Simonsen, A.H.; Hasselbalch, S.G.; Zetterberg, H.; Brinkmalm, A.; Blennow, K. A Parallel Reaction Monitoring Mass Spectrometric Method for Analysis of Potential CSF Biomarkers for Alzheimer's Disease. *PROTEOMICS—Clin. Appl.* **2018**, *12*, 1700131. [CrossRef] [PubMed]

103. Teunissen, C.E.; Elias, N.; Koel-Simmelink, M.J.A.; Durieux-Lu, S.; Malekzadeh, A.; Pham, T.V.; Piersma, S.R.; Beccari, T.; Meeter, L.H.H.; Dopper, E.G.P.; et al. Novel diagnostic cerebrospinal fluid biomarkers for pathologic subtypes of frontotemporal dementia identified by proteomics. *Alzheimer's Dement.* **2016**, *2*, 86–94. [CrossRef] [PubMed]
104. Oeckl, P.; Steinacker, P.; von Arnim, C.A.F.; Straub, S.; Nagl, M.; Feneberg, E.; Weishaupt, J.H.; Ludolph, A.C.; Otto, M. Intact Protein Analysis of Ubiquitin in Cerebrospinal Fluid by Multiple Reaction Monitoring Reveals Differences in Alzheimer's Disease and Frontotemporal Lobar Degeneration. *J. Proteome Res.* **2014**, *13*, 4518–4525. [CrossRef] [PubMed]
105. Sjödin, S.; Hansson, O.; Öhrfelt, A.; Brinkmalm, G.; Zetterberg, H.; Brinkmalm, A.; Blennow, K. Mass Spectrometric Analysis of Cerebrospinal Fluid Ubiquitin in Alzheimer's Disease and Parkinsonian Disorders. *PROTEOMICS—Clin. Appl.* **2017**, *11*, 1700100. [CrossRef] [PubMed]
106. Uhlen, M.; Bandrowski, A.; Carr, S.; Edwards, A.; Ellenberg, J.; Lundberg, E.; Rimm, D.L.; Rodriguez, H.; Hiltke, T.; Snyder, M.; et al. A proposal for validation of antibodies. *Nat. Methods* **2016**, *13*, 823. [CrossRef] [PubMed]
107. Guglielmo-Viret, V.; Attrée, O.; Blanco-Gros, V.; Thullier, P. Comparison of electrochemiluminescence assay and ELISA for the detection of Clostridium botulinum type B neurotoxin. *J. Immunol. Methods* **2005**, *301*, 164–172. [CrossRef] [PubMed]
108. Oh, E.S.; Mielke, M.M.; Rosenberg, P.B.; Jain, A.; Fedarko, N.S.; Lyketsos, C.G.; Mehta, P.D. Comparison of conventional ELISA with electrochemiluminescence technology for detection of amyloid-β in plasma. *J. Alzheimer's Dis.* **2010**, *21*, 769–773. [CrossRef] [PubMed]
109. Swartzman, E.E.; Miraglia, S.J.; Mellentin-Michelotti, J.; Evangelista, L.; Yuan, P.-M. A Homogeneous and Multiplexed Immunoassay for High-Throughput Screening Using Fluorometric Microvolume Assay Technology. *Anal. Biochem.* **1999**, *271*, 143–151. [CrossRef] [PubMed]
110. Martins, T.B. Development of internal controls for the Luminex instrument as part of a multiplex seven-analyte viral respiratory antibody profile. *Clin. Diagn. Lab. Immunol.* **2002**, *9*, 41–45. [CrossRef] [PubMed]
111. Christiansson, L.; Mustjoki, S.; Simonsson, B.; Olsson-Strömberg, U.; Loskog, A.S.I.; Mangsbo, S.M. The use of multiplex platforms for absolute and relative protein quantification of clinical material. *EuPA Open Proteom.* **2014**, *3*, 37–47. [CrossRef]
112. Kupcova Skalnikova, H.; Cizkova, J.; Cervenka, J.; Vodicka, P. Advances in Proteomic Techniques for Cytokine Analysis: Focus on Melanoma Research. *Int. J. Mol. Sci.* **2017**, *18*, 2697. [CrossRef] [PubMed]
113. Assarsson, E.; Lundberg, M.; Holmquist, G.; Björkesten, J.; Thorsen, S.B.; Ekman, D.; Eriksson, A.; Rennel Dickens, E.; Ohlsson, S.; Edfeldt, G.; et al. Homogenous 96-plex PEA immunoassay exhibiting high sensitivity, specificity, and excellent scalability. *PLoS ONE* **2014**, *9*, e95192. [CrossRef] [PubMed]
114. Lundberg, M.; Eriksson, A.; Tran, B.; Assarsson, E.; Fredriksson, S. Homogeneous antibody-based proximity extension assays provide sensitive and specific detection of low-abundant proteins in human blood. *Nucleic Acids Res.* **2011**, *39*, e102. [CrossRef] [PubMed]
115. Hensley, P. SOMAmers and SOMAscan—A Protein Biomarker Discovery Platform for Rapid Analysis of Sample Collections From Bench Top to the Clinic. *J. Biomol. Tech.* **2013**, *24*, S5.
116. Dencker, M.; Björgell, O.; Hlebowicz, J. Effect of food intake on 92 neurological biomarkers in plasma. *Brain Behav.* **2017**, *7*, e00747. [CrossRef] [PubMed]
117. Mattsson, N.; Andreasson, U.; Zetterberg, H.; Blennow, K. Alzheimer's Disease Neuroimaging Initiative Association of Plasma Neurofilament Light With Neurodegeneration in Patients With Alzheimer Disease. *JAMA Neurol.* **2017**, *74*, 557–566. [CrossRef] [PubMed]
118. Breen, E.C.; Reynolds, S.M.; Cox, C.; Jacobson, L.P.; Magpantay, L.; Mulder, C.B.; Dibben, O.; Margolick, J.B.; Bream, J.H.; Sambrano, E.; et al. Multisite comparison of high-sensitivity multiplex cytokine assays. *Clin. Vaccine Immunol.* **2011**, *18*, 1229–1242. [CrossRef] [PubMed]
119. Muszyński, P.; Groblewska, M.; Kulczyńska-Przybik, A.; Kułakowska, A.; Mroczko, B. YKL-40 as a Potential Biomarker and a Possible Target in Therapeutic Strategies of Alzheimer's Disease. *Curr. Neuropharmacol.* **2017**, *15*, 906–917. [CrossRef] [PubMed]
120. Craig-Schapiro, R.; Perrin, R.J.; Roe, C.M.; Xiong, C.; Carter, D.; Cairns, N.J.; Mintun, M.A.; Peskind, E.R.; Li, G.; Galasko, D.R.; et al. YKL-40: A novel prognostic fluid biomarker for preclinical Alzheimer's disease. *Biol. Psychiatry* **2010**, *68*, 903–912. [CrossRef] [PubMed]

121. Hye, A.; Riddoch-Contreras, J.; Baird, A.L.; Ashton, N.J.; Bazenet, C.; Leung, R.; Westman, E.; Simmons, A.; Dobson, R.; Sattlecker, M.; et al. Plasma proteins predict conversion to dementia from prodromal disease. *Alzheimer's Dement.* **2014**, *10*, 799–807. [CrossRef] [PubMed]
122. Ray, S.; Britschgi, M.; Herbert, C.; Takeda-Uchimura, Y.; Boxer, A.; Blennow, K.; Friedman, L.F.; Galasko, D.R.; Jutel, M.; Karydas, A.; et al. Classification and prediction of clinical Alzheimer's diagnosis based on plasma signaling proteins. *Nat. Med.* **2007**, *13*, 1359–1362. [CrossRef] [PubMed]
123. Kiddle, S.J.; Thambisetty, M.; Simmons, A.; Riddoch-Contreras, J.; Hye, A.; Westman, E.; Pike, I.; Ward, M.; Johnston, C.; Lupton, M.K.; et al. Plasma based markers of [^{11}C] PiB-PET brain amyloid burden. *PLoS ONE* **2012**, *7*, e44260. [CrossRef] [PubMed]
124. Baird, A.L.; Westwood, S.; Lovestone, S. Blood-Based Proteomic Biomarkers of Alzheimer's Disease Pathology. *Front. Neurol.* **2015**, *6*, 236. [CrossRef] [PubMed]
125. Voyle, N.; Baker, D.; Burnham, S.C.; Covin, A.; Zhang, Z.; Sangurdekar, D.P.; Tan Hehir, C.A.; Bazenet, C.; Lovestone, S.; Kiddle, S.; et al. AIBL research group, and the A. research Blood Protein Markers of Neocortical Amyloid-β Burden: A Candidate Study Using SOMAscan Technology. *J. Alzheimer's Dis.* **2015**, *46*, 947–961. [CrossRef] [PubMed]
126. Del Campo, M.; Mollenhauer, B.; Bertolotto, A.; Engelborghs, S.; Hampel, H.; Simonsen, A.H.; Kapaki, E.; Kruse, N.; Le Bastard, N.; Lehmann, S.; et al. Recommendations to standardize preanalytical confounding factors in Alzheimer's and Parkinson's disease cerebrospinal fluid biomarkers: An update. *Biomark. Med.* **2012**, *6*, 419–430. [CrossRef] [PubMed]
127. Leitão, M.J.; Baldeiras, I.; Herukka, S.-K.; Pikkarainen, M.; Leinonen, V.; Simonsen, A.H.; Perret-Liaudet, A.; Fourier, A.; Quadrio, I.; Veiga, P.M.; et al. Chasing the Effects of Pre-Analytical Confounders—A Multicenter Study on CSF-AD Biomarkers. *Front. Neurol.* **2015**, *6*, 153. [CrossRef] [PubMed]
128. Fourier, A.; Portelius, E.; Zetterberg, H.; Blennow, K.; Quadrio, I.; Perret-Liaudet, A. Pre-analytical and analytical factors influencing Alzheimer's disease cerebrospinal fluid biomarker variability. *Clin. Chim. Acta* **2015**, *449*, 9–15. [CrossRef] [PubMed]
129. Livesey, J.H.; Ellis, M.J.; Evans, M.J. Pre-analytical requirements. *Clin. Biochem. Rev.* **2008**, *29* (Suppl. 1), S11–S15.
130. Le Bastard, N.; De Deyn, P.P.; Engelborghs, S. Importance and Impact of Preanalytical Variables on Alzheimer Disease Biomarker Concentrations in Cerebrospinal Fluid. *Clin. Chem.* **2015**, *61*, 734–743. [CrossRef] [PubMed]
131. Vanderstichele, H.M.J.; Janelidze, S.; Demeyer, L.; Coart, E.; Stoops, E.; Herbst, V.; Mauroo, K.; Brix, B.; Hansson, O. Optimized Standard Operating Procedures for the Analysis of Cerebrospinal Fluid Aβ42 and the Ratios of Aβ Isoforms Using Low Protein Binding Tubes. *J. Alzheimer's Dis.* **2016**, *53*, 1121–1132. [CrossRef] [PubMed]
132. Vallabh, S.M.; Nobuhara, C.K.; Llorens, F.; Zerr, I.; Parchi, P.; Capellari, S.; Kuhn, E.; Klickstein, J.; Safar, J.; Nery, F.; et al. Prion protein quantification in cerebrospinal fluid as a tool for prion disease drug development. *bioRxiv* **2018**, 295063. [CrossRef]
133. Comstock, G.W.; Burke, A.E.; Norkus, E.P.; Gordon, G.B.; Hoffman, S.C.; Helzlsouer, K.J. Effects of repeated freeze-thaw cycles on concentrations of cholesterol, micronutrients, and hormones in human plasma and serum. *Clin. Chem.* **2001**, *47*, 139–142. [CrossRef] [PubMed]
134. Jani, D.; Allinson, J.; Berisha, F.; Cowan, K.J.; Devanarayan, V.; Gleason, C.; Jeromin, A.; Keller, S.; Khan, M.U.; Nowatzke, B.; et al. Recommendations for Use and Fit-for-Purpose Validation of Biomarker Multiplex Ligand Binding Assays in Drug Development. *AAPS J.* **2016**, *18*, 1–14. [CrossRef] [PubMed]
135. Rosenberg-Hasson, Y.; Hansmann, L.; Liedtke, M.; Herschmann, I.; Maecker, H.T. Effects of serum and plasma matrices on multiplex immunoassays. *Immunol. Res.* **2014**, *58*, 224–233. [CrossRef] [PubMed]
136. Tate, J.; Ward, G. Interferences in immunoassay. *Clin. Biochem. Rev.* **2004**, *25*, 105–120. [PubMed]
137. Martins, T.B.; Pasi, B.M.; Litwin, C.M.; Hill, H.R. Heterophile antibody interference in a multiplexed fluorescent microsphere immunoassay for quantitation of cytokines in human serum. *Clin. Diagn. Lab. Immunol.* **2004**, *11*, 325–329. [CrossRef] [PubMed]
138. Sharma, V.; Eckels, J.; Schilling, B.; Ludwig, C.; Jaffe, J.D.; MacCoss, M.J.; MacLean, B. Panorama Public: A public repository for quantitative data sets processed in Skyline. *Mol. Cell. Proteom.* **2018**, *17*, 1239–1244. [CrossRef] [PubMed]

139. Khoonsari, P.E.; Häggmark, A.; Lönnberg, M.; Mikus, M.; Kilander, L.; Lannfelt, L.; Bergquist, J.; Ingelsson, M.; Nilsson, P.; Kultima, K.; et al. Analysis of the Cerebrospinal Fluid Proteome in Alzheimer's Disease. *PLoS ONE* **2016**, *11*, e0150672. [CrossRef] [PubMed]
140. The UniProt Consortium. UniProt: The universal protein knowledgebase. *Nucleic Acids Res.* **2017**, *45*, D158–D169. [CrossRef] [PubMed]
141. Birney, E.; Andrews, T.D.; Bevan, P.; Caccamo, M.; Chen, Y.; Clarke, L.; Coates, G.; Cuff, J.; Curwen, V.; Cutts, T.; et al. An overview of Ensembl. *Genome Res.* **2004**, *14*, 925–928. [CrossRef] [PubMed]
142. Guldbrandsen, A.; Farag, Y.; Kroksveen, A.C.; Oveland, E.; Lereim, R.R.; Opsahl, J.A.; Myhr, K.-M.; Berven, F.S.; Barsnes, H. CSF-PR 2.0: An Interactive Literature Guide to Quantitative Cerebrospinal Fluid Mass Spectrometry Data from Neurodegenerative Disorders. *Mol. Cell. Proteom.* **2017**, *16*, 300–309. [CrossRef] [PubMed]

proteomes

MDPI

Review

Deep Profiling of the Aggregated Proteome in Alzheimer's Disease: From Pathology to Disease Mechanisms

Brianna M. Lutz and Junmin Peng *

Departments of Structural Biology and Developmental Neurobiology, Center for Proteomics and Metabolomics, St. Jude Children's Research Hospital, Memphis, TN 38105, USA; Brianna.lutz@stjude.org
* Correspondence: junmin.peng@stjude.org

Received: 22 September 2018; Accepted: 7 November 2018; Published: 12 November 2018

Abstract: Hallmarks of Alzheimer's disease (AD), a progressive neurodegenerative disease causing dementia, include protein aggregates such as amyloid beta plaques and tau neurofibrillary tangles in a patient's brain. Understanding the complete composition and structure of protein aggregates in AD can shed light on the as-yet unidentified underlying mechanisms of AD development and progression. Biochemical isolation of aggregates coupled with mass spectrometry (MS) provides a comprehensive proteomic analysis of aggregates in AD. Dissection of these AD-specific aggregate components, such as U1 small nuclear ribonucleoprotein complex (U1 snRNP), provides novel insights into the deregulation of RNA splicing in the disease. In this review, we summarize the methodologies of laser capture microdissection (LCM) and differential extraction to analyze the aggregated proteomes in AD samples, and discuss the derived novel insights that may contribute to AD pathogenesis.

Keywords: proteomics; proteome; mass spectrometry; Alzheimer's disease; protein aggregation; laser capture microdissection; splicing; U1 snRNP

1. Introduction

Alzheimer's Disease (AD) is a progressive neurodegenerative disease and the most common form of dementia, listed as the sixth leading cause of death [1,2]. AD represents a major economic burden predicted to surpass one trillion dollars worldwide in 2018 [3]. The cause of AD, however, is still not fully understood. There is no cure for AD, and current therapeutic strategies cannot hinder cognitive decline in AD [4].

The pathogenesis of AD has been extensively investigated by genetic and biochemical approaches. Genetic analysis of AD patients established three causative genes (*APP*, *PSEN1* and *PSEN2*) and a high-risk allele (*ApoE ε4*) [5,6], whereas genome-wide association studies led to the discovery of more than 20 low-risk genetic loci [7–9], and more recently, high-throughput sequencing identified rare, medium-risk genes, such as *TREM2* [10] and *UNC5C* [11]. Despite the genetic contributions, the vast majority of AD cases are sporadic, which may be attributed to the combination of genetic susceptibility and environmental factors [5], such as Herpesvirus infection [12,13] and environmental pollutants [14]. Biochemical dissection of AD brain tissue identified pathological hallmarks of amyloid-β (Aβ)-containing amyloid plaques, and neurofibrillary tangles (NFT) comprising hyperphosphorylated Tau in both familial and sporadic patients [15], although Tau mutations were identified in other forms of dementia, collectively termed tauopathy [16]. These results lead to the proposed amyloid cascade and Tau hypotheses [16,17] dominating AD research.

In the amyloid cascade and Tau hypotheses, the accumulation of Amyloid Precursor Protein (APP)-derived Aβ peptide is assumed to be the main cause of AD. Toxic Aβ species in the brain trigger a cascade that leads to inflammation, tau hyperphosphorylation and deposition, synaptic loss and neuronal degeneration, which eventually leads to dementia in AD. Based on the hypotheses, numerous animal models (largely mouse models) have been developed to mimic some phenotypes observed in AD patients, but these models cannot fully recapitulate human AD symptoms [18,19].

In addition, there is a lack of concordance between these models and clinical trials [18,19]. Given the amyloid hypothesis, targeting the cleavage of APP or the accumulation of Aβ has long been a goal for a pharmacological treatment for AD [20]. Unfortunately, clinical trials implementing Aβ antibody therapy or pharmacological intervention of APP cleavage have not yet been successful [21,22]. There is an urgent need for a broad understanding of synergistic interactions of molecular and cellular components in the brain, at asymptomatic—when the pathological hallmarks of AD are present but cognitive dysfunction is not evident [23]—and symptomatic stages during AD progression [24].

We believe that deep analysis of protein deposition in AD has the potential to discover novel disease mechanisms, considering the profound impact of the previous identification of Aβ and tau aggregation on our understanding of AD. In addition, protein aggregation is commonly observed in other neurodegenerative disorders, such as α-synuclein in Parkinson disease (PD) [25], and TDP-43 in ubiquitin-positive frontotemporal lobar degeneration (FTLD-U) and amyotrophic lateral sclerosis (ALS) [26]. In this review, we summarize the approaches toward profiling protein aggregates in AD, with a discussion of the benefits and pitfalls of the approaches, as well as potential novel AD mechanisms revealed by these analyses.

2. Proteomic Characterization of AD Amyloid Plaques and Neurofibrillary Tangles by Laser Capture Microdissection

Extracellular amyloid plaques consist of aggregated Aβ peptides entangled with microglial, neuronal, and vasculature components. Intracellular neurofibrillary tangles are also complex structures marked by anti-tau and anti-ubiquitin immunohistochemistry (IHC) [27,28]. The antibody-based IHC method is a targeted approach for detecting known aggregated proteins in the plaques and NFT of brain tissue, but the exact composition of the aggregated structures could not be uncovered.

Integration of laser capture microdissection (LCM) [29] with highly sensitive mass spectrometry (MS) [30] enables direct dissection of protein components in these AD aggregated structures [31,32]. In a pioneer study, Liao et al. isolated thioflavin-S-labeled senile plaques from frozen sections of human post-mortem brain tissue, and compared the plaque protein composition with the non-plaque regions by label-free quantification [33]. The analysis was performed with nanoscale liquid chromatography-tandem mass spectrometry (LC-MS/MS) on an LCQ ion trap mass spectrometer, identifying 488 proteins in the isolated plaques, in which 26 proteins were significantly enriched in the plaques compared to non-plaque regions. These proteins were classified into a variety of functional groups, including cell adhesion, cytoskeleton and membrane trafficking, chaperones and inflammation, kinase/phosphatase and regulators, and proteolysis, consistently with diverse cellular components in the plaque area [33]. Notably, the membrane trafficking protein dynein was enriched in the isolated plaques, and its localization was further validated by IHC in a transgenic AD mouse model. This study demonstrates the feasibility of proteomic analysis of minute amounts of LCM-isolated AD samples (Figure 1).

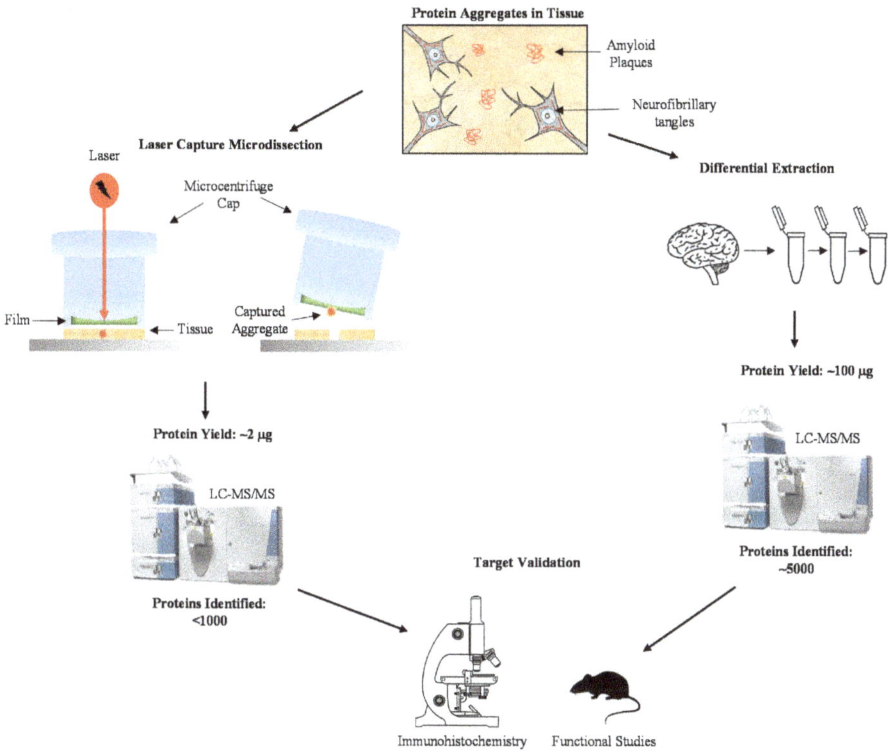

Figure 1. Methods for profiling the aggregated proteome in Alzheimer's Disease (AD). Isolation of protein aggregates in AD brain can be accomplished using laser capture microdissection or differential extraction. Laser capture microdissection specifically captures protein aggregates, resulting in a protein yield of around 2 μg for 1000 plaques. Using this minute amount, less than 1000 proteins were identified using LC-MS/MS. Differential extraction, the process of isolating insoluble aggregates through repeated centrifugation in varying reagents, yields around 100 μg of protein and around 5000 proteins identified using LC-MS/MS. Regardless of the method of aggregate isolation, protein targets need to be validated using specific immunohistochemical techniques and their function can be determined using comparable research models.

LCM was also used to isolate NFTs in AD brain for proteomic analysis. Wang et al. isolated NFTs from AD hippocampus samples and performed LC-MS/MS to determine NFT-associated proteins [34]. Out of 155 identified proteins, 63 novel proteins were found to be associated with NFT, including glyceraldehyde-3-phosphate dehydrogenase (GAPDH). The association of GAPDH with NFT was further supported by immunohistochemistry in AD brain samples, as well as biochemical fractionation of detergent-insoluble samples of AD brain lysate.

More recently, Drummond et al. implemented a method to extract proteins from archived, formalin-fixed paraffin-embedded (FFPE) human tissue slides, and analyzed amyloid plaques and NFT from FFPE AD brain tissue using LCM-LC-MS/MS [35]. The FFPE samples were extracted by formic acid and deparaffinized, followed by protein digestion. Using an Orbitrap Q-Exactive mass spectrometer, the group analyzed approximately 900 proteins in the plaques and 500 proteins in NFT with an FDR of 1%, deepening the understanding of neuropathological hallmarks in AD.

LCM allows for the specific isolation of plaques and NFT tissue which can lead to the identification of hundreds of proteins; however, these proteins only represent the most abundant components in the captured tissue areas. Another major drawback of the use of LCM for plaque and NFT isolation is the minute amount of sample that can be collected. For instance, using 10 μm thick sections to capture amyloid plaques, which are heterogeneous in size and about 60 μm in average diameter [31], the protein yield is approximately 2 ng per plaque and 2 μg from 1000 plaques. To address this drawback, protein differential extraction has been developed to increase the protein yield for deep proteome profiling (Table 1).

3. Deep Analysis of Aggregated Proteome in AD by Differential Extraction

Differential extraction has long been used for the enrichment of aggregated proteins in neurodegenerative diseases [36], as exemplified by biochemical purification of Aβ and tau in AD [37,38], α-synuclein in PD [25], and TDP-43 in FTLD-U and ALS [26]. Differential extraction is based on the principle that aggregated proteins usually display low solubility and are thus enriched in the pellet after detergent extraction (e.g., sarkosyl) as a detergent-insoluble fraction (Figure 1) [39].

The insolubility of amyloid plaque and NFT components provides an avenue for isolation and subsequent proteomic characterization. Insoluble aggregates can be isolated from whole homogenates of AD brain through sequential extraction. Gozal et al. isolated detergent-insoluble lysate from the frontal cortex of control, AD, and FTLD cases [40]. Label-free LC-MS/MS quantification identified 512 proteins, in which 11 proteins were significantly elevated in AD compared to FTLD and control cases. As expected, tau, Aβ, apolipoprotein E [41], and serum amyloid P [42] were enriched in the AD samples. The alteration of several proteins including serine protease 15, ankyrin B, and 14-3-3 eta, were validated by immunoblotting analysis.

Following the pilot study [40], Bai et al. performed a comprehensive profiling of aggregate-enriched, detergent-insoluble fractions from all major neurodegenerative diseases, including AD, PD, FTLD-U, ALS, corticobasal degeneration (CBD), and control samples [39]. To identify if proteins change early in the development of AD, mild cognitive impairment (MCI), a prodromal stage of AD, was also analyzed. This large-scale profiling was based on label-free quantification by gel-enhanced LC-MS/MS (gelLC-MS/MS)–protein separation by 1D SDS gel followed by in-gel digestion and LC-MS/MS, leading to the identification of 4216 proteins. After stringent statistical analysis and manual evaluation, a total of 36 proteins were shown to accumulate in AD. In addition to the known aggregate components such as Aβ, tau, ApoE, and complement proteins, the enriched proteins are involved in Aβ clearance [43], phosphorylation networks [16], synaptic plasticity [44], and mitochondrial regulation [45]. Interestingly, several U1 small nuclear ribonucleoprotein (U1 snRNP) spliceosome subunits (U1-70K and U1A) and the interacting RNA helicase Prp5 [46] were found to be highly increased in AD, leading to a novel U1 snRNP pathology, and implicating RNA splicing dysfunction in AD [39]. In addition to late onset sporadic AD cases, the U1 snRNP components were also found to aggregate in early onset genetic cases (e.g., mutations in APP and PS-1), as well as in trisomy 21 (the APP gene is in chromosome 21) [47].

To track the process of protein insolubility during the course of AD development, Hales et al. continued to quantify the detergent-insoluble brain proteome, and correlated them with Aβ and tau proteins in 35 cases of control, asymptomatic phase of AD (AsymAD), MCI, and AD [48]. Among 2711 proteins, six U1 snRNP subunits (U1-70K, U1A, SmD1, SmD2, SmD3, and SmB) are in the top 10 Aβ-correlated proteins, whereas three U1 snRNP subunits (U1-70K, U1A, and SmD) are also correlated with tau insolubility. These results suggest a possible link of these AD aggregated proteins during disease progression.

Table 1. A comparison of approaches for protein aggregate isolation for proteomic profiling.

Technique	Protein Yield	Instruments Required	Number of Proteins Identified *	Advantages	Disadvantages
LCM	~2 µg from 1000 plaques	Fluorescent Microscope with Laser Capture capability LC-MS/MS	155–900 [33–35]	(1) Precise collection of cellular components (2) Conservation of tissue integrity (3) Cellular region comparison within the same tissue	(1) Small amount of protein recovery (2) Extensive time required for LCM
Differential fractionation	1% of total protein input (e.g., 100 µg from 10 mg of tissue)	Centrifuge LC-MS/MS	512–4216 [39,40,49]	(1) A sufficient amount of protein can be extracted from individual samples (2) Flexible extraction methods using different combinations of detergents	(1) Detergent soluble aggregate proteins may not be included in the MS analysis (2) Contamination of the aggregated proteome by other detergent insoluble components

* The number of proteins identified may increase with the use of more sensitive instrumentation.

4. Implication of Disease Mechanisms by Aggregated Proteins in AD

Specifically analyzing the aggregate proteome in AD can be used to identify potential mechanisms of disease progression or development (Table 2). Since Aβ and tau are considered pathological hallmarks of AD, it is expected that these proteins would not only be identified in protein aggregates in AD, but also enriched in the AD aggregates compared to control patient aggregates. Consistently, Aβ and tau proteins are identified in the aggregate proteome in all AD patient samples [33,34,40,48]. While the exact molecular mechanisms of AD remain to be understood, aggregated Aβ can contribute to AD progression through neurotoxic effects including disruption of synaptic communication, free radical production, and disrupted calcium homeostasis [49]. The relationship between tau and Aβ is supported by in vitro studies that show Aβ-induced tau-dependent microtubule dysfunction, synaptic damage, and excitotoxicity [50], as well as in vivo studies that indicate Aβ-induced tau-mediated axonal transport defects [51]. Microtubules are key components of intracellular transport that exhibit reduced stability and subsequent reduced axonal transport in AD [52]. Loss of microtubules in AD has been attributed to aggregated tau-induced polyglutamylation of microtubules [52]. Additionally, Aβ oligomers can trigger tau-induced microtubule decay through elevated intracellular calcium, suggesting that Aβ aggregation may be an upstream event of tau-induced microtubule loss [52]. The loss of microtubules leads to impaired axonal transport which leads to dendritic spine decay and subsequent neuronal dysfunction [53].

Inflammatory proteins, including high-temperature requirement serine protease A1 (HTRA1) and complement C3, were found to associate with Aβ and tau aggregates in AD patient brain samples [39,48]. HTRA1 is a secreted serine protease that can bind tumor growth factor-β proteins, inhibiting their anti-inflammatory actions [54]. The correlation of HTRA1 and aggregated Aβ and tau in AD samples suggests possible upregulation of HTRA1 in AD, which could have implications in the inflammation associated with AD [48]. Complement C3 is released from microglia and is involved in phagocytosis [55]. In AD, Aβ initiates a complement cascade in which C3 production increases leading to phagocytosis of not only Aβ plaques, but also synapses [56]. This aberrant activation of microglia may contribute to the neuronal degeneration and synaptic dysfunction associated with AD. The association of inflammatory proteins with Aβ and tau aggregates in AD brain samples further exemplifies an inflammatory component to AD pathology.

U1 snRNP subunits (notably, U1-70K, SmD, and U1A) are highly correlated with insoluble tau and Aβ, suggesting a possible role in tau aggregation and AD pathogenesis [39,48]. U1 snRNP protein subunits are coupled with small nuclear RNAs (snRNAs) to form spliceosomes, which remove introns from mRNA transcripts in a process known as mRNA splicing [57]. The identification of multiple U1 snRNP subunits in the detergent-insoluble AD proteome strongly suggests the precipitation of the entire U1 snRNP complex. Indeed, IHC staining indicated tangle-like aggregates of snRNA in AD cases, and transmission electron microscopy showed snRNA co-localization with tau NFT [58]. The aggregation of U1-70K, a U1 snRNP, occurred in the form of cytoplasmic tangles in AD brain slices [39]. This localization was later confirmed using electron microscopy in which immunogold-labeled U1-70K co-localized with structures resembling NFT in AD frontal cortex samples [59]. This abnormal localization and enrichment of a U1 snRNP could play a role in AD. Consistently, deep RNA-sequencing revealed impaired RNA splicing in AD cortical samples [39]. This functional deficit could be the result of aggregation of spliceosome components and a loss-of-function effect in the AD brain [39,48].

The aggregation of U1-70K in AD has been confirmed in multiple studies, yet the cause of this abnormal aggregation in AD brain samples is still unclear. The presence of two specific low complexity (LC) domains in U1-70K protein suggests an inherent tendency for U1-70K aggregation [60]. Low complexity domains are repetitive sequences of amino acids that display a tendency to aggregate at high concentrations [61]. Recombinant protein studies concluded that one C-terminal LC domain in U1-70K contributed to its aggregation [59]. In AD brain homogenates, endogenous U1-70K aggregates formed direct interactions with recombinant U1-70K that was prone to aggregation via the incorporation of an LC domain. These results suggest that U1-70K aggregation in AD is the result of both an inherent potential for U1-70K to aggregate and co-aggregate with pre-existing seeds.

In addition to the aggregation hypothesis, the U1-70K loss-of-function may be the result of abnormal cleavage and peptide truncation in AD. Bai et al. showed that U1-70K can be cleaved to generate an N-terminal truncation identified as N40K [62]. This truncation occurred in about 50% of the 17 AD brain samples studied [62]. In these cases, the expression of N40K inversely correlated with the expression of U1-70K [62], suggesting that U1-70K loss-of-function could be due to truncation. Functionally, N40K displayed toxic pro-apoptotic effects in primary rat neurons [62].

Table 2. A comparison of significant AD-specific proteins identified in the insoluble fractions collected from two differential fractionation LC-MS/MS studies.

Protein	GeneBank™ Accession Number	Association with AD
Identified by Bai, B., et al., PNAS, 2013 [39]		
Collagen Type XXV, alpha 1 isoform 2	NP_000032.1	[63]
Cellular retinoic acid binding protein	NP_004369.1	[48]
Dystrobrevin alpha	NP_009224.2	[48]
Complement component 4a preproprotein	NP_116757.2	[64]
Complement component 3	NP_000055.2	[65]
Cyclin G-associated kinase	NP_005246.2	Not Found
Protein tyrosine phosphatase, zeta1	NP_002842.2	[66]
T-cell activation protein phosphatase 2C	NP_644812.1	Not Found
Synaptojanin 1	NP_982271.1	[67]
Amphiphysin	NP_001626.1	[68]
Syntaxin binding protein 5	NP_640337.3	[69]
Regulating synaptic membrane exocytosis 1	NP_055804.2	Not Found
Neuroblastoma-amplified protein (with a Sec39 domain)	NP_056993.2	Not Found
Glutamate receptor interacting protein 1	NP_066973.1	[70]
Mitochondrial nicotinamide nucleotide transhydrogenase	NP_892022.2	[71]
Mitochondrial NFS1 nitrogen fixation 1	NP_066923.3	Not Found
Mitochondrial fumarate hydratase	NP_000134.2	[72]
Optic atrophy 1	NP_570847.1	[73]
Mitochondrial processing peptidase	NP_004270.2	Not Found
U1 small nuclear ribonucleoprotein 70 kDa	NP_003080.2	[74]
U1 small nuclear ribonucleoprotein A	NP_004587.1	[39]
ATP-dependent RNA helicase DDX46, Prp5	NP_055644.2	Not Found
4-Aminobutyrate aminotransferase	NP_001120920.1	[75]
10-Formyltetrahydrofolate dehydrogenase	NP_036322.2	Not Found
Phytanoyl-CoA dioxygenase domain containing protein 1	NP_001094346.1	Not Found
Nicotinamide nucleotide adenylyltransferase 3	NP_835471.1	[76]
Asparagine-linked glycosylation 2	NP_149078.1	Not Found
GTPase activating protein and VPS9 domains 1	NP_056450.2	[77]

Table 2. *Cont.*

Protein	GeneBank™ Accession Number	Association with AD
Identified by Bai, B., et al., PNAS, 2013 [39]		
Phosphatidylinositol-dependent Rac exchanger 1	NP_065871.2	Not Found
Aminophospholipid transporter	NP_006086.1	[78]
RAN binding protein 16 (exportin 7)	NP_055839.3	[79]
ALFY, involved in macroautophagy	NP_055806.2	Not Found
Identified by Gozal, Y., et al., J. Proteome Res., 2009 [40]		
serum amyloid P component precursor	NP_001630.1	[42]
serine protease 15	NP_004784.2	Not Found
14-3-3, eta polypeptide	NP_003396.1	Not Found
14-3-3, zeta polypeptide	NP_663723.1	Not Found
ankyrin B	NP_066187.2	Not Found
dynamin 1	NP_004399.2	[80]
aquaporin 1	NP_000376.1	[81]
Identified in both studies		
Apolipoprotein E	NP_000032.1	[41]
Microtubule-associated protein tau	NP_058519.2	[16]
Amyloid β peptide	NP_000475.1	[82]
Complement component 4b	NP_001002029.3	[64]

It should be mentioned that N40K also contains a low complexity domain to form aggregates [59]. More recently, Bishof et al. extended the concept and proposed that a large number of RNA binding proteins containing basic-acidic dipeptide (BAD) domains may co-aggregate in Alzheimer's disease [74]. It will be highly interesting to further study if these RNA binding proteins contribute to AD pathogenesis.

5. Conclusions

Protein aggregation is a hallmark of AD typically associated with Aβ and hyperphosphorylated tau, however, other proteins can also self-aggregate or co-aggregate with amyloid plaques and NFT. Identifying this aggregated proteome could provide insight into the underlying mechanisms of AD development and progression. MS techniques coupled with plaque and NFT isolation allow for the analysis of the aggregate proteome in human AD samples. LCM and detergent-insoluble fractionation techniques have been successfully applied to isolate amyloid plaques and NFTs directly from AD brain samples for MS analysis. These techniques have identified novel aggregate proteins including U1-snRNP, a member of the spliceosome necessary for RNA splicing. Further studies have identified splicing loss-of-function in human AD samples. Additionally, comprehensive RNA-seq analyses from multiple cohorts implicate the role of RNA splicing dysfunction in AD [83]. Although further functional studies are needed to determine the exact role of aggregate-associated proteins in AD, MS proves to be an invaluable tool for dissecting AD pathology and pathogenesis.

Author Contributions: Conceptualization, J.P.; Manuscript Preparation, B.M.L. and J.P.

Funding: This work was partially supported by National Institutes of Health grants [R01AG047928, R01AG053987, R01GM114260], and ALSAC (American Lebanese Syrian Associated Charities).

Conflicts of Interest: The authors declared no conflict of interest.

References

1. James, B.D.; Leurgans, S.E.; Hebert, L.E.; Scherr, P.A.; Yaffe, K.; Bennett, D.A. Contribution of alzheimer disease to mortality in the united states. *Neurology* **2014**, *82*, 1045–1050. [CrossRef] [PubMed]
2. Scheltens, P.; Blennow, K.; Breteler, M.M.; de Strooper, B.; Frisoni, G.B.; Salloway, S.; Van der Flier, W.M. Alzheimer's disease. *Lancet* **2016**, *388*, 505–517. [CrossRef]

3. Wimo, A.; Guerchet, M.; Ali, G.C.; Wu, Y.T.; Prina, A.M.; Winblad, B.; Jönsson, L.; Liu, Z.; Prince, M. The worldwide costs of dementia 2015 and comparisons with 2010. *Alzheimers Dement.* **2017**, *13*, 1–7. [CrossRef] [PubMed]
4. Graham, W.V.; Bonito-Oliva, A.; Sakmar, T.P. Update on alzheimer's disease therapy and prevention strategies. *Annu. Rev. Med.* **2017**, *68*, 413–430. [CrossRef] [PubMed]
5. Tanzi, R.E. The genetics of alzheimer disease. *Cold Spring Harb. Perspect. Med.* **2012**, *2*. [CrossRef] [PubMed]
6. Guerreiro, R.; Bras, J.; Hardy, J. Snapshot: Genetics of alzheimer's disease. *Cell* **2013**, *155*, 968. [CrossRef] [PubMed]
7. Lambert, J.C.; Ibrahim-Verbaas, C.A.; Harold, D.; Naj, A.C.; Sims, R.; Bellenguez, C.; DeStafano, A.L.; Bis, J.C.; Beecham, G.W.; Grenier-Boley, B.; et al. Meta-analysis of 74,046 individuals identifies 11 new susceptibility loci for alzheimer's disease. *Nat. Genet.* **2013**, *45*, 1452–1458. [CrossRef] [PubMed]
8. Steinberg, S.; Stefansson, H.; Jonsson, T.; Johannsdottir, H.; Ingason, A.; Helgason, H.; Sulem, P.; Magnusson, O.T.; Gudjonsson, S.A.; Unnsteinsdottir, U.; et al. Loss-of-function variants in abca7 confer risk of alzheimer's disease. *Nat. Genet.* **2015**, *47*, 445–447. [CrossRef] [PubMed]
9. Sims, R.; van der Lee, S.J.; Naj, A.C.; Bellenguez, C.; Badarinarayan, N.; Jakobsdottir, J.; Kunkle, B.W.; Boland, A.; Raybould, R.; Bis, J.C.; et al. Rare coding variants in plcg2, abi3, and trem2 implicate microglial-mediated innate immunity in alzheimer's disease. *Nat. Genet.* **2017**, *49*, 1373–1384. [CrossRef] [PubMed]
10. Colonna, M.; Wang, Y. Trem2 variants: New keys to decipher alzheimer disease pathogenesis. *Nat. Rev. Neurosci.* **2016**, *17*, 201–207. [CrossRef] [PubMed]
11. Wetzel-Smith, M.K.; Hunkapiller, J.; Bhangale, T.R.; Srinivasan, K.; Maloney, J.A.; Atwal, J.K.; Sa, S.M.; Yaylaoglu, M.B.; Foreman, O.; Ortmann, W.; et al. A rare mutation in unc5c predisposes to late-onset alzheimer's disease and increases neuronal cell death. *Nat. Med.* **2014**, *20*, 1452–1457. [CrossRef] [PubMed]
12. Eimer, W.A.; Kumar, D.K.V.; Shanmugam, N.K.N.; Rodriguez, A.S.; Mitchell, T.; Washicosky, K.J.; Gyorgy, B.; Breakefield, X.O.; Tanzi, R.E.; Moir, R.D. Alzheimer's disease-associated beta-amyloid is rapidly seeded by herpesviridae to protect against brain infection. *Neuron* **2018**, *99*, 56–97. [CrossRef] [PubMed]
13. Readhead, B.; Haure-Mirande, J.V.; Funk, C.C.; Richards, M.A.; Shannon, P.; Haroutunian, V.; Sano, M.; Liang, W.S.; Beckmann, N.D.; Price, N.D.; et al. Multiscale analysis of independent alzheimer's cohorts finds disruption of molecular, genetic, and clinical networks by human herpesvirus. *Neuron* **2018**, *99*, 64–82.e67. [CrossRef] [PubMed]
14. Chin-Chan, M.; Navarro-Yepes, J.; Quintanilla-Vega, B. Environmental pollutants as risk factors for neurodegenerative disorders: Alzheimer and parkinson diseases. *Front. Cell. Neurosci.* **2015**, *9*, 124. [CrossRef] [PubMed]
15. Hyman, B.T.; Phelps, C.H.; Beach, T.G.; Bigio, E.H.; Cairns, N.J.; Carrillo, M.C.; Dickson, D.W.; Duyckaerts, C.; Frosch, M.P.; Masliah, E.; et al. National institute on aging-alzheimer's association guidelines for the neuropathologic assessment of alzheimer's disease. *Alzheimers Dement.* **2012**, *8*, 1–13. [CrossRef] [PubMed]
16. Ballatore, C.; Lee, V.M.; Trojanowski, J.Q. Tau-mediated neurodegeneration in alzheimer's disease and related disorders. *Nat. Rev. Neurosci.* **2007**, *8*, 663–672. [CrossRef] [PubMed]
17. Hardy, J.; Selkoe, D.J. The amyloid hypothesis of alzheimer's disease: Progress and problems on the road to therapeutics. *Science* **2002**, *297*, 353–356. [CrossRef] [PubMed]
18. Ashe, K.H.; Zahs, K.R. Probing the biology of alzheimer's disease in mice. *Neuron* **2010**, *66*, 631–645. [CrossRef] [PubMed]
19. LaFerla, F.M.; Green, K.N. Animal models of alzheimer disease. *Cold Spring Harb. Perspect. Med.* **2012**, *2*. [CrossRef] [PubMed]
20. Cummings, J.; Lee, G.; Ritter, A.; Zhong, K. Alzheimer's disease drug development pipeline: 2018. *Alzheimers Dement. (N Y)* **2018**, *4*, 195–214. [CrossRef] [PubMed]
21. Volloch, V.; Rits, S. Results of beta secretase-inhibitor clinical trials support amyloid precursor protein-independent generation of beta amyloid in sporadic alzheimer's disease. *Med. Sci.* **2018**, *6*. [CrossRef] [PubMed]
22. Honig, L.S.; Vellas, B.; Woodward, M.; Boada, M.; Bullock, R.; Borrie, M.; Hager, K.; Andreasen, N.; Scarpini, E.; Liu-Seifert, H.; et al. Trial of solanezumab for mild dementia due to alzheimer's disease. *N. Eng. J. Med.* **2018**, *378*, 321–330. [CrossRef] [PubMed]

23. Driscoll, I.; Troncoso, J. Asymptomatic alzheimer's disease: A prodrome or a state of resilience? *Curr. Alzheimer Res.* **2011**, *8*, 330–335. [CrossRef] [PubMed]
24. De Strooper, B.; Karran, E. The cellular phase of alzheimer's disease. *Cell* **2016**, *164*, 603–615. [CrossRef] [PubMed]
25. Spillantini, M.G.; Crowther, R.A.; Jakes, R.; Hasegawa, M.; Goedert, M. Alpha-synuclein in filamentous inclusions of lewy bodies from parkinson's disease and dementia with lewy bodies. *Proc. Natl. Acad. Sci. USA* **1998**, *95*, 6469–6473. [CrossRef] [PubMed]
26. Neumann, M.; Sampathu, D.M.; Kwong, L.K.; Truax, A.C.; Micsenyi, M.C.; Chou, T.T.; Bruce, J.; Schuck, T.; Grossman, M.; Clark, C.M.; et al. Ubiquitinated tdp-43 in frontotemporal lobar degeneration and amyotrophic lateral sclerosis. *Science* **2006**, *314*, 130–133. [CrossRef] [PubMed]
27. Serrano-Pozo, A.; Frosch, M.P.; Masliah, E.; Hyman, B.T. Neuropathological alterations in alzheimer disease. *Cold Spring Harb. Perspect. Med.* **2011**, *1*, a006189. [CrossRef] [PubMed]
28. Liebmann, T.; Renier, N.; Bettayeb, K.; Greengard, P.; Tessier-Lavigne, M.; Flajolet, M. Three-dimensional study of alzheimer's disease hallmarks using the idisco clearing method. *Cell Rep.* **2016**, *16*, 1138–1152. [CrossRef] [PubMed]
29. Emmert-Buck, M.R.; Bonner, R.F.; Smith, P.D.; Chuaqui, R.F.; Zhuang, Z.; Goldstein, S.R.; Weiss, R.A.; Liotta, L.A. Laser capture microdissection. *Science* **1996**, *274*, 998–1001. [CrossRef] [PubMed]
30. Aebersold, R.; Mann, M. Mass-spectrometric exploration of proteome structure and function. *Nature* **2016**, *537*, 347–355. [CrossRef] [PubMed]
31. Gozal, Y.M.; Cheng, D.; Duong, D.M.; Lah, J.J.; Levey, A.I.; Peng, J. Merger of laser capture microdissection and mass spectrometry: A window into the amyloid plaque proteome. *Methods Enzymol.* **2006**, *412*, 77–93. [CrossRef] [PubMed]
32. Nijholt, D.A.T.; Stingl, C.; Luider, T.M. Laser capture microdissection of fluorescently labeled amyloid plaques from alzheimer's disease brain tissue for mass spectrometric analysis. *Methods Mol. Biol.* **2015**, *1243*, 165–173. [CrossRef] [PubMed]
33. Liao, L.; Cheng, D.; Wang, J.; Duong, D.M.; Losik, T.G.; Gearing, M.; Rees, H.D.; Lah, J.J.; Levey, A.I.; Peng, J. Proteomic characterization of postmortem amyloid plaques isolated by laser capture microdissection. *J. Biol. Chem.* **2004**, *279*, 37061–37068. [CrossRef] [PubMed]
34. Wang, Q.; Woltjer, R.L.; Cimino, P.J.; Pan, C.; Montine, K.S.; Zhang, J.; Montine, T.J. Proteomic analysis of neurofibrillary tangles in alzheimer disease identifies gapdh as a detergent-insoluble paired helical filament tau binding protein. *FASEB J.* **2005**, *19*, 869–871. [CrossRef] [PubMed]
35. Drummond, E.; Nayak, S.; Pires, G.; Ueberheide, B.; Wisniewski, T. Isolation of amyloid plaques and neurofibrillary tangles from archived alzheimer's disease tissue using laser-capture microdissection for downstream proteomics. *Methods Mol. Biol.* **2018**, *1723*, 319–334. [CrossRef] [PubMed]
36. Taylor, J.P.; Hardy, J.; Fischbeck, K.H. Toxic proteins in neurodegenerative disease. *Science* **2002**, *296*, 1991–1995. [CrossRef] [PubMed]
37. Glenner, G.G.; Wong, C.W. Alzheimer's disease: Initial report of the purification and characterization of a novel cerebrovascular amyloid protein. *Biochem. Biophys. Res. Commun.* **1984**, *120*, 885–890. [CrossRef]
38. Guo, J.L.; Narasimhan, S.; Changolkar, L.; He, Z.; Stieber, A.; Zhang, B.; Gathagan, R.J.; Iba, M.; McBride, J.D.; Trojanowski, J.Q.; et al. Unique pathological tau conformers from alzheimer's brains transmit tau pathology in nontransgenic mice. *J. Exp. Med.* **2016**, *213*, 2635–2654. [CrossRef] [PubMed]
39. Bai, B.; Hales, C.M.; Chen, P.C.; Gozal, Y.; Dammer, E.B.; Fritz, J.J.; Wang, X.; Xia, Q.; Duong, D.M.; Street, C.; et al. U1 small nuclear ribonucleoprotein complex and rna splicing alterations in alzheimer's disease. *Proc. Natl. Acad. Sci. USA* **2013**, *110*, 16562–16567. [CrossRef] [PubMed]
40. Gozal, Y.M.; Duong, D.M.; Gearing, M.; Cheng, D.; Hanfelt, J.J.; Funderburk, C.; Peng, J.; Lah, J.J.; Levey, A.I. Proteomics analysis reveals novel components in the detergent-insoluble subproteome in alzheimer's disease. *J. Proteome Res.* **2009**, *8*, 5069–5079. [CrossRef] [PubMed]
41. Strittmatter, W.J.; Roses, A.D. Apolipoprotein e and alzheimer disease. *Proc. Natl. Acad. Sci. USA* **1995**, *92*, 4725–4727. [CrossRef] [PubMed]
42. Rostagno, A.; Lashley, T.; Ng, D.; Meyerson, J.; Braendgaard, H.; Plant, G.; Bojsen-Moller, M.; Holton, J.; Frangione, B.; Revesz, T.; et al. Preferential association of serum amyloid p component with fibrillar deposits in familial british and danish dementias: Similarities with alzheimer's disease. *J. Neurol. Sci.* **2007**, *257*, 88–96. [CrossRef] [PubMed]

43. Weiner, H.L.; Frenkel, D. Immunology and immunotherapy of alzheimer's disease. *Nat. Rev. Immunol.* **2006**, *6*, 404–416. [CrossRef] [PubMed]
44. Selkoe, D.J. Alzheimer's disease is a synaptic failure. *Science* **2002**, *298*, 789–791. [CrossRef] [PubMed]
45. Lin, M.T.; Beal, M.F. Mitochondrial dysfunction and oxidative stress in neurodegenerative diseases. *Nature* **2006**, *443*, 787–795. [CrossRef] [PubMed]
46. Staley, J.P.; Guthrie, C. Mechanical devices of the spliceosome: Motors, clocks, springs, and things. *Cell* **1998**, *92*, 315–326. [CrossRef]
47. Hales, C.M.; Seyfried, N.T.; Dammer, E.B.; Duong, D.; Yi, H.; Gearing, M.; Troncoso, J.C.; Mufson, E.J.; Thambisetty, M.; Levey, A.I.; et al. U1 small nuclear ribonucleoproteins (snrnps) aggregate in alzheimer's disease due to autosomal dominant genetic mutations and trisomy 21. *Mol. Neurodegener.* **2014**, *9*, 15. [CrossRef] [PubMed]
48. Hales, C.M.; Dammer, E.B.; Deng, Q.; Duong, D.M.; Gearing, M.; Troncoso, J.C.; Thambisetty, M.; Lah, J.J.; Shulman, J.M.; Levey, A.I.; et al. Changes in the detergent-insoluble brain proteome linked to amyloid and tau in alzheimer's disease progression. *Proteomics* **2016**, *16*, 3042–3053. [CrossRef] [PubMed]
49. Sadigh-Eteghad, S.; Sabermarouf, B.; Majdi, A.; Talebi, M.; Farhoudi, M.M.; Ahmoudi, J. Amyloid-beta: A crucial factor in alzheimer's disease. *Med. Princ. Pract.* **2015**, *24*, 1–10. [CrossRef] [PubMed]
50. Bloom, G.S. Amyloid-beta and tau: The trigger and bullet in alzheimer disease pathogenesis. *JAMA Neurol.* **2014**, *71*, 505–508. [CrossRef] [PubMed]
51. Vossel, K.A.; Zhang, K.; Brodbeck, J.; Daub, A.C.; Sharma, P.; Finkbeiner, S.; Cui, B.; Mucke, L. Tau reduction prevents abeta-induced defects in axonal transport. *Science* **2010**, *330*, 198. [CrossRef] [PubMed]
52. Zempel, H.; Luedtke, J.; Kumar, Y.; Biernat, J.; Dawson, H.; Mandelkow, E.; Mandelkow, E.M. Amyloid-beta oligomers induce synaptic damage via tau-dependent microtubule severing by ttll6 and spastin. *EMBO J.* **2013**, *32*, 2920–2937. [CrossRef] [PubMed]
53. Ebneth, A.; Godemann, R.; Stamer, K.; Illenberger, S.; Trinczek, B.; Mandelkow, E.-M.; Mandelkow, E. Overexpression of tau protein inhibits kinesin-dependent trafficking of vesicles, mitochondria, and endoplasmic reticulum: Implications for alzheimer's disease. *J. Cell Biol.* **1998**, *143*, 777–794. [CrossRef] [PubMed]
54. Lorenzi, M.; Lorenzi, T.; Marzetti, E.; Landi, F.; Vetrano, D.L.; Settanni, S.; Antocicco, M.; Bonassi, S.; Valdiglesias, V.; Bernabei, R.; et al. Association of frailty with the serine protease htra1 in older adults. *Exp. Gerontol.* **2016**, *81*, 8–12. [CrossRef] [PubMed]
55. Hong, S.; Beja-Glasser, V.F.; Nfonoyim, B.M.; Frouin, A.; Li, S.; Ramakrishnan, S.; Merry, K.M.; Shi, Q.; Rosenthal, A.; Barres, B.A.; et al. Complement and microglia mediate early synapse loss in alzheimer mouse models. *Science* **2016**, *352*, 712–716. [CrossRef] [PubMed]
56. Stephan, A.H.; Barres, B.A.; Stevens, B. The complement system: An unexpected role in synaptic pruning during development and disease. *Annu. Rev. Neurosci.* **2012**, *35*, 369–389. [CrossRef] [PubMed]
57. Nilsen, T.W. The spliceosome: The most complex macromolecular machine in the cell? *BioEssays* **2003**, *25*, 1147–1149. [CrossRef] [PubMed]
58. Hales, C.M.; Dammer, E.B.; Diner, I.; Yi, H.; Seyfried, N.T.; Gearing, M.; Glass, J.D.; Montine, T.J.; Levey, A.I.; Lah, J.J. Aggregates of small nuclear ribonucleic acids (snrnas) in alzheimer's disease. *Brain Pathol.* **2014**, *24*, 344–351. [CrossRef] [PubMed]
59. Diner, I.; Hales, C.M.; Bishof, I.; Rabenold, L.; Duong, D.M.; Yi, H.; Laur, O.; Gearing, M.; Troncoso, J.; Thambisetty, M.; et al. Aggregation properties of the small nuclear ribonucleoprotein u1-70k in alzheimer disease. *J. Biol. Chem.* **2014**, *289*, 35296–35313. [CrossRef] [PubMed]
60. Kato, M.; Han, T.W.; Xie, S.; Shi, K.; Du, X.; Wu, L.C.; Mirzaei, H.; Goldsmith, E.J.; Longgood, J.; Pei, J.; et al. Cell-free formation of rna granules: Low complexity sequence domains form dynamic fibers within hydrogels. *Cell* **2012**, *149*, 753–767. [CrossRef] [PubMed]
61. Kwon, I.; Kato, M.; Xiang, S.; Wu, L.; Theodoropoulos, P.; Mirzaei, H.; Han, T.; Xie, S.; Corden, J.L.; McKnight, S.L. Phosphorylation-regulated binding of rna polymerase ii to fibrous polymers of low-complexity domains. *Cell* **2013**, *155*, 1049–1060. [CrossRef] [PubMed]
62. Bai, B.; Chen, P.C.; Hales, C.M.; Wu, Z.; Pagala, V.; High, A.A.; Levey, A.I.; Lah, J.J.; Peng, J. Integrated approaches for analyzing u1-70k cleavage in alzheimer's disease. *J. Proteome Res.* **2014**, *13*, 4526–4534. [CrossRef] [PubMed]
63. Parmar, A.S.; Nunes, A.M.; Baum, J.; Brodsky, B. A peptide study of the relationship between the collagen triple-helix and amyloid. *Biopolymers* **2012**, *97*, 795–806. [CrossRef] [PubMed]

64. Kolev, M.V.; Ruseva, M.M.; Harris, C.L.; Morgan, B.P.; Donev, R.M. Implication of complement system and its regulators in alzheimer's disease. *Curr. Neuropharmacol.* **2009**, *7*, 1–8. [CrossRef] [PubMed]
65. Bonham, L.W.; Desikan, R.S.; Yokoyama, J.S. The relationship between complement factor c3, apoe epsilon4, amyloid and tau in alzheimer's disease. *Acta Neuropathol. Commun.* **2016**, *4*, 65. [CrossRef] [PubMed]
66. Xu, J.; Chatterjee, M.; Baguley, T.D.; Brouillette, J.; Kurup, P.; Ghosh, D.; Kanyo, J.; Zhang, Y.; Seyb, K.; Ononenyi, C.; et al. Inhibitor of the tyrosine phosphatase step reverses cognitive deficits in a mouse model of alzheimer's disease. *PLoS Biol.* **2014**, *12*, e1001923. [CrossRef] [PubMed]
67. McIntire, L.B.; Berman, D.E.; Myaeng, J.; Staniszewski, A.; Arancio, O.; Di Paolo, G.; Kim, T.W. Reduction of synaptojanin 1 ameliorates synaptic and behavioral impairments in a mouse model of alzheimer's disease. *J. Neurosci.* **2012**, *32*, 15271–15276. [CrossRef] [PubMed]
68. De Jesús-Cortés, H.J.; Nogueras-Ortiz, C.J.; Gearing, M.; Arnold, S.E.; Vega, I.E. Amphiphysin-1 protein level changes associated with tau-mediated neurodegeneration. *Neuroreport* **2012**, *23*, 942–946. [CrossRef] [PubMed]
69. Hernández-Zimbrón, L.F.; Rivas-Arancibia, S. Syntaxin 5 overexpression and β-amyloid 1-42 accumulation in endoplasmic reticulum of hippocampal cells in rat brain induced by ozone exposure. *BioMed. Res. Inter.* **2016**, *2016*, 2125643. [CrossRef] [PubMed]
70. Chatterjee, P.; Roy, D. Structural insight into grip1-pdz6 in alzheimer's disease: Study from protein expression data to molecular dynamics simulations. *J. Biomol. Struct. Dyn.* **2017**, *35*, 2235–2247. [CrossRef] [PubMed]
71. Ghosh, D.; Levault, K.R.; Brewer, G.J. Relative importance of redox buffers gsh and nad(p)h in age-related neurodegeneration and alzheimer disease-like mouse neurons. *Aging Cell* **2014**, *13*, 631–640. [CrossRef] [PubMed]
72. Shaerzadeh, F.; Motamedi, F.; Minai-Tehrani, D.; Khodagholi, F. Monitoring of neuronal loss in the hippocampus of abeta-injected rat: Autophagy, mitophagy, and mitochondrial biogenesis stand against apoptosis. *Neuromol. Med.* **2014**, *16*, 175–190. [CrossRef] [PubMed]
73. Alavi, M.V.; Fuhrmann, N. Dominant optic atrophy, opa1, and mitochondrial quality control: Understanding mitochondrial network dynamics. *Mol. Neurodegener.* **2013**, *8*, 32. [CrossRef] [PubMed]
74. Bishof, I.; Dammer, E.B.; Duong, D.M.; Kundinger, S.R.; Gearing, M.; Lah, J.J.; Levey, A.I.; Seyfried, N.T. Rna-binding proteins with basic-acidic dipeptide (bad) domains self-assemble and aggregate in alzheimer's disease. *J. Biol. Chem.* **2018**, *293*, 11047–11066. [CrossRef] [PubMed]
75. Aoyagi, T.; Wada, T.; Nagai, M.; Kojima, F.; Harada, S.; Takeuchi, T.; Takahashi, H.; Hirokawa, K.; Tsumita, T. Increased gamma-aminobutyrate aminotransferase activity in brain of patients with alzheimer's disease. *Chem. Pharm. Bull.* **1990**, *38*, 1748–1749. [CrossRef] [PubMed]
76. Musiek, E.S.; Xiong, D.D.; Patel, T.; Sasaki, Y.; Wang, Y.; Bauer, A.Q.; Singh, R.; Finn, S.L.; Culver, J.P.; Milbrandt, J.; et al. Nmnat1 protects neuronal function without altering phospho-tau pathology in a mouse model of tauopathy. *Ann. Clin. Transl. Neurol.* **2016**, *3*, 434–442. [CrossRef] [PubMed]
77. Mahlapuu, R.; Viht, K.; Balaspiri, L.; Bogdanovic, N.; Saar, K.; Soomets, U.; Land, T.; Zilmer, M.; Karelson, E.; Langel, U. Amyloid precursor protein carboxy-terminal fragments modulate g-proteins and adenylate cyclase activity in alzheimer's disease brain. *Mol. Brain Res.* **2003**, *117*, 73–82. [CrossRef]
78. Tong, Y.; Sun, Y.; Tian, X.; Zhou, T.; Wang, H.; Zhang, T.; Zhan, R.; Zhao, L.; Kuerban, B.; Li, Z.; et al. Phospholipid transfer protein (pltp) deficiency accelerates memory dysfunction through altering amyloid precursor protein (app) processing in a mouse model of alzheimer's disease. *Hum. Mol. Genet.* **2015**, *24*, 5388–5403. [CrossRef] [PubMed]
79. Mastroeni, D.; Chouliaras, L.; Grover, A.; Liang, W.S.; Hauns, K.; Rogers, J.; Coleman, P.D. Reduced ran expression and disrupted transport between cytoplasm and nucleus; a key event in alzheimer's disease pathophysiology. *PLoS ONE* **2013**, *8*, e53349. [CrossRef] [PubMed]
80. Zhu, L.; Su, M.; Lucast, L.; Liu, L.; Netzer, W.J.; Gandy, S.E.; Cai, D. Dynamin 1 regulates amyloid generation through modulation of bace-1. *PLoS ONE* **2012**, *7*, e45033. [CrossRef] [PubMed]
81. Misawa, T.; Arima, K.; Mizusawa, H.; Satoh, J. Close association of water channel aqp1 with amyloid-beta deposition in alzheimer disease brains. *Acta Neuropathol.* **2008**, *116*, 247–260. [CrossRef] [PubMed]

82. Murphy, M.P.; LeVine, H. Alzheimer's disease and the amyloid-beta peptide. *J. Alzheimers Dis.* **2010**, *19*, 311–323. [CrossRef] [PubMed]
83. Raj, T.; Li, Y.I.; Wong, G.; Humphrey, J.; Wang, M.; Ramdhani, S.; Wang, Y.C.; Ng, B.; Gupta, I.; Haroutunian, V.; et al. Integrative transcriptome analyses of the aging brain implicate altered splicing in alzheimer's disease susceptibility. *Nat. Genet.* **2018**. [CrossRef] [PubMed]

proteomes

MDPI

Review

From Synapse to Function: A Perspective on the Role of Neuroproteomics in Elucidating Mechanisms of Drug Addiction

Luis A. Natividad [1,*,†], Matthew W. Buczynski [2,†], Daniel B. McClatchy [3] and John R. Yates III [1,3,*]

1 Department of Neuroscience, The Scripps Research Institute, La Jolla, CA 92037, USA
2 School of Neuroscience, Virginia Polytechnic Institute and State University, Blacksburg, VA 24061, USA; mwb@vt.edu
3 Department of Molecular Medicine, The Scripps Research Institute, La Jolla, CA 92037, USA; dmcclat@scripps.edu
* Correspondence: lnativi@scripps.edu (L.A.N.); jyates@scripps.edu (J.R.Y.III); Tel.: +01-858-784-8862 (J.R.Y.III)
† These authors contributed equally to this work.

Received: 13 November 2018; Accepted: 7 December 2018; Published: 9 December 2018

Abstract: Drug addiction is a complex disorder driven by dysregulation in molecular signaling across several different brain regions. Limited therapeutic options currently exist for treating drug addiction and related psychiatric disorders in clinical populations, largely due to our incomplete understanding of the molecular pathways that influence addiction pathology. Recent work provides strong evidence that addiction-related behaviors emerge from the convergence of many subtle changes in molecular signaling networks that include neuropeptides (neuropeptidome), protein-protein interactions (interactome) and post-translational modifications such as protein phosphorylation (phosphoproteome). Advancements in mass spectrometry methodology are well positioned to identify these novel molecular underpinnings of addiction and further translate these findings into druggable targets for therapeutic development. In this review, we provide a general perspective of the utility of novel mass spectrometry-based approaches for addressing critical questions in addiction neuroscience, highlighting recent innovative studies that exemplify how functional assessments of the neuroproteome can provide insight into the mechanisms of drug addiction.

Keywords: neuroproteome; drug abuse; neuropeptidomics; phosphorylation; interactome

1. Drug Addiction: A Dysregulation of Plasticity in Motivational Circuitry

Drug addiction is a chronic relapsing disorder characterized by compulsive drug-seeking, the loss of control in limiting drug intake and the emergence of negative emotional states during drug abstinence [1]. In the past few decades, substantial research has contributed to our understanding of the underlying circuitry that influences addictive behaviors. The transition from casual use to dependence is mediated by changes in multiple interconnected brain systems involved in the processing of reward, stress, hormonal regulation and cognitive function [2]. Collectively, these molecular changes drive synaptic plasticity and alter connectivity between these brain regions, thus influencing a multitude of maladaptive behaviors that have come to characterize substance use disorders.

Drugs of abuse are widely recognized to facilitate dopamine transmission in brain reward systems, contributing to a powerful hedonic and euphoric response that reinforces drug-taking behavior. As drug use continues, increased dosage and frequency of use induces pivotal changes in brain reward pathways (e.g., the mesolimbic dopamine system) such that the same circuits recruited initially respond differently upon re-exposure. Pharmacological manipulation of the dopamine system has long been recognized to reduce drug intake in preclinical models of addiction [3,4]; however, there are multiple reasons why blocking this system is problematic in the clinical setting [5]. More generally, due to the disruptive side effects in mood (dysphoria) and motor-based (tardive dyskinesia) function of dopamine receptor antagonists, the lack of patient compliance remains a substantial problem. Conversely, dopamine receptor agonists reduce drug intake but also facilitate signs of drug-seeking in preclinical models, underscoring the potential for provoking drug relapse. The identification of non-dopaminergic targets may therefore provide an alternative for therapeutic treatment of addictive disorders.

Although the deregulation of brain reward remains a critical symptom of addiction, the maladaptive behaviors exhibited by addicts during drug abstinence are indicative of a more complex pathology driven by changes in additional brain structures. While a diverse set of hypotheses have been developed to explain the transition from casual use to dependence [2,5–8], "neuroplasticity" often emerges as a common theme linking these ideas. From this perspective, addictive behaviors emerge from a collection of neuroadaptations in specific neuronal circuits and neuroanatomical regions. By understanding the molecular changes that influence drug-induced neuroplasticity, it may be possible to stall or reverse these changes through a combination of cognitive behavioral therapy and small molecule treatments. For these reasons, developing a more comprehensive understanding of the molecular changes that occur in the brain following chronic drug use remains an important step in treating addiction.

2. Identification of Druggable Targets for Treating Addiction Using a "Neuroproteomics" Approach

While the population of drug-addicted individuals continues to grow in the United States and worldwide, the small number of available treatment options has failed to address this growing burden. Drugs that are FDA-approved for other diseases have been evaluated off-label to treat addiction, yet have been met with limited success, highlighting a critical need to identify addiction-specific changes in the central nervous system (CNS) that can be harnessed for clinical therapeutics.

From a pre-clinical perspective, a better understanding of the molecular alterations in the CNS would offer a wealth of information regarding the mechanistic underpinnings of disease, products of disease pathology (biomarkers) and the identification of high-value targets for the development of precision medicine. In the simplest case, a small molecule therapeutic would selectively bind a protein target in order to intercept a molecular change that underlies an important behavioral construct, such as drug reinforcement. Here, we depict the brain reward pathway (Figure 1) consisting of dopamine neurons in the ventral regions of the midbrain that project onto medium spiny neurons in the nucleus accumbens, where an increase in dopamine release is associated with the positive hedonic qualities of abused drugs. A broad-scale investigation of these regions would be useful not only for identifying changes in protein expression (e.g., proteomics) but also for capturing aspects of the proteome that confer functional changes. This review will discuss the following neuroproteomic strategies that are currently being employed to elucidate this dynamic signaling network: (1) changes in protein-protein interactions (affinity-purification proteomics), (2) post-translational modifications (e.g., phosphoproteomics) that alter cellular signaling pathways and (3) *in vivo* monitoring of signaling peptides (neuropeptidomics). In each of these cases, proteomics offers a valuable means to identify targets in a more unbiased manner than conventional protein assays, and together they establish a foundation for the "hit-to-lead" optimization of novel druggable targets for addiction and related psychiatric disorders. Here, we demonstrate the value of applying a proteomics approach to the CNS by highlighting recent studies that utilize novel methods for elucidating the mechanisms of addictive disorders.

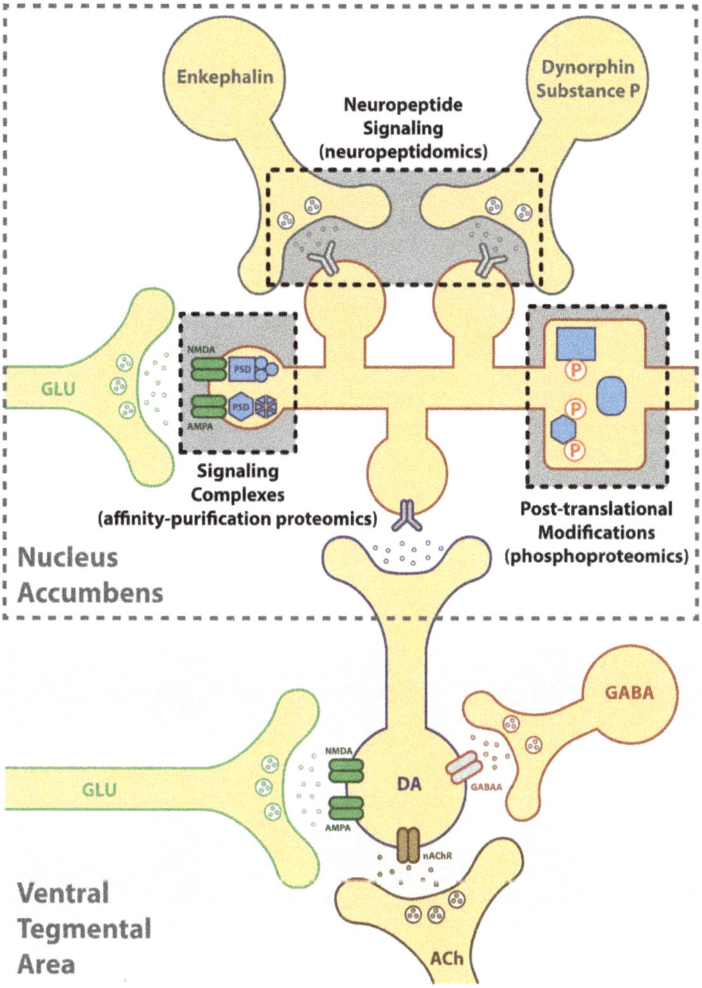

Figure 1. Application of neuroproteomic assessments in the study of addictive disorders. Synaptic plasticity underlying addiction-related behaviors can result from changes in (1) neuropeptide signaling (neuropeptidomics), (2) signaling protein complexes (affinity-based proteomics), (3) post-translational modifications such as phosphorylation (phosphoproteomics). DA: dopamine, GLU: glutamate, ACh: acetylcholine, GABA: gamma-aminobutyric acid.

3. Proteomics: Identifying Druggable Targets from Changes in Protein Expression

Traditional proteomic approaches have evaluated broad-scale changes in protein abundance in the brain following chronic drug exposure. The results have yielded a plethora of information on candidate targets, summarized elegantly in review articles for alcohol [9], morphine [10], and other psychostimulants [11]. Bottom-up proteomic strategies have greatly expanded the ability to identify the proteins in complex sample mixtures via the enzymatic digestion of proteins to generate peptides which are fragmented in the mass spectrometer [12]. Search algorithms (e.g., SEQUEST, Mascot) then match the fragmentation patterns of the peptides against theoretical spectra generated from protein databases, controlling for false positives with decoy strategies [13]. Thus, to maximize the number of peptides analyzed in biological mixtures, several components are often emphasized in method development: (a) sample protein preparation, (b) peptide fractionation (c) mass spectrometer acquisition, and (d) bioinformatic processing of the generated spectra. These approaches have been described at length in many excellent neuroproteomic reviews [14–18], and below we will mention a few issues regarding their implementation in addiction studies.

3.1. Current Limitations of Proteomics in Addiction Research

Obtaining a viable sample that is likely to contain relevant targets of addiction poses a critical challenge for neuroscientists. While anatomical structures provide some level of specificity, there is substantial heterogeneity at the molecular and behavioral level. An interesting example of this involves the dorsal striatum known to contain afferent dopaminergic terminals that are activated by drugs of abuse. While this structure is often dissected and analyzed as a singular region, site-specific inactivation has informed the rationale for distinguishing critical areas. In this regard, the dorsal medial region is important during acute drug exposure given its role in influencing goal-directed behavior that establishes early drug-cue associations; however, the dorsal lateral region becomes increasingly important as addictive behaviors become more automated, thus reflecting one of the core symptoms of addiction that influence compulsive intake [19–21]. Relatedly, methamphetamine only activates about 5–10% of neurons, suggesting that there may be a diluting component by the inactive majority when assessing whole tissue [22]. Moreover, subcellular locations within the brain are more likely to contain the sites of action of drugs of abuse (e.g., membrane/synaptic proteins, synaptosomes), suggesting that the enrichment of these compartments may result in a more viable fraction for further study. These technical aspects, as well as novel methods for evaluating the synaptosome [23], have been reviewed in much detail, yet the practical consideration remains that fractioning the proteome reduces the amount of starting material available. Thus, implementing a proteomics assessment often requires a delicate balance between the amount of protein required to identify a significant change and the amount that may be feasibly collected from an experimental preparation. For these reasons, the number of proteomics studies evaluating the effects of drug dependence remains relatively small when compared with other biological fields such as cancer and inflammation.

3.2. Identification of Important Changes in Expression

Identifying a druggable target from the many changes observed in the proteome remains a persistent challenge in neuroproteomics research. One strategy is to employ pathway analyses that can distill large amounts of information into known signaling networks, biological functions and associated disease states. A notable example from Salling et al. identified 29 proteins that were dysregulated by moderate alcohol drinking in the mouse amygdala using two-dimensional difference gel electrophoresis (2-DIGE) [24]. Ingenuity Pathway Analyses (IPA) revealed that many of these proteins annotated to neuronal signaling (CNS cell signaling, 14 proteins) and morphology (cell morphology, 8 proteins), while others were annotated to synaptic (excitatory plasticity, 7 proteins) and neurobehavioral (psychiatric disorders, 6 proteins) changes that were dysregulated by alcohol exposure. Of the total proteins identified, only calcium/calmodulin-dependent protein kinase II (CaMKII) was detected in each of these clusters, providing powerful support for the suggestion that this kinase plays a critical role during the early stages of alcohol dependence. In a separate study, Reissner et al. [25] used a BisoGenet plugin in Cytoscape to elucidate a glutamatergic signaling network based on the quantification of 42 proteins obtained from the nucleus accumbens of cocaine self-administering rats. Notably, the A-kinase anchoring protein 5 (AKAP5) located in the postsynaptic density was upregulated along with parallel changes in membrane-associated guanylate kinase markers such as discs large homolog-associated protein 3 (PSD-95). Follow-up studies applied the use of interfering peptide constructs to dissociate the interaction between AKAP and binding sequences for protein kinases that normally promote the insertion of ionotropic glutamate receptors, such as α-amino-3-hydroxy-5-methyl-4-isoxazolepropionic acid (AMPA).

The emergence of sample-labeling and multiplexing procedures provides an added dimension of research analysis of the brain proteome. Several strategies currently exist and are well documented in the literature (e.g., fluorophore, isobaric and metabolic labeling), each displaying their own unique advantage in terms of reducing the technical variation between mass spectrometric analysis and other aspects of labeling efficiency and sample preparation [26,27]. The recent emergence of multichannel labeling kits (e.g., iTRAQ and TMT) offer an attractive feature for neuroscientists who commonly employ more than two experimental groups. From an etiopathological perspective, it is feasible to gain insight into conditions that may result in discrepant changes at different stages of disease progression. For example, Lull et al. [28] utilized fluorophore labeling to compare the prefrontal cortical proteome in cocaine self-administering rats with 2-DIGE methods. The analysis comparing naïve versus cocaine-exposed rats experiencing acute or chronic abstinence yielded a total of 20 significant changes (e.g., synaptosomal-associated protein 25, dynamin-1), revealing a temporal pattern of protein expression classified as either (a) drug-induced changes that persisted into abstinence, (b) drug-induced changes that did not persist or (c) unique changes attributed exclusively to the abstinence period. The distributed pattern of expression argues against comprehensive treatment strategies that may miss a critical therapeutic window in which addicted individuals may be more or less prone to displaying altered cortical mechanisms. Nimitvilai et al. also investigated changes in the synaptosome in the orbitofrontal cortex of heavy alcohol-drinking macaque monkeys, a cortical region strongly implicated in decision-making and relapse [29]. iTRAQ labeling procedures identified 57 distinct protein changes in the synapse resulting from chronic alcohol exposure. Similar to Salling et al. [24], IPA analysis of all the differentially expressed proteins indicated a strong impact of alcohol on networks involved in cell-to-cell signaling, including a number of proteins that overlapped with glutamatergic function. Collectively, these studies demonstrate that synaptic transmission and excitatory signaling complexes represent important targets for understanding and treating dependence and addiction.

3.3. Confirmation of Targets Identified by Proteomics

A common strategy in proteomics research is to seek confirmatory evidence/validation of the targets derived from large-scale analyses with antibody-based molecular approaches. For instance, Nimitvilai et al. confirmed changes in a subtype of glutamate AMPA receptors (GluA1) in alcohol-dependent macaques using traditional western blot techniques [29], as validated antibodies exist for this target. However, confounds associated with commercial antibodies for western blot analysis, including issues with reliability (batch-dependent variations) and selectivity (cross-reactivity and antigen/epitope binding) [30], often render this technique as somewhat restricted to previously implicated targets with a limited capacity for identifying novel targets. To validate potential hits, many groups also rely on corresponding alterations in transcript levels as a correlate of central dogmatic principles. However, transcriptional changes often fail to reflect protein levels in CNS tissue and thus offer less predictable face validity than direct protein measurements. For this reason, targeted mass spectrometry approaches such as selected reaction monitoring (SRM) analysis have gained popularity for the confirmation and quantification of specific changes in the brain proteome [15,31].

To realize the ultimate goal of identifying druggable targets, the proteomics field would benefit from employing multiple strategies in elucidating proteome-derived targets. Exploring functional relevance with respect to whole organism behavior would enhance the translational value and application to the clinical population in question. For example, Salling et al. expanded on the initial proteome work in alcohol-drinking mice to elucidate a functional role for increased CaMKII subunit α expression in the amygdala, providing molecular and electrophysiological evidence of the strengthening of long-term potentiation (LTP) signaling ostensibly driven by CaMKII binding sites on GluA1 receptors [24]. These findings served as an impetus for site-anatomical procedures in which CaMKII inhibitors were injected directly into the amygdala and shown to have an ameliorating effect on symptoms of alcohol reinforcement, escalated drinking and relapse sensitivity [32–34]. Reissner et al. and other work from the Kalivas group have also applied the use of small-peptide inhibitors [25] and antisense oligonucleotides [35] to elegantly demonstrate both biological and behavioral roles in the tempering of cocaine reinstatement and drug-seeking behavior. Employing behavioral approaches may also distinguish between unforeseen predictions of applied treatment. For example, Chen et al. identified a network of proteins in the hippocampus that annotated to cyclin-dependent kinase 5 (CDK5) and ras homolog family member B (RhoB) signaling, both of which were upregulated in heroin self-administering rats [36]. Interestingly, the local infusion of a CDK5 inhibitor enhanced heroin intake, whereas a RhoB inhibitor reduced indices of heroin-seeking behavior as opposed to direct effects on self-administration. The authors concluded that CDK5 upregulation may have been a compensatory effect of repeated heroin exposure, whereas RhoB is likely to contribute to the sensitization of environmental cues that influence relapse. Collectively, this body of work highlights the inherent value of implementing multiple strategies for elucidating downstream pathways and target mechanisms of addiction.

4. Affinity-Based Proteomics: Protein Interactome as an Approach for Targeting Mechanisms of Addiction

Recent advancements in the isolation of neuronal circuits using chemogenetic or optogenetic tools clearly demonstrate the heterogeneity of brain structures involved in addiction [37,38]. Even within classifications of neuroanatomical structures, multiple overlaid circuits modulate distinct, and often opposing, behavioral responses. Synaptic activity-induced changes in receptor signaling complexes can drive neuroplasticity, which underlies behavioral abnormalities. Importantly, the discrete disruption of signaling complexes may circumvent the inherent side-effects often produced by pharmacotherapies that globally affect excitatory or inhibitory neural processing. For this reason, identifying relevant changes in protein signaling complexes represents an important approach in the optimization of potential druggable targets.

While substantial work has demonstrated an important role for nicotinic receptors in the development of nicotine dependence by targeting the ligand binding site, the proteins they interact with (i.e., the interactome) also participate in many important synaptic and immunological processes. Thus, drug-selective signaling complexes may represent unique opportunities for pharmacological intervention to disrupt addiction-related behaviors while minimizing therapeutic side-effects inherent in directly targeting nicotinic receptors. One such receptor complex in the hippocampus facilitates nicotine relapse-like behavior in rats [39]. In this study, Lui et al. demonstrated the formation of a complex between nicotinic α7 receptors and the glutamate receptor *N*-methyl-D-aspartate (NMDA) that is abundantly expressed in the hippocampus. To functionally evaluate this complex, they developed and validated small peptides that interfered with the formation of this complex without altering receptor-mediated signaling by either nicotinic α7 and NMDA. The intracerebroventricular delivery of this interfering peptide to the brain prevented the reinstatement of nicotine self-administration in rodents, providing evidence that this complex represents a viable druggable target for nicotine addiction.

Affinity-based enrichment can also provide insight into the network of protein interactors that may alter cellular signaling. For example, Wills et al. immunoprecipitated the NMDA receptor NR2B in hippocampal preparations to reveal a network of interacting proteins including scaffolding and PDZ-domain binding proteins that were dysregulated by chronic alcohol exposure [40]. Of the 64 proteins identified in synaptic fractions, the long-term depression (LTD) markers Arc and Homer 1 were observed to be upregulated, while the AMPA receptor GluA2 was downregulated in alcohol-dependent mice. The findings provided a basis for follow-up work elucidating a unique LTD mechanism in hippocampal electrophysiological recordings ostensibly driven by upstream increases in stress signaling molecules (i.e., adrenaline/noradrenaline) that are known to be amplified in the dependent state. Other work from Paulo et al. utilized a high affinity ligand for nicotinic α7 receptors (α-bgtx-conjugated beads) to identify 55 interacting proteins that were not present in α7-KO negative control samples [41]. The majority of the α7 receptor interactome were annotated to two pathways: (1) cell structure/protein trafficking and (2) signal transduction. Of interest, they identified multiple proteins related to GPCR signaling and phosphorylation, suggesting a more complex role for the α7 receptor in nicotine dependence beyond its canonical function as a calcium channel. McClure-Begley et al. performed similar analyses of the nicotinic α4β2 receptor using a β2 subunit antibody for immunopurification in conjunction with receptor subunit knockout mice [42] and identified 208 proteins in the α4β2 interactome. Subsequent use of this approach in post-mortem cortical tissue from nicotine-dependent mice and human subjects revealed 17 dose-dependent nicotine-induced changes in protein interactions in mice, with eight of these, including CaMKIIα, recapitulated in tissue from human smokers. The molecular relevance of the proposed interaction was recently elucidated by the Picciotto group using both *in vitro* and *in vivo* preparations [43]. These studies provide an excellent example of the utility of interactome studies in identifying novel protein interactions that may serve as the basis for developing precision medicine. Ongoing work in our laboratory is exploring the means by which immunoprecipitation methods may be applied towards the study of protein-protein interactors in mechanisms that mediate post-translational modifications (e.g., protein kinases) [44].

5. Phosphoproteomics: Signaling-Driven Phosphorylation States Underlying Addictive Behaviors

The detection of phosphoproteins has provided addiction scientists with a useful tool for measuring changes in activated states that may be devoid of changes in respective protein levels. Protein phosphorylation constitutes one of the most common post-translational modifications in protein biology, whereby the enzymatic addition or subtraction of a phosphate group onto nucleophilic residues (serine, threonine, or tyrosine) can alter the structural conformation of a protein, rendering it active, inactive or otherwise modifying its function. Though not exclusively linked to conformational changes, protein phosphorylation has been shown to serve as a molecular switch influencing a wide range of biological activity including signal transduction, cell differentiation/proliferation, protein-protein/-gene interactions and subcellular localization. Indeed, a large number of hypotheses recognize the importance of protein phosphorylation in directing the flow of molecular signaling, converging on key regulators of gene transcription (e.g., the cAMP response element-binding protein, delta fosB), membrane receptors (e.g., GluA1) and other important binding partners (e.g., transmembrane AMPA receptor regulatory proteins) that modulate neuroplasticity [45–47]. In this sense, several hundred kinases and phosphatases are encoded in the human genome and display a plethora of substrate targets [48]. A substantial component of receptor-mediated neuronal signaling involves the modulation of kinases and phosphatases, and in this regard, broad-scale approaches to the phosphoproteome are poised to contribute unique information into the role of phosphorylation states in addiction pathology.

Several aspects of the phosphoproteome are conducive to the identification of novel addiction-related targets. First, global assessments of the phosphoproteome often reveal an abundance of phosphoproteins involved in the regulation of phosphorylation states [49]. Indeed, numerous studies have elucidated the strong therapeutic potential of kinase inhibitors in oncology by targeting the phosphorylation-related constructs that drive cancer malignancy and metastasis [50]. Structurally, the conserved catalytic domain can be targeted with small-molecule inhibitors that exploit the biochemical features of the kinase core, leading to the generation of both reversible and irreversible inhibitors [50,51]. Likewise, protein phosphatases provide a complementary approach for addressing dysregulated phosphoproteins, and a number of successful drugs have been designed as substrate mimetics to target the active site of tyrosine phosphatase. Finally, a phosphoproteomics approach identifies unique peptide sequences in downstream protein substrates that are targeted by these mechanisms. The combined analysis of phospho-enriched proteins together with the unmodified proteome extrapolate well with identified nodes and canonical pathways, increasing the confidence of the candidate target [52,53]. Classical or allosteric approaches to modify the function of proteins by manipulating their phosphorylation state also provides an alternative approach to therapeutic development. In this sense, phosphoproteomics provides multiple avenues to gain insight into drug-related mechanisms of neuroplasticity and the development of precision medicine [54].

Presently, there are only a few studies that have applied a discovery-based phosphoproteomics approach in addiction models. Our laboratory recently utilized metabolic labeling procedures to quantify phosphopeptides enriched from the prefrontal cortex of rats receiving the psychedelic compound phencyclidine (PCP) [49]. We applied the analysis across experimental groups, allowing for the comparison of drug-induced effects relative to those incurred by an additional assessment of sensorimotor gating (i.e., prepulse inhibition) for evaluating schizophrenic-like phenotypes. In total, we identified approximately 120,000 phosphopeptides across experimental groups, 99,810 of which were confidently quantified using a nitrogen-heavy feeding protocol as an internal standard [55]. Overall, PCP treatment resulted in the downregulation of phosphorylation events that were enriched for LTP signaling. While consistent with the drug's mechanism of action (i.e., NMDA receptor blockade), the comparison of individual phosphosites revealed increased phosphorylation of proteins that regulate glutamatergic tone. Notably, the Serine 26 phosphosite on the light chain of the cysteine/glutamate transporter (SLC7A11) was hyperphosphorylated in PCP-treated rats undergoing prepulse inhibition testing, and site mutagenic procedures confirmed the role of this phosphosite in reducing glutamatergic uptake. This suggests that multiple levels of analysis (pathway, phosphosites and site mutagenics) are necessary to provide optimal insight into the molecular mechanisms influencing drug-induced glutamatergic dysregulation.

Another exceptional study by Rich et al. [56] utilized a label-free discovery approach to compare the amygdalar phosphoproteome in cocaine self-administering rats experiencing either drug-cue extinction or reconsolidation procedures. Microwave irradiation and phospho-enrichment procedures led to the quantification of phosphopeptides from 355 unique proteins, of which approximately 80 were compared across treatment conditions using SRM analysis. Interestingly, the authors reported 5 phosphopeptides that were regulated in the opposite manner by extinction and reconsolidation. This presents an attractive pattern of activation given that memories enter a labile state in which protein synthesis is required to re-stabilize drug-cue associations into long-term memory [57]. As extinction itself is driven by independent learning processes [58], molecules displaying less overlap between these conditions were hypothesized to serve as better therapeutic targets for ameliorating the effects of drug-cue memories. In this regard, a novel phosphosite (Serine 331) in the LTP-associated molecule CaMKIIα was examined further with site mutagenic procedures, ultimately showing that activation of this site reduced CaMKII catalytic activity. The localized infusion of CaMKII inhibitors into the amygdala prevented the reinstatement of cocaine-seeking, providing support for the assertion that CaMKIIα is critically involved in the retention of drug-cue memories that can trigger relapse. These findings are also in agreement with our recent work, displaying a similar pattern of upregulated CaMKIIα activity, albeit with the autophosphorylation site Threonine 286 in the dorsomedial prefrontal cortex of alcohol-dependent rats that may influence cognitive deficits during withdrawal [59]. Indeed, recent studies have demonstrated that genetic abrogation of Threonine 286 dysregulates the phosphorylation of other sites on CaMKII and further alters synaptic protein-protein interactions associated with neurodevelopmental disorders [60]. Pharmacological approaches may then target these regulatory phosphosites with peptidomimetic approaches that are shown to uniquely interface phosphoproteins [61].

6. Neuropeptides: Peptide Signals Driving Neuroplasticity in Addiction

Neuropeptides play a central role in the development and persistence of addictive behaviors [62]. For example, exogenous opiates such as oxycodone and heroin directly act on neuropeptide receptors in the brain to produce robust dopamine release, which is characteristic of drugs of abuse. Thus, both antagonists (naloxone) and partial agonists (buprenorphine) of these opiate receptors remain some of the few effective FDA-approved therapeutics for clinical use in treating drug addiction [63].

Research efforts studying the role of neuropeptides in addiction have not been limited to the opioid class, and a critical role has been established for a number of distinct neuropeptides, including (but not limited to) oxytocin [64], neuropeptide Y [65], substance P [66], and corticotropin releasing factor [67]. Notably, cocaine and amphetamine-regulated transcript (CART) represents a neuropeptide that was discovered due to its substantially enhanced expression following exposure to specific drugs of abuse [68].

Neuropeptides are short sequences of amino acids that can act like neurotransmitters, but with some critical distinctions that underlie their importance as potential druggable targets. While both neurotransmitters and neuropeptides are packaged in vesicles and released during neuronal activity, neuropeptides are expressed discretely throughout the CNS to facilitate specific behavioral responses. Neurons and glia have robust, rapid reuptake systems for most neurotransmitters, allowing the recycling of these molecular signals to reduce energy consumption in the CNS. In contrast, substantially more time and energy is required to produce a neuropeptide [69]; first, the pro-neuropeptide gene must be transcribed and translated into a pro-peptide sequence, then it is processed into active neuropeptides at consensus KK/KR sites by specific serine hydrolases. While neuropeptides are typically packaged into large dense-core vesicles that exhibit a longer latency for release and require prolonged stimulation [70], the mechanisms that regulate the vesicular loading and activity-dependent release of specific neuropeptides remain under investigation. Neuropeptides almost universally act at G-protein coupled receptors that do not directly produce action potentials [71], but instead modify neuronal responsivity by altering second messenger signaling (e.g., cAMP, IP3), phosphorylation states (via kinases, phosphatases), and protein levels (through changes in transcription, translation). In many cases, these factors cause neuropeptides to operate on longer time scales than typical neurotransmitters and allow them to exert hormone-like effects in the CNS that facilitate long-term behavioral changes.

Most established neuropeptides currently implicated in addiction were discovered prior to the development of modern proteomic approaches [72]. The idea that hormones acting as chemical messengers could transmit information over long distances originated over a hundred years ago, and the chemical identities of neuropeptide hormones implicated in addiction were discovered in the last century: oxytocin (1953), substance P (1971), enkephalins (1975), dynorphin (1979), and corticotrophin releasing factor (1981) [73–77]. In each case, the purification of each neuropeptide was performed from bulk tissue homogenates using sequential fractionation approaches, with a subsequent evaluation of each fraction for activity in basic *ex vivo* bioassays including the induction of uterine contractions, intestinal contractions, and secretion of other hormones. Likewise, their localization within the CNS was evaluated using antibody-based immunohistochemical techniques. The technical limitations of these approaches have biased studies toward high abundance neuropeptides, and as a result low abundance signals potentially dysregulated by abused drugs and other psychiatric disorders remain understudied.

Modern genetic and proteomic technologies have facilitated more comprehensive studies of neuropeptides. The human genome project (and corollary projects in related species) has identified over 70 genes that contain sequences capable of producing more than 1000 potential neuropeptides [78]. After considering the multitude of potential post-translational modifications, the technical challenges facing neuropeptidomics research becomes evident. The identification of these theoretical neuropeptides using matrix assisted laser desorption ionization (MALDI) or high performance liquid chromatography/electrospray ionization (HPLC/ESI) mass spectrometry has proven to be challenging, with many outstanding reviews outlining the technical issues with neuropeptide mass spectrometry [79–83]. Here, we highlight a few studies that have utilized a peptidomic approach to address critical biological questions regarding the role of neuropeptides in addiction and other psychiatric disorders.

6.1. Where Are Neuropeptides Located in the CNS?

Neuropeptide expression in discrete regions allows them to exert their effects on specific circuits to produce distinct behavioral outcomes. In the striatum, dopaminergic neurons regulate excitatory and inhibitory signaling to facilitate drug reward. Neuropeptides produce maladaptive effects by acting on specific neurons in distinct locations in the striatum and other limbic regions, ultimately contributing to drug craving and compulsive behaviors that influence relapse. Thus, the spatial location of precise neuropeptides plays a critical role in the development of addiction.

Hishimoto et al. investigated the localization of neuropeptides in the striatum using MALDI imaging mass spectrometry (MALDI-IMS) [84]. Mice exposed to acute nicotine were evaluated for changes in two distinct circuits in the striatum: (1) projections to the substantia nigra pars compacta (direct pathway) containing substance P and dynorphin, and (2) projections to the substantia nigra pars reticulate (indirect pathway) containing enkephalins. Using traditional immunohistochemical staining techniques, it is possible to distinguish (but not physically separate) these two pathways *in situ*. Hishimoto used MALDI-IMS with sufficient resolution (200 μm) to identify these structural features using substance P and enkephalin as molecular markers, and thereby demonstrated that nicotine administration decreased nigral substance P levels while increasing enkephalin levels. In total, 768 features were identified from mouse striatal spectra, and more than half of these were significantly regulated by nicotine. More importantly, they found that nicotine produces a negative correlation between substance P and other m/z species identified in those samples, whereas a positive correlation emerged between m/z identified in regions with enkephalin. These results suggest a more broad-scale remodeling of the direct (substance P) and indirect (enkephalin) pathways in the striatum, working in concert to facilitate behavioral changes produced by nicotine exposure.

Recent work has also identified the habenula as an important regulator of nicotine dependence. The habenular nuclei are morphologically and biochemically distinct structures that innervate reward circuitry: the medial habenula is thought to play a critical role in the maintenance of nicotine consumption, while the lateral habenula drives the aversive effects of the drug [85]. Immunohistochemical studies suggest that these regions have distinct repertoires of densely expressed neuropeptide receptors. For these reasons, Yang et al. utilized a tissue-stabilized liquid chromatography tandem mass spectrometry (LC-MS/MS) approach to identify the complimentary neuropeptidome in the habenula [86]. They identified 331 potential neuropeptides in this region, including many that were exclusive to either the medial (136 peptides) or lateral (51 peptides) habenula. While some of these peptides have well established roles in reward circuitry, the functions of others have not been widely investigated, suggesting that further studies investigating their involvement in addiction are warranted. Some of the neuropeptides identified remain more nebulous, as their putative receptors have not been fully established (e.g., secretogranins). Collectively, these results suggest that a comprehensive analysis of neuropeptide changes in emerging addiction structures such as the habenula will uncover new druggable targets for altering addictive behaviors.

6.2. Which Neuropeptides Get Released for Extracellular Signaling in the CNS?

In addition to exhibiting differences between brain structures, neuropeptide levels vary dramatically within brain structures at the subcellular level. Genetic sequence analyses predict that over 1000 potential neuropeptides may exist, with many having already been identified in CNS tissue using peptidomic approaches. In the striatum, Ye et al. identified 419 m/z that correspond to potential neuropeptides and discovered that many of these peptides were regulated by behavioral manipulation [87]. Specifically, they found that proSAAS derived peptides (big LEN, PEN, and little SAAS) were decreased in mice that were unfed (hungry). The microinjection of big LEN dramatically decreased food intake in regularly fed mice, demonstrating a clear functional link between big LEN and satiety, and suggesting that big LEN acts in the striatum through an extracellular, receptor-mediated mechanism.

For most studies, the analysis of neuropeptide content has typically been performed on samples from bulk tissue. As illustrated by analogous work studying neurotransmitters and endocannabinoid lipids in the CNS, there are some potential limitations to using this approach [88,89]. Briefly, it is not easy to distinguish neuropeptides found in vesicles and extracellular spaces (active signals) from those found in the endoplasmic reticulum and lysosomes (deactivated). Even with advanced sample handling technology [90,91], post-mortem peptide degradation remains a significant concern. In addition to these technical issues, environmental factors such as stress and circadian rhythm have a substantial impact on peptide hormonal levels. Given the large number of potential neuropeptides identified in comprehensive neuropeptidomics approaches, there is an unmet need for techniques that refine the population of peptides for identifying high-value targets.

Sampling using *in vivo* microdialysis addresses many of these concerns and provides a useful approach for identifying signaling-competent peptides in the CNS during drug exposure [88,92] (Figure 2). Through the implantation of a semi-permeable probe located within a specific brain structure, microdialysis allows for repeated sampling from awake and behaving animals. Moreover, microdialysate samples contain molecules in equilibrium with the extracellular space, providing an index for signaling-competent molecules available for binding to cell surface receptors. Due to the nature of this sampling approach, mass spectrometry analyses of microdialysates typically contain a lower neuropeptide diversity and content than traditional bulk tissue analyses. However, the neuropeptides identified using this technique have a greater likelihood of producing receptor-mediated effects, and they can be readily assessed for bioactivity using this approach. For example, Haskins et al. performed a peptidomic analysis of microdialysate samples collected from the striatum [93]. In this study, they used an untargeted LC-MS/MS approach and identified 3349 m/z features released in the rat striatum. From these data, they identified 29 potential neuropeptides from 6 different genes, including two peptides from enkephalin (PENK 198–207 and BAM 8–22). A similar study by Bernay et al. used multiple microdialysis and mass spectrometry approaches to more comprehensively investigate the striatal neuropeptidome and identified 97 peptides from these samples, including additional pro-enkephalin peptides PENK 114–133 and PENK 239–260 [94]. Given their lack of homology with met-enkephalin, these PENK peptides are unlikely to activate traditional opiate receptors and likely would exert their bioactivity through an alternative mechanism. Taken together, these studies indicate that pro-enkephalin biology may extend beyond traditional opioid receptor signaling, thereby warranting further investigation.

The reverse-dialysis of neuropeptides, a complementary technique to microdialysis, provides a rapid and effective readout for establishing *in vivo* bioactivity. Whereas microdialysis facilitates the diffusion of peptides in the brain to dialysate fluid in the probe, reverse-dialysis takes advantage of the ability to introduce potentially active substances into the dialysate fluid (e.g., neuropeptides) for subsequent distribution in the brain to their biological targets [89]. Thus, reverse-dialysis allows for the site-specific application of a potential neuropeptide to produce local effects on the CNS that can be assessed by changes in release of traditional neurotransmitters (e.g., glutamate, GABA, dopamine) implicated in addiction circuitry. Both Haskins et al. and Bernay et al. effectively used this approach to evaluate the bioactivity of their respective PENK peptides [93,94], demonstrating that both PENK 114–133 and PENK 198–207 increased glutamate release while suppressing GABA. Alternatively, PENK 239–260 produces robust increases in glutamate with no significant effect on GABA, whereas BAM8–22 activates GABA release. Collectively, these results suggest greater complexity in pro-enkephalin signaling than has been appreciated previously and implicate alternative strategies for targeting reward circuitry. Given the importance of the opioid system as a treatment for drug addiction, these pro-enkephalin peptides remain an important unanswered area in the study of addiction. More broadly, many of the neuropeptide genes implicated in addiction contain multiple peptides that can be identified *in vivo* using peptidomic strategies, and these peptides should be explored further as potential druggable targets.

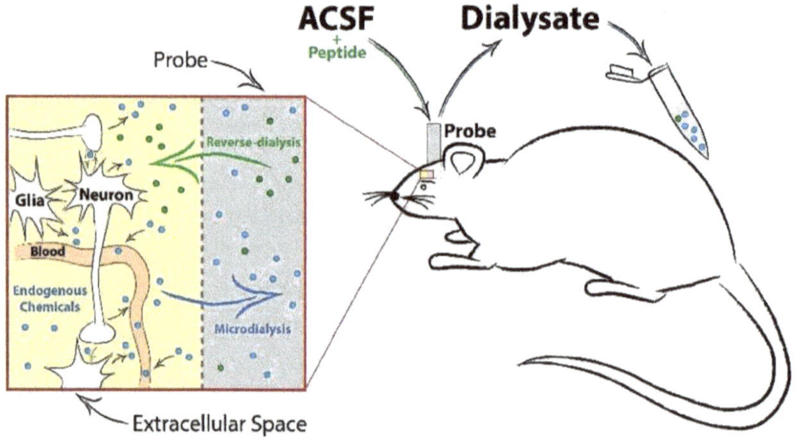

Figure 2. Schematic of an *in vivo* microdialysis probe setup, neurochemical diffusion and sample collection. Microdialysis sampling involves the implantation of a small-diameter probe into the brain region of interest. Artificial cerebrospinal fluid (ACSF) is perfused continuously into the probe, creating a concentration gradient at the semi-permeable membrane tip. This allows for the passive diffusion of extracellular transmitters (conventional dialysis) or solubilized compounds or drugs (reverse-dialysis) in the ACSF to freely enter or exit the probe. A timeline of collection can then be implemented to explore changes in neurotransmission before and after an experimental manipulation.

6.3. How do We Elucidate Neuropeptide Bioactivity?

The transition from peptide identification to target discovery remains a critical bottleneck in neuropeptidomics. As illustrated in the studies above, traditional behavioral and molecular biology approaches can establish a functional role for neuropeptides identified using mass spectrometry. As reported by Ye et al., peptides can be site-specifically infused into brain structures to evaluate behavioral outcomes and establish a functional link with the underlying biology [87]. Likewise, peptides can be site-specifically administered by reverse-dialysis to investigate changes in levels of traditional neurotransmitters such as glutamate and GABA [93,94]. Both approaches provide invaluable information about the bioactivity of a specific neuropeptide, yet fall short of elucidating specific receptor-mediated mechanisms. The following studies provide examples of different approaches for identifying specific receptors for orphan neuropeptides.

Given the importance of big LEN in food intake [95], Gomes et al. sought to identify the receptor responsible for its behavioral effects. Using the criteria of targeted receptor subclass (GPCR, $G_{i/o}$) and gene expression information (enriched in hypothalamus and Neuro2A cells), they limited the possibilities to four likely targets, ultimately revealing GPR171 as the endogenous big LEN receptor. Cells overexpressing GPR171 showed characteristic $G_{i/o}$-mediated decreases in cAMP levels following exposure to big LEN, and GPR171 siRNA delivered to the hypothalamus led to increased food intake in mice. Subsequent studies have identified a role for big LEN and GPR171 in anxiety-like behavior and fear conditioning [96], suggesting this system may influence addiction pathology. A small molecule agonist of GPR171 is now available [97], and future studies will evaluate its viability as a therapeutic target.

The proenkephalin-derived peptide BAM 8–22 has previously been identified as an endogenous anti-nociceptive peptide [98], and thus may counteract the actions of other opioid peptides. To identify receptor(s) that mediate the effects of BAM 8–22 and other orphan peptides, Kroeze et al. codon-optimized the Tango assay of Barnea et al. [99] and developed a method for the simultaneous and parallel interrogation of the entire human nonolfactory GPCRome (parallel receptorome expression and screening via transcriptional output, with transcriptional activation following arrestin translocation;

PRESTO-Tango). Although GPCRs can signal through a multitude of intracellular second messenger systems, thw sustained activation of nearly all GPCRs leads to the binding of β-arrestin and internalization, thereby providing a universal assay platform for screening a variety of receptors concurrently. Each Tango construct is engineered to promote β-arrestin recruitment and ultimately drive the expression of luciferase upon the binding of the ligand. Kroeze et al. designed and validated Tango constructs for nearly the entire human GPCRome (83%), enabling them to screen a single peptide against a wide range of potential receptors, including over 100 orphan GPCRs. Using this technology, BAM 8–22 was identified as a potent activator of Mas-related G-protein coupled receptor member X1 (MRGPRX1) and other members of this receptor family. This work helped facilitate the development of potential analgesics targeting MRGPRX1 [100], which may have reduced abuse liabilities as they do not activate the mu-opioid receptor. Further exploration of the MRGPRX family suggests that many opioid scaffolds can potently activate MRGPRX receptors [101], and that these receptors may be responsible for some of the undesirable side effects such as opioid tolerance [102]. More broadly, this technology provides a unique approach for uncovering links between peptidomics and orphan receptors and establishes a foundation for the hit-to-lead optimization of novel druggable targets for addiction and other psychiatric disorders.

7. Conclusions and Future Directions

The population of drug-addicted individuals continues to expand in the United States and worldwide, but the dearth of available treatment options has impeded progress in addressing this escalating burden. As our understanding of the neurobiology of addiction has evolved, it has become increasingly apparent that subtle changes in discrete brain circuits facilitate maladaptive behaviors associated with mood and cognition that ultimately lead to relapse and sustained drug use. Thus, the continued development of novel mass spectrometric-based methods such as cell-specific labeling [103] and single cell proteomics [104] in combination with currently available neuroscientific tools is critical for the detection of molecular drivers of addiction and their subsequent translation into viable clinical treatments. Future studies utilizing these approaches will undoubtedly improve our ability to interrogate brain circuitry, ultimately forging a unique path for identifying novel druggable targets of addiction and related psychiatric disorders.

Author Contributions: All authors contributed significantly to the conceptualization and framing of this review. L.A.N. and M.W.B. wrote the main text under the critical guidance and assistance of D.B.M. and J.R.Y.III.

Funding: This work was supported by the National Institutes of Health via the following mechanisms: From Alcohol Abuse and Alcoholism (K99-AA025393, L.A.N.); from Drug Abuse (R00-DA035865, M.W.B.); and from NIH (R01-MH067880, P41-GM103533: J.R.Y.III).

Acknowledgments: We are grateful for the advice, feedback and support from members of John Yates' and Marisa Roberto's laboratories at the Scripps Research Institute. We are especially grateful to Claire Delahunty for her contributions in editing this manuscript.

Conflicts of Interest: The authors declare no conflict of interest.

References

1. Hasin, D.S.; O'Brien, C.P.; Auriacombe, M.; Borges, G.; Bucholz, K.; Budney, A.; Compton, W.M.; Crowley, T.; Ling, W.; Petry, N.M.; et al. DSM-5 criteria for substance use disorders: recommendations and rationale. *Am. J. Psychiatry* **2013**, *170*, 834–851. [CrossRef] [PubMed]
2. Koob, G.F.; Volkow, N.D. Neurobiology of addiction: a neurocircuitry analysis. *Lancet Psychiatry* **2016**, *3*, 760–773. [CrossRef]
3. Awasaki, Y.; Nishida, N.; Sasaki, S.; Sato, S. Dopamine D(1) antagonist SCH23390 attenuates self-administration of both cocaine and fentanyl in rats. *Environ. Toxicol. Pharmacol.* **1997**, *3*, 115–122. [CrossRef]
4. Koob, G.F.; Le, H.T.; Creese, I. The D1 dopamine receptor antagonist SCH 23390 increases cocaine self-administration in the rat. *Neurosci. Lett.* **1987**, *79*, 315–320. [CrossRef]

5. Wise, R.A. D1- and D2-Type Contributions to Psychomotor Sensitization and Reward: Implications for Pharmacological Treatment Strategies. *Clin. Neuropharmacol.* **1995**, *18*, S74–S83. [CrossRef]
6. Robinson, T.E.; Berridge, K.C. The incentive sensitization theory of addiction: some current issues. *Philos. Trans. R. Soc. B Biol. Sci.* **2008**, *363*, 3137–3146. [CrossRef]
7. Russo, S.J.; Nestler, E.J. The brain reward circuitry in mood disorders. *Nat. Rev. Neurosci.* **2013**, *14*, 609–625. [CrossRef]
8. Scofield, M.D.; Heinsbroek, J.A.; Gipson, C.D.; Kupchik, Y.M.; Spencer, S.; Smith, A.C.; Roberts-Wolfe, D.; Kalivas, P.W. The Nucleus Accumbens: Mechanisms of Addiction across Drug Classes Reflect the Importance of Glutamate Homeostasis. *Pharmacol. Rev.* **2016**, *68*, 816–871. [CrossRef]
9. Gorini, G.; Harris, R.A.; Mayfield, R.D. Proteomic approaches and identification of novel therapeutic targets for alcoholism. *Neuropsychopharmacology* **2014**, *39*, 104–130. [CrossRef] [PubMed]
10. Abul-Husn, N.S.; Devi, L.A. Neuroproteomics of the synapse and drug addiction. *J. Pharmacol. Exp. Ther.* **2006**, *318*, 461–468. [CrossRef] [PubMed]
11. Kobeissy, F.H.; Zhang, Z.; Sadasivan, S.; Gold, M.S.; Wang, K.K. Methods in drug abuse neuroproteomics: methamphetamine psychoproteome. *Methods Mol. Biol.* **2009**, *566*, 217–228. [CrossRef] [PubMed]
12. Wolters, D.A.; Washburn, M.P.; Yates, J.R., 3rd. An automated multidimensional protein identification technology for shotgun proteomics. *Anal. Chem.* **2001**, *73*, 5683–5690. [CrossRef]
13. Zhang, Y.; Fonslow, B.R.; Shan, B.; Baek, M.C.; Yates, J.R., 3rd. Protein analysis by shotgun/bottom-up proteomics. *Chem. Rev.* **2013**, *113*, 2343–2394. [CrossRef]
14. Andrade, E.C.; Krueger, D.D.; Nairn, A.C. Recent advances in neuroproteomics. *Curr. Opin. Mol. Ther.* **2007**, *9*, 270–281.
15. Craft, G.E.; Chen, A.; Nairn, A.C. Recent advances in quantitative neuroproteomics. *Methods* **2013**, *61*, 186–218. [CrossRef] [PubMed]
16. Grant, K.J.; Wu, C.C. Advances in neuromembrane proteomics: efforts towards a comprehensive analysis of membrane proteins in the brain. *Brief. Funct. Genomic.* **2007**, *6*, 59–69. [CrossRef] [PubMed]
17. Hosp, F.; Mann, M. A Primer on Concepts and Applications of Proteomics in Neuroscience. *Neuron* **2017**, *96*, 558–571. [CrossRef]
18. Lull, M.E.; Freeman, W.M.; VanGuilder, H.D.; Vrana, K.E. The use of neuroproteomics in drug abuse research. *Drug Alcohol Depend.* **2010**, *107*, 11–22. [CrossRef]
19. Graybiel, A.M.; Grafton, S.T. The striatum: where skills and habits meet. *Cold Spring Harb. Perspect. Biol.* **2015**, *7*, a021691. [CrossRef]
20. Yin, H.H.; Knowlton, B.J.; Balleine, B.W. Lesions of dorsolateral striatum preserve outcome expectancy but disrupt habit formation in instrumental learning. *Eur. J. Neurosci.* **2004**, *19*, 181–189. [CrossRef]
21. Yin, H.H.; Ostlund, S.B.; Knowlton, B.J.; Balleine, B.W. The role of the dorsomedial striatum in instrumental conditioning. *Eur. J. Neurosci.* **2005**, *22*, 513–523. [CrossRef] [PubMed]
22. Liu, Q.R.; Rubio, F.J.; Bossert, J.M.; Marchant, N.J.; Fanous, S.; Hou, X.; Shaham, Y.; Hope, B.T. Detection of molecular alterations in methamphetamine-activated Fos-expressing neurons from a single rat dorsal striatum using fluorescence-activated cell sorting (FACS). *J. Neurochem.* **2014**, *128*, 173–185. [CrossRef] [PubMed]
23. Wang, Y.Z.; Savas, J.N. Uncovering Discrete Synaptic Proteomes to Understand Neurological Disorders. *Proteomes* **2018**, *6*. [CrossRef] [PubMed]
24. Salling, M.C.; Faccidomo, S.P.; Li, C.; Psilos, K.; Galunas, C.; Spanos, M.; Agoglia, A.E.; Kash, T.L.; Hodge, C.W. Moderate Alcohol Drinking and the Amygdala Proteome: Identification and Validation of Calcium/Calmodulin Dependent Kinase II and AMPA Receptor Activity as Novel Molecular Mechanisms of the Positive Reinforcing Effects of Alcohol. *Biol. Psychiatry* **2016**, *79*, 430–442. [CrossRef] [PubMed]
25. Reissner, K.J.; Uys, J.D.; Schwacke, J.H.; Comte-Walters, S.; Rutherford-Bethard, J.L.; Dunn, T.E.; Blumer, J.B.; Schey, K.L.; Kalivas, P.W. AKAP signaling in reinstated cocaine seeking revealed by iTRAQ proteomic analysis. *J. Neurosci.* **2011**, *31*, 5648–5658. [CrossRef]
26. Rauniyar, N.; Yates, J.R., 3rd. Isobaric labeling-based relative quantification in shotgun proteomics. *J. Proteome Res.* **2014**, *13*, 5293–5309. [CrossRef]
27. Rauniyar, N.; McClatchy, D.B.; Yates, J.R., 3rd. Stable isotope labeling of mammals (SILAM) for in vivo quantitative proteomic analysis. *Methods* **2013**, *61*, 260–268. [CrossRef]

28. Lull, M.E.; Erwin, M.S.; Morgan, D.; Roberts, D.C.; Vrana, K.E.; Freeman, W.M. Persistent proteomic alterations in the medial prefrontal cortex with abstinence from cocaine self-administration. *Proteomics Clin. Appl.* **2009**, *3*, 462–472. [CrossRef]
29. Nimitvilai, S.; Uys, J.D.; Woodward, J.J.; Randall, P.K.; Ball, L.E.; Williams, R.W.; Jones, B.C.; Lu, L.; Grant, K.A.; Mulholland, P.J. Orbitofrontal Neuroadaptations and Cross-Species Synaptic Biomarkers in Heavy-Drinking Macaques. *J. Neurosci.* **2017**, *37*, 3646–3660. [CrossRef]
30. Baker, M. Reproducibility crisis: Blame it on the antibodies. *Nature* **2015**, *521*, 274–276. [CrossRef]
31. MacLean, B.; Tomazela, D.M.; Shulman, N.; Chambers, M.; Finney, G.L.; Frewen, B.; Kern, R.; Tabb, D.L.; Liebler, D.C.; MacCoss, M.J. Skyline: an open source document editor for creating and analyzing targeted proteomics experiments. *Bioinformatics* **2010**, *26*, 966–968. [CrossRef] [PubMed]
32. Salling, M.C.; Hodge, C.J.; Psilos, K.E.; Eastman, V.R.; Faccidomo, S.P.; Hodge, C.W. Cue-induced reinstatement of alcohol-seeking behavior is associated with increased CaMKII T286 phosphorylation in the reward pathway of mice. *Pharmacol. Biochem. Behav.* **2017**, *163*, 20–29. [CrossRef] [PubMed]
33. Faccidomo, S.; Reid, G.T.; Agoglia, A.E.; Ademola, S.A.; Hodge, C.W. CaMKII inhibition in the prefrontal cortex specifically increases the positive reinforcing effects of sweetened alcohol in C57BL/6J mice. *Behav. Brain. Res.* **2016**, *298*, 286–290. [CrossRef]
34. Cannady, R.; Fisher, K.R.; Graham, C.; Crayle, J.; Besheer, J.; Hodge, C.W. Potentiation of amygdala AMPA receptor activity selectively promotes escalated alcohol self-administration in a CaMKII-dependent manner. *Addict. Biol.* **2017**, *22*, 652–664. [CrossRef] [PubMed]
35. Bowers, M.S.; McFarland, K.; Lake, R.W.; Peterson, Y.K.; Lapish, C.C.; Gregory, M.L.; Lanier, S.M.; Kalivas, P.W. Activator of G protein signaling 3: a gatekeeper of cocaine sensitization and drug seeking. *Neuron* **2004**, *42*, 269–281. [CrossRef]
36. Chen, Z.G.; Liu, X.; Wang, W.; Geng, F.; Gao, J.; Gan, C.L.; Chai, J.R.; He, L.; Hu, G.; Zhou, H.; et al. Dissociative role for dorsal hippocampus in mediating heroin self-administration and relapse through CDK5 and RhoB signaling revealed by proteomic analysis. *Addict. Biol.* **2017**, *22*, 1731–1742. [CrossRef] [PubMed]
37. Stamatakis, A.M.; Stuber, G.D. Optogenetic strategies to dissect the neural circuits that underlie reward and addiction. *Cold Spring Harb Perspect Med.* **2012**, *2*. [CrossRef]
38. Saunders, B.T.; Richard, J.M.; Janak, P.H. Contemporary approaches to neural circuit manipulation and mapping: focus on reward and addiction. *Philos. Trans. R. Soc. Lond. B Biol. Sci.* **2015**, *370*, 20140210. [CrossRef]
39. Li, S.; Li, Z.; Pei, L.; Le, A.D.; Liu, F. The alpha7nACh-NMDA receptor complex is involved in cue-induced reinstatement of nicotine seeking. *J. Exp. Med.* **2012**, *209*, 2141–2147. [CrossRef]
40. Wills, T.A.; Baucum, A.J., 2nd; Holleran, K.M.; Chen, Y.; Pasek, J.G.; Delpire, E.; Tabb, D.L.; Colbran, R.J.; Winder, D.G. Chronic intermittent alcohol disrupts the GluN2B-associated proteome and specifically regulates group I mGlu receptor-dependent long-term depression. *Addict. Biol.* **2017**, *22*, 275–290. [CrossRef]
41. Paulo, J.A.; Brucker, W.J.; Hawrot, E. Proteomic analysis of an α7 nicotinic acetylcholine receptor interactome. *J. Proteome Res.* **2009**, *8*, 1849–1858. [CrossRef] [PubMed]
42. McClure-Begley, T.D.; Esterlis, I.; Stone, K.L.; Lam, T.T.; Grady, S.R.; Colangelo, C.M.; Lindstrom, J.M.; Marks, M.J.; Picciotto, M.R. Evaluation of the Nicotinic Acetylcholine Receptor-Associated Proteome at Baseline and Following Nicotine Exposure in Human and Mouse Cortex. *eNeuro* **2016**, *3*. [CrossRef] [PubMed]
43. Miller, M.B.; Wilson, R.S.; Lam, T.T.; Nairn, A.C.; Picciotto, M.R. Evaluation of the Phosphoproteome of Mouse Alpha 4/Beta 2-Containing Nicotinic Acetylcholine Receptors In Vitro and In Vivo. *Proteomes* **2018**, *6*. [CrossRef] [PubMed]
44. McClatchy, D.B.; Yu, N.K.; Martinez-Bartolome, S.; Patel, R.; Pelletier, A.R.; Lavalle-Adam, M.; Powell, S.B.; Roberto, M.; Yates, J.R. Structural Analysis of Hippocampal Kinase Signal Transduction. *ACS Chem. Neurosci.* **2018**. [CrossRef] [PubMed]
45. Nestler, E.J. Molecular basis of long-term plasticity underlying addiction. *Nat. Rev. Neurosci.* **2001**, *2*, 119–128. [CrossRef] [PubMed]
46. Edwards, S.; Graham, D.L.; Whisler, K.N.; Self, D.W. Phosphorylation of GluR1, ERK, and CREB during spontaneous withdrawal from chronic heroin self-administration. *Synapse* **2009**, *63*, 224–235. [CrossRef] [PubMed]
47. Park, J. Phosphorylation of the AMPAR-TARP Complex in Synaptic Plasticity. *Proteomes* **2018**, *6*. [CrossRef]

48. Ardito, F.; Giuliani, M.; Perrone, D.; Troiano, G.; Lo Muzio, L. The crucial role of protein phosphorylation in cell signaling and its use as targeted therapy (Review). *Int. J. Mol. Med.* **2017**, *40*, 271–280. [CrossRef]
49. McClatchy, D.B.; Savas, J.N.; Martinez-Bartolome, S.; Park, S.K.; Maher, P.; Powell, S.B.; Yates, J.R., 3rd. Global quantitative analysis of phosphorylation underlying phencyclidine signaling and sensorimotor gating in the prefrontal cortex. *Mol. Psychiatry* **2016**, *21*, 205–215. [CrossRef]
50. Ferguson, F.M.; Gray, N.S. Kinase inhibitors: the road ahead. *Nat. Rev. Drug Discov.* **2018**, *17*, 353–377. [CrossRef]
51. Wu, P.; Nielsen, T.E.; Clausen, M.H. FDA-approved small-molecule kinase inhibitors. *Trends Pharmacol. Sci.* **2015**, *36*, 422–439. [CrossRef] [PubMed]
52. Ruse, C.I.; McClatchy, D.B.; Lu, B.; Cociorva, D.; Motoyama, A.; Park, S.K.; Yates, J.R., 3rd. Motif-specific sampling of phosphoproteomes. *J. Proteome Res.* **2008**, *7*, 2140–2150. [CrossRef] [PubMed]
53. Wu, R.; Dephoure, N.; Haas, W.; Huttlin, E.L.; Zhai, B.; Sowa, M.E.; Gygi, S.P. Correct interpretation of comprehensive phosphorylation dynamics requires normalization by protein expression changes. *Mol. Cell. Proteomics* **2011**, *10*, M111–009654. [CrossRef] [PubMed]
54. Ubersax, J.A.; Ferrell, J.E., Jr. Mechanisms of specificity in protein phosphorylation. *Nat. Rev. Mol. Cell. Biol.* **2007**, *8*, 530–541. [CrossRef] [PubMed]
55. McClatchy, D.B.; Dong, M.Q.; Wu, C.C.; Venable, J.D.; Yates, J.R., 3rd. 15N metabolic labeling of mammalian tissue with slow protein turnover. *J. Proteome Res.* **2007**, *6*, 2005–2010. [CrossRef] [PubMed]
56. Rich, M.T.; Abbott, T.B.; Chung, L.; Gulcicek, E.E.; Stone, K.L.; Colangelo, C.M.; Lam, T.T.; Nairn, A.C.; Taylor, J.R.; Torregrossa, M.M. Phosphoproteomic Analysis Reveals a Novel Mechanism of CaMKIIalpha Regulation Inversely Induced by Cocaine Memory Extinction versus Reconsolidation. *J. Neurosci.* **2016**, *36*, 7613–7627. [CrossRef]
57. Tronson, N.C.; Taylor, J.R. Molecular mechanisms of memory reconsolidation. *Nat. Rev. Neurosci.* **2007**, *8*, 262–275. [CrossRef]
58. Quirk, G.J.; Pare, D.; Richardson, R.; Herry, C.; Monfils, M.H.; Schiller, D.; Vicentic, A. Erasing fear memories with extinction training. *J. Neurosci.* **2010**, *30*, 14993–14997. [CrossRef]
59. Natividad, L.A.; Steinman, M.Q.; Laredo, S.A.; Irimia, C.; Polis, I.Y.; Lintz, R.; Buczynski, M.W.; Martin-Fardon, R.; Roberto, M.; Parsons, L.H. Phosphorylation of calcium/calmodulin-dependent protein kinase II in the rat dorsal medial prefrontal cortex is associated with alcohol-induced cognitive inflexibility. *Addict. Biol.* **2017**. [CrossRef]
60. Baucum, A.J., 2nd; Shonesy, B.C.; Rose, K.L.; Colbran, R.J. Quantitative proteomics analysis of CaMKII phosphorylation and the CaMKII interactome in the mouse forebrain. *ACS Chem. Neurosci.* **2015**, *6*, 615–631. [CrossRef]
61. Cai, D.; Lee, A.Y.; Chiang, C.M.; Kodadek, T. Peptoid ligands that bind selectively to phosphoproteins. *Bioorg Med. Chem. Lett.* **2011**, *21*, 4960–4964. [CrossRef] [PubMed]
62. Thiele, T.E. Neuropeptides and Addiction: An Introduction. *Int. Rev. Neurobiol.* **2017**, *136*, 1–3. [CrossRef] [PubMed]
63. Volkow, N.D.; Frieden, T.R.; Hyde, P.S.; Cha, S.S. Medication-assisted therapies–tackling the opioid-overdose epidemic. *N. Engl. J. Med.* **2014**, *370*, 2063–2066. [CrossRef] [PubMed]
64. Sarnyai, Z.; Kovacs, G.L. Oxytocin in learning and addiction: From early discoveries to the present. *Pharmacol. Biochem. Behav.* **2014**, *119*, 3–9. [CrossRef] [PubMed]
65. Goncalves, J.; Martins, J.; Baptista, S.; Ambrosio, A.F.; Silva, A.P. Effects of drugs of abuse on the central neuropeptide Y system. *Addict. Biol.* **2016**, *21*, 755–765. [CrossRef] [PubMed]
66. Schank, J.R.; Heilig, M. Substance P and the Neurokinin-1 Receptor: The New CRF. *Int. Rev. Neurobiol.* **2017**, *136*, 151–175. [CrossRef] [PubMed]
67. Roberto, M.; Spierling, S.R.; Kirson, D.; Zorrilla, E.P. Corticotropin-Releasing Factor (CRF) and Addictive Behaviors. *Int. Rev. Neurobiol.* **2017**, *136*, 5–51. [CrossRef]
68. Douglass, J.; McKinzie, A.A.; Couceyro, P. PCR differential display identifies a rat brain mRNA that is transcriptionally regulated by cocaine and amphetamine. *J. Neurosci.* **1995**, *15*, 2471–2481. [CrossRef]
69. Hook, V.; Funkelstein, L.; Lu, D.; Bark, S.; Wegrzyn, J.; Hwang, S.R. Proteases for processing proneuropeptides into peptide neurotransmitters and hormones. *Annu. Rev. Pharmacol. Toxicol.* **2008**, *48*, 393–423. [CrossRef]
70. Park, Y.; Kim, K.T. Short-term plasticity of small synaptic vesicle (SSV) and large dense-core vesicle (LDCV) exocytosis. *Cell. Signal.* **2009**, *21*, 1465–1470. [CrossRef]

71. Fricker, L.D.; Devi, L.A. Orphan neuropeptides and receptors: Novel therapeutic targets. *Pharmacol. Ther.* **2018**, *185*, 26–33. [CrossRef] [PubMed]
72. Burbach, J.P. What are neuropeptides? *Methods Mol. Biol.* **2011**, *789*, 1–36. [CrossRef] [PubMed]
73. Tuppy, H. The amino-acid sequence in oxytocin. *Biochim. Biophys. Acta* **1953**, *11*, 449–450. [CrossRef]
74. Chang, M.M.; Leeman, S.E.; Niall, H.D. Amino-acid sequence of substance P. *Nat. New Biol.* **1971**, *232*, 86–87. [CrossRef] [PubMed]
75. Hughes, J.; Smith, T.W.; Kosterlitz, H.W.; Fothergill, L.A.; Morgan, B.A.; Morris, H.R. Identification of two related pentapeptides from the brain with potent opiate agonist activity. *Nature* **1975**, *258*, 577–580. [CrossRef] [PubMed]
76. Goldstein, A.; Tachibana, S.; Lowney, L.I.; Hunkapiller, M.; Hood, L. Dynorphin-(1-13), an extraordinarily potent opioid peptide. *Proc. Natl. Acad. Sci. USA* **1979**, *76*, 6666–6670. [CrossRef]
77. Spiess, J.; Rivier, J.; Rivier, C.; Vale, W. Primary structure of corticotropin-releasing factor from ovine hypothalamus. *Proc. Natl. Acad. Sci. USA* **1981**, *78*, 6517–6521. [CrossRef]
78. Burbach, J.P. Neuropeptides from concept to online database www.neuropeptides.nl. *Eur. J. Pharmacol.* **2010**, *626*, 27–48. [CrossRef]
79. OuYang, C.; Liang, Z.; Li, L. Mass spectrometric analysis of spatio-temporal dynamics of crustacean neuropeptides. *Biochim. Biophys. Acta* **2015**, *1854*, 798–811. [CrossRef]
80. Romanova, E.V.; Sweedler, J.V. Peptidomics for the discovery and characterization of neuropeptides and hormones. *Trends Pharmacol. Sci.* **2015**, *36*, 579–586. [CrossRef]
81. Verdonck, R.; De Haes, W.; Cardoen, D.; Menschaert, G.; Huhn, T.; Landuyt, B.; Baggerman, G.; Boonen, K.; Wenseleers, T.; Schoofs, L. Fast and Reliable Quantitative Peptidomics with labelpepmatch. *J. Proteome Res.* **2016**, *15*, 1080–1089. [CrossRef] [PubMed]
82. Secher, A.; Kelstrup, C.D.; Conde-Frieboes, K.W.; Pyke, C.; Raun, K.; Wulff, B.S.; Olsen, J.V. Analytic framework for peptidomics applied to large-scale neuropeptide identification. *Nat. Commun.* **2016**, *7*, 11436. [CrossRef] [PubMed]
83. Kim, Y.G.; Lone, A.M.; Saghatelian, A. Analysis of the proteolysis of bioactive peptides using a peptidomics approach. *Nat. Protoc.* **2013**, *8*, 1730–1742. [CrossRef] [PubMed]
84. Hishimoto, A.; Nomaru, H.; Ye, K.; Nishi, A.; Lim, J.; Aguilan, J.T.; Nieves, E.; Kang, G.; Angeletti, R.H.; Hiroi, N. Molecular Histochemistry Identifies Peptidomic Organization and Reorganization Along Striatal Projection Units. *Biol. Psychiatry* **2016**, *79*, 415–420. [CrossRef] [PubMed]
85. Mathis, V.; Kenny, P.J. From controlled to compulsive drug-taking: The role of the habenula in addiction. *Neurosci. Biobehav. Rev.* **2018**. [CrossRef] [PubMed]
86. Yang, N.; Anapindi, K.D.B.; Rubakhin, S.S.; Wei, P.; Yu, Q.; Li, L.; Kenny, P.J.; Sweedler, J.V. Neuropeptidomics of the Rat Habenular Nuclei. *J. Proteome Res.* **2018**, *17*, 1463–1473. [CrossRef] [PubMed]
87. Ye, H.; Wang, J.; Tian, Z.; Ma, F.; Dowell, J.; Bremer, Q.; Lu, G.; Baldo, B.; Li, L. Quantitative Mass Spectrometry Reveals Food Intake-Induced Neuropeptide Level Changes in Rat Brain: Functional Assessment of Selected Neuropeptides as Feeding Regulators. *Mol. Cell. Proteomics* **2017**, *16*, 1922–1937. [CrossRef] [PubMed]
88. Buczynski, M.W.; Parsons, L.H. Quantification of brain endocannabinoid levels: methods, interpretations and pitfalls. *Br. J. Pharmacol.* **2010**, *160*, 423–442. [CrossRef]
89. De Luca, M.A.; Buczynski, M.W.; Di Chiara, G. Loren Parsons' contribution to addiction neurobiology. *Addict. Biol.* **2018**. [CrossRef]
90. Fridjonsdottir, E.; Nilsson, A.; Wadensten, H.; Andren, P.E. Brain Tissue Sample Stabilization and Extraction Strategies for Neuropeptidomics. *Methods Mol. Biol* **2018**, *1719*, 41–49. [CrossRef]
91. Boren, M. Sample preservation through heat stabilization of proteins: principles and examples. *Methods Mol. Biol.* **2015**, *1295*, 21–32. [CrossRef] [PubMed]
92. Van Wanseele, Y.; De Prins, A.; De Bundel, D.; Smolders, I.; Van Eeckhaut, A. Challenges for the in vivo quantification of brain neuropeptides using microdialysis sampling and LC-MS. *Bioanalysis* **2016**, *8*, 1965–1985. [CrossRef] [PubMed]
93. Haskins, W.E.; Watson, C.J.; Cellar, N.A.; Powell, D.H.; Kennedy, R.T. Discovery and neurochemical screening of peptides in brain extracellular fluid by chemical analysis of in vivo microdialysis samples. *Anal. Chem.* **2004**, *76*, 5523–5533. [CrossRef] [PubMed]

94. Bernay, B.; Gaillard, M.C.; Guryca, V.; Emadali, A.; Kuhn, L.; Bertrand, A.; Detraz, I.; Carcenac, C.; Savasta, M.; Brouillet, E.; et al. Discovering new bioactive neuropeptides in the striatum secretome using in vivo microdialysis and versatile proteomics. *Mol. Cell. Proteomics* **2009**, *8*, 946–958. [CrossRef] [PubMed]
95. Wardman, J.H.; Berezniuk, I.; Di, S.; Tasker, J.G.; Fricker, L.D. ProSAAS-derived peptides are colocalized with neuropeptide Y and function as neuropeptides in the regulation of food intake. *PLoS ONE* **2011**, *6*, e28152. [CrossRef] [PubMed]
96. Bobeck, E.N.; Gomes, I.; Pena, D.; Cummings, K.A.; Clem, R.L.; Mezei, M.; Devi, L.A. The BigLEN-GPR171 Peptide Receptor System Within the Basolateral Amygdala Regulates Anxiety-Like Behavior and Contextual Fear Conditioning. *Neuropsychopharmacology* **2017**, *42*, 2527–2536. [CrossRef] [PubMed]
97. Wardman, J.H.; Gomes, I.; Bobeck, E.N.; Stockert, J.A.; Kapoor, A.; Bisignano, P.; Gupta, A.; Mezei, M.; Kumar, S.; Filizola, M.; et al. Identification of a small-molecule ligand that activates the neuropeptide receptor GPR171 and increases food intake. *Sci. Signal.* **2016**, *9*, ra55. [CrossRef]
98. Chen, T.; Cai, Q.; Hong, Y. Intrathecal sensory neuron-specific receptor agonists bovine adrenal medulla 8-22 and (Tyr6)-gamma2-MSH-6-12 inhibit formalin-evoked nociception and neuronal Fos-like immunoreactivity in the spinal cord of the rat. *Neuroscience* **2006**, *141*, 965–975. [CrossRef]
99. Barnea, G.; Strapps, W.; Herrada, G.; Berman, Y.; Ong, J.; Kloss, B.; Axel, R.; Lee, K.J. The genetic design of signaling cascades to record receptor activation. *Proc. Natl. Acad. Sci. USA* **2008**, *105*, 64–69. [CrossRef]
100. Li, Z.; Tseng, P.Y.; Tiwari, V.; Xu, Q.; He, S.Q.; Wang, Y.; Zheng, Q.; Han, L.; Wu, Z.; Blobaum, A.L.; et al. Targeting human Mas-related G protein-coupled receptor X1 to inhibit persistent pain. *Proc. Natl. Acad. Sci. USA* **2017**, *114*, E1996–E2005. [CrossRef]
101. Lansu, K.; Karpiak, J.; Liu, J.; Huang, X.P.; McCorvy, J.D.; Kroeze, W.K.; Che, T.; Nagase, H.; Carroll, F.I.; Jin, J.; et al. In silico design of novel probes for the atypical opioid receptor MRGPRX2. *Nat. Chem. Biol.* **2017**, *13*, 529–536. [CrossRef] [PubMed]
102. Cai, Q.; Jiang, J.; Chen, T.; Hong, Y. Sensory neuron-specific receptor agonist BAM8-22 inhibits the development and expression of tolerance to morphine in rats. *Behav. Brain Res.* **2007**, *178*, 154–159. [CrossRef] [PubMed]
103. Cai, H.L.; Zhang, Y.; Fu, D.W.; Zhang, W.; Liu, T.; Yoshikawa, H.; Awaga, K.; Xiong, R.G. Above-room-temperature magnetodielectric coupling in a possible molecule-based multiferroic: triethylmethylammonium tetrabromoferrate(III). *J. Am. Chem. Soc.* **2012**, *134*, 18487–18490. [CrossRef] [PubMed]
104. Thompson, J.A.; Pande, H.; Paxton, R.J.; Shively, L.; Padma, A.; Simmer, R.L.; Todd, C.W.; Riggs, A.D.; Shively, J.E. Molecular cloning of a gene belonging to the carcinoembryonic antigen gene family and discussion of a domain model. *Proc. Natl. Acad. Sci. USA* **1987**, *84*, 2965–2969. [CrossRef] [PubMed]

proteomes

MDPI

Review

Exploring Morphine-Triggered PKC-Targets and Their Interaction with Signaling Pathways Leading to Pain via TrkA

Darlene A. Pena [1], Mariana Lemos Duarte [2], Dimitrius T. Pramio [1], Lakshmi A. Devi [2,*] and Deborah Schechtman [1,*]

[1] Department of Biochemistry, Chemistry Institute, University of São Paulo, Sao Paulo 05508-220, Brazil; darlenebqi@yahoo.com.br (D.A.P.); dimibiomed@hotmail.com (D.T.P.)

[2] Department of Pharmacological Sciences, Icahn School of Medicine at Mount Sinai, New York, NY 10029, USA; mlduarte@gmail.com

* Correspondence: lakshmi.devi@mssm.edu (L.A.D.); deborah@iq.usp.br (D.S.)

Received: 24 August 2018; Accepted: 2 October 2018; Published: 6 October 2018

Abstract: It is well accepted that treatment of chronic pain with morphine leads to μ opioid receptor (MOR) desensitization and the development of morphine tolerance. MOR activation by the selective peptide agonist, D-Ala2, N-MePhe4, Gly-ol]-enkephalin(DAMGO), leads to robust G protein receptor kinase activation, β-arrestin recruitment, and subsequent receptor endocytosis, which does not occur in an activation by morphine. However, MOR activation by morphine induces receptor desensitization, in a Protein kinase C (PKC) dependent manner. PKC inhibitors have been reported to decrease receptor desensitization, reduce opiate tolerance, and increase analgesia. However, the exact role of PKC in these processes is not clearly delineated. The difficulties in establishing a particular role for PKC have been, in part, due to the lack of reagents that allow the selective identification of PKC targets. Recently, we generated a conformation state-specific anti-PKC antibody that preferentially recognizes the active state of this kinase. Using this antibody to selectively isolate PKC substrates and a proteomics strategy to establish the identity of the proteins, we examined the effect of morphine treatment on the PKC targets. We found an enhanced interaction of a number of proteins with active PKC, in the presence of morphine. In this article, we discuss the role of these proteins in PKC-mediated MOR desensitization and analgesia. In addition, we posit a role for some of these proteins in mediating pain by TrKA activation, via the activation of transient receptor potential cation channel subfamily V member 1 (TRPV1). Finally, we discuss how these new PKC interacting proteins and pathways could be targeted for the treatment of pain.

Keywords: morphine; opioid receptors; conformational antibody; analgesia; GPCR signaling

---

1. Introduction

Morphine-Mediated Signal Transduction Pathways and Receptor Desensitization

Treatment of chronic pain has been a challenge as the most effective treatment that uses opiates has many unwanted side effects; for example, chronic exposure leads to desensitization of opioid receptors, development of tolerance, and addiction [1]. One of the alarming effects, reported in 2016, is that more than 100 people die daily due to opioid-related overdose (CDC/NCHS, National Vital Statistics System, Mortality, CDC Wonder, Atlanta, GA: US Department of Health and Human Services, CDC; 2017).

Opiates, such as morphine and heroin, interact with opioid receptors and it is generally thought that they function primarily via the activation of the μ opioid receptor (MOR), although, at high concentrations, they can activate δ and κ opioid receptors [2]. Opioid receptors are located both pre-and-post-synaptically and are coupled to the Gi/Go proteins. Upon ligand binding, Gi/Go-coupled receptors, acutely inhibit adenylyl cyclase (AC) activity, decreasing the levels of the cyclic AMP (cAMP) and decreasing the activity of the protein kinase A (PKA) [3], or of the exchange protein directly activated by cAMP (EPAC) [4]. Opioid receptor activation also leads to the stimulation of inward rectifying potassium channels and the inhibition of voltage-gated calcium channels, causing a decreased neurotransmitter release from the pre-synaptic nerve terminal. Thus, the net effect of acute opiate administration is to inhibit neuronal transmission, and this is thought to lead to analgesia [5].

Chronic opiate administration, on the other hand, has been shown to upregulate the activity of AC and PKA [6]. This upregulation of the cAMP pathway has been reported to occur in several regions of the brain, reduce analgesia, and is thought to contribute to opiate addiction [7].

In addition to PKA, opioid receptors have been shown to regulate a number of other kinases. Activation of opioid receptors leads to the activation of G protein-coupled receptor kinases (GRK), mitogen-activated protein kinase (MAPK), protein kinase B (PKB/AKT), calcium/calmodulin-dependent kinase II (CAMKII), and protein kinase C (PKC) [2,8–13]. Some of these kinases are thought to play a role in opiate-mediated tolerance, dependence [14], and addiction [15]. In the case of PKC, studies show that PKC inhibitors decrease receptor desensitization, development of opiate tolerance, and opiate addiction [16,17], (reviewed in [18]).

PKC is a family of serine/threonine kinases, composed of eleven different isoenzymes, divided into three sub-families. These include, (i) classical PKCs (cPKCs) including α, Bi, βII, and γ, which are calcium-dependent and are activated by phosphatidyl serine (PS) and diacylglycerol (DAG), (ii) novel PKCs including δ, ε, η, and θ, which are calcium-independent, but depend on PS and DAG for their activation, and (iii) atypical PKCs including ζ and λ/ι, which are calcium-independent [19] and are thought to be activated by protein–protein interactions [20]. Different PKC isoenzymes are expressed at different subcellular locations. For example, PKCα is found in both pre-and-post-synaptic sites, at the outer surface of synaptic vessels. However, PKCγ in adult rats is only expressed in postsynaptic dendrites, perikaryal cytoplasm, and postsynaptic densities. On the other hand, PKCε is found only in small and medium-sized dorsal root ganglion (DRG) neuronal soma, and presynaptic terminals of nociceptive neurons in the dorsal spinal horn (reviewed in [18]). It has not yet been determined if these PKCs are active, and what proteins they are interacting with, or are being phosphorylated by them.

Distinct PKC isoenzymes have been implicated in opioid receptor desensitization and addiction. Protein levels and activity of PKCα and γ are increased in the dorsal spinal cord, during chronic exposure to morphine [21,22]. Selective inhibitors for PKCα, γ, and ε completely reverse morphine-tolerance [18]. In particular, PKCγ has been suggested to play a central role in morphine tolerance, both in the spinal cord and the nucleus accumbens (NAc), having a role in sensory signal processing [23]. Determining the exact role of PKC, in addiction, has been difficult, due to the fact that PKC also plays a critical role in the formation and maintenance of memory [24,25], including drug-induced memory [26]. However, it is not clear how PKC is activated, following the activation of MOR, by morphine. Following are the possible mechanisms that follow from the MOR activation:

(i) MOR activation enables the Gβγ subunit to activate PLC which then would lead to PKC activation [27],
(ii) MOR activation leads to an activation of a Gq-coupled receptor that, in turn, leads to PLCβ activation, as seen in the case of M3 muscarinic receptor activation-mediated increase in the MOR desensitization [28].
(iii) MOR activation leads to activation of the receptor-coupled and non-coupled tyrosine kinases, which in turn lead to PLCγ activation.
(iv) MOR activation leads to the activation of a small G protein which would then activate PLCε and subtypes of PLCβ and γ [29].
(v) MOR activation leads to activation of PI3K which then activates PKC.

However, this last scenario has been shown to occur only in the case of atypical PKCs that are insensitive to DAG [30]. In intestinal epithelial cells, MOR has been shown to activate PI3K, via Gβγ, leading to a decrease in cell death [31]). Not much is also known about which PKC isoenzymes are activated by morphine. A recent study with DRG neurons and HEK-293 cells that were overexpressing MOR, showed that both PKCα and ε were activated at the plasma membrane within the first minute of the receptor activation by morphine, and that this activation was sustained for at least 20 min. This was specific to morphine, since MOR activation by DAMGO did not activate PKC, in this time frame. The authors also demonstrated that PKCα was activated by Gβγ, and led to MOR phosphorylation at specific sites, that restricted the plasma membrane localization of MOR and inhibited subsequent nuclear activation of the extracellular signal-regulated kinase (ERK) [32]. Sequential activation of PKCα and ε has been previously shown to be responsible for sustained ERK1/2 activation, upon ymechanical stress [33]. If this also happens in the case of MOR activation by morphine, or whether both PKCs are activated simultaneously, remains to be determined. Therefore, understanding the spatial and temporal dynamics of the PKC signaling can help us elucidate the mechanisms that lead to MOR desensitization, mediated by these kinases.

It is clear that PKC has an important role not only in receptor desensitization but also in inhibition of receptor recycling [18,32]. One of the main features that distinguishes morphine from other potent MOR agonists, such as DAMGO and fentanyl, is that morphine activates PKC signaling (and minimally activates GRK), whereas, DAMGO robustly activates GRK which phosphorylates the receptor and recruits β-arrestin, leading to a receptor endocytosis and recycling [32]. Even though PKC has been demonstrated to be a key target in morphine-mediated receptor desensitization, the mechanism by which PKC mediates this process is still not clear. Targeting PKC itself to decrease receptor desensitization could be problematic, as PKC is involved in several processes, including mediating immunological responses specifically against viral infections [34]. Thus, identifying PKC targets can be useful in elucidating the signal transduction processes involved in MOR desensitization and opioid-tolerance. MOR itself is a PKC target [35]. The carboxy-terminal tail of MOR contains 12 serine/threonine residues and two of them have the consensus sequence for phosphorylation by PKC. Mutations of eleven of these phosphorylation sites (including the two PKC sites) led to a functional receptor that was not desensitized or internalized, indicating that phosphorylation of MOR is important for receptor recycling [36,37]. A possibility that MOR activation of PKC leads to the phosphorylation of proteins other than the receptor, and that these PKC targets participate in desensitization, has not been well explored [35,38,39]. In order to address this issue, we developed a new strategy to identify the PKC interacting proteins/substrates within the context of an acute morphine treatment. For this, we used an antibody that specifically recognizes the active state of cPKCs (anti-C2Cat) [40]. Using this anti-C2Cat antibody, we immunoprecipitated active PKC-associated proteins from Neuro-2A cells treated with acute morphine, and identified the associated proteins by mass spectrometry. A number of proteins were identified, including a few known PKC targets.

In this article, we describe these proteins, discussing them in the context of pain mediated by nerve growth factor (NGF) signaling. In nociceptive neurons, NGF has a central role in pain. Inflammation leads to the release of NGF and activation of tropomyosin receptor kinase A (TrkA), a tyrosine kinase coupled to the NGF receptor [41]. NGF-binding leads to TrkA dimerization, auto-phosphorylation and subsequent binding and activation of PLCγ [42]. Amongst the several pathways activated by TrkA, PLCγ activation causes DAG generation and opening of an ion channel, the transient receptor potential cation channel subfamily V member 1 (TRPV1), a nonselective cation channel involved in a variety of nociceptive processes and activated by several stimuli (including acidic pH, heat, endocannabinoids, endogenous lipids, and capsaicin). Activation of TRPV1 causes a cation influx followed by depolarization and pain [43]. PKA and PKC bind to AKAP79/150, and this complex can then phosphorylate and activate TRPV1 [44] (Figure 1). One of morphine's targets is TRPV1 (reviewed in [45]). Blocking PKA-signaling by MOR, inhibits the TRPV1 channel activity and the TRPV1 active multimer-translocation to the membrane [46]. Moreover, a cAMP analog, 8Br-cAMP, can reverse the opioid-mediated inhibition of TRPV1, in DRG neurons [47]. Furthermore, blocking TRPV1 decreases morphine tolerance [48]. In DRG neurons and the spinal cord, TRPV1 and MOR are co-localized and their expression increases, upon inflammation [48,49]. These observations suggest that TRPV1 [50] and TrkA [51,52] could be drug targets for the development of non-opioid analgesics. Therefore, understanding the interaction between morphine-mediated analgesia and TrkA mediated pain could lead to the development of analgesics with lesser side-effects than the currently used drugs.

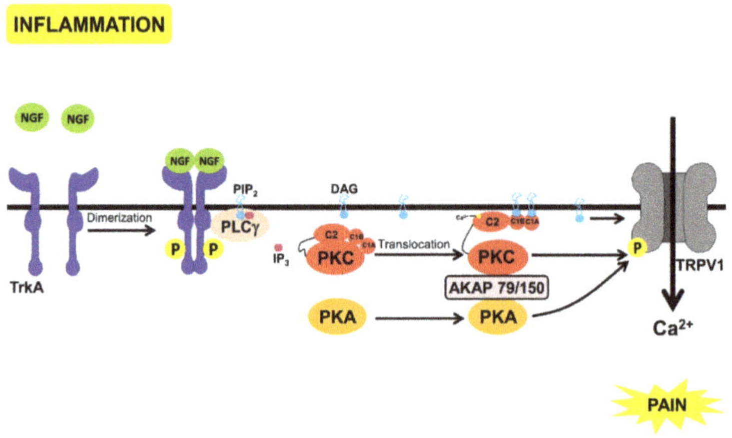

Figure 1. Schematic representation of the pathways that lead to pain through NGF-binding to TrkA. Inflammation leads to the secretion of NGF that binds to TrkA, leading to receptor dimerization and auto-phosphorylation, followed by PLCγ binding and activation. PLCγ cleaves inositol bisphosphate (PtdIns(4,5)P2) to inositol (1,4,5) trisphosphate and DAG. DAG directly activates TRPV1 [43] and the complex of PKC, PKA, and AKAP 79/150 that phosphorylates and also activates TRPV1 [42,44]).

2. Methods

2.1. Cell Culture

Mouse neuroblastoma (Neuro 2A) cell line was obtained from American Type Culture Collection (Manassas, VA, USA) and maintained in DMEM high Glucose, supplemented with 10% fetal bovine serum (FBS), penicillin/streptomycin [50 U/mL and 50 μg/mL, respectively, (Gibco-BRL®)], at 37 °C, under 5% CO_2.

2.2. Immunofluorescence

Neuro 2A cells were cultured on 13 mm glass coverslips, at 60% confluency, and treated with 50 nM PMA or 1 μM of Morphine or ATP for 1, 3, and 30 min. Cells were then fixed with 4% PFA, permeabilized with Phosphate buffered saline (PBS), 0.1% Triton X-100. Next, the cells were blocked in PBS, 0.1% Triton-X100, 1% normal goat serum, for 40 min, at room temperature. Cells were subsequently incubated overnight at 4 °C, with anti-C2Cat serum [40], diluted 1:100 in blocking solution, or anti PKCα (4 μg/mL, Santa Cruz Biotechnology®) and incubated for 1 h, at room temperature (RT), with anti-rabbit conjugated to Alexa 555 (4 μg/mL), diluted as above. As a negative control, secondary antibodies were incubated alone (without prior incubation with primary antibodies) to assess the nonspecific-binding. Coverslips were mounted with Vectashield®/DAPI, and Immunofluorescence staining was analyzed using a Leica DM6000 fluorescent microscope. The level of fluorescence intensity from 12-15 images, for each condition, with an average of 60 cells per field, was quantified using ImageJ® software and the amount of cPKC activity in each treatment was normalized to the control levels (fluorescence detected in unstimulated cells), which was set to 100. Statistical analyses of ANOVA (Dunnett's test) was done.

2.3. Western Blot

To analyze the cPKC expression in Neuro2A cells, cells were cultured in 25-cm^2 flasks in DMEM high Glucose, supplemented with FBS and antibiotics, until they reached 60% to 80% confluence. The total cells were counted using a Neubauer chamber, to get 1×10^6 cells. The cells were then lysed in Laemmli buffer, run on 10% SDS-PAGE, and the separated polypeptide chains were transferred to nitrocellulose membranes, as described previously [40]. Membranes were incubated for 2 h, at RT, with anti-PKCα, βI, βII or γ (0.4 µg/mL, Santa Cruz Biotechnology®), and anti-α-tubulin (1:5000, Sigma-Aldrich®) as primary antibodies diluted in PBS/0.1% Tween-20, and 10% non-fat milk. Goat secondary antibodies, anti-rabbit IgG, and anti-mouse IgG, conjugated to horseradish peroxidase (GE Healthcare Life Science®), were diluted 1:1000 in PBS/0.1% Tween-20. For the negative control, membranes were incubated with the secondary antibodies only (without prior incubation with primary antibodies). Immunodetection was performed by chemiluminescence and the quantification was performed using ImageJ® software.

2.4. Real-Time PCR

Total RNA was extracted from 1×10^6 Neuro 2A cells with TRIzol® (ThermoFisher Scientific, Waltham, MA, USA), following the manufacturer's instructions. cDNA was synthesized with iScript cDNA synthesis kit (Bio-Rad®, Hercules, CA, USA). Real-time PCR reactions were performed with the PowerUp SYBR® Green Master Mix (ThermoFisher Scientific) on a 7500 Real-Time PCR System (Applied Biosystems, Foster City, CA, USA), using the reaction default parameters. Primers used on the reactions were the following: PKCα Fwd: 5′CTGGAGAACAGGGAGATCCA3′, Rev: 5′ ACTGGGGGTTGACATACGAG3′; PKCβI Fwd: 5′AGAAACTCGAACGCAAGGAG3′, Rev: 5′CGAGAAGCCAGCAAACTCAT3′; PKCβII Fwd: 5′ AGAAACTCGAACGCAAGGAG3′, Rev: 5′TCCTGATGACTTCCTGGTCA3′; PKCγ Fwd: 5′GGAAATTGCACCTCCTTTCA3′, Rev: 5′ACGAAGTCCGGGTTCACATA3′; GAPDH Fwd: 5′AGGTCGGTGTGAACGGATTTG3′, Rev: 5′GGGGTCGTTGATGGCAACA3′. Relative abundance of the transcripts was normalized by the expression of the *GAPDH* gene and calculated with the equation: $2^{20} - [Ct(PKC) - Ct(GAPDH)]$, where 2^{20} would be an arbitrary number of copies of GAPDH [53].

2.5. Cross-Linking Antibodies to Beads

The antibodies anti-C2Cat (5 µL) or pre-immune serum (5 µL) [40] were crosslinked to protein G beads [(50 µL) (Invitrogen®)], for use in immunoprecipitation assays. Briefly, beads were washed, by centrifugation, with 1 mg/mL BSA, in PBS. Anti-C2Cat (5 µL) or pre-immune serum (5 µL) were incubated with 50 µL of packed volume of protein G beads, for 1 h at 4 °C. Antibody-bound beads were washed with 1 mg/mL BSA, in PBS, and resuspended in PBS. Beads were incubated under rotation with dimethylpimelimidate [(DMP) (Sigma®)] solution, freshly made up in 0.2 M triethanolamine, with the pH readjusted to pH 8.2, for 30 min at RT. Subsequently, beads were washed with 0.2 M of triethanolamine, in PBS, by incubation for 5 min, at RT, under rotation and then incubated twice with an equal volume of DMP solution, for 5 min at RT, under rotation. The reaction was stopped by an addition of an equal volume of 50 mM ethanolamine, prepared in PBS. After 5 min of incubation at RT under rotation, the excess (non-bounded) antibody was removed by washing with 1M glycine pH 3.0, and beads were resuspended in a PBS buffer.

2.6. Immunoprecipitation Assays and iTRAQ®

For the immunoprecipitation assays, Neuro 2A cells were cultured in 75-cm^2 flasks, in DMEM high Glucose supplemented with FBS and antibiotics, until they reached 60% to 80% confluence (containing approximately 5 × 10^6 cells). Neuro 2A cells were then treated with either saline or morphine (1 μM), for 3 min, and lysed in PBS containing 1% Triton X-100, protease (Sigma-Aldrich®), and phosphatase [PhosphoStop™ (Sigma®)] inhibitor cocktails, followed by three freeze–thaw cycles. Cells were sonicated for 30 min, at 80 Hz (output), with a probe sonicator (Branson Sonifier 250). Cell lysates were precleared with protein G beads, for 1 h at 4 °C, incubated with crosslinked antibody-bound beads, overnight at 4 °C, and subsequently washed with PBS. Experimental triplicates of immunoprecipitated proteins with anti-C2Cat from control (vehicle-treated) or morphine-treated cells were performed and individual samples from the cells immunoprecipitated with pre-immune serum, in the presence or absence of morphine, were prepared, totaling 8 samples. Proteins were eluted with 0.1 M glycine pH 3.0 and subject to SDS-PAGE, followed by in-gel trypsin digestion. The resulting peptides were subjected to 8-plex iTRAQ® (isobaric tags for relative and absolute quantitation), essentially, as previously described [54]. Labeling was performed according to the manufacturer's protocol, using the following isobaric iTRAQ® tags: 113, 114, and 115 for saline-treated samples, immunoprecipitated with anti-C2Cat, 116, 117, and 118 for morphine-treated samples, immunoprecipitated with anti-C2Cat, 119, and 121 for saline and morphine, respectively, both immunoprecipitated with pre-immune serum. After that, the iTRAQ® labeled peptides from saline and morphine-treated samples were combined and subjected to strong cation exchange liquid chromatography (SCXLC). Fractions containing the labeled peptides were analyzed by RPLC-MS/MS on Obitrap Velos mass spectrometer. MS/MS spectra were searched against UniRef 100 mouse database, using Mascot search engine. A total of 2889 unique peptides, corresponding to 956 proteins, were identified. Homologous protein redundancy was reduced by Scaffold software (http://www.proteomesoftware.com/QPlus/ScaffoldQ+.html) to a minimum. The false discovery rate was less than 1%. Each protein was identified with at least one unique peptide, with FDR less than 1%.

To identify the proteins that preferably interacted more with cPKC, upon morphine treatment, proteins were analyzed using the Scaffold software, as discussed in the results section and Figure 3D. Further network analysis of PKC interaction proteins was performed using Strings (https://string-db.org/) webtool. Immunoprecipitated proteins that interacted more with anti-C2Cat, in the presence of morphine and PKCα, were analyzed for their interaction with each other, according to the following criteria. (i) text mining, (ii) experiments, and (iii) databases. The analysis was performed with a minimum confidence of 0.4. Seventeen proteins were shown to interact with PKCα, according to these criteria, and a network of the interaction amongst these proteins was made. The analysis was performed with a minimum confidence of 0.4.

3. Results

Previously, we have developed and characterized an antibody that preferentially recognizes active cPKC. This antibody was used to study the spatial and temporal dynamics of active cPKC, in SK-N-SH cells, activated by the DAG analog, Phorbol myristate acetate (PMA), or activated by ATP and glutamate [40]. The anti-active state cPKC antibody was named anti-C2Cat, since this antibody is directed towards an intramolecular interaction between the C2 and the catalytic domains of PKC that occurs only in an inactive kinase, and thus the epitope recognized by the antibody was only exposed upon activation [40]. Previously, we have used this antibody to immunoprecipitate cPKC from breast cancer cell lines and found higher levels of active cPKC in a metastatic breast cancer cell line (MDA-MB-231), compared to a non-metastatic cell line (MCF-7). We also, showed a significantly higher level of active cPKC in metastatic, triple negative breast cancer samples, as compared to estrogen receptor positive (ER+) samples [40].

In this study, we used anti-C2Cat to study the temporal dynamics of PKC activation, after stimulation of Neuro 2A with morphine, ATP, and PMA. As can be seen in Figure 2, PKC was active after 3 min of treatment, with 1 μM morphine, and its activation was faster (1 min) when ATP was used. Treatment with PMA led to the PKC activation, by 1 min, and this was sustained for up to 30 min, after treatment. Of the classical PKCs, PKCα was expressed the most, in Neuro2A cells, as seen in Figure 3A.

cPKCs were detected, both, by real-time PCR and Western blot with isoenzyme specific antibodies. Immunofluorescence with anti-PKCα antibodies, following treatment with 1 μM morphine, for 3 min, detected an increase in PKCα at the cell membrane (indicative of active PKC) (Figure 3A, bottom).

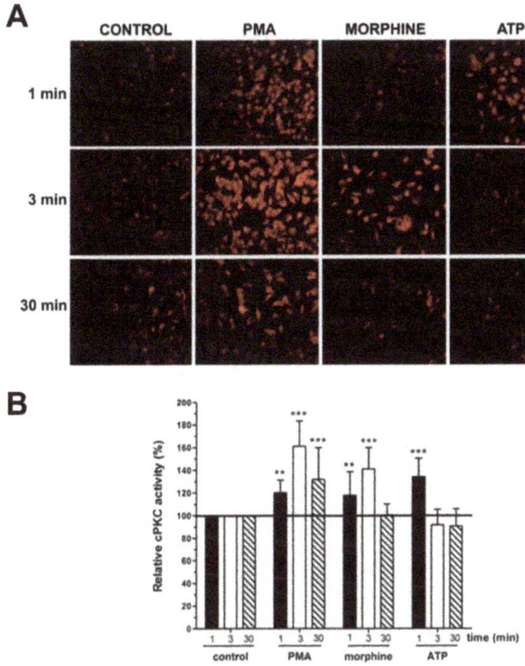

Figure 2. Morphine and Phorbol ester triggered cPKC activation with anti-C2Cat. (**A**) Neuro-2A cells were treated with 1 μM Morphine, ATP, or 100 nM PMA, for the indicated periods. Activators were removed by washing with PBS and the cells were fixed for immunofluorescence analysis. cPKC activation was performed using anti-C2Cat [40] as the primary antibody, and as a secondary antibody, when conjugated with Alexa 568. (**B**) The level of fluorescence intensity was quantified using ImageJ® software and the amount of cPKC activity, in each treatment, was normalized to the control levels (fluorescence detected in unstimulated cells), which was set to 100. Results represent the average ± SD of measurements from twelve to fifteen different images, for each condition, statistical significance was determined by ANOVA–Dunnett's test where *** represents $p < 0.001$ and ** represents $p < 0.01$.

To detect proteins and substrates with enhanced interaction with PKC, in response to MOR activation by morphine, immunoprecipitation experiments with anti-C2Cat (and pre-immune serum as a control) were performed with Neuro 2A cells, treated with and without 1μM morphine, for 3 min (Figure 3B). Immunoprecipitated proteins were labeled by iTRAQ® and submitted to mass spectrometry, as described in Methods. We identified 757 proteins, excluding putative and uncharacterized proteins. Based on the quantitative data of the immunoprecipitated proteins (analyzed using Scaffold 4 software), the following criteria were used to select proteins that exhibited enhanced interaction with PKC, in response to MOR activation by morphine:

(i) To detect the proteins that interacted with active PKC upon morphine treatment, we eliminated proteins that immunoprecipitated from cells not treated with morphine. Six hundred proteins met this criterion ($p \leq 0.05$, per t-test analysis) and these proteins exhibited a ratio of ≥ 2.5, when proteins immunoprecipitated from morphine-treated cells were compared with vehicle-treated cells.

(ii) To exclude proteins that bound non-specifically to the anti-C2Cat antibody, we eliminated proteins that interacted with pre-immune serum in the presence of morphine. This reduced the list to 557 proteins and these proteins exhibited a ratio ≥2.5, when proteins that were immunoprecipitated by anti-C2Cat antibody were compared to proteins immunoprecipitated with pre-immune serum (from cells treated with morphine).

(iii) To exclude proteins that interacted with PKC under basal conditions (absence of morphine), we only considered proteins that had a ratio ≤1.0, when comparing proteins immunoprecipitated with anti-C2Cat, to those with pre-immune serum, in vehicle-treated cells. This further reduced the list to 434 proteins.

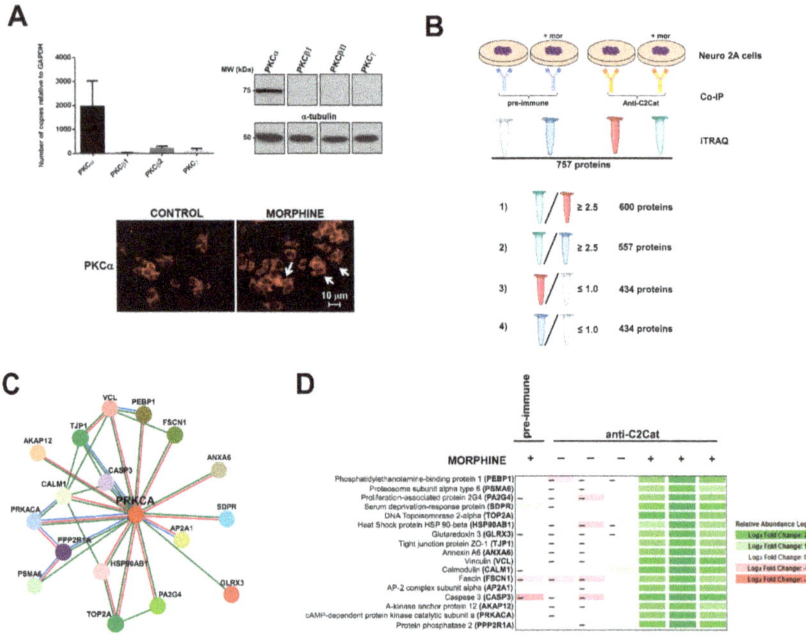

Figure 3. cPKC binding partners found in Neuro 2A cells that interacted more with anti-C2Cat, upon treatment with 1 μM morphine, for 3 min. (**A**) cPKC expression in Neuro2A cells analyzed by Real-time PCR (average of $n = 3$), and Western blot, with isoenzyme specific antibodies (top panels), immunofluorescence of PKCα, in control or morphine-treated cells (1 μM morphine for 3 min), white arrows indicate that PKCα was present in the membrane (indicative of PKC activation). (**B**) A representative diagram of how the proteins that interacted more with anti-C2Cat, upon morphine activation of MOR, were selected to eliminate non-specific binding (as discussed in the results section). (**C**) Using Strings (https://string-db.org/), a network with 17 PKCα interacting proteins was created (as discussed in the text). (1) Red lines indicate PKCα interacting proteins obtained from both experimental data and text mining (proteins that were co-mentioned). (2) Green lines indicate proteins that were co-mentioned. (3) Blue lines indicate putative homologs co-mentioned in other species. (**D**) A Heat map indicating a quantitative analysis of the intensities of the peptides found for each of the 17 proteins, which were found to interact more with anti-C2Cat, in the presence of morphine.

Of these, approximately 20% of the identified proteins were ribosomal proteins or proteins involved with RNA processing and 20% involved in metabolism. We did not immunoprecipitate PKC, because of limitations of mass spectrometry. However, when the immunoprecipitated proteins were subjected to Western Blotting, using an anti-PKCα antibody, we found higher levels of PKCα signal in the anti- C2Cat immunoprecipitate of cells treated with morphine (Supplementary Figure

S1). Next, we used Strings (https://string-db.org/) software to generate a network of the anti-C2Cat immunoprecipitated proteins, from cells treated with morphine; of these seventeen proteins were shown to directly interact with PKCα (Figure 3C). As can be seen in this figure some proteins interacted with each other, suggesting they could be in a complex of PKCα and associated binding proteins (Figure 3D). Other proteins in the list also interacted with the known PKCα-binding proteins previously identified, for example, eleven proteins have been previously reported to interact with the catalytic domain of PKA (PRKACA), including the regulatory domain. Below we discuss the role for some of these proteins in the context of the pain-signaling pathway, activated by the Nerve Growth Factor (NGF) and the Tropomyosin Receptor Kinase A (TrkA).

A heat map shown in Figure 3D summarizes the relative quantification of the 17 proteins immunoprecipitated with anti-C2Cat, from cells treated with morphine (and compares it to those immunoprecipitated following vehicle treatment or immunoprecipitated with pre-immune serum from morphine-treated cells). Data were analyzed using Scaffold 4 (Proteome Software, Inc.). It was clearly seen that these 17 proteins interacted more with anti-C2Cat antibody i.e., active cPKC, upon MOR stimulation with morphine, and that our strategy with activation-specific antibodies enriched the PKC binding partners (Figure 3D).

Below, we discuss some of the identified PKCα interacting proteins, detected with the anti-active-state specific cPKC antibody and a PKC substrate (neurogranin), previously validated in the literature [55], that could be regulating spatial and temporal dynamics, in the context of TrkA and MOR signaling.

3.1. *Phosphatidylethanolamine Binding Protein 1 (PEBP1)*

As discussed above, one of the main features that distinguish DAMGO from morphine is that MOR activated by DAMGO leads to robust GRK activation, while MOR activated by morphine activates PKC signaling, without significant GRK activation [32]. One of the mechanisms for this could be through the Phosphatidylethanolamine-binding protein 1 (PEBP1) phosphorylation by PKC. PEBP1, also known as Raf kinase inhibitory protein (RKIP), which inhibits Raf1-signaling and thus the MAPK activation. PKC phosphorylates Serine 153 of PEBP1, leading to its dimerization, as dimerized PEBP1 is known to interact with and inhibit GRK2 [56] (Figure 4). GRK2 phosphorylation of MOR was essential for arrestin-binding and endocytosis [18,57]. Interestingly, non-phosphorylated PEBP1 has also been shown to bind to δ-opioid receptors (DOR), and inhibit Gαβγ binding to DOR. Upon PKC activation, phosphorylated PEBP1 binds to GRK2, enabling G αβγ-binding and signaling, via DOR [58].

3.2. *Scaffolds (Annexin 6 and AKAP12) and PKA*

Spatial dynamics of signaling pathways is essential to position proteins at specific subcellular localizations. Adaptor proteins often mediate interactions between kinases and substrates, and help create microdomains. Of these, we found two scaffolds that have been shown to interact with PKC, annexin 6 [59], and AKAP12, also known as, Gravin, SSeCKS, or AKAP250 [60].

The role of PKC activation/inhibition of ERK by PEBP1 and Annexin 6 should be more carefully examined. Annexin 6 binds Ras-GTPase activating protein p120GAP, leading to Ras-GDP and consequently inhibiting ERK activation, besides being a scaffold protein for PKC [61]. Recently, Halls and collaborators (2016) showed that the spatial and temporal dynamics of ERK-activation is regulated by morphine versus DAMGO. They also showed that PKCα is a key player for this regulation, since it inhibits the receptor translocation within the plasma membrane which triggers a transient nuclear activation of the ERK, while sustaining cytosolic ERK-activation [32]. This study suggests that modulators of ERK-signaling, such as PEBP1 and Annexin 6, should be analyzed within the context of spatial-signaling, triggered by morphine.

AKAP12 is a PKA substrate and scaffold [62] that has been shown to interact with a β-adrenergic receptor and mediate its phosphorylation and desensitization. AKAP12 is localized at the plasma membrane and relocalizes to the cytoplasm, upon an increase in calcium in a PKC-dependent manner [60]. Indeed, we found AKAP12 and PKA (both regulatory and catalytic domains), interacting with active PKC, upon morphine treatment. The redistribution of PKA, by AKAP12, to the cytoplasm, upon PKC activation, could be a mechanism to inhibit the PKA membrane localization and the binding of other AKAPS (such as 79/50). Whether AKAP12 interacts with PKC, at the membrane, and is phosphorylated and relocalized to the cytoplasm, or whether PKCα relocalizes with AKAP12, upon activation, remains to be investigated. Interestingly, both, PKC and PKA phosphorylate TRPV1, at specific residues, increasing the channel activity and decreasing the threshold for channel activation (reviewed by [63]). These phosphorylations only occur upon binding of the kinases, PKA and PKC, to AKAP79/150 [44]. It is possible that morphine could not only be inhibiting the PKA activity through Gαi, but could also be mediating PKA (regulatory and catalytic domains) redistribution via PKC, and this should be further investigated.

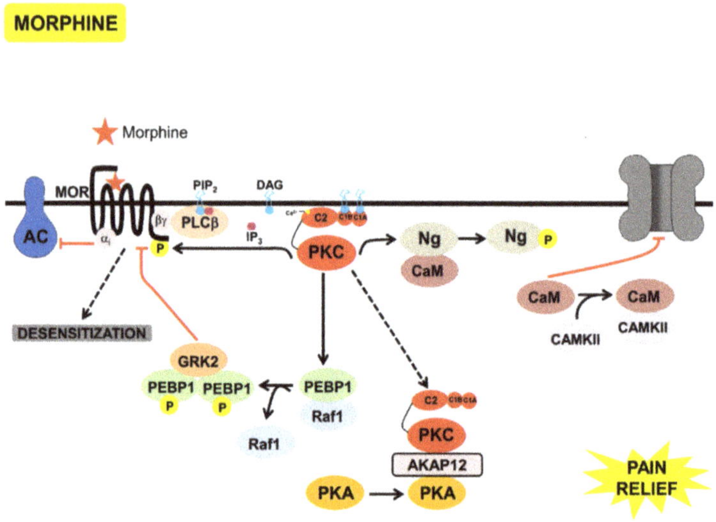

Figure 4. Model of the signaling processes mediated by acute morphine, that lead to desensitization of MOR and pain-relief, through inhibition of TRPV1. Morphine bound to MOR, leading to Gαi-mediated inhibition of AC and Gbγ-mediated activation of PKC. PKC phosphorylated neurogranin bound to calmodulin, calmodulin was then released and inhibited TRPV1. PKC phosphorylated PEBP1, releasing it from Raf1. Phosphorylated PEBP1 dimerized, bound to, and inhibited GRK2, inhibiting β-arrestin-mediated MOR endocytosis, leading to MOR desensitization. Active PKC bound to AKAP12, leading to PKA relocalization to the cytoplasm, and inhibition of TRPV1 phosphorylation by PKC/PKA.

3.3. Neurogranin and Calmodulin

Neurogranin (Ng) is one of the few well-characterized PKC substrates, in the context of morphine treatment [55]. Neurogranin is expressed in post-synaptic neuronal cell bodies and dendrites of the hippocampus, amygdala, striatum, and olfactory bulb, and interacts both with Ca^{2+}/calmodulin dependent protein kinase II (CAMKII) and PKC, and has been shown to be a regulator of CAMKII activity. At low levels of calcium, neurogranin binds Calmodulin (CaM), sequestering it and inhibiting it from binding to CAMKII. An increase in cytosolic calcium releases Ng from CaM, enabling it to bind to and activate CAMKII. Phosphorylation of neurogranin, by PKC, can also release CaM bound to Ng and lead to CAMKII activation. Free CaM can also bind directly to TRPV1, and prevent AKAP79/150 bound to PKC and PKA, from binding to TRPV1, consequently inhibiting the channel and causing a decrease in pain sensation (Figure 4).

An increase in neurogranin phosphorylation and in CaMKII activity was observed in the brains and spinal cord of opioid-tolerant mice [55]. Inhibition of neurogranin, through antisense oligodeoxynucleotides, decreased morphine dependence, and activation of CaMKII and of the transcription factor cAMP response element-binding protein (CREB) [64]. Further, neurogranin phosphorylation potentiates synaptic transmission and long-term potentiation (LTP) processes [65], which are important for the development of opioid tolerance. Neurogranin is an important link between PKC and CAMKII, which are two key proteins involved in the development of tolerance [55].

3.4. Morphine Inhibition of TrKA-Signaling Pathways via PKC

Signaling networks in the context of specific processes are complex and often involve cross-talk between different pathways and receptors. As discussed above, besides activation of PKC by Gβγ, in Gι coupled receptors, activation/sustained activation of PKC could possibly be mediated by subsequent activation of Gαq coupled receptors, tyrosine kinases, and small G proteins via PLC activation [28,29]

In the context of morphine signaling, we proposed a cross-talk between MOR, PKC, TrkA, and TRPV1, which is summarized below.

Upon inflammation, NGF is secreted and, via PLCγ, it activates TRPV1 leading to pain (Figure 1). On the other hand, attenuation of pain by acute morphine administration leads to the following:

(a) Inhibition of PKA by Gαi.
(b) Transient activation of cPKC, initially via Gβγ. Sustained PKC activation, possibly of nPKCs, could be mediated through other mechanisms discussed above.
(c) PKC-mediated phosphorylation of neurogranin, making CAM available to activate CAMKII and to bind to TRPV1, and inhibit the channel and nociception.
(d) Displacement of PKC and PKA, through scaffold proteins, annexin 6, and AKAP 12, inhibiting these kinases from binding to AKAP79/150 and activating TRPV1 (Figure 4).

In part, morphine-mediated desensitization could be due to the inhibition of endocytosis and receptor recycling via β-arrestin. Phosphorylation of PEBP1 by PKC leads to localized MAPK-activation and inactivation of GRK2, thereby, inhibiting MOR phosphorylation, internalization by β-arrestin, and receptor recycling. Upon chronic morphine treatment, PKA was activated [45] leading to an increase in NGF secretion [66]. PKC and PKA activation led to TRPV1 activation and pain, these mechanisms should be further explored in future studies. Other mechanisms of PKC activation upon chronic morphine exposure, such as EPAC-mediated PKCε activation, should also be explored [67]. MOR activation, in fact, has been shown to cause a decrease in EPAC activation [68].

4. Concluding Remarks

Despite the fact that PKC plays a role in MOR desensitization, the exact role and mechanisms that lead to receptor desensitization, via this kinase family, are still unclear. One of the reasons is the difficulty in identifying kinase interaction proteins. Using an active-state specific PKC antibody we identified proteins that interact with PKC upon MOR-activation and we discuss pathways that could be activated by these PKC interacting proteins, in the context of acute morphine treatment. Future studies will validate the proposed interactions. These proteins should be validated within the context of pain, in animal models.

Pain is discussed in the context of TrKA and TRPV1 activation, and its attenuation by morphine. Other cation channels involved in depolarization and pain could be affected in a similar manner. Due to the opioid epidemic, it is important to integrate signaling pathways and understand the molecular mechanisms involved in MOR desensitization, as this may lead to new strategies to modulate morphine-signaling and decreasing tolerance. Mutations in TrKA lead to *Congenital Insensitivity to Pain with Anhidrosis* (CIPA) [69,70], and a decrease in kinase activity observed in the naked mole rat leads to a reduction in pain sensitivity [71]. Efforts to develop selective inhibitors, for both TRPV1 [50] and TrKA [51,52], are underway. New and more specific therapeutic approaches towards pain can be developed, upon understanding the molecular mechanisms of these signaling pathways.

Supplementary Materials: The following are available online at http://www.mdpi.com/2227-7382/6/4/39/s1, Figure S1: PKCα was immunoprecipitated with anti-C2Cat antibody following morphine treatment. Western blot probing for PKCα in two of the three samples processed for protein quantification and identification experiments by Mass Spectrometry, in the presence and absence of 1 μM Morphine for three minutes. No reactivity with anti PKCα was found in samples immunoprecipitated with Pre-immune serum (data not shown).

Author Contributions: Conceptualization, L.A.D. and D.S.; Data curation, D.A.P.; Formal analysis, D.T.P. and D.S.; Funding acquisition, D.A.P. and D.S.; Investigation, M.L.D. and D.T.P.; Methodology, D.A.P., M.L.D. and D.T.P.; Project administration, L.A.D.; Resources, L.A.D.; Supervision, D.S.; Writing—original draft, L.A.D.; Writing—review & editing, M.L.D., L.A.D. and D.S.

Funding: This research was funded by NIH-NIDA, Fundação de Amparo Pesquisa do Estado de São Paulo (FAPESP) Grant 2015/21786-8 to DS and Coordenação de Aperfeiçoamento de Pessoal de Nível Superior (CAPES) Grant 88881.0622072014-01 to LAD & DS and NIH DA-08863 to LAD.

Acknowledgments: We would like to thank Ivone Gomes for critical reading of this manuscript and for the constructive comments.

Conflicts of Interest: The authors declare no conflict of interest.

References

1. Koob, G.F.; Sanna, P.P.; Bloom, F.E. Neuroscience of addiction. *Neuron* **1998**, *21*, 467–476. [CrossRef]
2. Williams, J.T.; Ingram, S.L.; Henderson, G.; Chavkin, C.; von Zastrow, M.; Schulz, S.; Koch, T.; Evans, C.J.; Christie, M.J. Regulation of mu-opioid receptors: desensitization, phosphorylation, internalization, and tolerance. *Pharmacol. Rev.* **2013**, *65*, 223–254. [CrossRef] [PubMed]
3. Koch, W.J.; Hawes, B.E.; Allen, L.F.; Lefkowitz, R.J. Direct evidence that Gi-coupled receptor stimulation of mitogen-activated protein kinase is mediated by G beta gamma activation of p21ras. *Proc. Natl. Acad. Sci. USA* **1994**, *91*, 12706–12710. [CrossRef] [PubMed]
4. Cheng, X.; Ji, Z.; Tsalkova, T.; Mei, F. Epac and PKA: a tale of two intracellular cAMP receptors. *Acta Biochim Biophys. Sin. (Shanghai)* **2008**, *40*, 651–662. [CrossRef] [PubMed]
5. Guitart, X.; Nestler, E.J. Second messenger and protein phosphorylation mechanisms underlying opiate addiction: Studies in the rat locus coeruleus. *Neurochem. Res.* **1993**, *18*, 5–13. [CrossRef] [PubMed]
6. Duman, R.S.; Tallman, J.F.; Nestler, E.J. Acute and chronic opiate-regulation of adenylate cyclase in brain: specific effects in locus coeruleus. *J. Pharmacol. Exp. Ther.* **1988**, *246*, 1033–1039. [PubMed]
7. Terwilliger, R.Z.; Beitner-Johnson, D.; Sevarino, K.A.; Crain, S.M.; Nestler, E.J. A general role for adaptations in G-proteins and the cyclic AMP system in mediating the chronic actions of morphine and cocaine on neuronal function. *Brain Res.* **1991**, *548*, 100–110. [CrossRef]
8. Fukuda, K.; Kato, S.; Morikawa, H.; Shoda, T.; Mori, K. Functional coupling of the delta-, mu-, and kappa-opioid receptors to mitogen-activated protein kinase and arachidonate release in Chinese hamster ovary cells. *J. Neurochem.* **1996**, *67*, 1309–1316. [CrossRef] [PubMed]
9. Law, P.Y.; Wong, Y.H.; Loh, H.H. Molecular mechanisms and regulation of opioid receptor signaling. *Annu. Rev. Pharmacol. Toxicol.* **2000**, *40*, 389–430. [CrossRef] [PubMed]
10. Narita, M.; Akai, H.; Nagumo, Y.; Sunagawa, N.; Hasebe, K.; Nagase, H.; Kita, T.; Hara, C.; Suzuki, T. Implications of protein kinase C in the nucleus accumbens in the development of sensitization to methamphetamine in rats. *Neuroscience* **2004**, *127*, 941–948. [CrossRef] [PubMed]
11. Polakiewicz, R.D.; Schieferl, S.M.; Dorner, L.F.; Kansra, V.; Comb, M.J. A mitogen-activated protein kinase pathway is required for mu-opioid receptor desensitization. *J. Biol. Chem.* **1998**, *273*, 12402–12406. [CrossRef] [PubMed]
12. Pak, Y.; O'Dowd, B.F.; George, S.R. Agonist-induced desensitization of the mu opioid receptor is determined by threonine 394 preceded by acidic amino acids in the COOH-terminal tail. *J. Biol. Chem.* **1997**, *272*, 24961–24965. [CrossRef] [PubMed]
13. Pei, G.; Kieffer, B.L.; Lefkowitz, R.J.; Freedman, N.J. Agonist-dependent phosphorylation of the mouse delta-opioid receptor: involvement of G protein-coupled receptor kinases but not protein kinase C. *Mol. Pharmacol.* **1995**, *48*, 173–177. [PubMed]
14. Liu, J.G.; Anand, K.J. Protein kinases modulate the cellular adaptations associated with opioid tolerance and dependence. *Brain Res. Brain. Res. Rev.* **2001**, *38*, 1–19. [CrossRef]

15. Ferrer-Alcon, M.; Garcia-Fuster, M.J.; La Harpe, R.; Garcia-Sevilla, J.A. Long-term regulation of signalling components of adenylyl cyclase and mitogen-activated protein kinase in the pre-frontal cortex of human opiate addicts. *J. Neurochem.* **2004**, *90*, 220–230. [CrossRef] [PubMed]
16. Smith, F.L.; Javed, R.R.; Elzey, M.J.; Dewey, W.L. The expression of a high level of morphine antinociceptive tolerance in mice involves both PKC and PKA. *Brain Res.* **2003**, *985*, 78–88. [CrossRef]
17. Hull, L.C.; Llorente, J.; Gabra, B.H.; Smith, F.L.; Kelly, E.; Bailey, C.; Henderson, G.; Dewey, W.L. The effect of protein kinase C and G protein-coupled receptor kinase inhibition on tolerance induced by mu-opioid agonists of different efficacy. *J. Pharmacol. Exp. Ther.* **2010**, *332*, 1127–1135. [CrossRef] [PubMed]
18. Bailey, C.P.; Smith, F.L.; Kelly, E.; Dewey, W.L.; Henderson, G. How important is protein kinase C in mu-opioid receptor desensitization and morphine tolerance? *Trends Pharmacol. Sci.* **2006**, *27*, 558–565. [CrossRef] [PubMed]
19. Nishizuka, Y. The molecular heterogeneity of protein kinase C and its implications for cellular regulation. *Nature* **1988**, *334*, 661–665. [CrossRef] [PubMed]
20. Newton, A.C. Protein kinase C: Poised to signal. *Am. J. Physiol. Endocrinol. Metab.* **2010**, *298*, E395–402. [CrossRef] [PubMed]
21. Bailey, C.P.; Oldfield, S.; Llorente, J.; Caunt, C.J.; Teschemacher, A.G.; Roberts, L.; McArdle, C.A.; Smith, F.L.; Dewey, W.L.; Kelly, E.; et al. Involvement of PKC alpha and G-protein-coupled receptor kinase 2 in agonist-selective desensitization of mu-opioid receptors in mature brain neurons. *Br. J. Pharmacol.* **2009**, *158*, 157–164. [CrossRef] [PubMed]
22. Malmberg, A.B.; Chen, C.; Tonegawa, S.; Basbaum, A.I. Preserved acute pain and reduced neuropathic pain in mice lacking PKCgamma. *Science* **1997**, *278*, 279–283. [CrossRef] [PubMed]
23. Song, Z.; Zou, W.; Liu, C.; Guo, Q. Gene knockdown with lentiviral vector-mediated intrathecal RNA interference of protein kinase C gamma reverses chronic morphine tolerance in rats. *J. Gene Med.* **2010**, *12*, 873–880. [CrossRef] [PubMed]
24. Sacktor, T.C.; Hell, J.W. The genetics of PKMzeta and memory maintenance. *Sci. Signal.* **2017**, *10*. [CrossRef] [PubMed]
25. Sun, M.K.; Alkon, D.L. Protein kinase C activators as synaptogenic and memory therapeutics. *Arch. Pharm. (Weinheim)* **2009**, *342*, 689–698. [CrossRef] [PubMed]
26. Li, Z.; Wu, C.F.; Pei, G.; Xu, N.J. Reversal of morphine-induced memory impairment in mice by withdrawal in Morris water maze: Possible involvement of cholinergic system. *Pharmacol. Biochem. Behav.* **2001**, *68*, 507–513. [CrossRef]
27. Zhu, X.; Birnbaumer, L. G protein subunits and the stimulation of phospholipase C by Gs-and Gi-coupled receptors: Lack of receptor selectivity of Galpha(16) and evidence for a synergic interaction between Gbeta gamma and the alpha subunit of a receptor activated G protein. *Proc. Natl. Acad. Sci. USA* **1996**, *93*, 2827–2831. [CrossRef] [PubMed]
28. Bailey, C.P.; Kelly, E.; Henderson, G. Protein kinase C activation enhances morphine-induced rapid desensitization of mu-opioid receptors in mature rat locus ceruleus neurons. *Mol. Pharmacol.* **2004**, *66*, 1592–1598. [CrossRef] [PubMed]
29. Gresset, A.; Sondek, J.; Harden, T.K. The phospholipase C isozymes and their regulation. *Subcell. Biochem.* **2012**, *58*, 61–94. [PubMed]
30. Bekhite, M.M.; Finkensieper, A.; Binas, S.; Muller, J.; Wetzker, R.; Figulla, H.R.; Sauer, H.; Wartenberg, M. VEGF-mediated PI3K class IA and PKC signaling in cardiomyogenesis and vasculogenesis of mouse embryonic stem cells. *J. Cell Sci.* **2011**, *124*, 1819–1830. [CrossRef] [PubMed]
31. Goldsmith, J.R.; Perez-Chanona, E.; Yadav, P.N.; Whistler, J.; Roth, B.; Jobin, C. Intestinal epithelial cell-derived mu-opioid signaling protects against ischemia reperfusion injury through PI3K signaling. *Am. J. Pathol.* **2013**, *182*, 776–785. [CrossRef] [PubMed]
32. Halls, M.L.; Yeatman, H.R.; Nowell, C.J.; Thompson, G.L.; Gondin, A.B.; Civciristov, S.; Bunnett, N.W.; Lambert, N.A.; Poole, D.P.; Canals, M. Plasma membrane localization of the mu-opioid receptor controls spatiotemporal signaling. *Sci. Signal.* **2016**, *9*, ra16. [CrossRef] [PubMed]
33. Cheng, J.J.; Wung, B.S.; Chao, Y.J.; Wang, D.L. Sequential activation of protein kinase C (PKC)-alpha and PKC-epsilon contributes to sustained Raf/ERK1/2 activation in endothelial cells under mechanical strain. *J. Biol. Chem.* **2001**, *276*, 31368–31375. [CrossRef] [PubMed]

34. Wang, W.; Wang, Y.; Debing, Y.; Zhou, X.; Yin, Y.; Xu, L.; Herrera Carrillo, E.; Brandsma, J.H.; Poot, R.A.; Berkhout, B.; et al. Biological or pharmacological activation of protein kinase C alpha constrains hepatitis E virus replication. *Antivir. Res.* **2017**, *140*, 1–12. [CrossRef] [PubMed]
35. Johnson, E.A.; Oldfield, S.; Braksator, E.; Gonzalez-Cuello, A.; Couch, D.; Hall, K.J.; Mundell, S.J.; Bailey, C.P.; Kelly, E.; Henderson, G. Agonist-selective mechanisms of mu-opioid receptor desensitization in human embryonic kidney 293 cells. *Mol. Pharmacol.* **2006**, *70*, 676–685. [CrossRef] [PubMed]
36. El Kouhen, R.; Burd, A.L.; Erickson-Herbrandson, L.J.; Chang, C.Y.; Law, P.Y.; Loh, H.H. Phosphorylation of Ser363, Thr370, and Ser375 residues within the carboxyl tail differentially regulates mu-opioid receptor internalization. *J. Biol. Chem.* **2001**, *276*, 12774–12780. [CrossRef] [PubMed]
37. Arttamangkul, S.; Heinz, D.A.; Bunzow, J.R.; Song, X.; Williams, J.T. Cellular tolerance at the micro-opioid receptor is phosphorylation dependent. *eLife* **2018**, *7*, e34989. [CrossRef] [PubMed]
38. Feng, B.; Li, Z.; Wang, J.B. Protein kinase C-mediated phosphorylation of the mu-opioid receptor and its effects on receptor signaling. *Mol. Pharmacol.* **2011**, *79*, 768–775. [CrossRef] [PubMed]
39. Lau, E.K.; Trester-Zedlitz, M.; Trinidad, J.C.; Kotowski, S.J.; Krutchinsky, A.N.; Burlingame, A.L.; von Zastrow, M. Quantitative encoding of the effect of a partial agonist on individual opioid receptors by multisite phosphorylation and threshold detection. *Sci. Signal.* **2011**, *4*, ra52. [CrossRef] [PubMed]
40. Pena, D.A.; Andrade, V.P.; Silva, G.A.; Neves, J.I.; Oliveira, P.S.; Alves, M.J.; Devi, L.A.; Schechtman, D. Rational design and validation of an anti-protein kinase C active-state specific antibody based on conformational changes. *Sci. Rep.* **2016**, *6*, 22114. [CrossRef] [PubMed]
41. Hefti, F.F.; Rosenthal, A.; Walicke, P.A.; Wyatt, S.; Vergara, G.; Shelton, D.L.; Davies, A.M. Novel class of pain drugs based on antagonism of NGF. *Trends Pharmacol. Sci.* **2006**, *27*, 85–91. [CrossRef] [PubMed]
42. Obermeier, A.; Lammers, R.; Wiesmuller, K.H.; Jung, G.; Schlessinger, J.; Ullrich, A. Identification of Trk binding sites for SHC and phosphatidylinositol 3′-kinase and formation of a multimeric signaling complex. *J. Biol. Chem.* **1993**, *268*, 22963–22966. [PubMed]
43. Chuang, H.H.; Prescott, E.D.I; Kong, H.; Shields, S.; Jordt, S.E.; Basbaum, A.I.; Chao, M.V.; Julius, D. Bradykinin and nerve growth factor release the capsaicin receptor from PtdIns(4,5)P2-mediated inhibition. *Nature* **2001**, *411*, 957–962. [CrossRef] [PubMed]
44. Zhang, X.; Li, L.; McNaughton, P.A. Proinflammatory mediators modulate the heat-activated ion channel TRPV1 via the scaffolding protein AKAP79/150. *Neuron* **2008**, *59*, 450–461. [CrossRef] [PubMed]
45. Bao, Y.; Gao, Y.; Yang, L.; Kong, X.; Yu, J.; Hou, W.; Hua, B. The mechanism of mu-opioid receptor (MOR)-TRPV1 crosstalk in TRPV1 activation involves morphine anti-nociception, tolerance and dependence. *Channels (Austin)* **2015**, *9*, 235–243. [CrossRef] [PubMed]
46. Vetter, I.; Cheng, W.; Peiris, M.; Wyse, B.D.; Roberts-Thomson, S.J.; Zheng, J.; Monteith, G.R.; Cabot, P.J. Rapid, opioid-sensitive mechanisms involved in transient receptor potential vanilloid 1 sensitization. *J. Biol. Chem.* **2008**, *283*, 19540–19550. [CrossRef] [PubMed]
47. Endres-Becker, J.; Heppenstall, P.A.; Mousa, S.A.; Labuz, D.; Oksche, A.; Schafer, M.; Stein, C.; Zollner, C. Mu-opioid receptor activation modulates transient receptor potential vanilloid 1 (TRPV1) currents in sensory neurons in a model of inflammatory pain. *Mol. Pharmacol.* **2007**, *71*, 12–18. [CrossRef] [PubMed]
48. Chen, Y.; Geis, C.; Sommer, C. Activation of TRPV1 contributes to morphine tolerance: involvement of the mitogen-activated protein kinase signaling pathway. *J. Neurosci.* **2008**, *28*, 5836–5845. [CrossRef] [PubMed]
49. Chen, S.R.; Pan, H.L. Blocking mu opioid receptors in the spinal cord prevents the analgesic action by subsequent systemic opioids. *Brain Res.* **2006**, *1081*, 119–125. [CrossRef] [PubMed]
50. Marwaha, L.; Bansal, Y.; Singh, R.; Saroj, P.; Bhandari, R.; Kuhad, A. TRP channels: Potential drug target for neuropathic pain. *Inflammopharmacology* **2016**, *24*, 305–317. [CrossRef] [PubMed]
51. Hirose, M.; Kuroda, Y.; Murata, E. NGF/TrkA Signaling as a Therapeutic Target for Pain. *Pain Pract.* **2016**, *16*, 175–182. [CrossRef] [PubMed]
52. Kumar, V.; Mahal, B.A. NGF–The TrkA to successful pain treatment. *J. Pain Res.* **2012**, *5*, 279–287. [CrossRef] [PubMed]
53. Livak, K.J.; Schmittgen, T.D. Analysis of relative gene expression data using real-time quantitative PCR and the 2(-Delta Delta C(T)) Method. *Methods* **2001**, *25*, 402–408. [CrossRef] [PubMed]
54. Stockton, S.D., Jr.; Gomes, I.; Liu, T.; Moraje, C.; Hipolito, L.; Jones, M.R.; Ma'ayan, A.; Moron, J.A.; Li, H.; Devi, L.A. Morphine Regulated Synaptic Networks Revealed by Integrated Proteomics and Network Analysis. *Mol. Cell Proteom.* **2015**, *14*, 2564–2576. [CrossRef] [PubMed]

55. Shukla, P.K.; Tang, L.; Wang, Z.J. Phosphorylation of neurogranin, protein kinase C, and Ca^{2+}/calmodulin dependent protein kinase II in opioid tolerance and dependence. *Neurosci. Lett.* **2006**, *404*, 266–269. [CrossRef] [PubMed]
56. Deiss, K.; Kisker, C.; Lohse, M.J.; Lorenz, K. Raf kinase inhibitor protein (RKIP) dimer formation controls its target switch from Raf1 to G protein-coupled receptor kinase (GRK) 2. *J. Biol. Chem.* **2012**, *287*, 23407–23417. [CrossRef] [PubMed]
57. Kelly, E.; Bailey, C.P.; Henderson, G. Agonist-selective mechanisms of GPCR desensitization. *Br. J. Pharmacol.* **2008**, *153*, S379–S388. [CrossRef] [PubMed]
58. Brackley, A.D.; Gomez, R.; Akopian, A.N.; Henry, M.A.; Jeske, N.A. GRK2 Constitutively Governs Peripheral Delta Opioid Receptor Activity. *Cell Rep.* **2016**, *16*, 2686–2698. [CrossRef] [PubMed]
59. Koese, M.; Koese, M.; Rentero, C.; Kota, B.P.; Hoque, M.; Cairns, R.; Wood, P.; Vila de Muga, S.; Reverter, M.; Alvarez-Guaita, A.; Monastyrskaya, K.; et al. Annexin A6 is a scaffold for PKCalpha to promote EGFR inactivation. *Oncogene* **2013**, *32*, 2858–2872. [CrossRef] [PubMed]
60. Schott, M.B.; Grove, B. Receptor-mediated Ca^{2+} and PKC signaling triggers the loss of cortical PKA compartmentalization through the redistribution of gravin. *Cell Signal.* **2013**, *25*, 2125–2135. [CrossRef] [PubMed]
61. Grewal, T.; Evans, R.; Rentero, C.; Tebar, F.; Cubells, L.; de Diego, I.; Kirchhoff, M.F.; Hughes, W.E.; Heeren, J.; Rye, K.A.; et al. Annexin A6 stimulates the membrane recruitment of p120GAP to modulate Ras and Raf-1 activity. *Oncogene* **2005**, *24*, 5809–5820. [CrossRef] [PubMed]
62. Tao, J.; Wang, H.Y.; Malbon, C.C. Protein kinase A regulates AKAP250 (gravin) scaffold binding to the beta2-adrenergic receptor. *EMBO J.* **2003**, *22*, 6419–6429. [CrossRef] [PubMed]
63. Palazzo, E.; Luongo, L.; de Novellis, V.; Rossi, F.; Marabese, I.; Maione, S. Transient receptor potential vanilloid type 1 and pain development. *Curr. Opin. Pharmacol.* **2012**, *12*, 9–17. [CrossRef] [PubMed]
64. Tang, L.; Shukla, P.K.; Wang, Z.J. Disruption of acute opioid dependence by antisense oligodeoxynucleotides targeting neurogranin. *Brain Res.* **2007**, *1143*, 78–82. [CrossRef] [PubMed]
65. Zhong, L.; Cherry, T.; Bies, C.E.; Florence, M.A.; Gerges, N.Z. Neurogranin enhances synaptic strength through its interaction with calmodulin. *EMBO J.* **2009**, *28*, 3027–3039. [CrossRef] [PubMed]
66. Cheppudira, B.P.; Trevino, A.V.; Petz, L.N.; Christy, R.J.; Clifford, J.L. Anti-nerve growth factor antibody attenuates chronic morphine treatment-induced tolerance in the rat. *BMC Anesthesiol.* **2016**, *16*. [CrossRef] [PubMed]
67. Wang, W.; Ma, X.; Luo, L.; Huang, M.; Dong, J.; Zhang, X.; Jiang, W.; Xu, T. Exchange factor directly activated by cAMP-PKCepsilon signalling mediates chronic morphine-induced expression of purine P2X3 receptor in rat dorsal root ganglia. *Br. J. Pharmacol.* **2018**, *175*, 1760–1769. [CrossRef] [PubMed]
68. Storch, U.; Straub, J.; Erdogmus, S.; Gudermann, T.; Mederos, Y.; Schnitzler, M. Dynamic monitoring of Gi/o-protein-mediated decreases of intracellular cAMP by FRET-based Epac sensors. *Pflugers Arch.* **2017**, *469*, 725–737. [CrossRef] [PubMed]
69. Mardy, S.; Miura, Y.; Endo, F.; Matsuda, I.; Indo, Y. Congenital insensitivity to pain with anhidrosis (CIPA), effect of TRKA (NTRK1) missense mutations on autophosphorylation of the receptor tyrosine kinase for nerve growth factor. *Hum. Mol. Genet.* **2001**, *10*, 179–188. [CrossRef] [PubMed]
70. Miura, Y.; Mardy, S.; Awaya, Y.; Nihei, K.; Endo, F.; Matsuda, I.; Indo, Y. Mutation and polymorphism analysis of the TRKA (NTRK1) gene encoding a high-affinity receptor for nerve growth factor in congenital insensitivity to pain with anhidrosis (CIPA) families. *Hum. Genet.* **2000**, *106*, 116–124. [CrossRef] [PubMed]
71. Omerbasic, D.; Smith, E.S.; Moroni, M.; Homfeld, J.; Eigenbrod, O.; Bennett, N.C.; Reznick, J.; Faulkes, C.G.; Selbach, M.; Lewin, G.R. Hypofunctional TrkA Accounts for the Absence of Pain Sensitization in the African Naked Mole-Rat. *Cell Rep.* **2016**, *17*, 748–758. [CrossRef] [PubMed]

proteomes

MDPI

Article

Granulocyte-Colony-Stimulating Factor Alters the Proteomic Landscape of the Ventral Tegmental Area

Nicholas L. Mervosh [1,2], Rashaun Wilson [3], Navin Rauniyar [3], Rebecca S. Hofford [1,2], Munir Gunes Kutlu [4], Erin S. Calipari [4], TuKiet T. Lam [3,5,6] and Drew D. Kiraly [1,2,7,*]

1 Department of Psychiatry, Icahn School of Medicine at Mount Sinai, New York, NY 10029, USA; nicholasmervosh@gmail.com (N.L.M.); rebecca.hofford@mssm.edu (R.S.H.)
2 Fishberg Department of Neuroscience, Icahn School of Medicine at Mount Sinai, New York, NY 10029, USA
3 Yale/NIDA Neuroproteomics Center, New Haven, CT 06511, USA; rashaun.wilson@yale.edu (R.W.); navin.rauniyar@yale.edu (N.R.); tukiet.lam@yale.edu (T.T.L.)
4 Department of Pharmacology, Vanderbilt Center for Addiction Research, Vanderbilt University School of Medicine, Nashville, TN 37232, USA; gunes.kutlu@vanderbilt.edu (M.G.K.); erin.calipari@vanderbilt.edu (E.S.C.)
5 Department of Molecular Biophysics & Biochemistry, New Haven, CT 06510, USA
6 Yale MS & Proteomics Resource, New Haven, CT 06510, USA
7 Seaver Autism Center for Research and Treatment, Icahn School of Medicine at Mount Sinai, New York, NY 10029, USA
* Correspondence: drew.kiraly@mssm.edu; Tel.: +1-212-824-8973

Received: 19 July 2018; Accepted: 20 September 2018; Published: 23 September 2018

Abstract: Cocaine addiction is characterized by aberrant plasticity of the mesolimbic dopamine circuit, leading to dysregulation of motivation to seek and take drug. Despite the significant toll that cocaine use disorder exacts on society, there are currently no available pharmacotherapies. We have recently identified granulocyte-colony stimulating factor (G-CSF) as a soluble cytokine that alters the behavioral response to cocaine and which increases dopamine release from the ventral tegmental area (VTA). Despite these known effects on behavior and neurophysiology, the molecular mechanisms by which G-CSF affects brain function are unclear. In this study mice were treated with repeated injections of G-CSF, cocaine or a combination and changes in protein expression in the VTA were examined using an unbiased proteomics approach. Repeated G-CSF treatment resulted in alterations in multiple signaling pathways related to synaptic plasticity and neuronal morphology. While the treatment groups had marked overlap in their effect, injections of cocaine and the combination of cocaine and G-CSF lead to distinct patterns of significantly regulated proteins. These experiments provide valuable information as to the molecular pathways that G-CSF activates in an important limbic brain region and will help to guide further characterization of G-CSF function and evaluation as a possible translational target.

Keywords: cocaine; addiction; cytokine; neuroimmune; ventral tegmental area

1. Introduction

Pathological substance use disorders are a group of recalcitrant, relapsing and remitting conditions that have deleterious effects on the patient, their family, and society at large. While there have been attempts made to mitigate the prevalence of substance abuse disorders, the incidences of illicit substance abuse and misuse has remained steady or increased since 1990 [1], and the economic burden created by substance use disorders is tremendous with a societal cost of over 500 billion dollars per year in the United States alone [2]. Of these conditions, pathological use of psychostimulants such as cocaine and amphetamine account for a significant portion of the morbidity and mortality. However, there are currently no FDA-approved pharmacological treatments for cocaine use disorder [3,4]. Previous drug discovery attempts in this arena have generally failed due to lack of efficacy, intolerable side effects, or both [5–7].

In recent years there has been growing interest in the role that neuroimmune interactions play in the development of psychiatric illness, including addictive disorders [8–10]. This raises the intriguing possibility that targeting neuroimmune signaling pathways may be a viable translational treatment strategy to reduce the persistence of pathological substance use disorders. Our lab recently discovered granulocyte-colony stimulating factor (G-CSF) as a cytokine that is up-regulated both centrally and peripherally after chronic cocaine treatment [11]. Peripheral injections of G-CSF potentiated the development of locomotor sensitization, conditioned place preference, and self-administration of cocaine, and blockade of G-CSF function in the mesolimbic dopamine system abrogated the formation of conditioned place preference.

While the behavioral effects of G-CSF on cocaine-induced behavioral plasticity are known, the cellular and molecular mechanisms underlying these effects remain to be identified. We have recently found that acute treatment with G-CSF enhances release of dopamine from the ventral tegmental area (VTA) into the nucleus accumbens (NAc) [12]. Previous work has found that the G-CSF receptor is robustly expressed on dopamine expressing neurons of the midbrain [13,14]. G-CSF has been found to be a potent neurotrophic and neuroprotective factor in response to stroke or other insults [15–17]. Importantly, G-CSF is also neuroprotective in the midbrain where treatment with G-CSF reduces neuronal death in the MPTP model of Parkinson's disease [18]. Additionally, within these midbrain neurons, G-CSF has been found to induce activity of the immediate-early gene *Cfos* and acute treatments upregulate tyrosine hydroxylase—the rate limiting step in dopamine synthesis [13]. Moving forward, it will be critical to determine the molecular signaling cascades that control the effects of G-CSF on behavior.

Given the known effects of G-CSF within the midbrain and the importance of the VTA in the development and persistence of substance use disorders [19,20] we characterized the effect of G-CSF and its interaction with cocaine on the proteomic makeup of the VTA. Via an unbiased quantitative proteomics approach, we identified and characterized the regulation pattern of more than two thousand proteins in the VTA. We found that G-CSF treatment on its own regulated many of the same signaling pathways that are regulated by cocaine and induced numerous factors important for neurite and dendritic spine plasticity. Specifically, we found significant regulation of proteins predicted to be downstream from Fragile X mental retardation (FMRP) and mammalian target of rapamycin (mTOR). Additionally, we report multiple intracellular signaling cascades that are differentially regulated by combined cocaine and G-CSF treatment, suggesting future targets for study on the effects of G-CSF on the behavioral response to cocaine.

2. Materials and Methods

2.1. Animals and Drug Treatments

Male C57BL/6J mice (7 weeks old ~20–25 g; Jackson Laboratories, Bar Harbor, ME, USA) were housed in the animal facilities at Icahn School of Medicine at Mount Sinai. Mice were maintained on a 12:12 h light/dark cycle with lights on at 0700 and lights off at 1900. Mice had food and water available *ad libitum* throughout the experiments. Drug treatments were performed in a 2×2 design with the first group receiving phosphate buffered saline vehicle, followed by saline (PBS/Sal), the second group was injected with G-CSF 50 μg/kg (GenScript Biotech, Piscataway, NJ—G-CSF/Sal) followed by saline, the third group was injected with PBS followed by cocaine hydrochloride 7.5 mg/kg (NIDA—PBS/Coc), and the fourth group with both G-CSF and cocaine (G-CSF/Coc). Injections were performed once daily for 7 days and the animals were euthanized 24 h after the final injection. All animals were maintained according to the National Institutes of Health guidelines in Association for Assessment and Accreditation of Laboratory Animal Care accredited facilities. All experimental protocols were approved by the Institutional Animal Care and Use Committee at Mount Sinai.

2.2. Protein Preparation

For each mouse the VTA was dissected from fresh tissue on ice using a reference brain atlas and anatomical landmarks to guide dissection. Tissue from each animal was then sonicated into 50 µL of ice-cold RIPA buffer (50 mM Tris [pH 8.0], 1 mM EDTA, 1% Triton X-100, 0.1% sodium deoxycholate, 0.1% SDS, 110 mM NaCl & Halt Protease and Phosphatase Inhibitor Cocktails [Fisher]). Protein concentrations were determined by Bradford colorimetric assay according to manufacturer protocols (Thermo Fisher, Waltham, MA, USA). For these analyses tissue from individual animals was used as distinct data points. There was no pooling of samples between animals other than to make the master mix for cross-assay normalization as described below.

2.3. Tandem Mass Tag (TMT) Labeling

TMT samples were prepared according to the manufacturer's instructions. Briefly, 50 µg proteins per condition were reduced by incubating the samples with TCEP (tris(2-carboxyethyl)phosphine) at 55 °C for 1 h and alkylated by incubating with iodoacetamide at room temperature in the dark for 30 min. The proteins were precipitated by the acetone precipitation, resuspended in 25 mM Triethyl ammonium bicarbonate (TEAB) and digested with trypsin at 37 °C overnight. The peptide concentrations of the tryptic digests were measured by Amino Acid Analysis method using a Hitachi L-8900 Amino Acid Analyzer. Equal amount (30 µg) of peptides were labeled with TMT reagents from the TMT-10plex kit (ThermoFisher Scientific). The samples were labeled by distributing them into two experimental groups. Each TMT experimental setup has two TMT tags (126 and 129N) that labeled the two pooled samples, which were created by collecting and combining an equal amount of peptides from each sample. The pooled samples served as a global internal standard for normalizing the data across the two experimental setups and is henceforth referred to as the Master Mix. The remaining 8 TMT reagents in each experimental setup were used for labeling the two biological replicates for each of the four conditions. The TMT labels carried by each sample and the mixing design is shown in Figure 1B. For labeling, the peptides were incubated with TMT reagents for 1 h at room temperature. The labeling reaction was quenched by adding 5% hydroxylamine to the sample and incubating for 15 min. Before combining the labeled samples for mass spectrometry analysis, an aliquot was combined and analyzed by LC-MS/MS to ensure the labeling was complete and also that the mixing generated a ratio of 1. Eventually, all ten labeled samples were combined and fractionated offline by high pH reversed-phase fractionation.

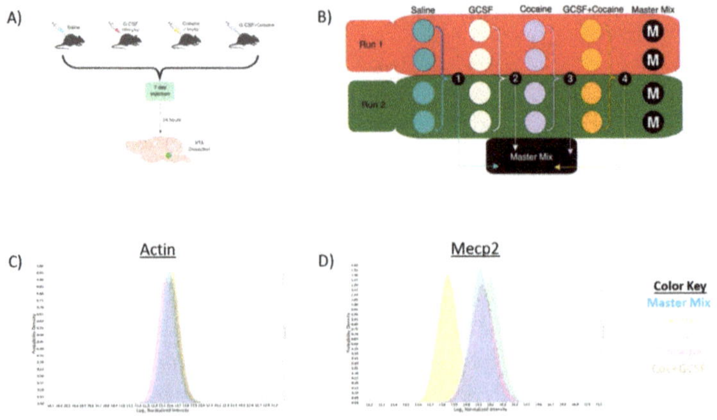

Figure 1. Experimental design and validation. (**A**) Animals were injected with ± G-CSF

(50 μg/kg) ± Cocaine (7.5 mg/kg) a 2 × 2 design. Injections were done once daily for 7 days and animals sacrificed 24 h after the final injection and the VTA dissected out for analysis. (**B**) To allow for significant power, two runs of the TMT 10-plex were run with two samples from each group per run (total 4/group) with a mix comprised of an equal amount of each sample run as a Master Mix run to allow normalization between runs. Median intensity values of actin, which was not significantly changed in any group show near complete overlap (**C**) whereas Mecp2 shows increase in all non-saline groups with the expected change in median intensity (**D**).

2.4. High-pH Reversed-Phase C18 Peptide Fractionation

High-pH reversed-phase C18 peptide fractionation was performed on an ACQUITY UPLC H-class system (Waters Corporation, Milford, MA, USA) on ACQUITY UPLC BEH C18 column, 1.7 μm, 2.1 mm × 50 mm. Elution was performed at a flowrate of 0.4 mL/min using a gradient of mobile phase A (10 mM ammonium acetate) and B (10 mM ammonium acetate in 90% acetonitrile). The gradient extended from 2% to 37% mobile phase B in 17.6 min and then to 75% mobile phase B in another 8.8 min. The collected pooled 10 fractions were dried in a speed-vac centrifuged and reconstituted in buffer A (0.1% formic acid in water); 400 ng digests from each fraction were used for reversed-phase liquid chromatography-tandem mass spectrometry (RP-LC-MS/MS/MS) analysis.

2.5. SPS-MS3 TMT Data Acquisition on an Orbitrap Fusion Tribrid Mass Spectrometer

RP-LC-MS/MS/MS was performed using a nanoACQUITY UPLC system (Waters Corporation, Milford, MA, USA) connected to an Orbitrap Fusion Tribrid (ThermoFisher Scientific, San Jose, CA, USA) mass spectrometer. After injection, samples were loaded into a trapping column (nanoACQUITY UPLC Symmetry C18 Trap column, 180 μm × 20 mm) at a flowrate of 5 μL/min and separated with a C18 column (nanoACQUITY column Peptide BEH C18, 75 μm × 250 mm). The compositions of mobile phases A and B were 0.1% formic acid in water and 0.1% formic acid in acetonitrile, respectively. Peptides were eluted with a gradient extending from 6% to 20% mobile phase B in 120 min and then to 40% mobile phase B in another 50 min at a flowrate of 300 nL/min and a column temperature of 37 °C. The data were acquired with the mass spectrometer operating in a top speed data-dependent mode with multinotch synchronous precursor selection (SPS)-MS3 scanning for TMT tags. The full scan was performed in the range of 380–1580 m/z at an Orbitrap resolution of 120,000 at 200 m/z and automatic gain control (AGC) target value of 2 × 10^5, followed by selection of ions above an intensity threshold of 5000 for collision-induced dissociation (CID)-MS fragmentation in the linear ion trap with collision energy of 35%. The isolation width was set to 1.6 m/z. The top 10 fragment ions for each peptide MS2 were notched out with an isolation width of 2 m/z and co-fragmented with higher-energy collision dissociation (HCD) at a collision energy of 65% to produce MS3 scans which were analyzed in the Orbitrap at a resolution of 60,000.

2.6. Protein Identification and Quantification

Raw data from the Orbitrap Fusion were processed using Proteome Discoverer software (version 2.1, ThermoFisher Scientific, San Jose, CA, USA). MS2 spectra were searched using Sequest HT which was set up to search against the SwissProt mouse database (downloaded on 06292017). The search criteria included 10 ppm precursor mass tolerance, 0.6 Da fragment mass tolerance, trypsin enzyme and maximum missed cleavage sites of two. Static modification included carbidomethylation (+57.02146 Da) on cysteine and TMT labels (+229.16293 Da) on lysine and peptide N-terminus. Dynamic modifications included oxidation (+15.99492 Da) on methionine, deamidation (+0.98402 Da) on asparagine and glutamine, and acetylation (+42.01057 Da) on protein N-terminus. Peptide spectral match (PSM) error rates were determined using the target-decoy strategy coupled to Percolator modeling of true and false matches [21]. Reporter ions were quantified from MS3 scans using an integration tolerance of 20 ppm and the most confident centroid as the integration method in the Reporter Ions Quantifier node.

2.7. Mass Spec Data Analysis

Scaffold Q+ (version Scaffold_4.8.5, Proteome Software Inc., Portland, OR, USA) was used for label-based TMT10-plex quantitation of peptide and protein identifications. Peptide identifications were accepted if they could be established at greater than 95.0% probability by the Scaffold Local FDR algorithm. Protein identifications were accepted if they could be established at greater than 99.0% probability and contained at least 2 identified peptides. Peptide probabilities were calculated by the Scaffold Local FDR algorithm, and protein probabilities were assigned using the Protein Prophet algorithm [22]. Proteins identified with fewer than two peptides were excluded from quantitation. Proteins sharing redundant peptides were grouped into clusters. Normalization was performed iteratively (across samples and spectra) on intensities, as previously described [23]. After setting the minimum dynamic range to 5%, removing spectra that were missing a reference value and those that arose from degenerate peptides that match to more than one protein, the remaining log-transformed spectra were weighted by an adaptive intensity weighting algorithm. Of 71,507 spectra in the experiment, 59,861 (84%) met the threshold criteria and were included in quantitation. Statistical testing was performed using uncorrected Student's t-test between groups. p-values < 0.05 were considered statistically significant. Volcano plots were created using GraphPad Prism version 7 (La Jolla, CA, USA). Pathway analyses to determine specifically regulated pathways were created using Ingenuity Pathway Analysis software from Qiagen. The network diagrams depicted in Figure 3 were created using significantly regulated proteins from our dataset that were predicted to be directly downstream of the hub genes, and then up to 5 genes predicted to be downstream of each of those was added to the outer layer. There were no additional filters applied. Predicted targets downstream from activity-dependent transcription factors was performed using the Enrichr analysis suite (http://amp.pharm.mssm.edu/Enrichr/). Full methodology for the Enrichr analyses is described in detail in the original Chen et al. paper [24]. Heatmaps were created using the freely available Morpheus software from the Broad Institute (https://software.broadinstitute.org/morpheus).

2.8. Western Blot Analysis

For Western blot analysis animals were treated identically to those above, and VTA tissue was fresh dissected and frozen on ice until further processing. Samples were thoroughly sonicated into SDS lysis buffer (1% SDS, 50 mM Tris [pH 8.0], 130 mM NaCl, 5 mM EDTA, 50 mM NaF, 1 mM PMSF, protease and phosphatase inhibitor cocktails from ThermoFisher) according to previously published procedures [25]. Sample concentrations were determined using a Bradford colorimetric assay (ThermoFisher) according to manufacturer protocols, and 10µg of protein was run on a 4–12% gradient gel. Proteins were transferred to PVDF membranes using standard techniques. Membranes were blocked using LiCor blocking buffer with TBS based mixed 1:1 with standard TBS for one hour at room temperature. Primary antibodies were incubated with mixing at 4 °C overnight with constant agitation. Primary antibodies used were tyrosine hydroxylase (AbCam #ab112, 1:1000), Mecp2 (Cell Signaling #3456, 1:1000) & actin (Cell Signaling #3700, 1:10,000). Membranes were washed with TBS + Tween-20 before incubation with secondary antibodies raised against the appropriate species (LiCor, 1:10,000) for one hour at room temperature. Membranes were then washed with TBS + Tween-20, rinsed with TBS without Tween, and imaged using a LiCor Fluorescent imager. Image quantification was performed using freely available ImageJ software. Representative images shown in Figure 8 were flipped horizontally to achieve representative bands in the correct order but were not otherwise altered or retouched.

3. Results

3.1. Experimental Design

We have previously demonstrated that peripheral injections of G-CSF alter gene expression in the NAc in response to cocaine [11]. More recently, we have identified G-CSF as a potent regulator of dopamine release from the VTA into the NAc [12]. These data lead us to the hypothesis that G-CSF may be inducing changes in VTA function that lead to downstream alterations in neuronal responsiveness in the NAc. To assess the effects of G-CSF alone and in combination with cocaine a 2×2 experimental design was utilized in which animals were being injected with vehicle, G-CSF (50 μg/kg), cocaine (7.5 mg/kg), or both—with the appropriate additional vehicle controls (Figure 1A). Animals received 7 daily injections as this treatment paradigm leads to significant alterations of important synaptic plasticity pathways and protein changes [26].

To allow for sufficient power to detect protein changes in a complex mixture using this 2×2 experimental design, two parallel runs were performed utilizing the TMT-10-plex labeling method as described in the Methods section. To allow for quantitative comparisons between the two runs we pooled an equal amount of each of the experimental samples and ran it in duplicate as the "Master Mix" in each run (Figure 1B). This allowed for a standard for normalizing protein expression between runs and allowed for an N of 4 for each experimental group in the discovery proteomics analysis. Figure 1C,D provides examples of median intensity plots for a regulated (MeCP2, Figure 1D) and non-regulated (Actin, Figure 1C) protein. In Figure 1C the median intensity values for actin-derived peptides are presented with the colored-in peaks representing the median Log2 normalized intensity, and the corresponding lines representing the full range. All groups including the master mix show alignment of their median intensities. In Figure 1D we provide an example of a protein that was shown to be up-regulated relative to Saline in all other treatment groups, Mecp2. The median intensities of the other treatment groups are increased relative to that of Saline. Additionally, within this group the median intensity of the Master Mix is shifted towards the up-regulated groups but is somewhat downshifted compared to the three experimental conditions, suggesting that the lower levels of Mecp2 in the saline samples caused a shift in the Master Mix graph, as would be expected.

3.2. Proteomic Effects of G-CSF in the VTA

For our initial analyses we queried the effects of chronic G-CSF alone on the VTA proteome. While the G-CSF receptor has been shown to be robustly expressed in the midbrain [27,28], there has not yet been a detailed molecular analysis of the effects of chronically increased G-CSF signaling. Figure 2A is a volcano plot of the fold-change and p-value of regulation for each protein that was detected in the proteomics analysis. There were 2353 reliably detected proteins, 475 met a threshold of $p < 0.05$ when the normalized mean intensity of detected peptides was compared to those from the saline group (colored dots on volcano plot). Of these 475, we found that 121 were down-regulated and 354 were up-regulated, suggesting that repeated treatment with G-CSF was more likely to upregulate protein networks in the VTA. To look more stringently at proteins that were regulated by repeated G-CSF, we identified proteins that were up or down-regulated by more than 20%. By applying this criterion, we identified 184 proteins 51 were down-regulated and 153 were up-regulated (green dots on volcano plot). A full list of proteins significantly regulated by G-CSF with corresponding fold-change information and p values is available as Table S1.

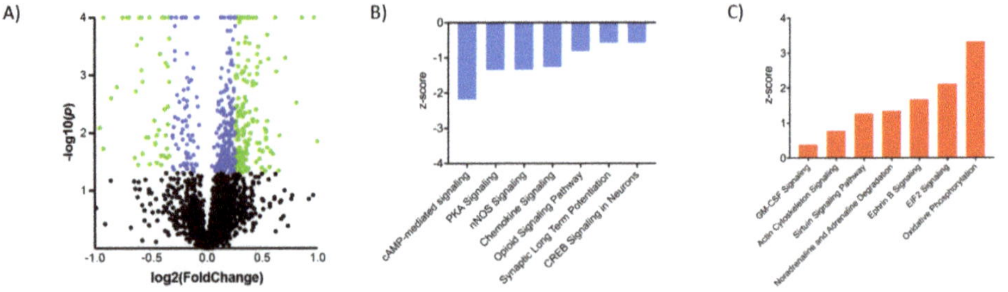

Figure 2. G-CSF regulated proteins and signaling pathways. (**A**) Volcano plot demonstrating proteins in the G-CSF group relative to saline controls with Log2 Fold change on the x-axis and Log10 p value on the y-axis. Proteins that were significantly changed with a nominal p value of <0.05 are represented by blue dots, and those with a ±20% change and a $p < 0.05$ are represented by green dots. Ingenuity pathway analysis demonstrated that amongst the significantly regulated proteins there were multiple canonical signaling pathways that were found to be down-regulated (**B**) as well as up-regulated (**C**) relative to saline controls.

It should be noted that the primary purpose of these experiments was to gain further insight into the effect of G-CSF on the proteomic landscape of the VTA, and to identify important signaling networks for future more mechanistic studies into the effects of G-CSF in the brain. Given this, and that our study was not powered to allow for statistical correction of multiple tests, uncorrected p values were utilized in this figure and throughout the manuscript. Additionally, all proteins that were found to be significantly regulated were included in subsequent pathway and network analysis, regardless of the fold change. While this methodology may bias reported results towards an increased number of false positives, we feel that it is appropriate for a discovery analysis such as this one.

To provide a context for how these large-scale protein changes induced by G-CSF might be affecting neuronal function in the VTA we analyzed the subset of proteins found to be significantly regulated to look for changes in intracellular signaling networks. Analysis of canonical signaling pathways identified multiple that were significantly up or down-regulated (Table 1 and Figure 2B,C). Among the seven most significantly down-regulated pathways (Figure 2B), there were multiple that relate to signaling downstream of cyclic-adenosine monophosphate (cAMP)—a second-messenger signaling system heavily implicated in response to drugs of abuse [20]. We see down-regulation specifically of the cAMP-mediated signaling pathway, the protein kinase A (PKA) signaling pathway which is downstream of activated cAMP, and the CREB1 signaling pathway. When looking at pathways that were significantly upregulated (Figure 2C) we see increases in pathways related to transcriptional and translational control. This includes marked increases in the eukaryotic initiation factor 2 (EIF2) pathway which is critical for the initiation of translation from mRNA to protein [29]. Furthermore, sirtuin and granulocyte-macrophage colony stimulating factor (GM-CSF) are also increased. Sirtuins are a class of histone deacetylase enzymes and changes in their function have previously been shown to be important for behavioral response to cocaine and opiates [30,31]. GM-CSF is another colony stimulating factor molecule that shares some signaling pathways with G-CSF, and this increase in this signaling pathway may be due to the overlap in the signaling between the two sets of proteins.

Table 1. Canonical signaling pathways altered by G-CSF treatment. Ingenuity pathway analysis of proteins significantly altered by repeated G-CSF treatment reveals multiple signaling networks that are up and downregulated.

	Ingenuity Canonical Pathways	−log(*p*-value)	z-score
Downregulated	cAMP-mediated signaling	2.95	−2.183
	Protein Kinase A Signaling	6.62	−1.347
	nNOS Signaling in Neurons	3.2	−1.342
	Chemokine Signaling	4.48	−1.265
	Opioid Signaling Pathway	6.77	−0.816
	Synaptic Long-Term Potentiation	4.1	−0.577
	CREB Signaling in Neurons	3.37	−0.577
	RhoA Signaling	2.56	−0.333
	Dopamine-DARPP32 in cAMP Signaling	2.83	−0.277
Upregulated	Oxidative Phosphorylation	3.78	3.317
	EIF2 Signaling	5.19	2.111
	Ephrin B Signaling	3.24	1.667
	Noradrenaline/Adrenaline Degradation	2.87	1.342
	Sirtuin Signaling Pathway	3.07	1.265
	Tryptophan Degradation	4.93	0.816
	Actin Cytoskeleton Signaling	3.15	0.775
	Thrombin Signaling	3.4	0.535
	GM-CSF Signaling	3.28	0.378
	14-3-3-mediated Signaling	3.93	0.333
	Neuropathic Pain Signaling in Dorsal Horn Neurons	3.07	0.302
	Calcium Signaling	3.46	0.277

Given that G-CSF signals through multiple intracellular signaling pathways the data were also analyzed to identify key signaling molecules that might serve as signaling hubs upstream of proteins regulated by G-CSF. These analyses provide information on the specific intracellular signaling networks driven by effects of G-CSF in the VTA. Ingenuity Pathway Analysis (IPA) revealed multiple key regulators predicted to be upstream of proteins regulated by G-CSF as shown in Table 2. The top protein predicted to be an upstream regulator of proteins altered by G-CSF was fragile X mental retardation protein (FMRP), a key protein in translation initiation and the site of the most common mutation seen in Fragile X syndrome. Based on these analyses it was estimated that at least 26 of the proteins that were altered with chronic G-CSF treatment are known to be downstream of FMRP. The data from these samples was used to create a network diagram of all proteins significantly regulated by G-CSF that were predicted to be downstream of FMRP were added. To show the complexity of this network, up to 5 downstream targets of each of the proteins directly downstream of FMRP. This is displayed as a network diagram in Figure 3A. This analysis revealed a total of 157 G-CSF regulated proteins (37 down-regulated and 120 up-regulated—*p* values and fold change values in Table S2) are predicted to be downstream of FMRP.

Table 2. Predicted upstream regulators of G-CSF affected proteins. Data from Ingenuity Pathway Analysis predicting the regulator genes with the greatest influence on significantly regulated proteins from animals treated with daily G-CSF.

Master Regulator	Molecule Type	Participating Regulators	Activation z-score	*p*-Value	Direct Targets
FMR1	Translation Regulator	Akt1, FMR1, MAPT, **MTOR**	1.4	7.83×10^{-14}	26
MMP9	Peptidase	AKT1, FMR1, GRIN1, MAPT, MMP9, **MTOR**	−1.8	1.51×10^{-13}	25
CDK5	Kinase	CDK5, FMR1, MAP2, MAP2K1, MAPK1, MAPT, **MTOR**, RPS6KB1, STAT3, TRPV1	−0.78	1.49×10^{-10}	25
SLC6A3	Transporter	CDK5, GSK3B, MAP2, MAP2K1, MAPK10, MAPT, **MTOR**, RPS6KB1, SLC6A3	1.09	3.76×10^{-10}	21
EGR1	Transcription Regulator	CDK5, EGR1, GSK3B, MAP2, MAP2K1, MAPK10, MAPT, **MTOR**, RPS6KB1	−1.34	1.05×10^{-9}	20

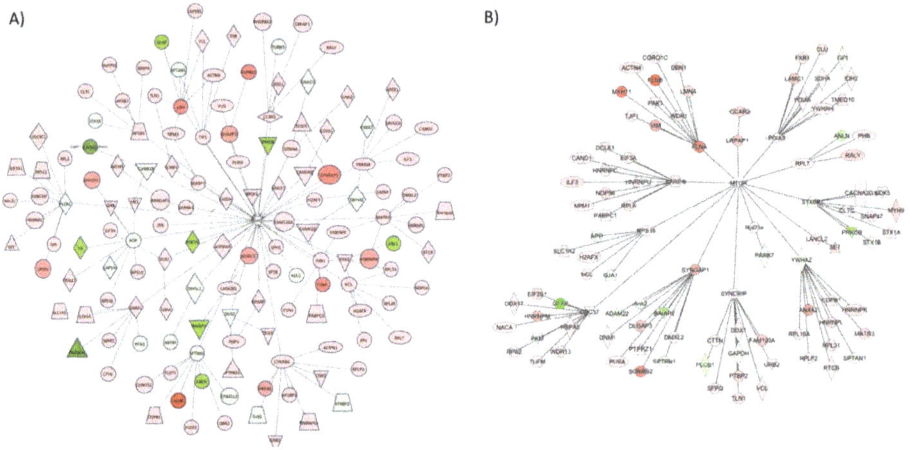

Figure 3. Key upstream regulators of proteins altered by G-CSF. From the proteins identified as significantly altered by repeated G-CSF treatment, we used IPA analysis to identify key upstream regulators. Two of the most robust were FMRP (**A**) and mTOR (**B**). These dendrograms represent all proteins that were significantly changed in this dataset that are predicted to be directly downstream of these regulators, and those that are predicted to be directly downstream of those (two degrees of regulation). Proteins visualized in red are significantly increased, and those in green were significantly decreased.

While FMRP was predicted to be the top upstream regulator of G-CSF-altered signaling networks, it was also noted in our analyses that mTOR was the only protein that was predicted to be a participating regulator in each of the top master regulatory networks identified. Given its apparent broad involvement in those proteins that were regulated by prolonged G-CSF exposure, a network diagram of significantly regulated proteins from our dataset that would be predicted to be downstream of mTOR and its direct effectors was created. This is illustrated in Figure 3B, and from these analyses we see that 101 proteins predicted to be downstream of mTOR are significantly regulated by chronic G-CSF treatment (18 down-regulated, 83 up-regulated—*p* values and fold change values in Table S3).

Examination of the most significantly regulated disease and function changes predicted by IPA in the G-CSF treated samples revealed networks related to changes in neuronal morphology, with the most significantly regulated network being "Morphology of Neurons" ($p = 1.22 \times 10^{-13}$). The significantly regulated proteins belonging to this network are illustrated according to their predicted subcellular distribution in Figure 4—which demonstrates that G-CSF had significant effects on nuclear, cytosolic and cell membrane proteins known to affect the morphological structure of neurons. In sum this network of significantly regulated proteins was comprised of 64 proteins 23 down-regulated and 41 up-regulated (*p* values and fold change values in Table S4).

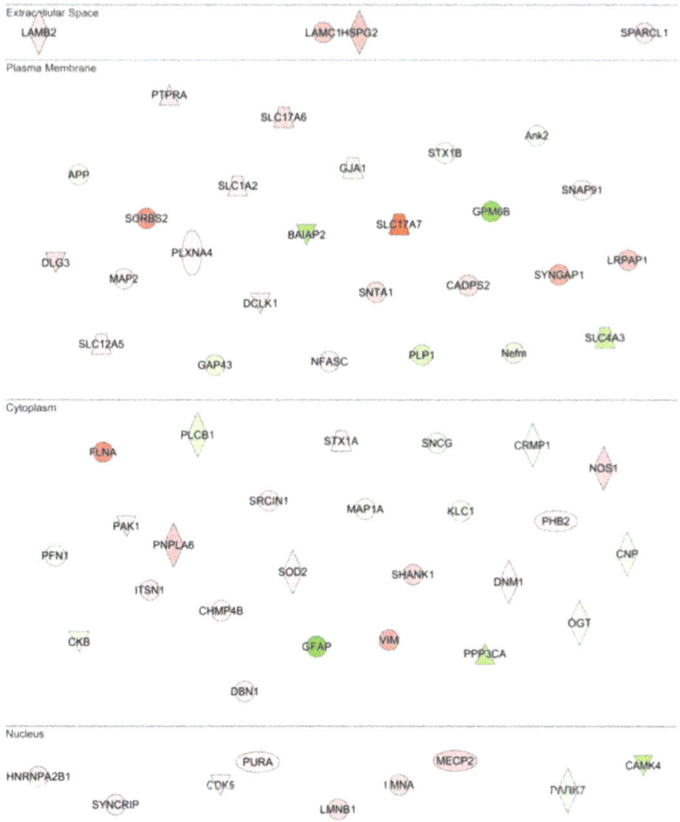

Figure 4. Drivers of neuronal morphology affected by G-CSF. IPA analysis of the most significantly altered cellular functions following G-CSF treatment revealed that proteins involved in altering neuronal morphology were significantly changed. This diagram shows all significantly regulated proteins predicted to be involved in affecting neuronal morphology, and their corresponding predicted subcellular distribution. Proteins visualized in red are significantly increased, and those in green were significantly decreased.

3.3. *Interaction Effects of G-CSF & Cocaine in the VTA*

Following the analyses of G-CSF treated animals, the effects of cocaine and the combination of cocaine and G-CSF on VTA proteomics were examined (Figure 1A). Cocaine treatment significantly altered 422 with an uncorrected *p* value of <0.05 (Figure 5A—blue dots). Of these, there were 152 that also exhibited a >20% increase or decrease in expression from the saline group (Figure 5A—green dots). Treatment with a combination of G-CSF plus cocaine resulted in 327 proteins that were significantly regulated, 99 of which were increased or decreased by 20% or more compared to saline treatment (Figure 5B). Combination treatment of G-CSF plus cocaine significantly altered 195 proteins, 63 of which were up or down-regulated more than 20% compared to cocaine alone (Figure 5C). A list of all significantly regulated proteins from each pairwise comparison is available as Table S1.

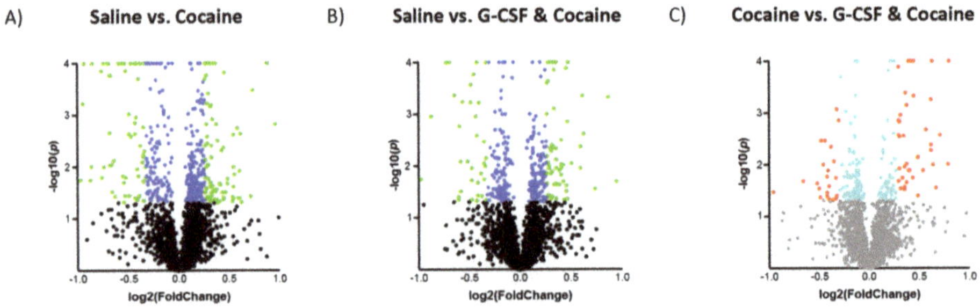

Figure 5. Changes in VTA protein expression in cocaine-treated groups. Volcano plots of proteins in the Cocaine (**A**) and Cocaine + G-CSF (**B**) groups relative to saline controls. Log2 Fold change on the *x*-axis and Log10 *p* value on the *y*-axis. Proteins that were significantly changed with a nominal *p* value of <0.05 are represented by blue dots, and those with a $\pm 20\%$ change and a $p < 0.05$ are represented by green dots. (**C**) Demonstrates the changes in the Cocaine + G-CSF group relative to the Cocaine group. Proteins with a nominal *p* value of <0.05 are represented by turquoise dots, and those with a $\pm$ 20% change and a $p < 0.05$ are represented by red dots.

Analysis of all proteins that were regulated compared to saline revealed that there was considerable overlap in changes in protein expression between the groups, but also significant subsets of proteins that were only regulated by one treatment group. Treatment with G-CSF only had the highest number of uniquely regulated proteins (Figure 6A). A breakdown of all proteins in each segment of the Venn diagram is available as Table S5. To further illustrate the differential expression patterns between the treatment groups we measured the mean fold saline expression level of all significantly regulated proteins (N = 789 unique proteins) relative to saline controls. Expression levels were then z-score normalized and sorted using *k*-means clustering ($k = 5$, Figure 6B). Functional characterization of these protein clusters will be the subject of future analyses.

As was done for the G-CSF only group, we also performed IPA analysis of the two cocaine treatment groups to identify canonical signaling pathways that were altered compared to Saline. The *p* values of nine of the most significantly regulated pathways are presented in Figure 7A. These analyses demonstrate that there is indeed a good degree of commonality in the regulated proteins in all three treatment groups compared to the control group. Notably, the tryptophan degradation pathway was significantly regulated in the G-CSF and in the G-CSF + Cocaine groups, but not in the cocaine only group. Full data for these pathways with regulated protein lists and directional z-scores are available as Table S6.

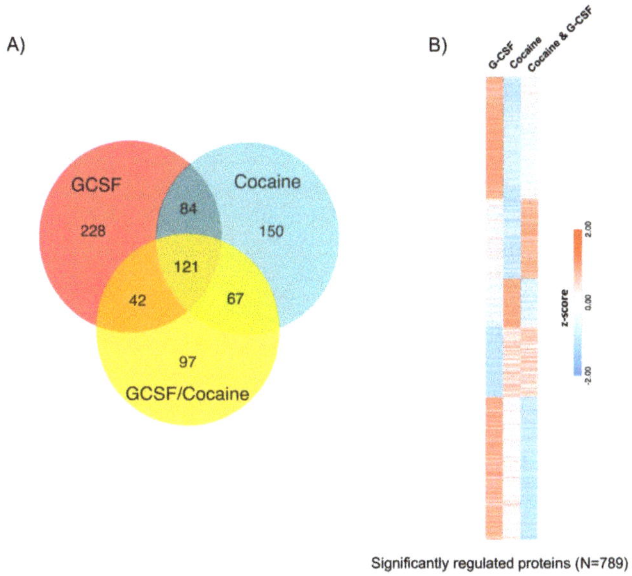

Figure 6. Comparison of significantly-regulated proteins between all treatment groups. (**A**) Venn diagram demonstrating overlap and differences of proteins changed between the three treatment groups relative to saline controls. (**B**) Heatmap visualization of the 789 proteins that were significantly regulated in any treatment group demonstrates clusters of proteins that are differentially affected based on the three treatment groups. K-means clustering ($k = 5$) used to create heatmap of z-scored mean fold-change from saline.

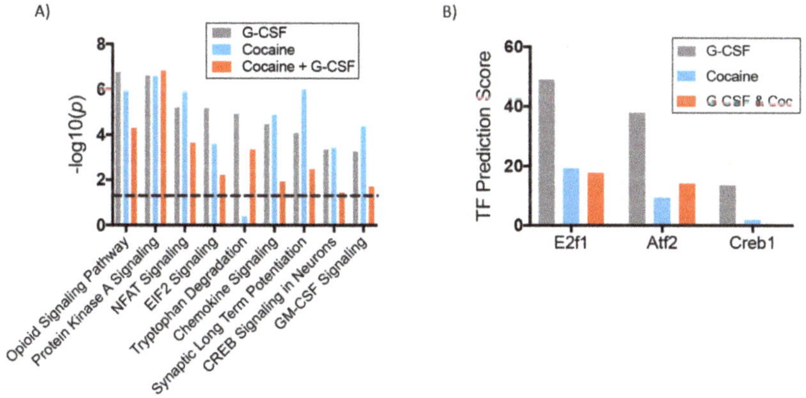

Figure 7. Canonical pathways regulated in all groups and predicted transcription factors of upregulated proteins. (**A**) Ingenuity Pathway Analysis software was used to compare significantly altered canonical signaling pathways amongst all treatment groups. The height of the bars represents the statistical strength of the change but does not represent directionality of change. Directional data available in Table S6. (**B**) Using Enrichr software we identified transcription factors with the highest predicted number of targets in our datasets. This graph demonstrates the calculated transcription factor (TF) prediction score for the three chosen transcription factors. Significantly regulated targets are available in Table S9.

Further pathway analyses was performed utilizing Gene Ontology enrichment analysis (geneontology.org) to assess for specific molecular functions altered in each treatment group relative to Saline controls [32–34]. For these analyses, only the upregulated proteins from each treatment group were included. The top 12 significantly regulated molecular processes (as defined by lowest FDR-corrected *p* value) from each treatment group are presented in Table 3. Similar to what was seen with the IPA analyses, we found that 7/12 predicted changes in molecular function were common amongst the three treatment groups. The full list of all significantly changed molecular function pathways is available as Table S7 and the full list of all significantly regulated cellular component pathways is available as Table S8.

Table 3. Gene Ontology analysis of significantly regulated molecular functions in each treatment group. Date representing the top 12 significantly regulated molecular functions from each treatment group sorted from smallest to largest FDR-corrected p value. Only proteins significantly upregulated relative to saline were included in these analyses. Bolded GO terms are those that were significantly regulated in all three treatment groups.

	GO Molecular Function Complete	REFLIST (22262)	Upload Match	Upload Expected	Upload +/−	Fold Enrichment	Raw p-Value	FDR
G-CSF vs. Saline	**binding (GO:0005488)**	13001	296	200.9	+	1.47	2.49×10^{-28}	1.13×10^{-24}
	RNA binding (GO:0003723)	1019	73	15.75	+	4.64	2.86×10^{-27}	6.48×10^{-24}
	protein binding (GO:0005515)	8900	227	137.53	+	1.65	6.02×10^{-22}	9.09×10^{-19}
	heterocyclic compound binding (GO:1901363)	4917	152	75.98	+	2	1.82×10^{-19}	2.07×10^{-16}
	organic cyclic compound binding (GO:0097159)	5006	152	77.35	+	1.96	1.08×10^{-18}	9.77×10^{-16}
	mRNA binding (GO:0003729)	243	28	3.75	+	7.46	1.56×10^{-15}	1.18×10^{-12}
	protein-containing complex binding (GO:0044877)	1163	58	17.97	+	3.23	9.27×10^{-15}	6.01×10^{-12}
	enzyme binding (GO:0019899)	2260	84	34.92	+	2.41	7.23×10^{-14}	4.10×10^{-11}
	signaling receptor activity (GO:0038023)	2271	2	35.09	−	0.06	1.56×10^{-13}	7.87×10^{-11}
	structural molecule activity (GO:0005198)	613	39	9.47	+	4.12	3.37×10^{-13}	1.53×10^{-10}
	molecular transducer activity (GO:0060089)	2324	3	35.91	−	0.08	1.18×10^{-12}	4.88×10^{-10}
	transmembrane signaling receptor activity (GO:0004388)	2082	2	32.17	−	0.06	3.02×10^{-12}	1.14×10^{-9}
Coc. vs. Saline	***protein binding (GO:0005515)***	*8900*	*170*	*96.35*	+	*1.76*	2.40×10^{-21}	1.09×10^{-17}
	binding (GO:0005488)	*13001*	*206*	*140.74*	+	*1.46*	2.12×10^{-19}	4.82×10^{-16}
	structural molecule activity (GO:0005198)	*613*	*32*	*6.64*	+	*4.82*	5.44×10^{-13}	8.23×10^{-10}
	RNA binding (GO:0003723)	*1019*	*41*	*11.03*	+	*3.72*	7.98×10^{-13}	9.05×10^{-10}
	organic cyclic compound binding (GO:0097159)	*5006*	*100*	*54.19*	+	*1.85*	6.20×10^{-13}	5.62×10^{-8}
	heterocyclic compound binding (GO:1901363)	*4917*	*98*	*53.23*	+	*1.84*	1.34×10^{-10}	1.02×10^{-7}
	molecular transducer activity (GO:0060089)	*2324*	*1*	*25.16*	−	*0.04*	1.66×10^{-10}	1.07×10^{-7}
	identical protein binding (GO:0042802)	*1840*	*52*	*19.92*	+	*2.61*	1.98×10^{-10}	1.12×10^{-7}
	cytoskeletal protein binding (GO:0008092)	*936*	*35*	*10.13*	+	*3.45*	3.12×10^{-10}	1.57×10^{-7}
	signaling receptor activity (GO:0038023)	*2271*	*1*	*24.58*	−	*0.04*	4.19×10^{-10}	1.90×10^{-7}
	mRNA binding (GO:0003729)	*243*	*18*	*2.63*	+	*6.84*	4.91×10^{-10}	2.02×10^{-7}
	actin binding (GO:0003779)	*411*	*22*	*4.45*	+	*4.94*	1.85×10^{-9}	7.00×10^{-7}
G-CSF + Coc vs. Saline	**binding (GO:0005488)**	13001	167	108.62	+	1.54	7.88×10^{-21}	3.57×10^{-17}
	protein binding (GO:0005515)	8900	137	74.36	+	1.84	2.67×10^{-20}	6.06×10^{-17}
	identical protein binding (GO:0042802)	1840	46	15.37	+	2.99	1.67×10^{-11}	2.52×10^{-8}
	cytoskeletal protein binding (GO:0008092)	936	31	7.82	+	3.96	9.34×10^{-11}	1.06×10^{-7}
	RNA binding (GO:0003723)	1019	32	8.51	+	3.76	1.65×10^{-10}	1.50×10^{-7}
	structural molecule activity (GO:0005198)	613	24	5.12	+	4.69	6.58×10^{-10}	4.97×10^{-7}
	actin binding (GO:0003779)	411	19	3.43	+	5.53	3.48×10^{-9}	2.25×10^{-6}
	molecular transducer activity (GO:0060089)	2324	1	19.42	−	0.05	7.39×10^{-8}	3.72×10^{-5}
	heterocyclic compound binding (GO:1901363)	4917	74	41.08	+	1.8	6.61×10^{-8}	3.75×10^{-5}
	signaling receptor activity (GO:0038023)	2271	1	18.97	−	0.05	1.12×10^{-7}	5.07×10^{-5}
	organic cyclic compound binding (GO:0097159)	5006	74	41.83	+	1.77	1.38×10^{-7}	5.68×10^{-5}
	enzyme binding (GO:0019899)	2260	43	18.88	+	2.28	2.95×10^{-7}	1.11×10^{-4}

Given the substantial number of proteins altered in all treatment groups, analyses were performed to determine which transcription factors were predicted to be affecting the largest number of proteins in the samples. To do this, all of the proteins that were up-regulated relative to saline in each group were uploaded to the Enrichr software package (freely available: http://amp.pharm.mssm.edu/Enrichr/). Using inputs from an exhaustive list of published studies, this software predicts the transcription factors most likely to be upstream of regulated proteins and provides a transcription factor prediction score [24]. Based on this, the transcription factors likely to be responsible for the most changes in each group were identified. For these analyses focus was placed on two transcription factors that achieved statistically significant prediction value for each of the three treatment groups, as well as CREB1 which was significantly regulated only in the G-CSF group, but which has been broadly implicated in the neurobiology of addiction [35,36]. These analyses predicted the E2F1 transcription factor to be the strongest regulator of proteins in the G-CSF group, but is also a significant driver of transcription in the other two experimental groups (Figure 7B and Table S9). A similar pattern is seen for both Atf2 and CREB1. While not yet conclusive, these analyses identify potential hub molecules that are driven by G-CSF signaling to induce neuronal and potentially behavioral plasticity.

3.4. Protein Validation

Due to the relatively small sample size (N = 4) of each of the treatment groups and the relatively large number of proteins defined as significantly regulated by various treatments, we performed experiments to validate the scale and directionality of change of some key regulated proteins. For these analyses we chose tyrosine hydroxylase, the rate limiting enzyme in dopamine synthesis, and a protein predicted to be significantly decreased in all three treatment groups. We additionally examined changes in Mecp2, a methyl-DNA binding protein that has been shown to be important in numerous aspects of neuronal and behavioral response to cocaine [37,38], and which was predicted to be increased in all three treatment groups in our mass spec analyses. For these experiments animals received the same treatments as in Figure 1 and protein levels in the VTA were examined with quantitative Western blot analysis.

Analysis of tyrosine hydroxylase levels demonstrated changes similar in magnitude to those that were reported with the initial analyses. Two-way ANOVA demonstrated a main effect of G-CSF ($F_{(1,20)} = 11.63$; $p = 0.003$) and a significant G-CSF x cocaine interaction ($F_{(1,20)} = 7.612$; $p = 0.012$) but no main effect of cocaine ($F_{(1,20)} = 0.202$; $p = 0.66$). Post-hoc analyses (Fisher's LSD) demonstrated significant differences with all treatment groups compared to the Saline controls (Figure 8A—asterisks). To compare the results from the Western blots to the mass spec data from above, the fold-change from saline for all groups is marked on the graphs with a blue line. While the magnitude of the changes were not identical, they were quite similar and all in the same direction. A similar pattern for Mecp2 was also seen. We found a main effect of G-CSF ($F_{(1,20)} = 6.707$; $p = 0.018$) and a significant Cocaine x G-CSF interaction ($F_{(1,20)} = 11.19$; $p = 0.003$), but no main effect of cocaine ($F_{(1,20)} = 0.02$; $p = 0.88$). Post-hoc testing demonstrated significant differences between Saline and G-CSF and Saline and Cocaine (Figure 8B)

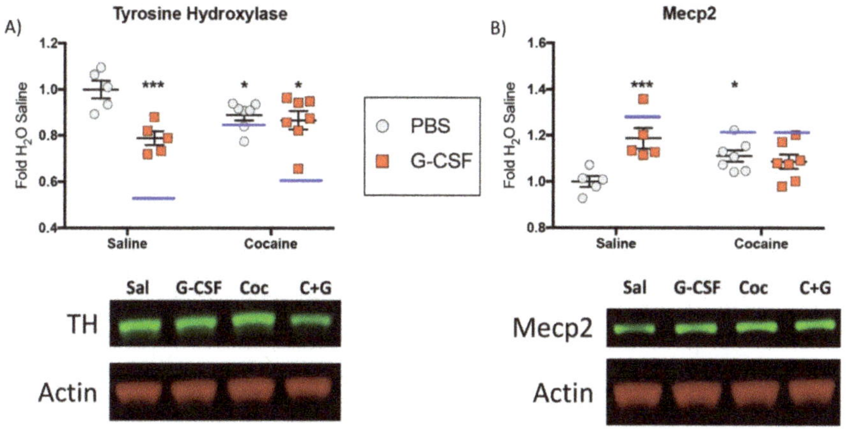

Figure 8. Western blot validations—To validate proteins identified as changed by mass spectrometry additional Western blot analysis of similarly treated tissue was performed. Graphical fold-change from saline control is shown for tyrosine hydroxylase (**A**) and Mecp2 (**B**) with corresponding representative images shown below. Blue lines on the graph represent the fold-change from saline that was seen in each of the treatment groups with mass spec analysis.

4. Discussion

We have recently identified G-CSF as a key mediator of neuronal and behavioral plasticity in response to cocaine [11]. In this manuscript an unbiased proteomics analysis is employed to identify protein changes induced in the VTA by G-CSF, both on its own and in combination with cocaine. In our original studies G-CSF signaling in the NAc was found to play a key role in the behavioral effects of G-CSF. Given that dopamine release from VTA terminals in the NAc is a crucial substrate of reward learning and the attribution of salience to rewarding stimuli, understanding changes in protein expression in the VTA is critical for understanding the neuroplasticity that occurs in response to drugs of abuse. Additionally, since the publication of our initial study we found that peripheral injections of G-CSF are capable of modulating dopamine signaling by enhancing release from VTA terminals in the NAc [12]. Given this and the fact that G-CSF receptors are densely expressed in the VTA [28] lead to these proteomic analyses of the VTA.

Review of the literature demonstrates that the exact intracellular signaling mechanisms of G-CSF in the brain are not fully clear and may be complex. G-CSF treatment has variously been shown to induce activity of the Jak-Stat, Erk, and CREB1 signaling cascades among others [28,39–42]. These results demonstrate that treatment with G-CSF decreased signaling in the CREB1 transcription factor signaling cascades, as well as the cAMP and PKA pathways which are well known to be upstream of CREB1 (Figure 2B) [43]. Increased expression of CREB1 in the NAc and in subregions of the VTA has been shown to decrease cocaine reward, and inhibition of CREB1 in these regions has been shown to enhance reward in a region-specific manner [35,44]. Analysis of significantly upregulated proteins in the G-CSF treatment group found that CREB1 was predicted to be one of the transcription factors driving gene expression (Figure 7). This apparent discrepancy in Figures 2 and 7 may be due to the fact that the IPA analysis looks at networks of proteins based on literature review, while the Enrichr software looks only at those proteins predicted to be directly downstream of the transcription factor. Since G-CSF enhances cocaine intake and place preference and alters CREB1-related signaling, it is possible that the behavioral effects of G-CSF are at least partially mediated through the CREB1 pathway.

We also observed regulation of proteins related to the maintenance of synapses and other cell-cell contacts in our G-CSF treated groups (Figures 2B and 4). This is of particular interest as numerous studies have demonstrated that changes in synapse density are induced by cocaine and are important for the behavioral response to drugs of abuse [45,46]. While most of these studies have focused on the NAc, there is also evidence for synaptic remodeling in the VTA [47,48]. These findings raise the possibility that G-CSF may participate in neurite remodeling, and may prime animals for further changes in synaptic structure in response to cocaine, thus leading to the potentiation of behavioral response induced by G-CSF [11].

The G-CSF-treated animals displayed significant changes in signaling cascades that are related to initiation of mRNA translation. IPA analyses predicted that one of the most up-regulated canonical signaling pathways is the eukaryotic initiation factor 2 (EIF2) pathway which is a critical mediator of protein translation initiation and has been implicated in synaptic plasticity and memory [49] (Figure 2C). Interestingly, EIF2 signaling has been shown to be inhibited by PKA signaling which is found to be decreased in our G-CSF-treated animals (Figure 2B). EIF2 is also known to be activated by the mTOR pathway which was predicted to be a key upstream regulator of the altered proteins in our dataset [50] (Figure 3B). Indeed, the two most highly predicted upstream regulators, mTOR and FMRP (Figure 3), have been shown to be critical regulators of translation of synaptic mRNAs and play key roles in synaptic plasticity [51].

There is a growing literature demonstrating the importance of regulators of synaptic translation regulators in the neuronal and behavioral plasticity in response to cocaine. Recently, an elegant study by the Wolf lab demonstrated increased protein translation during cue-induced drug seeking, and inhibition of mTOR or EIF2 could significantly attenuate cocaine seeking [52]. Studies of FMRP have shown that it is also critical for the rewarding effects of cocaine and changes in synapse structure in response to cocaine [53]. A number of studies have found roles for mTOR-mediated intracellular signaling cascades in NAc in response to cocaine [54–56]. Behaviorally it has been demonstrated that inhibition of mTOR with rapamycin can reduce locomotor sensitization, conditioned place preference, and cocaine seeking [57–59]. The role of mTOR in the VTA was recently interrogated by Liu and colleagues who found that deletion of mTOR reduced VTA dopamine release and decreased conditioned place preference for cocaine [60].

When examining the number of proteins that were significantly altered between the different treatment groups, it was found that treatment with G-CSF alone leads to changes in the largest number of proteins (Figures 5 and 6). This may be due to the fact that activation of the G-CSF receptor has been coupled to direct activation of transcription factors [39,40,61]. In contrast, cocaine directly leads to changes in multiple neurotransmitter systems, but its effects on gene expression are tightly coupled with context and behavior [62–64]. It is interesting that the combination of G-CSF and cocaine lead to the smallest number of regulated proteins of the three treatment groups (Figure 5B). This suggests the possibility that there are interactions between signaling pathways after G-CSF and cocaine in the two that temper changes in protein expression in the VTA.

One of the more surprising findings from these studies was the similarity in changes between treatment groups. Pathways that were regulated by G-CSF, Cocaine, or the combination were largely the same (Figure 7A and Table 3) despite some differences. Given that G-CSF enhances the behavioral effects of cocaine [11] and enhances dopamine release from the VTA [12] one might have suspected that the effects of G-CSF and cocaine on protein expression in the VTA would have been additive. Comparisons of levels of proteins relative to Saline revealed only 42 proteins in which Saline <G-CSF <Cocaine <G-CSF + Cocaine and 107 in which Saline > G-CSF > Cocaine > G-CSF + Cocaine (Table S10). This raises the possibility that the behavioral and physiological responses potentiated by G-CSF may be owing in part to this smaller subset of proteins, or, more likely, that the changes induced by G-CSF are complex and dependent on the function and response of multiple brain regions. Further examination of these clusters of regulated proteins will be important for understanding interactions between G-CSF and cocaine.

While these results have provided new and interesting findings related to the effects of G-CSF and cocaine on proteomic expression in the midbrain, there are important caveats to their interpretation. This study was designed as a discovery analysis to identify G-CSF and cocaine interactions in a 2 × 2 design, and while this allowed us to investigate effects and interactions it lead to a study with low power in terms of sample size (N = 4/group). While we were able to perform successful Western blot validation of several regulated targets (Figure 8) the low sample size and decision not to correct *p* values leads to a high likelihood that some of the reported changes are indeed false positives. Additionally, while the use of network and pathway analyses (IPA, GO, Enrichr) are very useful for the identification of potentially regulated pathways, it is important to note that none of these software packages are built on a comprehensive review of the entire scientific knowledge base, but rather large cross-sections of data that are available to be mined. Additionally, most of these software packages pool data across tissues to increase statistical power in the analyses. While this has utility, it is important to note that regulation of intracellular pathways in other tissues, or even in other brain regions, is likely to be different from that seen in the VTA and has the potential to lead to spurious conclusions.

In sum, we have identified G-CSF as a neuroimmune factor that significantly influences the behavioral and neuronal response to cocaine [11]. While this initial study established the possibility that G-CSF may be a translationally-relevant target for the treatment of cocaine abuse, there remains much to be done to establish its mechanism of action in the brain. Here we present an unbiased proteomic analysis of the VTA animals treated with G-CSF, cocaine, or both. This study identified key intracellular signaling pathways that are altered by systemic G-CSF treatment and lays the groundwork for future mechanistic studies into the effects of G-CSF in brain reward structures.

Supplementary Materials: The following are available online at http://www.mdpi.com/2227-7382/6/4/35/s1, Table S1: All significantly regulated proteins for each pairwise comparison with corresponding p-value and Log2 Fold Change; Table S2: Significantly regulated proteins predicted to be downstream of FMRP; Table S3: Significantly regulated proteins predicted to be downstream of mTOR; Table S4: Significantly regulated proteins predicted to be involved in neuronal morphology; Table S5: Breakdown of all significantly regulated proteins from all 7 portions of the Venn diagram presented in Figure 6; Table S6: Comparisons of significantly regulated canonical signaling pathways for each treatment group as predicted with Ingenuity Pathway Analysis; Table S7: Gene ontology analysis of predicted molecular function of all significantly upregulated proteins in each pairwise comparison; Table S8: Gene ontology analysis of predicted cellular component of all significantly upregulated proteins in each pairwise comparison; Table S9: Predicted transcription factors of upregulated proteins for each pairwise comparison; Table S10: Lists of significantly regulated proteins following the pattern of Saline < G-CSF < Cocaine < G-CSF + Cocaine and Saline > G-CSF > Cocaine > G-CSF + Cocaine.

Author Contributions: D.D.K. conceived and designed the study. N.R. and T.T.L. performed protein isolation, mass spectrometry analyses, and initial data processing. R.W. performed data analysis and statistical guidance. N.L.M., R.S.H., M.G.K., E.S.C. & D.D.K. performed detailed data analyses and created the figures. N.L.M. & D.D.K. wrote the manuscript. All authors provided critical feedback and edits on the final version of the manuscript.

Funding: Proteomic analysis for this study was supported by the Yale/NIDA Neuroproteomic Center Grant from NIDA (P30-DA018343) including pilot award funds from that grant to D.D.K. Additional support was provided from NIDA to D.D.K. (DA044308) and to E.S.C. (DA042111), from the Brain and Behavior Research Foundation to D.D.K. & E.S.C., funds from the Whitehall Foundation and the Edward Mallinckrodt Jr. Foundation to E.S.C., as well as funds from the Friedman Brain Institute, Leon Levy Foundation, and Seaver Family Foundation all to D.D.K.

Conflicts of Interest: The authors declare no conflict of interest. The funders had no role in the design of the study; in the collection, analyses, or interpretation of data; in the writing of the manuscript, and in the decision to publish the results.

References

1. Patel, V.; Araya, R.; Chatterjee, S.; Chisholm, D.; Cohen, A.; De Silva, M.; Hosman, C.; McGuire, H.; Rojas, G.; van Ommeren, M. Treatment and prevention of mental disorders in low-income and middle-income countries. *Lancet* **2007**, *370*, 991–1005. [CrossRef]
2. Patel, V.; Chisholm, D.; Parikh, R.; Charlson, F.J.; Degenhardt, L.; Dua, T.; Ferrari, A.J.; Hyman, S.; Laxminarayan, R.; Levin, C.; et al. Addressing the burden of mental, neurological, and substance use disorders: Key messages from Disease Control Priorities, 3rd edition. *Lancet* **2016**, *387*, 1672–1685. [CrossRef]
3. Castells, X.; Cunill, R.; Pérez-Mañá, C.; Vidal, X.; Capellà, D. Psychostimulant drugs for cocaine dependence. *Cochrane Database Syst. Rev.* **2016**, *9*, CD007380. [CrossRef] [PubMed]

4. Shorter, D.; Domingo, C.B.; Kosten, T.R. Emerging drugs for the treatment of cocaine use disorder: A review of neurobiological targets and pharmacotherapy. *Expert Opin. Emerg. Drugs* **2015**, *20*, 15–29. [CrossRef] [PubMed]
5. Platt, D.M.; Rowlett, J.K.; Spealman, R.D. Behavioral effects of cocaine and dopaminergic strategies for preclinical medication development. *Psychopharmacology* **2002**, *163*, 265–282. [CrossRef] [PubMed]
6. Preti, A. New developments in the pharmacotherapy of cocaine abuse. *Addict. Biol.* **2007**, *12*, 133–151. [CrossRef] [PubMed]
7. Shorter, D.; Kosten, T.R. Novel pharmacotherapeutic treatments for cocaine addiction. *BMC Med.* **2011**, *9*, 119. [CrossRef] [PubMed]
8. Hofford, R.S.; Russo, S.J.; Kiraly, D.D. Neuroimmune mechanisms of psychostimulant and opioid use disorders. *Eur. J. Neurosci.* **2018**. [CrossRef] [PubMed]
9. Lacagnina, M.J.; Rivera, P.D.; Bilbo, S.D. Glial and neuroimmune mechanisms as critical modulators of drug use and abuse. *Neuropsychopharmacology* **2017**, *42*, 156–177. [CrossRef] [PubMed]
10. Hodes, G.E.; Kana, V.; Menard, C.; Merad, M.; Russo, S.J. Neuroimmune mechanisms of depression. *Nat. Neurosci.* **2015**, *18*, 1386–1393. [CrossRef] [PubMed]
11. Calipari, E.S.; Godino, A.; Peck, E.G.; Salery, M.; Mervosh, N.L.; Landry, J.A.; Russo, S.J.; Hurd, Y.L.; Nestler, E.J.; Kiraly, D.D. Granulocyte-colony stimulating factor controls neural and behavioral plasticity in response to cocaine. *Nat. Commun.* **2018**, *9*, 9. [CrossRef] [PubMed]
12. Kutlu, M.G.; Brady, L.J.; Peck, E.G.; Hofford, R.S.; Yorgason, J.T.; Siciliano, C.A.; Kiraly, D.D.; Calipari, E.S. Granulocyte colony stimulating factor enhances reward learning through potentiation of mesolimbic dopamine system function. *J. Neurosci.* **2018**, 1116–1118. [CrossRef] [PubMed]
13. Kumar, A.S.; Jagadeeshan, S.; Subramanian, A.; Chidambaram, S.B.; Surabhi, R.P.; Singhal, M.; Bhoopalan, H.; Sekar, S.; Pitani, R.S.; Duvuru, P.; et al. Molecular Mechanism of Regulation of MTA1 Expression by Granulocyte Colony-stimulating Factor. *J. Biol. Chem.* **2016**, *291*, 12310–12321. [CrossRef] [PubMed]
14. Prakash, A.; Medhi, B.; Chopra, K. Granulocyte colony stimulating factor (GCSF) improves memory and neurobehavior in an amyloid-β induced experimental model of Alzheimer's disease. *Pharmacol. Biochem. Behav.* **2013**, *110*, 46–57. [CrossRef] [PubMed]
15. Minnerup, J.; Sevimli, S.; Schäbitz, W.-R. Granulocyte-colony stimulating factor for stroke treatment: Mechanisms of action and efficacy in preclinical studies. *Exp. Transl. Stroke Med.* **2009**, *1*, 2. [CrossRef] [PubMed]
16. Schäbitz, W.-R.; Schneider, A. New targets for established proteins: Exploring G-CSF for the treatment of stroke. *Trends Pharmacol. Sci.* **2007**, *28*, 157–161. [CrossRef] [PubMed]
17. Lu, C.Z.; Xiao, B.G. G-CSF and neuroprotection: A therapeutic perspective in cerebral ischaemia. *Biochem. Soc. Trans.* **2006**, *34*, 1327–1333. [CrossRef] [PubMed]
18. Meuer, K.; Pitzer, C.; Teismann, P.; Krüger, C.; Göricke, B.; Laage, R.; Lingor, P.; Peters, K.; Schlachetzki, J.C.M.; Kobayashi, K.; et al. Granulocyte-colony stimulating factor is neuroprotective in a model of Parkinson's disease. *J. Neurochem.* **2006**, *97*, 675–686. [CrossRef] [PubMed]
19. Kauer, J.A. Learning mechanisms in addiction: Synaptic plasticity in the ventral tegmental area as a result of exposure to drugs of abuse. *Annu. Rev. Physiol.* **2004**, *66*, 447–475. [CrossRef] [PubMed]
20. Nestler, E.J. Is there a common molecular pathway for addiction? *Nat. Neurosci.* **2005**, *8*, 1445–1449. [CrossRef] [PubMed]
21. Käll, L.; Canterbury, J.D.; Weston, J.; Noble, W.S.; MacCoss, M.J. Semi-supervised learning for peptide identification from shotgun proteomics datasets. *Nat. Methods* **2007**, *4*, 923–925. [CrossRef] [PubMed]
22. Nesvizhskii, A.I.; Keller, A.; Kolker, E.; Aebersold, R. A statistical model for identifying proteins by tandem mass spectrometry. *Anal. Chem.* **2003**, *75*, 4646–4658. [CrossRef] [PubMed]
23. Oberg, A.L.; Mahoney, D.W.; Eckel-Passow, J.E.; Malone, C.J.; Wolfinger, R.D.; Hill, E.G.; Cooper, L.T.; Onuma, O.K.; Spiro, C.; Therneau, T.M.; et al. Statistical analysis of relative labeled mass spectrometry data from complex samples using ANOVA. *J. Proteome Res.* **2008**, *7*, 225–233. [CrossRef] [PubMed]
24. Chen, E.Y.; Tan, C.M.; Kou, Y.; Duan, Q.; Wang, Z.; Meirelles, G.V.; Clark, N.R.; Ma'ayan, A. Enrichr: Interactive and collaborative HTML5 gene list enrichment analysis tool. *BMC Bioinform.* **2013**, *14*, 128. [CrossRef] [PubMed]

25. Kiraly, D.D.; Stone, K.L.; Colangelo, C.M.; Abbott, T.; Wang, Y.; Mains, R.E.; Eipper, B.A. Identification of kalirin-7 as a potential post-synaptic density signaling hub. *J. Proteome Res.* **2011**, *10*, 2828–2841. [CrossRef] [PubMed]
26. Lüscher, C. Cocaine-evoked synaptic plasticity of excitatory transmission in the ventral tegmental area. *Cold Spring Harb. Perspect. Med.* **2013**, *3*, a012013. [CrossRef] [PubMed]
27. Ridwan, S.; Bauer, H.; Frauenknecht, K.; Hefti, K.; von Pein, H.; Sommer, C.J. Distribution of the hematopoietic growth factor G-CSF and its receptor in the adult human brain with specific reference to Alzheimer's disease. *J. Anat.* **2014**, *224*, 377–391. [CrossRef] [PubMed]
28. Huang, H.-Y.; Lin, S.-Z.; Kuo, J.-S.; Chen, W.-F.; Wang, M.-J. G-CSF protects dopaminergic neurons from 6-OHDA-induced toxicity via the ERK pathway. *Neurobiol. Aging* **2007**, *28*, 1258–1269. [CrossRef] [PubMed]
29. Jennings, M.D.; Pavitt, G.D. A new function and complexity for protein translation initiation factor eIF2B. *Cell Cycle* **2014**, *13*, 2660–2665. [CrossRef] [PubMed]
30. Ferguson, D.; Koo, J.W.; Feng, J.; Heller, E.; Rabkin, J.; Heshmati, M.; Renthal, W.; Neve, R.; Liu, X.; Shao, N.; et al. Essential role of SIRT1 signaling in the nucleus accumbens in cocaine and morphine action. *J. Neurosci.* **2013**, *33*, 16088–16098. [CrossRef] [PubMed]
31. Renthal, W.; Kumar, A.; Xiao, G.; Wilkinson, M.; Covington, H.E.; Maze, I.; Sikder, D.; Robison, A.J.; LaPlant, Q.; Dietz, D.M.; et al. Genome-wide analysis of chromatin regulation by cocaine reveals a role for sirtuins. *Neuron* **2009**, *62*, 335–348. [CrossRef] [PubMed]
32. Ashburner, M.; Ball, C.A.; Blake, J.A.; Botstein, D.; Butler, H.; Cherry, J.M.; Davis, A.P.; Dolinski, K.; Dwight, S.S.; Eppig, J.T.; et al. Gene ontology: Tool for the unification of biology. The Gene Ontology Consortium. *Nat. Genet.* **2000**, *25*, 25–29. [CrossRef] [PubMed]
33. The Gene Ontology Consortium. Expansion of the Gene Ontology knowledgebase and resources. *Nucleic Acids Res.* **2017**, *45*, D331–D338. [CrossRef] [PubMed]
34. Mi, H.; Huang, X.; Muruganujan, A.; Tang, H.; Mills, C.; Kang, D.; Thomas, P.D. PANTHER version 11: Expanded annotation data from Gene Ontology and Reactome pathways, and data analysis tool enhancements. *Nucleic Acids Res.* **2017**, *45*, D183–D189. [CrossRef] [PubMed]
35. Carlezon, W.A.; Thome, J.; Olson, V.G.; Lane-Ladd, S.B.; Brodkin, E.S.; Hiroi, N.; Duman, R.S.; Neve, R.L.; Nestler, E.J. Regulation of cocaine reward by CREB. *Science* **1998**, *282*, 2272–2275. [CrossRef] [PubMed]
36. Brown, T.E.; Lee, B.R.; Mu, P.; Ferguson, D.; Dietz, D.; Ohnishi, Y.N.; Lin, Y.; Suska, A.; Ishikawa, M.; Huang, Y.H.; et al. A silent synapse-based mechanism for cocaine-induced locomotor sensitization. *J. Neurosci.* **2011**, *31*, 8163–8174. [CrossRef] [PubMed]
37. Deng, J.V.; Wan, Y.; Wang, X.; Cohen, S.; Wetsel, W.C.; Greenberg, M.E.; Kenny, P.J.; Calakos, N.; West, A.E. MeCP2 phosphorylation limits psychostimulant-induced behavioral and neuronal plasticity. *J. Neurosci.* **2014**, *34*, 4519–4527. [CrossRef] [PubMed]
38. Im, H.-I.; Hollander, J.A.; Bali, P.; Kenny, P.J. MeCP2 controls BDNF expression and cocaine intake through homeostatic interactions with microRNA-212. *Nat. Neurosci.* **2010**, *13*, 1120–1127. [CrossRef] [PubMed]
39. Tian, S.S.; Tapley, P.; Sincich, C.; Stein, R.B.; Rosen, J.; Lamb, P. Multiple signaling pathways induced by granulocyte colony-stimulating factor involving activation of JAKs, STAT5, and/or STAT3 are required for regulation of three distinct classes of immediate early genes. *Blood* **1996**, *88*, 4435–4444. [PubMed]
40. Marino, V.J.; Roguin, L.P. The granulocyte colony stimulating factor (G-CSF) activates Jak/STAT and MAPK pathways in a trophoblastic cell line. *J. Cell Biochem.* **2008**, *103*, 1512–1523. [CrossRef] [PubMed]
41. Cassinat, B.; Zassadowski, F.; Ferry, C.; Llopis, L.; Bruck, N.; Lainey, E.; Duong, V.; Cras, A.; Despouy, G.; Chourbagi, O.; et al. New role for granulocyte colony-stimulating factor-induced extracellular signal-regulated kinase 1/2 in histone modification and retinoic acid receptor α recruitment to gene promoters: Relevance to acute promyelocytic leukemia cell differentiation. *Mol. Cell. Biol.* **2011**, *31*, 1409–1418. [CrossRef] [PubMed]
42. Diederich, K.; Schäbitz, W.-R.; Kuhnert, K.; Hellström, N.; Sachser, N.; Schneider, A.; Kuhn, H.-G.; Knecht, S. Synergetic effects of granulocyte-colony stimulating factor and cognitive training on spatial learning and survival of newborn hippocampal neurons. *PLoS ONE* **2009**, *4*, e5303. [CrossRef] [PubMed]
43. Robison, A.J.; Nestler, E.J. Transcriptional and epigenetic mechanisms of addiction. *Nat. Rev. Neurosci.* **2011**, *12*, 623–637. [CrossRef] [PubMed]

44. Olson, V.G.; Zabetian, C.P.; Bolanos, C.A.; Edwards, S.; Barrot, M.; Eisch, A.J.; Hughes, T.; Self, D.W.; Neve, R.L.; Nestler, E.J. Regulation of drug reward by cAMP response element-binding protein: Evidence for two functionally distinct subregions of the ventral tegmental area. *J. Neurosci.* **2005**, *25*, 5553–5562. [CrossRef] [PubMed]
45. Kiraly, D.D.; Ma, X.-M.; Mazzone, C.M.; Xin, X.; Mains, R.E.; Eipper, B.A. Behavioral and morphological responses to cocaine require kalirin7. *Biol. Psychiatry* **2010**, *68*, 249–255. [CrossRef] [PubMed]
46. Dietz, D.M.; Sun, H.; Lobo, M.K.; Cahill, M.E.; Chadwick, B.; Gao, V.; Koo, J.W.; Mazei-Robison, M.S.; Dias, C.; Maze, I.; et al. Rac1 is essential in cocaine-induced structural plasticity of nucleus accumbens neurons. *Nat. Neurosci.* **2012**, *15*, 891–896. [CrossRef] [PubMed]
47. Sarti, F.; Borgland, S.L.; Kharazia, V.N.; Bonci, A. Acute cocaine exposure alters spine density and long-term potentiation in the ventral tegmental area. *Eur. J. Neurosci.* **2007**, *26*, 749–756. [CrossRef] [PubMed]
48. Ungless, M.A.; Whistler, J.L.; Malenka, R.C.; Bonci, A. Single cocaine exposure in vivo induces long-term potentiation in dopamine neurons. *Nature* **2001**, *411*, 583–587. [CrossRef] [PubMed]
49. Costa-Mattioli, M.; Sonenberg, N.; Richter, J.D. Translational regulatory mechanisms in synaptic plasticity and memory storage. *Prog. Mol. Biol. Transl. Sci.* **2009**, *90*, 293–311. [PubMed]
50. Taha, E.; Gildish, I.; Gal-Ben-Ari, S.; Rosenblum, K. The role of eEF2 pathway in learning and synaptic plasticity. *Neurobiol. Learn. Mem.* **2013**, *105*, 100–106. [CrossRef] [PubMed]
51. Khlebodarova, T.M.; Kogai, V.V.; Trifonova, E.A.; Likhoshvai, V.A. Dynamic landscape of the local translation at activated synapses. *Mol. Psychiatry* **2018**, *23*, 107–114. [CrossRef] [PubMed]
52. Werner, C.T.; Stefanik, M.T.; Milovanovic, M.; Caccamise, A.; Wolf, M.E. Protein Translation in the Nucleus Accumbens Is Dysregulated during Cocaine Withdrawal and Required for Expression of Incubation of Cocaine Craving. *J. Neurosci.* **2018**, *38*, 2683–2697. [CrossRef] [PubMed]
53. Smith, L.N.; Jedynak, J.P.; Fontenot, M.R.; Hale, C.F.; Dietz, K.C.; Taniguchi, M.; Thomas, F.S.; Zirlin, B.C.; Birnbaum, S.G.; Huber, K.M.; et al. Fragile X mental retardation protein regulates synaptic and behavioral plasticity to repeated cocaine administration. *Neuron* **2014**, *82*, 645–658. [CrossRef] [PubMed]
54. Cahill, M.E.; Bagot, R.C.; Gancarz, A.M.; Walker, D.M.; Sun, H.; Wang, Z.-J.; Heller, E.A.; Feng, J.; Kennedy, P.J.; Koo, J.W.; et al. Bidirectional Synaptic Structural Plasticity after Chronic Cocaine Administration Occurs through Rap1 Small GTPase Signaling. *Neuron* **2016**, *89*, 566–582. [CrossRef] [PubMed]
55. Shi, X.; Miller, J.S.; Harper, L.J.; Poole, R.L.; Gould, T.J.; Unterwald, E.M. Reactivation of cocaine reward memory engages the Akt/GSK3/mTOR signaling pathway and can be disrupted by GSK3 inhibition. *Psychopharmacology* **2014**, *231*, 3109–3118. [CrossRef] [PubMed]
56. Sutton, L.P.; Caron, M.G. Essential role of D1R in the regulation of mTOR complex1 signaling induced by cocaine. *Neuropharmacology* **2015**, *99*, 610–619. [CrossRef] [PubMed]
57. Bailey, J.; Ma, D.; Szumlinski, K.K. Rapamycin attenuates the expression of cocaine-induced place preference and behavioral sensitization. *Addict. Biol.* **2012**, *17*, 248–258. [CrossRef] [PubMed]
58. Wu, J.; McCallum, S.E.; Glick, S.D.; Huang, Y. Inhibition of the mammalian target of rapamycin pathway by rapamycin blocks cocaine-induced locomotor sensitization. *Neuroscience* **2011**, *172*, 104–109. [CrossRef] [PubMed]
59. James, M.H.; Quinn, R.K.; Ong, L.K.; Levi, E.M.; Smith, D.W.; Dickson, P.W.; Dayas, C.V. Rapamycin reduces motivated responding for cocaine and alters GluA1 expression in the ventral but not dorsal striatum. *Eur. J. Pharmacol.* **2016**, *784*, 147–154. [CrossRef] [PubMed]
60. Liu, X.; Li, Y.; Yu, L.; Vickstrom, C.R.; Liu, Q.-S. VTA mTOR Signaling Regulates Dopamine Dynamics, Cocaine-Induced Synaptic Alterations, and Reward. *Neuropsychopharmacology* **2018**, *43*, 1066–1077. [CrossRef] [PubMed]
61. Schäbitz, W.R.; Kollmar, R.; Schwaninger, M.; Juettler, E.; Bardutzky, J.; Schölzke, M.N.; Sommer, C.; Schwab, S. Neuroprotective effect of granulocyte colony-stimulating factor after focal cerebral ischemia. *Stroke* **2003**, *34*, 745–751. [CrossRef] [PubMed]
62. Uslaner, J.; Badiani, A.; Norton, C.S.; Day, H.E.; Watson, S.J.; Akil, H.; Robinson, T.E. Amphetamine and cocaine induce different patterns of c-fos mRNA expression in the striatum and subthalamic nucleus depending on environmental context. *Eur. J. Neurosci.* **2001**, *13*, 1977–1983. [CrossRef] [PubMed]

63. Lasseter, H.C.; Xie, X.; Arguello, A.A.; Wells, A.M.; Hodges, M.A.; Fuchs, R.A. Contribution of a mesocorticolimbic subcircuit to drug context-induced reinstatement of cocaine-seeking behavior in rats. *Neuropsychopharmacology* **2014**, *39*, 660–669. [CrossRef] [PubMed]
64. Stankeviciute, N.M.; Scofield, M.D.; Kalivas, P.W.; Gipson, C.D. Rapid, transient potentiation of dendritic spines in context-induced relapse to cocaine seeking. *Addict. Biol.* **2014**, *19*, 972–974. [CrossRef] [PubMed]

proteomes

MDPI

Review

Phosphorylation of the AMPAR-TARP Complex in Synaptic Plasticity

Joongkyu Park [1,2]

[1] Department of Pharmacology, Wayne State University School of Medicine, Detroit, MI 48201, USA; joongkyu.park@wayne.edu; Tel.: +1-303-577-1580

[2] Department of Neurology, Wayne State University School of Medicine, Detroit, MI 48201, USA

Received: 11 September 2018; Accepted: 6 October 2018; Published: 8 October 2018

Abstract: Synaptic plasticity has been considered a key mechanism underlying many brain functions including learning, memory, and drug addiction. An increase or decrease in synaptic activity of the α-amino-3-hydroxy-5-methyl-4-isoxazolepropionic acid receptor (AMPAR) complex mediates the phenomena as shown in the cellular models of synaptic plasticity, long-term potentiation (LTP), and depression (LTD). In particular, protein phosphorylation shares the spotlight in expressing the synaptic plasticity. This review summarizes the studies on phosphorylation of the AMPAR pore-forming subunits and auxiliary proteins including transmembrane AMPA receptor regulatory proteins (TARPs) and discusses its role in synaptic plasticity.

Keywords: phosphorylation; AMPA receptor complex; transmembrane AMPA receptor regulatory protein; synaptic plasticity

---

1. Introduction

Animal behavior is dynamic. One of the essential features of brain function is the ability to be dynamic in order to express various behaviors. Selective strengthening and weakening of synaptic transmission have been modeled as a critical mechanism for many brain functions including learning, memory, and drug addiction. Long-term potentiation (LTP) and depression (LTD) are well-characterized models of synaptic plasticity, and they can be regulated by changes at presynaptic (e.g., changes in the release of neurotransmitters) and postsynaptic (e.g., changes in the number and properties of neurotransmitter receptors) sites. Importance of the α-amino-3-hydroxy-5-methyl-4-isoxazolepropionic acid receptor (AMPAR) complex has emerged notably in LTP and LTD. In particular, protein phosphorylation is well known to play a pivotal role in the expression of synaptic plasticity, for example, Ca^{2+}/CaM-dependent protein kinase II (CaMKII) in hippocampal LTP [1–3]. This review gives an overview of the studies on phosphorylation of the AMPAR pore-forming subunits and auxiliary proteins including transmembrane AMPA receptor regulatory proteins (TARPs) and discusses its role in those plastic cellular phenomena.

2. AMPAR Complex

AMPARs are predominantly distributed at excitatory synapses and mediate the majority of fast transmission. The AMPAR complex consists of four pore-forming subunits (GluA1–4) and auxiliary proteins including TARP, cornichons-like (CNIH), and cysteine-knot AMPAR modulating protein (CKAMP)/Shisa family in the brain [4–9]. Knockout mice of GluA1, TARPγ-8, or CNIH-2/-3 show a substantial reduction in hippocampal LTP [10–12], and GluA2 knockout leads to LTD impairment in cultured cerebellar neurons and anterior cingulated cortex slices [13,14], highlighting their significance for synaptic plasticity.

3. Phosphorylation of the Pore-Forming Subunits: GluA1

To elucidate how phosphorylation regulates synaptic plasticity, extensive research has been conducted to identify phosphorylation sites on the pore-forming subunits [15]. Based on the revised topology [16], the C-terminal intracellular region of the GluA1 subunit has emerged as a potential phosphorylation target (Figure 1).

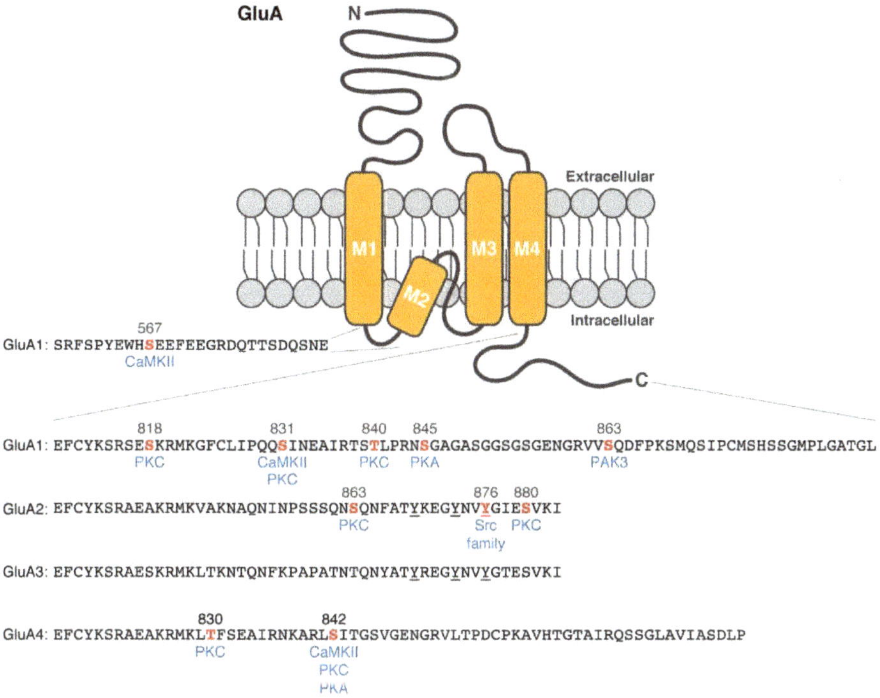

Figure 1. Schematic illustration of α-amino-3-hydroxy-5-methyl-4-isoxazolepropionic acid receptor (AMPAR) subunit structure and phosphorylation sites in the intracellular loop 1 and C-terminal region (referred to [15]). The identified phosphorylation sites are shown in red, and the protein kinases are listed below the sites in blue. The conserved three tyrosine residues on GluA2 and GluA3 subunits are underlined. M1–4 indicates transmembrane domains. CaMKII = Ca^{2+}/CaM-dependent protein kinase II; PKC = protein kinase C; PKA = cAMP-dependent protein kinase; PAK3 = p21-activated kinase-3.

A serine residue (Ser) at 831 in the intracellular region (Ser831) is one of the most attractive phosphorylation sites of the GluA1 subunit. In vitro phosphorylation assays reveal that Ser831 can be directly phosphorylated by CaMKII [17] or protein kinase C (PKC) [18]. Ser831 phosphorylation is also shown on transiently expressed GluA1 in a heterologous cell line (a quail-origin fibroblast, QT6) co-transfected with a constitutively active form of CaMKII [19] and in a PKC- and cAMP-dependent protein kinase (PKA)-activating condition of human embryonic kidney 293 (HEK293) cells (treated with phorbol 12-myristate 13-acetate, forskolin, and 3-isobutyl-1-methylxanthine) [18], respectively. Therefore, this single site can be a shared target that can be phosphorylated by two different kinases: CaMKII and PKC. Functional studies with whole-cell recording or outside-out membrane patches reveal that Ser831 phosphorylation is critical for CaMKII- or PKC-induced AMPAR potentiation and channel conductance enhancement in heterologous systems [17,20,21].

A serine residue at 845 (Ser845) is another phosphorylation target site that has been characterized well. In vitro phosphorylation assay with purified PKA identifies direct phosphorylation at Ser845 [18].

In heterologous cells, PKA activation (by forskolin and 3-isobutyl-1-methylxanthine) induces Ser845 phosphorylation on transiently expressed GluA1 [18,19]. Recording of whole-cell patches or single-channel currents with PKA infusion reveals that Ser845 is necessary for PKA-induced AMPAR potentiation and enhancement of channel open probability [18,22]. The positive effects of Ser831 and Ser845 phosphorylation on AMPAR potentiation and channel conductance may contribute to synaptic plasticity upon stimulation.

In ex vivo slices, phosphorylation at Ser831 and Ser845 of GluA1 highly correlates with LTP and LTD. To describe their phosphorylation states, phosphorylation site-specific antibodies have been developed and validated by activation of PKC (by phorbol dibutyrate) and PKA (by forskolin and 3-isobutyl-1-methylxanthine) in rat hippocampal slices [19]. Theta burst stimulation (TBS)-induced LTP increases Ser831 phosphorylation (but not at Ser845) in rat hippocampal slices, and the subsequent low-frequency stimulation (LFS)-induced depotentiation decreases the phosphorylation [23]. This is consistent with the notion that Ser831 residue of GluA1 subunit can be phosphorylated by CaMKII, which is required for LTP [3,17,24]. In contrast, Ser845 phosphorylation (not at Ser831) reduces in the LFS-induced LTD condition of rat hippocampal slices [23].

PKC is known to phosphorylate other sites on the GluA1 subunit. Autoradiograph and phosphorylation site-specific antibody combined with purified C-terminal mutants of GluA1 reveal that PKC (but not CaMKII) can directly phosphorylate Ser818 in vitro, and Ser818 phosphorylation increases upon chemical LTP and TBS [25]. Thr840 is another PKC target site that was uncovered by in vitro phosphorylation assays, a phosphorylation site-specific antibody, and PKC activation (by phorbol ester) or inhibition (by Gö6976 or chelerythrine) in hippocampal slices [26,27]. Interestingly, Thr840 phosphorylation inhibits a PKA-mediated increase in Ser845 phosphorylation and subsequent AMPAR potentiation whereas a phospho-mimetic aspartate mutation at Ser845 (S845D) inhibits PKC-mediated Thr840 phosphorylation in vitro and in hippocampal slices [27]. This suggests that GluA1 phosphorylation may regulate AMPAR channel properties dynamically depending on upstream kinase signaling pathways. This idea that the GluA1 subunit has a hyper-regulatory domain with multiple phosphorylation sites is also supported by a study that single or multiple aspartate mutations at Ser818, Thr840, and Ser831 residues enhance the weighted mean channel conductance [28]. In addition, surface expression and synaptic trafficking of AMPARs can be regulated by p21-activated kinase-3 (PAK3)-mediated Ser863 phosphorylation [29] and CaMKII-mediated phosphorylation of Ser567 residue in the loop 1 of GluA1 subunits [30]. This information is listed in Table 1.

Table 1. Phosphorylation of AMPAR pore-forming subunits and auxiliary proteins and their involvement in synaptic plasticity. LTP = long-term potentiation; LTD = long-term depression; TBS = theta burst stimulation; KO = knockout; TARP = transmembrane AMPA receptor regulatory proteins; CKAMP = cysteine-knot AMPAR modulating protein; GRIP = glutamate receptor-interacting protein; ERK2 = extracellular signal-regulated protein kinase 2; p38 MAPK = p38 mitogen-activated protein kinase.

Protein	Target Site	Kinase	Identification	Effect on the AMPAR Complex	Involvement in Synaptic Plasticity
GluA1	Ser567	CaMKII	In vitro; Phospho-specific antibody with rat hippocampal lysate [30]	Regulation of synaptic trafficking [30]	
	Ser818	PKC	In vitro; PKC activation of heterologous cells; Phospho-specific antibody with rat cortical lysate [25]	Enhancement of the weighted mean channel conductance [28]	A correlational increase upon chemical LTP and TBS [25]
	Ser831	CaMKII	In vitro [17]; Co-expression of a constitutively active CaMKII in heterologous cells [19]; Phospho-specific antibody with rat hippocampal lysate [19,23]	Potentiation of AMPAR current [17]; Enhancement of channel conductance [20,21]	A correlational increase upon LTP [23]
		PKC	In vitro; PKC- and PKA-activation of heterologous cells [18]		
	Thr840	PKC	In vitro [26,27]; Phosphopeptide mapping of hippocampal slices [26]; Phospho-specific antibody with mouse hippocampal lysate [26,27]	Inhibition of PKA-induced AMPAR potentiation [27]; Enhancement of the weighted mean channel conductance [28]	A correlational change upon PKC activity [26,27]
	Ser845	PKA	In vitro [18]; PKA activation of heterologous cells [18,19]; Phospho-specific antibody with rat hippocampal lysate [19,23]	Potentiation of AMPAR current [18]; Enhancement of channel opening probability [22]	A correlational decrease upon LTD [23]
	Ser863	PAK3	In vitro; Co-expression in heterologous cells; Phospho-specific antibody with cortical lysate [29]	Regulation of surface expression [29]	

Table 1. *Cont.*

Protein	Target Site	Kinase	Identification	Effect on the AMPAR Complex	Involvement in Synaptic Plasticity
GluA2	Ser863	PKC	In vitro; PKC activation of heterologous cells; Phospho-specific antibody with cortical lysate [31]		
	Tyr876	Lyn	Co-expression in heterologous cells [32]	Regulation of GRIP binding [32,33]	
	Tyr869, Tyr873, Tyr876			Internalization of GluA2 subunits [32,34]	A correlational increase upon LTD [34]
	Ser880	PKC	In vitro; PKC activation of heterologous cells [31,33]	Regulation of GRIP binding [33]	Contribution to LTD in GluA2 KO cerebellar Purkinje cell cultures [13]
GluA4	Thr830	PKC	In vitro [35]		
	Ser842	CaMKII	In vitro [35]	Synaptic incorporation of GluA4 subunits [36]	A correlational increase upon PKA activation [36]
		PKC	In vitro [35]		
		PKA	In vitro; PKC activation of heterologous cells [35]		
Stargazin/ TARPγ-2	Ser228, Ser237, Ser239, Ser240, Ser241, Ser243, Ser247, Ser249, Ser253	CaMKII, PKC	Phosphopeptide mapping of heterologous cells and cortical neurons [37]	Dissociation of the cytoplasmic domain from the plasma membrane [38,39]	Contribution to LTP in hippocampal slice culture [37]
	Thr321	PKA, ERK2, p38 MAPK	In vitro [40]	Regulation of PSD-95 binding [40]	
TARPγ-8	Ser277, Ser281	CaMKII	In vitro; Radio-Edman sequencing [41]	Enhancement of synaptic AMPAR activity [41]	A correlational increase upon chemical LTP [41]
CKAMP44/ Shisa9	N/A	PKC	In vitro; PKC activation of heterologous cells [42]		

4. Significance of GluA1 Phosphorylation for Synaptic Plasticity: Knock-In Mouse Studies

The phosphorylation studies using heterologous cell systems and overexpression of GluA1 mutants have provided us with valuable mechanistic information. However, it has to be validated in more physiological conditions such as targeted knock-in mouse models (Table 2). A 'Penta' phosphomutant mouse line is generated with alanine mutations at Ser831, Thr838, Ser839, Thr840, and Ser845 [26]. The 'Penta' phosphomutant mice show a reduction in LTP and LTD compared to that of wild-type mice at adult (~3-month-old) but not young (3–4-week-old) stage [26], suggesting that some or all of those five phosphorylation sites are necessary for synaptic plasticity and there are different mechanisms involved in those phenomena depending on age. The reduced LTP and LTD in adult mice are also shown in a 'Double' phosphomutant mouse line that lacks phosphorylation at Ser831 and Ser845 residues, suggesting that the other three sites including Thr840 may not contribute additionally to synaptic plasticity [43]. Interestingly, a single knock-in mouse that has an alanine mutation at Ser831 residue exhibits intact LTP and LTD at adult and young stages [44]. This may suggest that there are other CaMKII substrates that contribute to LTP more than Ser831 phosphorylation of GluA1 subunits because disrupting CaMKII shows impairment in LTP in a knockout mouse or a knock-in mouse of a mutant CaMKII [3,24]. One of the possible candidates for CaMKII substrates could be Ser567 residue in the loop 1 of GluA1 subunits as it is shown to be involved in the synaptic targeting of AMPARs [30], but Ser567 phosphorylation has not been validated yet by a knock-in study. On the other hand, a single knock-in mouse that lacks Ser845 phosphorylation displays an abolished LTD at adult and young stages [44], consistent with the previous finding that Ser845 phosphorylation correlates with LFS-induced LTD condition in rat hippocampal slices [23].

Table 2. Targeted knock-in mouse studies of AMPAR and TARP phosphorylation sites.

Protein	Phosphorylation Site (Mutated to Alanine Residues)	Effect on Synaptic Plasticity
GluA1	Ser831, Thr838, Ser839, Thr840, and Ser845	A reduction in LTP and LTD at adult stage [26]
	Ser831 and Ser845	A reduction in LTP and LTD at adult stage [43]
	Ser831	Normal LTP and LTD [44]
	Ser845	Abolished LTD [44]
Stargazin/TARPγ-2	Ser228, Ser237, Ser239, Ser240, Ser241, Ser243, Ser247, Ser249, and Ser253	Normal LTP [41]
TARPγ-8	Ser277 and Ser281	A reduction in LTP [41]

5. Phosphorylation of the Pore-Forming Subunits: GluA2 and GluA3

Unlike GluA1, no CaMKII phosphorylation is detected on transiently expressed GluA2 in HEK293 cells [17]. However, the C-terminal intracellular region of the GluA2 subunit serves phosphorylation substrate sites (Table 1) despite the limited homology to the GluA1's C-terminal region (Figure 1). In vitro phosphorylation and phosphopeptide mapping reveal that PKC can directly phosphorylate GluA2 subunits at Ser863 and Ser880 [31,33]. Phosphorylation site-specific antibodies also show that Ser863 and Ser880 residues are phosphorylated on transiently expressed GluA2 in a PKC-activating condition of HEK293 cells (treated by phorbol 12-myristate 13-acetate) [31,33]. The Ser863 phosphorylation likely exists in vivo brains as shown by an immunoblot of rat brain homogenates using the phosphorylation site-specific antibody [31].

LTD is abolished in cultured cerebellar Purkinje cells from GluA2 knockout mice, and the abolished LTD can be rescued by transient expression of the wild-type GluA2 subunit [13]. However, expression of a mutant form of GluA2 that lacks Ser880 phosphorylation fails to restore LTD whereas expression of its phospho-mimetic form with a glutamate mutation at Ser880 residue occludes LTD, suggesting the importance of Ser880 phosphorylation in cerebellar LTD [13].

The C-terminal ends of GluA2 and GluA3 subunits uniquely have three conserved tyrosine residues (Figure 1). Immunoblots with anti-phosphotyrosine (PY20) antibody of immunoprecipitated GluA2 or GluA3 from mouse brains show their tyrosine phosphorylation [32]. Tyrosine phosphorylation-specific antibody against the C-terminal part of GluA2 subunits and site-specific mutants further identify that a Src family protein tyrosine kinase Lyn phosphorylates Tyr876 of GluA2 subunits [32]. The phosphorylation at the C-terminal end of GluA2 subunits (i.e., Tyr876 and Ser880 adjacent to its PDZ-binding motif) negatively regulates GluA2 binding to glutamate receptor-interacting protein (GRIP) [32,33], which is a synaptic PDZ domain-containing protein [45]. The C-terminal tyrosine phosphorylation on GluA2 subunits (including Tyr876) is required for AMPA-, N-methyl-D-aspartate (NMDA)-, and insulin-induced internalization of GluA2 subunits in cultured cortical neurons [32,34]. LFS-induced LTD condition correlates with an increase in tyrosine phosphorylation of GluA2 subunits in homogenates from the stimulated rat hippocampal slices [34]. Also, the C-terminal peptide of wild-type GluA2 (but not a mutant with alanine substitution at Tyr869, Tyr873, and Tyr876) in an intracellular recording solution interferes with LFS-induced LTD, suggesting that tyrosine phosphorylation of GluA2 subunits is required for LFS-induced hippocampal LTD [34].

6. Phosphorylation of the Pore-Forming Subunits: GluA4

Transiently expressed GluA4 subunits are phosphorylated in a PKC-activating (by phorbol 12-myristate 13-acetate) or PKA-activating (by forskolin) condition of HEK293T cells [35]. Phosphopeptide mapping identifies Ser842 residue as the major phosphorylation site in the C-terminal intracellular region of GluA4 subunits, which can be phosphorylated by CaMKII, PKC, and PKA in vitro and in a PKA-activating condition of HEK293T cells [35] (Table 1). Also, PKC can phosphorylate Thr830 residue of GluA4 [35].

PKA activity is necessary and sufficient for synaptic incorporation of GluA4 subunits in hippocampal slices [36]. The PKA activation (by forskolin and 3-isobutyl-1-methylxanthine) leads to AMPAR potentiation and an increase in Ser842 phosphorylation of GluA4 subunits in hippocampal slices [36].

7. TARP Phosphorylation and Its Roles in LTP

AMPARs exist in the brain as a protein complex with auxiliary proteins (e.g., TARP, CNIH, and CKAMP/Shisa) [4,9,46–48]. Stargazin/TARPγ-2 is firstly focused, as its mutant mice (termed *stargazer*) show a loss of AMPAR-mediated transmission in cerebellar granule cells [49,50]. The TARP family comprises two classes, type I (stargazin/TARPγ-2, γ-3, γ-4, and γ-8) and type II (TARPγ-5 and γ-7) [9,51]. TARPs have four transmembrane domains (Figure 2) and regulate trafficking and channel properties of AMPARs [9,52–54]. TARP shows differential expression patterns in adult rodent brains, for example, stargazin/TARPγ-2 is the dominant isoform of TARPs in the cerebellum whereas TARPγ-8 is highly enriched in the hippocampus [51], suggesting distinct roles of each member in different brain regions.

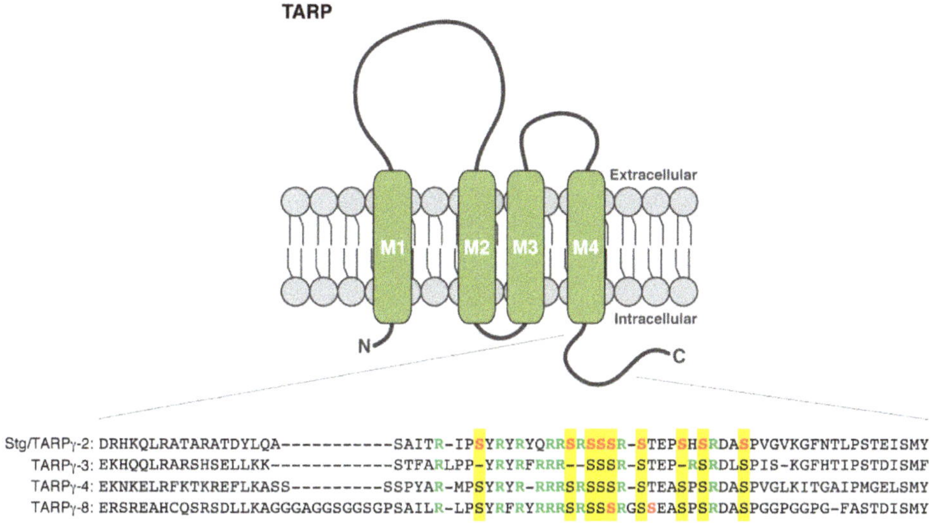

Figure 2. Schematic illustration of TARP isoform structure and phosphorylation sites in the part of the C-terminal cytoplasmic tail of each isoform (referred to [37]). The identified phosphorylation sites are shown in red, and adjacent arginine residues are indicated in green. The conserved serine residues of stargazin/TARPγ-2 phosphorylation sites are highlighted in yellow. M1–4 indicates transmembrane domains.

Phosphorylation of stargazin/TARPγ-2 was firstly described. Both extra-synaptic (Triton X-100-soluble) and post-synaptic density (Triton X-100-insoluble) fractions show phosphorylation of stargazin/TARPγ-2 [37]. Intriguingly, the post-synaptic density fraction dominantly displays the highest degree of stargazin/TARPγ-2 phosphorylation whereas the extra-synaptic fraction has multiple phosphorylation bands with variable degrees, suggesting that synaptic stargazin/TARPγ-2 is preferentially phosphorylated [37]. The phosphorylation sites of stargazin/TARPγ-2 are identified as nine serine residues at 228, 237, 239, 240, 241, 243, 247, 249, and 253 in the C-terminal cytoplasmic tail (mouse stargazin/TARPγ-2) by phosphopeptide mapping with primary cortical neurons and transfected Chinese hamster ovary (CHO) cells [37] (Table 1). CaMKII and PKC can be the kinases for stargazin/TARPγ-2 phosphorylation as shown by in vitro phosphorylation assays and inhibitor studies [37]. A stargazin/TARPγ-2 phospho-mimetic knock-in mouse line that has aspartate mutations at all those nine serine residues (S9D) (Stargazin^{SD}) shows an increase in synaptic AMPAR activity in cerebellar mossy fiber-granule cell synapses [38], consistent with the notion that phosphorylated forms of stargazin/TARPγ-2 are dominant in the post-synaptic density fraction [37]. In hippocampal slice cultures, LTP is occluded by expression of a phospho-mimetic form of stargazin/TARPγ-2 (S9D) and prevented by expression of a phospho-deficient stargazin/TARPγ-2 that has alanine mutations (S9A) [37]. Also, Thr321 of stargazin/TARPγ-2 can be phosphorylated by PKA, extracellular signal-regulated protein kinase 2, and p38 mitogen-activated protein kinase in vitro [40].

Phosphorylation of TARPγ-8 by CaMKII is shown by in vitro phosphorylation assays [41]. The nine phosphorylation sites of stargazin/TARPγ-2 are highly conserved in all four TARP isoforms including TARPγ-8 (e.g., serine at 264, 273, 275, 276, 277, 280, 284, 286, and 290 of mouse TARPγ-8), and TARPγ-8 uniquely has one more serine residue at 281 [37,41] (Figure 2). In vitro phosphorylation assay and radio-Edman sequencing identify Ser277 and Ser281 as CaMKII phosphorylation sites on TARPγ-8 [41] (Table 1). A TARPγ-8 knock-in mouse line containing alanine mutations at Ser277 and Ser281 residues (TARPγ-8^{Cm}) shows a substantial reduction in hippocampal LTP, suggesting that CaMKII phosphorylation of TARPγ-8 at these two sites is required for LTP [41] (Table 2). The possible contribution of stargazin/TARPγ-2 phosphorylation, TARPγ-3, and TARPγ-4 to hippocampal LTP may be excluded because hippocampal LTP is intact in both a stargazin/TARPγ-2 knock-in mouse line that lacks the nine phosphorylation sites (Stargazin^{SA}) and a triple mutant mouse with Stargazin^{SA} knock-in plus knockout of TARPγ-3 and TARPγ-4 [41]. Previously, TARPγ-8 knockout mice exhibited a substantial reduction in LTP, but not LTD, as well as altered expression of AMPAR subunits (i.e., GluA1 and GluA2) [11]. Possible secondary effect of the altered protein expression on the LTP impairment can be ruled out because the TARPγ-8^{Cm} knock-in mice with S277A and S281A show no obvious differences in AMPAR expression compared to wild-type mice [41]. In addition, overexpression of a TARPγ-8 phospho-mimetic form with aspartate mutations at Ser277 and Ser281 is sufficient to enhance synaptic AMPAR activity in the hippocampus [41].

The molecular mechanism for how TARP phosphorylation regulates synaptic AMPAR activity is one of the most interesting topics in synaptic plasticity. In the C-terminal cytoplasmic tail of TARPs, the serine sites are adjacent to many arginine residues (e.g., arginine residues at 225, 230, 232, 235, 236, 238, 242, and 250 of mouse stargazin/TARPγ-2) (Figure 2). The positive charges from these arginine residues of stargazin/TARPγ-2 can directly bind to negatively charged lipids in vitro, for example, phosphatidic acid, phosphatidylinositol-4-phosphate (PIP), phosphatidylinositol-4,5-bisphosphate (PIP_2), and phosphatidylinositol-3,4-5-triphosphate (PIP_3) [38]. Phospho-mimetic S9D mutations (aspartate substitution at nine serine residues of stargazin/TARPγ-2) disrupt the electrostatic interaction between the cytoplasmic domain of stargazin/TARPγ-2 and the negatively charged lipids [38]. Consistently, the charge-dependent dissociation of the TARP cytoplasmic domain from the plasma membrane occurs in cultured hippocampal neurons [39]. Fluorescence lifetime imaging microscopy reveals that the phospho-mimetic S9D mutant of stargazin/TARPγ-2 (GFP-tagged right after the arginine/serine-rich domain) exhibits a longer GFP lifetime than wild-type stargazin/TARPγ-2 due to being further from the plasma membrane (stained by a plasma membrane marker R18; octadecyl rhodamine B chloride), suggesting that the C-terminus of the phospho-mimetic form of stargazin/TARPγ-2 extends further into the cytoplasm than wild-type [39]. As discussed earlier, highly phosphorylated forms of stargazin/TARPγ-2 are dominant in the post-synaptic density fraction [37], and the Stargazin^{SD} knock-in mice show an increase in synaptic AMPAR activity [38]. Taken together, TARP phosphorylation may disrupt the membrane binding of the C-terminal cytoplasmic tail to trigger a synaptic enhancement of AMPAR activity.

Since all TARP isoforms commonly have the eight to nine arginine residues adjacent to the phosphorylation sites (Figure 2), it would not be surprising that other TARP isoforms behave in a similar way to stargazin/TARPγ-2 in terms of the binding to/dissociation from membranes. For example, it is possible that CaMKII phosphorylation at Ser277 and Ser281 of TARPγ-8 may lead to dissociation of the cytoplasmic tail from the plasma membrane to enhance synaptic AMPAR activity in the hippocampus as does stargazin/TARPγ-2 in cerebellar granule cells. This model is further supported by a study with a TARPγ-8 knock-in mouse line (TARPγ-8$^{\Delta 4}$) that lacks the last four amino acids, a PDZ-binding motif. The interaction between PDZ-binding motifs of TARPs and membrane-associated guanylate kinase family proteins (e.g., PSD-95) is proposed to stabilize AMPAR-TARP complexes at synapses. Consistent with this idea, TARPγ-8$^{\Delta 4}$ knock-in mice exhibit ~30% reduction in basal AMPAR transmission in the hippocampus [55]. However, hippocampal LTP is intact in the TARPγ-8$^{\Delta 4}$ knock-in mice [55], suggesting that PDZ binding of TARPγ-8 is not necessary for LTP expression.

8. Phosphorylation of Other Auxiliary Proteins of AMPAR

Accumulating reports have identified more auxiliary proteins of the AMPAR complex, such as CNIH, germ cell-specific gene 1-like protein (GSG1-L), and CKAMP/Shisa [9,46–48]. CNIH-2/-3 are identified by proteomic analysis of native AMPAR complexes from rat brains [56]. CNIHs regulate surface expression and channel properties of AMPARs in heterologous cells and mouse brain slices [12,

56,57]. GSG1-L also modulates AMPAR trafficking and desensitization in heterologous cells [58,59]. However, phosphorylation of CNIHs and GSG1-L remains unknown.

Among the binding proteins to AMPARs [47,58], CKAMP44/Shisa9 is known to be phosphorylated by PKC and protein interacting with C kinase 1 (PICK1) [42]. In vitro phosphorylation assay and Phos-tag polyacrylamide gel electrophoresis reveal that CKAMP44/Shisa9 can be phosphorylated by PKC in vitro and in a PKC-activating condition of COS-7 cells (treated by phorbol 12-myristate 13-acetate) [42]. However, the phosphorylation sites and their involvement in synaptic plasticity are unknown.

9. Closing Remarks

Phosphorylation of the pore-forming subunits and auxiliary proteins of the AMPAR complex has been extensively studied. The broad spectrum of biochemical approaches such as in vitro phosphorylation assay, phosphopeptide mapping, phosphorylation site-specific antibodies, and Phos-tag polyacrylamide gel electrophoresis has allowed us to identify the actual phosphorylation sites and to describe their states in synaptic plasticity. Combined with the biochemical approaches, in vivo and ex vivo studies including genetically modified mice and electrophysiological analyses have found compelling phenomena and their molecular and cellular mechanisms. However, many questions remain in the field of synaptic plasticity, in particular, in regards to proteomics. Although the native AMPAR complex constituents were recently profiled by proteomics [58], phosphoproteomics of the native complex has not been reported yet. Very little is known to date regarding phosphorylation of many AMPAR complex constituents including TARPs, CNIHs, GSG1-L, and CKAMPs. Also, roles of the previously identified phosphorylation sites of the complex in synaptic plasticity need to be investigated further in more physiological conditions (e.g., targeted knock-in mice). Accumulating knowledge from the phosphorylation studies of the native AMPAR complex will provide us with more insight into the activity-dependent dynamic changes at synapses that underlie various animal behaviors. Although this review is only limited to AMPAR complex proteins, other synaptic proteins including neurotransmitter receptors and scaffolding proteins may serve as phosphorylation substrates that contribute to synaptic plasticity.

Funding: This work was supported by the Yale/NIDA Neuroproteomics Center (P30 DA018343).

Acknowledgments: The author thanks Susumu Tomita and Jeeyun Chung for their valuable input to the manuscript.

Conflicts of Interest: The author declares no conflict of interest.

References

1. Malenka, R.C.; Kauer, J.A.; Perkel, D.J.; Mauk, M.D.; Kelly, P.T.; Nicoll, R.A.; Waxham, M.N. An essential role for postsynaptic calmodulin and protein kinase activity in long-term potentiation. *Nature* **1989**, *340*, 554–557. [CrossRef] [PubMed]
2. Malinow, R.; Schulman, H.; Tsien, R.W. Inhibition of postsynaptic PKC or CaMKII blocks induction but not expression of LTP. *Science* **1989**, *245*, 862–866. [CrossRef] [PubMed]
3. Silva, A.J.; Stevens, C.F.; Tonegawa, S.; Wang, Y. Deficient hippocampal long-term potentiation in alpha-calcium-calmodulin kinase II mutant mice. *Science* **1992**, *257*, 201–206. [CrossRef] [PubMed]
4. Nicoll, R.A.; Tomita, S.; Bredt, D.S. Auxiliary subunits assist AMPA-type glutamate receptors. *Science* **2006**, *311*, 1253–1256. [CrossRef] [PubMed]
5. Jackson, A.C.; Nicoll, R.A. The expanding social network of ionotropic glutamate receptors: TARPs and other transmembrane auxiliary subunits. *Neuron* **2011**, *70*, 178–199. [CrossRef] [PubMed]
6. Greger, I.H.; Watson, J.F.; Cull-Candy, S.G. Structural and functional architecture of AMPA-type glutamate receptors and their auxiliary proteins. *Neuron* **2017**, *94*, 713–730. [CrossRef] [PubMed]
7. Twomey, E.C.; Yelshanskaya, M.V.; Grassucci, R.A.; Frank, J.; Sobolevsky, A.I. Elucidation of AMPA receptor-stargazin complexes by cryo-electron microscopy. *Science* **2016**, *353*, 83–86. [CrossRef] [PubMed]

8. Chen, S.; Zhao, Y.; Wang, Y.; Shekhar, M.; Tajkhorshid, E.; Gouaux, E. Activation and desensitization mechanism of AMPA receptor-TARP complex by cryo-EM. *Cell* **2017**, *170*, 1234–1246. [CrossRef] [PubMed]
9. Yan, D.; Tomita, S. Defined criteria for auxiliary subunits of glutamate receptors. *J. Physiol.* **2012**, *590*, 21–31. [CrossRef] [PubMed]
10. Zamanillo, D.; Sprengel, R.; Hvalby, O.; Jensen, V.; Burnashev, N.; Rozov, A.; Kaiser, K.M.; Köster, H.J.; Borchardt, T.; Worley, P.; et al. Importance of AMPA receptors for hippocampal synaptic plasticity but not for spatial learning. *Science* **1999**, *284*, 1805–1811. [CrossRef] [PubMed]
11. Rouach, N.; Byrd, K.; Petralia, R.S.; Elias, G.M.; Adesnik, H.; Tomita, S.; Karimzadegan, S.; Kealey, C.; Bredt, D.S.; Nicoll, R.A. TARP gamma-8 controls hippocampal AMPA receptor number, distribution and synaptic plasticity. *Nat. Neurosci.* **2005**, *8*, 1525–1533. [CrossRef] [PubMed]
12. Herring, B.E.; Shi, Y.; Suh, Y.H.; Zheng, C.Y.; Blankenship, S.M.; Roche, K.W.; Nicoll, R.A. Cornichon proteins determine the subunit composition of synaptic AMPA receptors. *Neuron* **2013**, *77*, 1083–1096. [CrossRef] [PubMed]
13. Chung, H.J.; Steinberg, J.P.; Huganir, R.L.; Linden, D.J. Requirement of AMPA receptor GluR2 phosphorylation for cerebellar long-term depression. *Science* **2003**, *300*, 1751–1755. [CrossRef] [PubMed]
14. Toyoda, H.; Wu, L.J.; Zhao, M.G.; Xu, H.; Jia, Z.; Zhuo, M. Long-term depression requires postsynaptic AMPA GluR2 receptor in adult mouse cingulate cortex. *J. Cell. Physiol.* **2007**, *211*, 336–343. [CrossRef] [PubMed]
15. Wang, J.Q.; Arora, A.; Yang, L.; Parelkar, N.K.; Zhang, G.; Liu, X.; Choe, E.S.; Mao, L. Phosphorylation of AMPA receptors: mechanisms and synaptic plasticity. *Mol. Neurobiol.* **2005**, *32*, 237–249. [CrossRef]
16. Hollmann, M.; Maron, C.; Heinemann, S. N-Glycosylation site tagging suggests a three transmembrane domain topology for the glutamate receptor GluR1. *Neuron* **1994**, *13*, 1331–1343. [CrossRef]
17. Barria, A.; Derkach, V.; Soderling, T. Identification of the Ca^{2+}/calmodulin-dependent protein kinase II regulatory phosphorylation site in the alpha-amino-3-hydroxyl-5-methyl-4-isoxazole-propionate-type glutamate receptor. *J. Biol. Chem.* **1997**, *272*, 32727–32730. [CrossRef] [PubMed]
18. Roche, K.W.; O'Brien, R.J.; Mammen, A.L.; Bernhardt, J.; Huganir, R.L. Characterization of multiple phosphorylation sites on the AMPA receptor GluR1 subunit. *Neuron* **1996**, *16*, 1179–1188. [CrossRef]
19. Mammen, A.L.; Kameyama, K.; Roche, K.W.; Huganir, R.L. Phosphorylation of the alpha-amino-3-hydroxy-5-methylisoxazole4-propionic acid receptor GluR1 subunit by calcium/calmodulin-dependent kinase II. *J. Biol. Chem.* **1997**, *272*, 32528–32533. [CrossRef] [PubMed]
20. Derkach, V.; Barria, A.; Soderling, T.R. Ca^{2+}/calmodulin-kinase II enhances channel conductance of alpha-amino-3-hydroxy-5-methyl-4-isoxazolepropionate type glutamate receptors. *Proc. Natl. Acad. Sci. USA* **1999**, *96*, 3269–3274. [CrossRef] [PubMed]
21. Jenkins, M.A.; Traynelis, S.F. PKC phosphorylates GluA1-Ser831 to enhance AMPA receptor conductance. *Channels* **2012**, *6*, 60–64. [CrossRef] [PubMed]
22. Banke, T.G.; Bowie, D.; Lee, H.K.; Huganir, R.L.; Schousboe, A.; Traynelis, S.F. Control of GluR1 AMPA receptor function by cAMP-dependent protein kinase. *J. Neurosci.* **2000**, *20*, 89–102. [CrossRef] [PubMed]
23. Lee, H.K.; Barbarosie, M.; Kameyama, K.; Bear, M.F.; Huganir, R.L. Regulation of distinct AMPA receptor phosphorylation sites during bidirectional synaptic plasticity. *Nature* **2000**, *405*, 955–959. [CrossRef] [PubMed]
24. Giese, K.P.; Fedorov, N.B.; Filipkowski, R.K.; Silva, A.J. Autophosphorylation at Thr286 of the alpha calcium-calmodulin kinase II in LTP and learning. *Science* **1998**, *279*, 870–873. [CrossRef] [PubMed]
25. Boehm, J.; Kang, M.G.; Johnson, R.C.; Esteban, J.; Huganir, R.L.; Malinow, R. Synaptic incorporation of AMPA receptors during LTP is controlled by a PKC phosphorylation site on GluR1. *Neuron* **2006**, *51*, 213–225. [CrossRef] [PubMed]
26. Lee, H.K.; Takamiya, K.; Kameyama, K.; He, K.; Yu, S.; Rossetti, L.; Wilen, D.; Huganir, R.L. Identification and characterization of a novel phosphorylation site on the GluR1 subunit of AMPA receptors. *Mol. Cell. Neurosci.* **2007**, *36*, 86–94. [CrossRef] [PubMed]
27. Gray, E.E.; Guglietta, R.; Khakh, B.S.; O'Dell, T.J. Inhibitory interactions between phosphorylation sites in the C terminus of α-Amino-3-hydroxy-5-methyl-4-isoxazolepropionic acid-type glutamate receptor GluA1 subunits. *J. Biol. Chem.* **2014**, *289*, 14600–14611. [CrossRef] [PubMed]

28. Jenkins, M.A.; Wells, G.; Bachman, J.; Snyder, J.P.; Jenkins, A.; Huganir, R.L.; Oswald, R.E.; Traynelis, S.F. Regulation of GluA1 α-amino-3-hydroxy-5-methyl-4-isoxazolepropionic acid receptor function by protein kinase C at serine-818 and threonine-840. *Mol. Pharmacol.* **2014**, *85*, 618–629. [CrossRef] [PubMed]
29. Hussain, N.K.; Thomas, G.M.; Luo, J.; Huganir, R.L. Regulation of AMPA receptor subunit GluA1 surface expression by PAK3 phosphorylation. *Proc. Natl. Acad. Sci. USA* **2015**, *112*, E5883–E5890. [CrossRef] [PubMed]
30. Lu, W.; Isozaki, K.; Roche, K.W.; Nicoll, R.A. Synaptic targeting of AMPA receptors is regulated by a CaMKII site in the first intracellular loop of GluA1. *Proc. Natl. Acad. Sci. USA* **2010**, *107*, 22266–22271. [CrossRef] [PubMed]
31. McDonald, B.J.; Chung, H.J.; Huganir, R.L. Identification of protein kinase C phosphorylation sites within the AMPA receptor GluR2 subunit. *Neuropharmacology* **2001**, *41*, 672–679. [CrossRef]
32. Hayashi, T.; Huganir, R.L. Tyrosine phosphorylation and regulation of the AMPA receptor by SRC family tyrosine kinases. *J. Neurosci.* **2004**, *24*, 6152–6160. [CrossRef] [PubMed]
33. Matsuda, S.; Mikawa, S.; Hirai, H. Phosphorylation of serine-880 in GluR2 by protein kinase C prevents its C terminus from binding with glutamate receptor-interacting protein. *J. Neurochem.* **1999**, *73*, 1765–1768. [CrossRef] [PubMed]
34. Ahmadian, G.; Ju, W.; Liu, L.; Wyszynski, M.; Lee, S.H.; Dunah, A.W.; Taghibiglou, C.; Wang, Y.; Lu, J.; Wong, T.P.; et al. Tyrosine phosphorylation of GluR2 is required for insulin-stimulated AMPA receptor endocytosis and LTD. *EMBO J.* **2004**, *23*, 1040–1050. [CrossRef] [PubMed]
35. Carvalho, A.L.; Kameyama, K.; Huganir, R.L. Characterization of phosphorylation sites on the glutamate receptor 4 subunit of the AMPA receptors. *J. Neurosci.* **1999**, *19*, 4748–4754. [CrossRef] [PubMed]
36. Esteban, J.A.; Shi, S.H.; Wilson, C.; Nuriya, M.; Huganir, R.L.; Malinow, R. PKA phosphorylation of AMPA receptor subunits controls synaptic trafficking underlying plasticity. *Nat. Neurosci.* **2003**, *6*, 136–143. [CrossRef] [PubMed]
37. Tomita, S.; Stein, V.; Stocker, T.J.; Nicoll, R.A.; Bredt, D.S. Bidirectional synaptic plasticity regulated by phosphorylation of stargazing-like TARPs. *Neuron* **2005**, *45*, 269–277. [CrossRef] [PubMed]
38. Sumioka, A.; Yan, D.; Tomita, S. TARP phosphorylation regulates synaptic AMPA receptors through lipid bilayers. *Neuron* **2010**, *66*, 755–767. [CrossRef] [PubMed]
39. Hafner, A.S.; Penn, A.C.; Grillo-Bosch, D.; Retailleau, N.; Poujol, C.; Philippat, A.; Coussen, F.; Sainlos, M.; Opazo, P.; Choquet, D. Lengthening of the Stargazin cytoplasmic tail increases synaptic transmission by promoting interaction to deeper domains of PSD-95. *Neuron* **2015**, *86*, 475–489. [CrossRef] [PubMed]
40. Stein, E.L.; Chetkovich, D.M. Regulation of stargazin synaptic trafficking by C-terminal PDZ ligand phosphorylation in bidirectional synaptic plasticity. *J. Neurochem.* **2010**, *113*, 42–53. [CrossRef] [PubMed]
41. Park, J.; Chávez, A.E.; Mineur, Y.S.; Morimoto-Tomita, M.; Lutzu, S.; Kim, K.S.; Picciotto, M.R.; Castillo, P.E.; Tomita, S. CaMKII phosphorylation of TARPγ-8 is a mediator of LTP and learning and memory. *Neuron* **2016**, *92*, 75–83. [CrossRef] [PubMed]
42. Kunde, S.A.; Rademacher, N.; Zieger, H.; Shoichet, S.A. Protein kinase C regulates AMPA receptor auxiliary protein Shisa9/CKAMP44 through interactions with neuronal scaffold PICK1. *FEBS Open Bio* **2017**, *7*, 1234–1245. [CrossRef] [PubMed]
43. Lee, H.K.; Takamiya, K.; Han, J.S.; Man, H.; Kim, C.H.; Rumbaugh, G.; Yu, S.; Ding, L.; He, C.; Petralia, R.S.; et al. Phosphorylation of the AMPA receptor GluR1 subunit is required for synaptic plasticity and retention of spatial memory. *Cell* **2003**, *112*, 631–643. [CrossRef]
44. Lee, H.K.; Takamiya, K.; He, K.; Song, L.; Huganir, R.L. Specific roles of AMPA receptor subunit GluR1 (GluA1) phosphorylation sites in regulating synaptic plasticity in the CA1 region of hippocampus. *J. Neurophysiol.* **2010**, *103*, 479–489. [CrossRef] [PubMed]
45. Dong, H.; O'Brien, R.J.; Fung, E.T.; Lanahan, A.A.; Worley, P.F.; Huganir, R.L. GRIP: A synaptic PDZ domain-containing protein that interacts with AMPA receptors. *Nature* **1997**, *386*, 279–284. [CrossRef] [PubMed]
46. Tigaret, C.; Choquet, D. More AMPAR garnish. *Science* **2009**, *323*, 1295–1296. [CrossRef] [PubMed]
47. Farrant, M.; Cull-Candy, S.G. Neuroscience. AMPA receptors—Another twist? *Science* **2010**, *327*, 1463–1465. [CrossRef] [PubMed]
48. Haering, S.C.; Tapken, D.; Pahl, S.; Hollmann, M. Auxiliary subunits: shepherding AMPA receptors to the plasma membrane. *Membranes* **2014**, *4*, 469–490. [CrossRef] [PubMed]

49. Hashimoto, K.; Fukaya, M.; Qiao, X.; Sakimura, K.; Watanabe, M.; Kano, M. Impairment of AMPA receptor function in cerebellar granule cells of ataxic mutant mouse stargazer. *J. Neurosci.* **1999**, *19*, 6027–6036. [CrossRef] [PubMed]
50. Chen, L.; Chetkovich, D.M.; Petralia, R.S.; Sweeney, N.T.; Kawasaki, Y.; Wenthold, R.J.; Bredt, D.S.; Nicoll, R.A. Stargazin regulates synaptic targeting of AMPA receptors by two distinct mechanisms. *Nature* **2000**, *408*, 936–943. [CrossRef] [PubMed]
51. Tomita, S.; Chen, L.; Kawasaki, Y.; Petralia, R.S.; Wenthold, R.J.; Nicoll, R.A.; Bredt, D.S. Functional studies and distribution define a family of transmembrane AMPA receptor regulatory proteins. *J. Cell Biol.* **2003**, *161*, 805–816. [CrossRef] [PubMed]
52. Tomita, S.; Adesnik, H.; Sekiguchi, M.; Zhang, W.; Wada, K.; Howe, J.R.; Nicoll, R.A.; Bredt, D.S. Stargazin modulates AMPA receptor gating and trafficking by distinct domains. *Nature* **2005**, *435*, 1052–1058. [CrossRef] [PubMed]
53. Kato, A.S.; Zhou, W.; Milstein, A.D.; Knierman, M.D.; Siuda, E.R.; Dotzlaf, J.E.; Yu, H.; Hale, J.E.; Nisenbaum, E.S.; Nicoll, R.A.; et al. New transmembrane AMPA receptor regulatory protein isoform, gamma-7, differentially regulates AMPA receptors. *J. Neurosci.* **2007**, *27*, 4969–4977. [CrossRef] [PubMed]
54. Kato, A.S.; Siuda, E.R.; Nisenbaum, E.S.; Bredt, D.S. AMPA receptor subunit-specific regulation by a distinct family of type II TARPs. *Neuron* **2008**, *59*, 986–996. [CrossRef] [PubMed]
55. Sumioka, A.; Brown, T.E.; Kato, A.S.; Bredt, D.S.; Kauer, J.A.; Tomita, S. PDZ binding of TARPγ-8 controls synaptic transmission but not synaptic plasticity. *Nat. Neurosci.* **2011**, *14*, 1410–1412. [CrossRef] [PubMed]
56. Schwenk, J.; Harmel, N.; Zolles, G.; Bildl, W.; Kulik, A.; Heimrich, B.; Chisaka, O.; Jonas, P.; Schulte, U.; Fakler, B.; et al. Functional proteomics identify cornichon proteins as auxiliary subunits of AMPA receptors. *Science* **2009**, *323*, 1313–1319. [CrossRef] [PubMed]
57. Kato, A.S.; Gill, M.B.; Ho, M.T.; Yu, H.; Tu, Y.; Siuda, E.R.; Wang, H.; Qian, Y.W.; Nisenbaum, E.S.; Tomita, S.; et al. Hippocampal AMPA receptor gating controlled by both TARP and cornichon proteins. *Neuron* **2010**, *68*, 1082–1096. [CrossRef] [PubMed]
58. Schwenk, J.; Harmel, N.; Brechet, A.; Zolles, G.; Berkefeld, H.; Müller, C.S.; Bildl, W.; Baehrens, D.; Hüber, B.; Kulik, A.; et al. High-resolution proteomics unravel architecture and molecular diversity of native AMPA receptor complexes. *Neuron* **2012**, *74*, 621–633. [CrossRef] [PubMed]
59. Shanks, N.F.; Savas, J.N.; Maruo, T.; Cais, O.; Hirao, A.; Oe, S.; Ghosh, A.; Noda, Y.; Greger, I.H.; Yates, J.R., III; et al. Differences in AMPA and kainate receptor interactomes facilitate identification of AMPA receptor auxiliary subunit GSG1L. *Cell Rep.* **2012**, *1*, 590–598. [CrossRef] [PubMed]

proteomes

MDPI

Article

Phosphoproteomic Analysis of the Amygdala Response to Adolescent Glucocorticoid Exposure Reveals G-Protein Coupled Receptor Kinase 2 as a Target for Reducing Motivation for Alcohol

Megan L. Bertholomey [1], Kathryn Stone [2], TuKiet T. Lam [2,3], Seojin Bang [4], Wei Wu [4], Angus C. Nairn [5], Jane R. Taylor [5,6] and Mary M. Torregrossa [1,*]

1 Department of Psychiatry, University of Pittsburgh, 450 Technology Dr. Bridgeside Point II, Suite 223, Pittsburgh, PA 15219, USA; bertholomeym@upmc.edu

2 Keck Foundation Biotechnology Resource Laboratory, Yale University, New Haven, CT 06536, USA; gstonecreek@gmail.com (K.S.); tukiet.lam@yale.edu (T.T.L.)

3 Department of Molecular Biophysics and Biochemistry, Yale University, New Haven, CT 06508, USA

4 Computational Biology Department, Carnegie Mellon University, Pittsburgh, PA 15213, USA; seojinb@andrew.cmu.edu (S.B.); weiwu2@cs.cmu.edu (W.W.)

5 Department of Psychiatry, Yale University, New Haven, CT 06508, USA; angus.nairn@yale.edu (A.C.N.); jane.taylor@yale.edu (J.R.T.)

6 Department of Psychology, Yale University, New Haven, CT 06508, USA

* Correspondence: torregrossam@upmc.edu; Tel.: +1-412-624-5723; Fax: +1-412-624-5280

Received: 28 August 2018; Accepted: 9 October 2018; Published: 12 October 2018

Abstract: Early life stress is associated with risk for developing alcohol use disorders (AUDs) in adulthood. Though the neurobiological mechanisms underlying this vulnerability are not well understood, evidence suggests that aberrant glucocorticoid and noradrenergic system functioning play a role. The present study investigated the long-term consequences of chronic exposure to elevated glucocorticoids during adolescence on the risk of increased alcohol-motivated behavior, and on amygdalar function in adulthood. A discovery-based analysis of the amygdalar phosphoproteome using mass spectrometry was employed, to identify changes in function. Adolescent corticosterone (CORT) exposure increased alcohol, but not sucrose, self-administration, and enhanced stress-induced reinstatement with yohimbine in adulthood. Phosphoproteomic analysis indicated that the amygdala phosphoproteome was significantly altered by adolescent CORT exposure, generating a list of potential novel mechanisms involved in the risk of alcohol drinking. In particular, increased phosphorylation at serines 296–299 on the α_{2A} adrenergic receptor (α_{2A}AR), mediated by the G-protein coupled receptor kinase 2 (GRK2), was evident after adolescent CORT exposure. We found that intra-amygdala infusion of a peptidergic GRK2 inhibitor reduced alcohol seeking, as measured by progressive ratio and stress reinstatement tests, and induced by the α_{2A}AR antagonist yohimbine. These results suggest that GRK2 represents a novel target for treating stress-induced motivation for alcohol which may counteract alterations in brain function induced by adolescent stress exposure.

Keywords: adolescence; corticosterone; proteomics; yohimbine; progressive ratio; reinstatement; ethanol

1. Introduction

Chronic stress is an environmental factor known to increase the risk for psychiatric disorders, including alcohol use disorders. Importantly, chronic stress during critical developmental periods can have long-lasting effects on alcoholism risk [1–5]. Specifically, exposure to multiple adverse events in childhood, including adolescence, is associated with greater lifetime incidence and earlier onset of alcohol dependence [6,7]. Adolescence may be a period of particular vulnerability because of the ongoing development of brain circuits responsive to glucocorticoids during that time. Indeed, we have found that chronic exposure to the glucocorticoid stress hormone corticosterone (CORT) during adolescence, using an established procedure that produces a depression-like syndrome in adults [8,9], increases impulsivity on the delay-discounting test of impulsive choice, indicating that adolescent CORT has long-term effects on behavior [10]. Moreover, impulsivity on delay discounting tasks is frequently associated with presence of alcohol use disorders, and it is a possible risk factor for alcoholism [11].

Preclinical studies suggest that stress hormone exposure in adolescence may influence motivation for ethanol in adulthood, including reports demonstrating that post-weaning social isolation stress for either 42 [12,13] or 90 [14] days, which includes but is not limited to the adolescent period, can increase subsequent operant ethanol self-administration. Moreover, we recently reported that adolescent corticosterone exposure from postnatal day (PND) 30–50 can increase a variety of alcohol-motivated behaviors in male and female rats [15].

In Experiment 1 of the current study, we demonstrate that under certain training conditions, male rats exposed to CORT in adolescence demonstrate an increased motivation for alcohol, as evidenced by increased operant alcohol self-administration and yohimbine-induced reinstatement of alcohol-seeking behavior in adulthood, whereas responding for sucrose was unchanged. We then used a discovery-based phosphoproteomics approach to determine what signaling systems were persistently altered in the amygdala of rats exposed to CORT in adolescence, which might interact with the ethanol self-administration experience. The proteomics analysis revealed several persistent changes in protein phosphorylation based on adolescent experience that may represent potential mechanisms underlying increased motivation for alcohol, including increased phosphorylation of the α_{2A} adrenergic receptor (α_{2A}AR) by CORT. Consequently, Experiment 2 sought to determine the effects of direct manipulation of α_{2A}AR function on adult alcohol-motivated behavior in adolescent CORT-exposed male rats.

2. Materials and Methods

2.1. Subjects

Male Sprague Dawley rats aged 24–27 days upon delivery to the animal facility were used in all experiments. Experiment 1 was conducted at Yale University in the Connecticut Mental Health Center, and Experiment 2 was conducted at the University of Pittsburgh. Rats were obtained from Charles River (Kingston, NY, USA) at Yale and from Harlan/Envigo (Frederick, MD, USA) at the University of Pittsburgh. We used different vendors to minimize animal shipping time to both facilities. In addition, at Yale, rats were housed in shoebox cages with water bottles on standard racks, while at the University of Pittsburgh, rats were housed in individually ventilated caging (IVC) with an automated watering system. All other housing and procedural parameters were the same between the two universities, unless otherwise noted. In addition, all procedures were conducted in accordance with the National Institutes of Health *Guide for the Care and Use of Laboratory Animals* and were approved by each institution's Institutional Animal Care and Use Committee. Rats acclimated to the facility for 3–5 days before CORT exposure began on postnatal day (PND) 30. Rats were pair-housed and maintained on a 12:12 hour light-dark cycle in a temperature- and humidity-controlled environment. The rats were given ad libitum access to food and water except during periods of food restriction, as described below.

2.2. Drugs

Corticosterone hemisuccinate (CORT; 4-pregnen-11β,21-diol-3,20-dione21-hemisuccinate, Steraloids, Newport, RI, USA) was prepared fresh every three days. CORT was dissolved in tap water and stirred overnight at a pH of 10–11 and neutralized to a pH of 7.0–7.4 prior to use. Ethanol (EtOH; Decon Labs, King of Prussia, PA, USA) and saccharin (Acros Organics, Pittsburgh, PA, USA) were diluted in tap water to concentrations of 10% (*v*/*v*) and 0.1% (*w*/*v*), respectively, to make the sweetened EtOH solution. Yohimbine (Sigma, St. Louis, MO, USA) was dissolved in double distilled water to a concentration of 1.25 mg/mL. GRK2i (GRK2 inhibitory polypeptide; Tocris, Minneapolis, MN, USA) was dissolved in saline to a concentration of 2 mM.

2.3. Chronic Corticosterone Exposure

From postnatal day (PND) 30–50, a period which spans the majority of adolescence in rodents [1], rats received access to a bottle containing either water or a solution of CORT as their sole source of fluid (note: the automated watering system was disabled in Experiment 2). For the first 14 days of exposure, rats received a concentration of 50 μg/mL CORT, which was then reduced to 25 μg/mL and finally to 12.5 μg/mL for three days each. During the exposure period, CORT- and water (H_2O)-containing bottles were weighed daily, and rats were weighed every other day. Following cessation of CORT exposure, all rats were returned to normal tap water. Behavioral testing began following a 10-day washout period to allow for the re-establishment of endogenous hypothalamic-pituitary-adrenal (HPA) axis functioning [9] and for the rats to age into adulthood. The CORT exposure procedure was the same as that previously described and that has been reported to produce circulating CORT levels of greater than 800 ng/mL [9,16].

2.4. Operant Self-Administration

All rats remained CORT-free during behavioral testing. To facilitate acquisition of self-administration, rats were mildly food restricted during training. In Experiment 1, rats were maintained on this restriction, while in Experiment 2, the restriction was gradually eased such that rats were fed ad libitum by the end of training. Self-administration sessions were conducted in standard operant chambers (MedAssociates, St. Albans, VT, USA) housed in sound-attenuating cubicles. Rats were trained to respond for 10 s presentation of the reinforcer paired with a light + tone cue on a fixed ratio (FR)1 schedule of reinforcement. In both experiments, rats were trained to self-administer a solution of 10% (*v*/*v*) EtOH + 0.1% (*w*/*v*) saccharin, and in Experiment 1, a control group of rats were trained to self-administer a solution of 20% (*w*/*v*) sucrose. Sucrose was used for comparison as it contains calories similar to ethanol, and has a sweet taste, similar to saccharin. In Experiment 1, rats received 20 one-hour self-administration sessions during the light cycle. In Experiment 2, sessions were conducted in the dark cycle and were initially 30 min in length, and they were subsequently extended to 60 min to match the duration of Experiment 1. Rats received a total of 21 self-administration sessions prior to surgery, and an additional 10 sessions following surgery to re-establish baseline responding.

2.5. Experiment 1. Analysis of Adolescent CORT Effects on Adult Ethanol Self-Administration and the Amygdala Phosphoproteome

Quantitative Label-Free Phosphoproteomics

Fourteen to 16 days after the last day of self-administration, rats from Experiment 1 were euthanized by focused microwave irradiation, in order to analyze the amygdala phosphoproteome using high resolution tandem mass spectrometry. Focused microwave irradiation is known to maintain the post-translational modification state of proteins during the post-mortem period [17]. Rats were lightly anesthetized with isoflurane prior to euthanasia, to reduce the influence of acute stress on protein phosphorylation. After euthanasia, the amygdala was dissected and homogenized by sonication in a buffer containing urea (ThermoFisher, 8 M), ammonium bicarbonate (ThermoFisher, 0.4 M), and protease (Pierce, at 1% of lysis buffer) and phosphatase inhibitor cocktails (Pierce, at 2.5% of lysis buffer). Samples from two rats in each experimental group were randomly pooled to create a total of four biological samples per group. Pooled samples were then analyzed by the Yale/NIDA

Neuroproteomics Center as previously described [18]. Briefly, 20 µL of 45 mM dithiothreitol (DTT) was added to each sample and incubated at 37 °C for 20 min to reduce Cys residues. Samples were cooled and 20 µL of 100 mM iodoacetamide (IAM) was added to each sample and incubated at room temperature in the dark for 20 min for alkylation of the reactive free sulfhydro of the reduced Cys. Dual enzymatic digestion was carried out by adding 600 µL of dH_2O and 30 µL of 1 mg/mL Lys C followed by incubation at 37 °C for 4 h, with subsequent digestion by incubation with 30 µL of 1 mg/mL trypsin overnight at 37 °C. Samples were macrospin desalted and dried by a Speedvac. Pellets were dissolved in 50 µL of a solution containing 0.5% trifluoroacetic acid (TFA) and 50% acetonitrile. Samples were then subjected to titanium dioxide (TiO_2) phosphopeptide enrichment using TopTips (Glygen, Columiba, MD, USA). A three-step conditioning of the TopTip was utilized with 1 min at 2000 rpm on a bench top centrifuge (ThermoFisher) for each step. First, the TopTip was washed with 2 × 60 µL 100% acetonitrile, then with 2 × 60 µL 0.2 M sodium phosphate (pH 7.0), and finally with 2 × 60 µL 0.5% TFA in a 50% acetonitrile solution. The acidified digest supernatants were loaded into the TopTip, and bound phosphopeptides were washed with 2 × 40 µL of a buffer containing 0.5% TFA in 50% acetonitrile, spun at 1000 rpm for 1 min, and then at 3000 rpm for 2 min. Phosphopeptides were eluted from each TopTip by three aliquots of 30 µL of 28% high purity ammonium hydroxide (ThermoFisher). The eluted fraction was dried and re-dried with 2 × 30 µL water by speedvac. Enriched fractions were dissolved in 10 µL of 70% formic acid and 30 µL of 50 mM sodium phosphate. Peptide concentrations were determined by NanoDrop to load 0.3 µg/5 µL of each sample.

For LC/MS-MS, 5 µL of each sample was injected onto a LTQ Orbitrap XL LC-MS/MS system. Peptide separation was performed on the nanoACQUITY™ ultra-high pressure liquid chromatography (UPLC™) system (Waters, Milford, MA), using a Waters Symmetry® C18 180 µm × 20 mm trap column and a 1.7 µm, 75 µm × 250 mm nanoACQUITY™ UPLC™ column (35 °C). Trapping was done at 15 µL/min, with 99% Buffer A (0.1% formic acid in water) for 1 min. Peptide separation was performed over 120 min at a flow rate of 300 nL/min beginning with 95% Buffer A and 5% Buffer B (0.075% formic acid in acetonitrile) to 40% B from 1–9 min, to 85% B from 9–91 min, held at 85% B from 91–95 min, then returned to 5% B from 95–96 min. Two washes were made between each sample run to ensure no carryover (1. 100% acetonitrile, 2. Buffer A). The LC was in-line with an LTQ-Orbitrap XL mass spectrometer. MS was acquired in the Orbitrap using one microscan, and a maximum injection time of 900 msec followed by 3–6 data-dependent MS/MS acquisitions in the ion trap (with precursor ion threshold of >3000). The total cycle time for both MS and MS/MS fragmentation by collision induced dissociation (CID) were first isolated with a 2 Da window, followed by normalized collision energy of 35%. Dynamic exclusion was activated where former target ions were excluded for 30 s. Three technical replicates were injected for each sample and all samples and replicates were randomized across the entire run time.

2.6. *Experiment 2. Determining the Role of GRK2 in Regulating Ethanol-Motivated Behaviors*

The proteomics analysis revealed several changes in protein phosphorylation of potential interest for mediating the increase in motivation to respond for alcohol in adolescent CORT-exposed rats. One of particular relevance was the increase in phosphorylation of the α_{2A} adrenergic receptor (α_{2A}R) on serines (S) 296–299 in the amygdala of CORT-exposed animals relative to controls. Phosphorylation of these four neighboring residues is mediated by the G-protein coupled receptor kinase 2 (GRK2, also known as the β-adrenergic receptor kinase-βARK) [19]. Therefore, we sought to determine whether inhibition of GRK2 in the amygdala could prevent or reduce yohimbine-induced increases in alcohol-motivated behavior that were measured by progressive ratio responding for alcohol, and reinstatement of alcohol seeking.

2.6.1. Surgery

Rats were anesthetized with a combination of 87.5 mg/kg ketamine and 5 mg/kg xylazine, and were injected with 5 mg/kg of Rimadyl (NSAID analgesic) and 5 mL of lactated Ringers' solution prior to surgery. Rats were implanted bilaterally with intracranial guide cannulae (28 gauge, Plastics One) aimed 1 mm above the basolateral amygdala using the following coordinates: from bregma: AP −3.0 mm; ML ±5.3 mm; DV −7.9 mm. Cannulae were anchored with three stainless steel screws and dental cement, and obturators were placed within the cannulae to maintain patency. Rats were monitored for seven days post-surgery and then they resumed EtOH self-administration.

2.6.2. Progressive Ratio (PR)

The PR schedule began with a ratio of 1 that increased by 2 within each step, then it increased by 1 every four steps (e.g., 1, 3, 5, 7; 10, 13, 16, 19; 23, 27, 31, 35; etc.). Rats were first given three baseline PR sessions, the last of which included a sham injection to acclimate the rats to the infusion/injection procedure. On test days, rats were infused with 1 nmol/side of GRK2i (vs vehicle (saline), between-subjects) 10 min prior to an intraperitoneal injection of 1.25 mg/kg yohimbine (vs. vehicle (H_2O), within-subjects] given 10 min prior to each of two PR sessions. Yohimbine and vehicle PR sessions were given in a counterbalanced manner and were separated by a non-injection PR session.

2.6.3. Extinction/Yohimbine-Induced Reinstatement of EtOH Seeking

Following self-administration (Experiment 1) or PR testing (Experiment 2), rats underwent at least five days of extinction (no more than 10 days) until meeting the extinction criterion ($\leq$20 active lever presses over two consecutive days). In Experiment 1, rats were tested for stress-induced reinstatement of EtOH seeking following challenge with yohimbine and vehicle in two separate test days [15,20,21]. Similarly, in Experiment 2, rats were tested in two reinstatement sessions (yohimbine and vehicle) using the same infusion/injection parameters described above for PR testing. Assignment to GRK2i and vehicle groups was counterbalanced such that some rats received the same infusion for both PR and reinstatement (e.g., GRK2i, GRK2i or VEH, VEH) while others received the opposite infusion (e.g., GRK2i, VEH or VEH, GRK2i).

2.7. Statistical Analysis

2.7.1. Behavioral Data

Behavioral data were analyzed by multi factor ANOVA, including repeated measures where appropriate, or by Student's *t*-test. Significance was set at an alpha of 0.05 and any significant interaction effects were further analyzed by Bonferroni's post hoc test.

2.7.2. Proteomics Data

Chromatographic/spectral alignment, feature extraction, data filtering, and statistical analysis was carried out using Nonlinear Dynamics Progenesis LC-MS software (www.nonlinear.com). Raw data files were imported into the program and detected mass spectral features were aligned based on the retention time of the detected m/z peaks based on a randomly selected reference run. All other runs were automatically aligned to the reference run to minimize retention time variability between runs. No adjustments were necessary in the m/z dimension, due to the high mass accuracy of the spectrometer (typically <3 ppm). All runs were selected for detection with an automatic detection limit. Features within retention time ranges of 0–5 min were filtered out, as were features with charge state greater than +6 or singly charged peptides (as no MS/MS fragmentations were taken for these charge states during data collection) for reduction of false positive peptide assignments. A normalization factor based on the use of the median and the median absolute deviation was then calculated to account for the approximation of the variance to remove the influence of outliers, and to account for differences in sample load between injections. The experimental design grouped multiple injections from each condition. Stringent conditions were set in in-house MASCOT search engine (Matrix Science, Boston, MA) to filter out low scoring identified peptides by imposing a confidence probability score (p) of <0.05. A protein was quantified if it contained at least two unique identified peptides. The filtered MS/MS spectral features along with their precursor spectra were exported in the form of an .mgf file (Mascot generic file) for database searching using the Mascot algorithm [22]. These data were searched against the Uniprot (*Rattus norvegicus*) database. The confidence level was set to 95% within the MASCOT search engine for peptides assigned hits based on randomness. MS/MS analysis was based on the use of trypsin and the following variable modifications: carbamidomethyl (Cys), Oxidation (Met), Phospho (Ser, Thr, Tyr). Other search parameters included peptide mass tolerance of ±15 ppm, fragment mass tolerance of ±0.5 Da, and maximum missed cleavages of 3. A decoy search (based on the reverse sequence search) was performed to estimate the false discovery rate (FDR), with a setting of acceptable protein ID having FDR of 2%. A protein was considered to be positively identified if there were two or more significantly labeled unique peptides (bold red based on Mascot MOWSE scoring). The Mascot significance score match is based on a MOWSE score, and it relies on multiple matches to more than one peptide from the same protein. The Mascot search results were exported to an .xml file using a significance cutoff of <0.05, and ion score cutoff of 28, and a requirement of at least one bold (first time any match to the spectrum has appeared in the report) and red (top scoring peptide match for this spectrum) peptide. The .xml file was then imported into the Progenesis LCMS software, where search hits were assigned to corresponding detected features (post-translational modifications), identified as described above. Verification of phosphorylation site(s) was carried out using the PhosphoRS algorithm [23], and phosphorylated peptides with PhosphoRS probability greater than 0.7 (confidently assigned from the MS/MS fragmentation spectra) were considered in our analyses.

Once proteins and protein modifications for each peptide were determined, the normalized intensity values for each sample were averaged within each group. When the same modification was identified on a protein as separate peptides (due to charge state or cleavage differences), the peptide with the lowest coefficient of variation was chosen as the representative readout for that phosphorylation event. However, in cases where the multiple peptides did not show the same general statistical relationships between groups, then both peptides were kept in the analysis. The four groups were compared by ANOVA to identify *p*-values of potential differences between any of the four groups. The −log10 value of these *p*-values was then plotted against the log2 ratio of the average intensity for the CORT group relative to the water group, and the EtOH group relative to the sucrose group, in volcano plots. These data were also used to determine the significant main effects of relative phosphopeptide abundance. These data were also analyzed by two-way ANOVA to determine whether there were any interactions in phosphopeptide abundance based on adolescent treatment and reinforcer self-administered. Note that phosphopeptide abundance was not normalized to total protein levels, allowing the possibility that apparent differences in phosphorylation are driven by changes in total protein expression. Nevertheless, the data still indicate a change in the amount of signaling related to that phosphorylation event in the amygdala. Given that the purpose of this study was to discover potential new targets for treating stress-associated alcohol drinking, we present data with *p*-values of $p < 0.05$ for main effects (CORT vs H_2O treatment or sucrose vs EtOH self-administration) and $p < 0.1$ for interaction effects after correcting for multiple comparisons using the Benjamini–Hochberg procedure, assuming a 10% false discovery rate. Significant interactions were further examined using Tukey's post-hoc test.

Phosphopeptides showing significant interactions were further analyzed using a principal components analysis and a hierarchical clustering analysis to better visualize the relationship of phosphopeptide abundance between groups. The distance between two clusters was calculated using Ward's method.

3. Results

3.1. Experiment 1. Analysis of Adolescent CORT Effects on Adult Ethanol Self-Administration and the Amygdala Phosphoproteome

A total of 32 rats were exposed to CORT (n = 16) or normal tap water (n = 16) in adolescence. In adulthood, half of the rats were trained to self-administer either sucrose or ethanol + saccharin, with n = 8 in each treatment group (sucrose–H_2O, sucrose–CORT, EtOH–H_2O, EtOH–CORT). Figure 1A illustrates the experimental design.

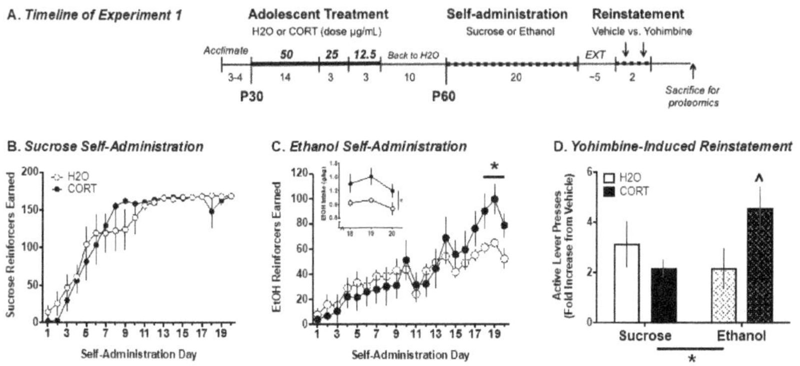

Figure 1. Ethanol and sucrose self-administration and yohimbine-induced reinstatement in H_2O- and CORT-exposed animals. (**A**) Experimental timeline. CORT exposure did not alter sucrose self-administration (**B**), but increased the response for ethanol during the last three days of self-administration (**C**) and ethanol (g/kg body weight) intake (**C**), inset. Pairwise comparisons following a significant reinforcer × exposure interaction showed a strong trend for increased yohimbine-induced reinstatement of ethanol, but not sucrose seeking in CORT-exposed rats (**D**). Thus, CORT exposure selectively alters ethanol-motivated behavior. Data are presented as the mean ± the standard error of the mean (SEM). * $p < 0.05$' ^ $p = 0.06$.

3.1.1. Ethanol Self-Administration

In adulthood, adolescent CORT-exposed and control rats were trained to self-administer ethanol or sucrose. Factorial ANOVA analysis of overall numbers of reinforcers earned, active lever presses, and magazine entries across self-administration sessions as a function of adolescent treatment (H_2O vs. CORT) and reinforcer type (sucrose vs. EtOH), indicated no treatment × reinforcer interactions (all $p > 0.05$). However, there were significant effects of time and time × reinforcer interactions for reinforcers earned ($F_{(19, 8)} = 58.22$, $p < 0.001$; $F_{(19, 8)} = 43.99$, $p < 0.001$), indicating that sucrose self-administration (Figure 1B) was acquired more quickly for both treatment groups. Inspection of the ethanol self-administration data indicated that once animals had acquired stable self-administration (e.g., the final 3 days of testing), there was a separation between the CORT-exposed and control groups in the number of ethanol reinforcers earned (Figure 1C). Analysis of the number of reinforcers earned on the last three days of self-administration revealed a significant reinforcer type by adolescent exposure interaction ($F_{(1, 26)} = 8.10$, $p = 0.009$), and subsequent analysis by separate two-way ANOVAs for each reinforcer type indicated that there was only a significant effect of adolescent CORT exposure on reinforcers earned in rats responding for ethanol ($F_{(1, 13)} = 17.14$, $p < 0.001$), but not for sucrose ($F_{(1, 13)} = 1.829$, $p > 0.05$). Therefore, adolescent CORT exposure resulted in a significant increase in the number of ethanol reinforcers earned once self-administration was acquired, but did not affect self-administration of sucrose (Figure 1B,C). Similarly, analysis of ethanol intake (g/kg) during the same time frame revealed a significant effect of adolescent condition ($F_{(1, 26)} = 6.94$, $p = 0.02$), but no interaction with session day ($F_{(2, 26)} = 0.32$, $p > 0.05$), indicating that the adolescent CORT-exposed animals self-administered significantly more ethanol as a g/kg dose during all three of the last self-administration sessions (Figure 1C, inset).

3.1.2. Yohimbine-Induced Reinstatement of EtOH vs. Sucrose Seeking

Following self-administration and subsequent extinction training, rats were tested for yohimbine-induced reinstatement. Data were expressed as a fold change in responses on the active lever following yohimbine relative to vehicle injection (Figure 1D). A two-way ANOVA revealed a significant reinforce type × adolescent exposure interaction ($F_{(1, 27)} = 4.9$, $p = 0.036$). Subsequent *t*-tests comparing the effects of adolescent exposure revealed a trend ($p = 0.06$) for increased yohimbine-induced reinstatement of ethanol, but not sucrose ($p = 0.33$) seeking. These results indicate that adolescent CORT exposure may enhance sensitivity to the ability of a pharmacological stressor to induce reward seeking, but that this effect is selective for responding for ethanol.

3.1.3. Phosphoproteomic Analysis

Two weeks after the reinstatement test, rats were euthanized and their brains were analyzed for changes in the levels of phosphorylated proteins in the amygdala. A discovery-based mass spectrometry approach was used to identify potentially novel biological signaling differences in the brains of rats that self-administered sucrose vs ethanol or that were exposed to CORT in adolescence, and their interaction. A total of 156 unique proteins were identified on which sites of phosphorylation could be resolved. Within these proteins, 478 unique phosphorylation patterns were identified, and of these, 270 phosphopeptides were significantly regulated in at least one of the experimental conditions (Suppl. Table S1, phosphopeptides above yellow row are significant). Next, volcano plots were created to compare the magnitude of change in phosphopeptide abundance based on the main effect of adolescent CORT exposure (Figure 2A) versus the main effect of ethanol self-administration (Figure 2B) relative to the −log10 of the *p*-value from the ANOVA to identify highly significant differences (*y*-axis) of large effect size (*x*-axis). This analysis revealed that adolescent CORT exposure produced 16 changes in protein phosphorylation (red dots in Figure 2A) that were both significantly different from H_2O exposure (points above gray line = $p < 0.05$ after correcting for multiple comparisons) and that were of large effect size (either increases or decreases with an effect size greater than a four-fold change from H_2O exposed control = log2(ratio) >2 or <−2). On the other hand, there was only one significant difference of large effect size identified, based on the reinforcer that was previously self-administered (Figure 2B), suggesting that adolescent CORT exposure had a larger long-term effect on the amygdala phosphoproteome than the prior ethanol self-administration experience. Indeed, the protein seemingly regulated by ethanol self-administration, microtubule-associated protein 2 (MAP2), was also regulated by CORT exposure, and both effects were likely driven by a few large values in the CORT–sucrose group.

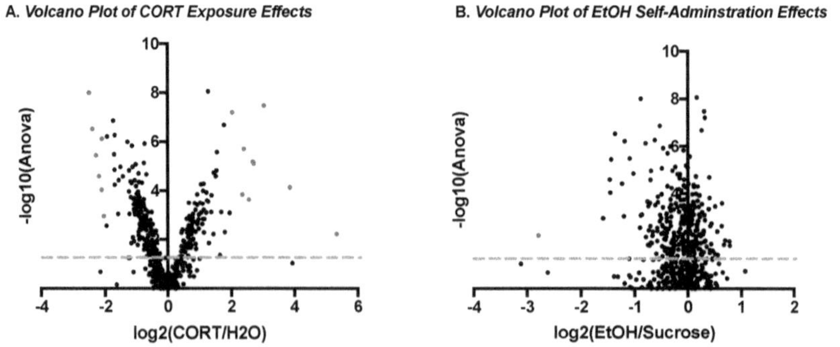

Figure 2. Volcano plots of main effects of adolescent exposure and adult self-administration. Plots are shown comparing main effect of adolescent CORT exposure (**A**) or the main effect of ethanol (EtOH) self-administration experience (**B**). Each plot shows the identified phosphopeptides based on the log2 effect size of adolescent CORT relative to H_2O exposure on the *x*-axis and the −log10 of the ANOVA *p*-value on the *y*-axis. The gray lines represent the Benjamini–Hochberg-corrected significance point for $p < 0.05$. Points in red represent phosphopeptides that are both highly significantly different and have a large effect size (>4-fold difference).

We went on to test for potential interaction effects between adolescent CORT exposure and ethanol self-administration. After correcting for multiple comparisons, significant interactions ($p < 0.1$) were identified for 10 phosphopeptides (Suppl. Table S2). Of these 10, seven were different phosphopeptides from the neurofilament heavy and medium chain proteins. Figure 3A shows the quantitative difference between groups from one of these neurofilament phosphopeptides, which was representative of the pattern of results observed for all of the neurofilament phosphopeptides. Tukey's post-hoc analysis revealed that phosphorylation of the neurofilament proteins was highest in the adolescent control group that self-administered sucrose ($p < 0.0001$ relative to all other groups). A similar pattern was observed for two of the other phosphopeptides identified, synaptotagmin 2 and Map 1a (all $p < 0.0001$ comparing H_2O–sucrose to all other groups; Figure 3B,C). Therefore, either prior adolescent CORT exposure or ethanol self-administration resulted in reduced phosphorylation of these peptides relative to controls. The only phosphopeptide to show a different pattern of results was IPP2 (protein phosphatase inhibitor 2, PPP1R2). Phosphorylation of IPP2 on serines 121 and 122 was reduced in the adolescent CORT-exposed rats that self-administered sucrose relative to H_2O–sucrose controls ($p = 0.012$), but CORT-exposed rats that self-administered ethanol showed a significant reversal of this effect ($p = 0.035$; Figure 3D). Thus, with the exception of IPP2, all significant interactions between adolescent exposure groups and reinforcer types indicated that self-administration of ethanol could reduce protein phosphorylation in the adolescent H_2O-exposed group to the levels of adolescent CORT-exposed rats, while ethanol produced no further effects beyond the CORT exposure.

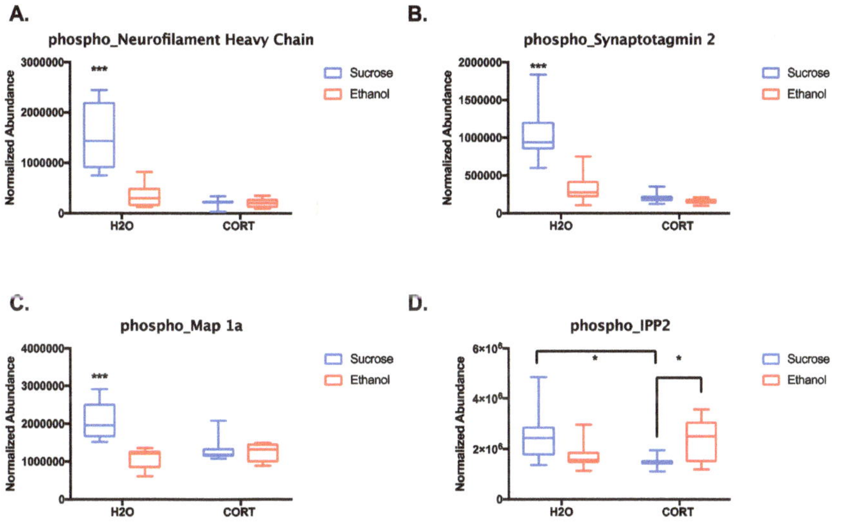

Figure 3. Box plots of phosphopeptides with significant interactions. Plots show the effect of adolescent exposure versus reinforcer self-administered in adulthood for four of the phosphopeptides found to have significant interaction effects after two-way ANOVA: (**A**) neurofilament heavy chain (SPAEAKpSPAEAKPPAEAK), (**B**) synaptotagmin 2 (GGQDDDDAETGLpTEGEGEGEEEKEPENLGK), (**C**) Map 1a (GFKpSPPCEDFSVTGESEK), and (**D**) IPP2 (EQEpSpSGEEDNDLSPEER). Significant interactions were followed by Tukey's post-hoc test, *** $p < 0.0001$, * $p < 0.05$.

These results were further supported by a principal component analysis (PCA) and a hierarchical clustering analysis. Figure 4 illustrates that the first principal component explained the majority of the variance, with the H_2O–sucrose group showing a concentration ellipse that did not overlap with the other three groups. Overlaying the PCA plot is a biplot indicating that the H_2O–sucrose group generally had higher values for each phosphopeptide relative to the other groups, suggestive of reduced phosphorylation in the experimental groups. In addition, hierarchical clustering analysis based on the abundance of the 10 phosphopeptides was significantly different among groups, showing that the adolescent H_2O- and CORT-exposed groups largely clustered separately, independent of the reinforcer self-administered, with the exception of some of the rats in the H_2O–ethanol group, which clustered more closely with the CORT groups (Figure 5).

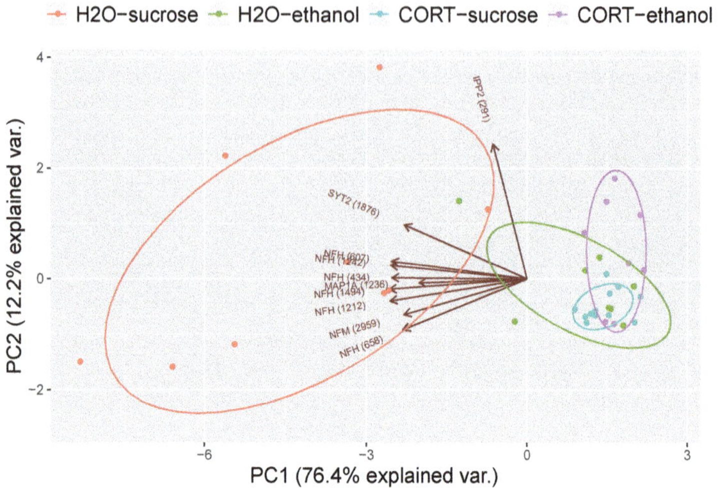

Figure 4. Principal components analysis of significant interactions. Plot shows the first principal component on the *x*-axis and second principal component on the *y*-axis. Each colored ellipsis represents a different group and the clustering of the CORT groups shows that much of the variance between groups could be explained by adolescent CORT exposure. The H_2O–sucrose group was the most different, suggesting that ethanol self-administration shifted the H_2O group to be more similar to adolescent CORT group. Overlaid is a biplot (brown circle and arrows) indicating the majority of the phosphopeptides in the H_2O–sucrose group are in greater abundance than the other three groups.

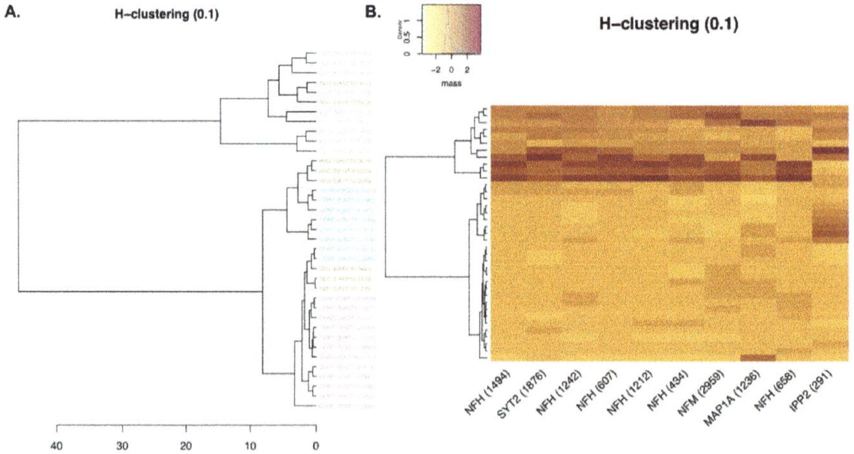

Figure 5. Hierarchical clustering analysis. (**A**) Hierarchical clustering of individual samples based on the abundance of the 10 phosphopeptides significantly different among groups showing general clustering of adolescent H_2O and CORT groups, with the H_2O–ethanol group showing mixed clustering between the two. (**B**). Heat map of clusters versus phosphopeptides with darker colors representing greater abundance of the phosphopeptide.

Next, due to the large effect of adolescent CORT exposure on the amygdala phosphoproteome, independent of self-administration condition, we focused our analysis on the 16 significantly regulated phosphopeptides shown in red in Figure 2A. The identity of each of the phosphopeptides is listed in Table 1, where the protein, modified peptide sequence, log2 magnitude of change, and *p*-value from the ANOVA are given. The phosphorylated residues are shown in red.

Table 1. Phosphopeptides significantly regulated by adolescent CORT exposure.

Protein	Peptide Sequence + Phosphorylation Sites (in Red)	Log2(CORT/H_2O)-Magnitude	*p*-Value (ANOVA)
Top Up-Regulated Phosphopeptides			
Microtubule-associated protein 2	RLSNVSSSGSINLLESPQLATLAEDVTAALAK	5.325915864	0.005821436
Microtubule-associated protein 2	RLSNVSSSGSINLLESPQLATLAEDVTAALAK	3.863577679	7.21×10^{-5}
Gap junction alpha-1 protein	VAAGHELQPLAIVDQRPSSRASSR	3.054725626	3.34×10^{-8}
Microtubule-associated protein 2	RLSNVSSSGSINLLESPQLATLAEDVTAALAK	2.726525438	7.69×10^{-6}
Protein phosphatase 1 regulatory subunit 1A	RRPTPATLVLTSDQSSPEVDEDRIPNPLLK	2.69773257	6.46×10^{-6}
Canalicular multispecific organic anion transporter 2	IPLNLLPQLISGMTQTSVSLK	2.568107399	0.000230896
Microtubule-associated protein tau	HLSNVSSTGSIDMVDSPQLATLADEVSASLAK	2.409694538	1.97×10^{-6}
Microtubule-associated protein 2	RLSNVSSSGSINLLESPQLATLAEDVTAALAK	2.359792109	0.000139602
Alpha-2A adrenergic receptor	DGDALDLEESSSSEHAERPQGPGKPER	2.043347801	6.23×10^{-8}
Top Down-Regulated Phosphopeptides			
Neurofilament light polypeptide	AEEAKDEPPSEGEAEEEEK	−2.48762892	9.91×10^{-9}
Neurofilament heavy polypeptide	TLDVKSPEAK	−2.38663396	2.96×10^{-7}
Neurofilament heavy polypeptide	SLAEAKSPEK	−2.276582671	3.53×10^{-6}
Neurofilament heavy polypeptide	SPAEAKSPAEAKPPAEAK	−2.178346111	2.49×10^{-5}
Neurofilament heavy polypeptide	SPVEVKSPEK	−2.100576776	9.08×10^{-5}
Neurofilament heavy polypeptide	SPAEAKSPAEVK	−2.093897637	7.44×10^{-7}
Neurofilament medium polypeptide	AEEEGGSEEEVGDKSPQESK	−2.035085213	0.001096364

Overall, adolescent CORT exposure appeared to produce increased phosphorylation of the microtubule-associated protein MAP2, particularly in the N-terminal domain, while the phosphorylation of neurofilament proteins was decreased. These data are suggestive of CORT-induced structural changes in the amygdala, though the exact functions of the phosphorylation sites identified are currently unknown. Of interest for alcohol use and other psychiatric disorders, adolescent CORT exposure also regulated the gap junction protein, connexin43, the protein phosphatase 1 regulatory subunit 1a (PPP1R1a), which is also known as inhibitor 1 (I-1), and the α_{2A}AR. In addition, the most highly statistically significant change in phosphopeptide abundance between groups was for the metabotropic glutamate receptor 5 (mGluR5), though the magnitude of effect was slightly less than 4-fold. Given the known relevance of mGluR5, particularly in the amygdala, for alcohol-motivated behaviors [24–28], we further inspected the two-way interaction between adolescent treatment and reinforcer self-administered for this receptor and the other highly regulated phosphopeptides. A Two-way ANOVA revealed the main effects of CORT in increasing the abundance of each of these phosphopeptides (connexin 43: ($F_{(1, 33)} = 14.44$, $p < 0.001$); I-1: ($F_{(1, 33)}=15.5$, $p < 0.001$); α_{2A}AR: ($F_{(1, 33)} = 24.17$, $p < 0.001$); mGluR5: ($F_{(1, 33)}=26.64$, $p < 0.001$)), with no effect of reinforcer consumed during self-administration (all $p > 0.25$; Figure 6A–D).

We next determined the potential functional implications of the phosphorylation events observed. The function of the phosphorylation sites on I-1 (Ser43, Ser46, and Ser47) and mGluR5 (Ser1014 and Ser1016) are unknown. On the other hand, increased phosphorylation of connexin43 was found on Ser365, Ser368, and Ser369, which have been described previously [29]. In particular, phosphorylation of Ser368 is known to decrease the permeability of gap junctions, and it is thought to be mediated by protein kinase C (PKC) [30,31]. Thus, adolescent CORT exposure may lead to long-lasting changes in neural signaling via gap junctions in the amygdala.

Finally, increased phosphorylation of the α_{2A}AR was found on four consecutive serines (366–369), which are a known substrate of the G-protein coupled receptor kinase 2 (GRK2) [19,32]. Phosphorylation at these sites mediates agonist-stimulated receptor desensitization, association with arrestin, decoupling from the G-protein, and clathrin-mediated endocytosis [19,32]. Thus, we predicted that the adolescent CORT-treated rats would have a reduced sensitivity to α_{2A}R-mediated signaling, which could result in an increase in norepinephrine release and post-synaptic signaling in the brain, as the normal autoreceptor-mediated brake on noradrenergic-transmission would be impaired. Given the large literature on the involvement of heightened noradrenergic signaling, particularly in the amygdala, for both stress- and alcohol-related behaviors, including the potential clinical use of α_{2A}AR agonists as a treatment for alcohol use disorders [33–35], GRK2-mediated phosphorylation of α_{2A}AR after adolescent CORT exposure is a strong candidate as a mediator of increased motivation for alcohol. Therefore, we decided to directly test if inhibition of GRK2 in the amygdala could reduce alcohol-motivated behaviors in adolescent CORT- or H_2O-exposed rats.

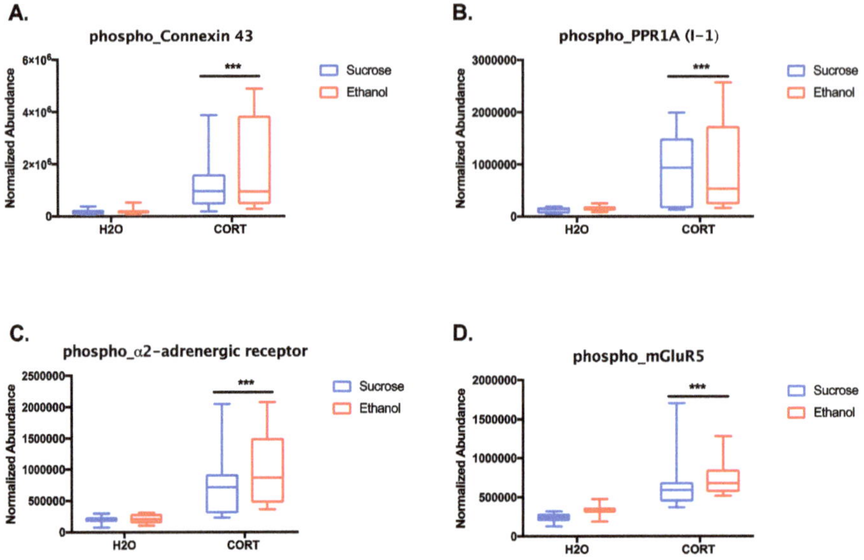

Figure 6. Box plots of phosphopeptides highly differentially regulated by adolescent CORT exposure. Plots show the effect of adolescent exposure group versus reinforcer self-administered in adulthood for four of the phosphopeptides found to have highly significant differences of large effect size based on adolescent CORT exposure with relevance to alcohol use disorders: (**A**) connexin 43 (VAAGHELQPLAIVDQRPSpSRApSpSR), (**B**) PPR1A (RRPTPATLVLTpSDQpSpSPEVDEDRIPNPLLK), (**C**) α_{2A}AR (DGDALDLEEpSpSpSpSEHAERPQGPGKPER), and (**D**) mGluR5 (pSPpSPISTLSHLAGSAGR). Significant main effects of CORT, *** $p < 0.0001$.

3.2. Experiment 2. Determining the Role of GRK2 in Regulating Ethanol Motivated Behaviors

A total of 94 rats were exposed to CORT (n = 46) or normal tap H_2O (n = 48) in adolescence at the University of Pittsburgh as in Experiment 1. Of these rats, 26 were ultimately excluded from data analysis due to failure to meet acquisition criteria, misplaced cannula, or death during surgery. Thus, the final sample sizes for this experiment were n = 34 for both CORT and H_2O groups (n = 16–17 each for GRK2i- and vehicle-treated rats). Figure 7A illustrates the experimental design.

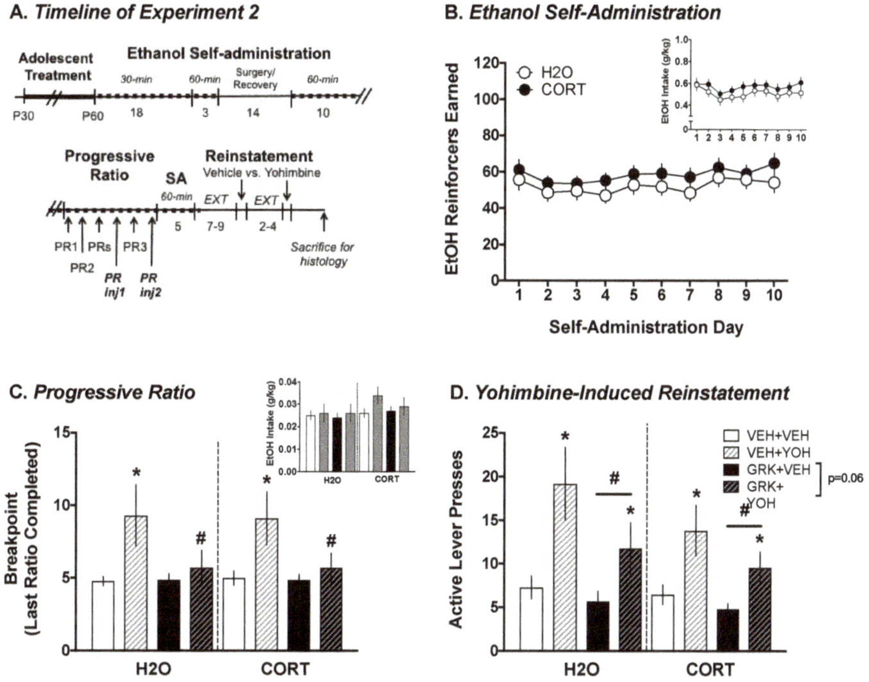

Figure 7. Inhibition of GRK2 attenuates yohimbine-induced increases in ethanol-motivated behavior. (**A**) Experimental timeline. (**B**) CORT exposure consistently, but not significantly, tended to increase reinforcers earned and intake (inset) during ethanol self-administration. (**C**) Intra-basolateral amygdala inhibition of GRK2 blocks yohimbine-induced increases in breakpoint during progressive ratio testing, though the overall low number of reinforcers earned precluded the detection of differences in ethanol intake (inset). (**D**) GRK2 inhibition significantly reduced the reinstatement of ethanol seeking and tended to reduce the effects of yohimbine in increasing reinstatement. * $p < 0.05$ yohimbine (YOH) vs. vehicle (VEH); # $p < 0.05$ GRK2i vs. vehicle.

3.2.1. Ethanol Self-Administration

To more closely equate the results of Experiment 2 to those of Experiment 1, analyses were conducted on the 10-day postoperative self-administration period, during which time rats were tested in 60 min sessions. Mixed factorial ANOVAs with adolescent exposure (CORT vs. H_2O) as the between-subjects factor and day (10) as the within-subject factor revealed significant main effects of day for reinforcers earned (($F_{(9, 702)} = 3.208$, $p < 0.001$); Figure 7B) and ethanol intake (g/kg; ($F_{(9, 702)} = 3.18$, $p = 0.001$); (Figure 7B, inset); however, day-to-day variability was not systematic. Similar patterns evident for all other self-administration parameters (e.g., active lever presses, magazine entries; data not shown). Though no statistically significant main effects of or interactions involving adolescent exposure were found for any outcome measure, CORT-exposed rats consistently showed greater ethanol-motivated behavior compared to H_2O-exposed controls.

3.2.2. Progressive Ratio Testing: Yohimbine vs. Vehicle

Mixed factorial ANOVAs with adolescent group (CORT vs. H_2O) and infusion (GRK2i vs. vehicle) as the between-subjects factors and injection (yohimbine vs vehicle) as the within-subject factor revealed an overall main effect of injection ($F_{(1, 64)} = 10.997$, $p = 0.002$), and an injection × infusion interaction ($F_{(1, 64)} = 5.085$, $p = 0.028$) for breakpoint (last ratio completed), with the yohimbine-injected rats showing greater "willingness to work" for ethanol than vehicle-injected rats, and a GRK2i-mediated reduction in these measures was only observed in yohimbine-injected rats (t(66) = 2.229, $p = 0.029$); Figure 7C. Similar patterns were evident for other self-administration parameters (e.g., reinforcers earned, active lever presses; data not shown). Though the analysis of ethanol intake did not reveal significant effects of injection or infusion, rats only earned roughly 3–4 reinforcers on average during PR testing, thus making it difficult to detect differences in very low levels of intake (Figure 7C, inset). No main effect of adolescent group was evident during PR.

3.2.3. Yohimbine-Induced Reinstatement

Mixed factorial ANOVAs with adolescent exposure (CORT vs. H_2O) and infusion (GRK2i vs. vehicle) as the between-subjects factors and injection (yohimbine vs. vehicle) as the within-subject factor revealed significant main effects of injection ($F_{(1, 63)} = 26.169$, $p < 0.001$) and infusion ($F_{(1, 63)} = 4.293$, $p = 0.042$) for active lever presses during reinstatement (Figure 7D). Yohimbine-injected rats responded more than vehicle-injected rats, and GRK2i-infused rats responded less on the active lever than rats infused with the GRK2i vehicle. An exploratory analysis indicated that like during PR testing, yohimbine-induced increases in ethanol-motivated behavior tended to be attenuated ($p = 0.06$) in GRK2i-infused rats relative to those that received vehicle infusion. No main effect of the adolescent group was evident during reinstatement. Taken together, these results indicate that intra-BLA inhibition of GRK2 reduces yohimbine-induced increases in ethanol-motivated behavior.

4. Discussion

In the present series of studies, we first examined the impact of chronic exposure to the glucocorticoid stress hormone corticosterone (CORT) during adolescence (PND 30–50) on ethanol-motivated behaviors and on the amygdala phosphoproteome (Experiment 1). We found that rats chronically exposed to CORT during adolescence self-administered significantly more of a sweetened ethanol solution than control rats, once self-administration was acquired. Further, CORT-exposed rats displayed enhanced yohimbine stress-induced reinstatement in ethanol-reinforced, but not sucrose-reinforced rats. Importantly, chronic CORT exposure increased phosphorylation of a series of serine residues in the α_{2A} adrenergic receptor protein in the amygdala, at which yohimbine exerts its pharmacological action. We then targeted the kinase that phosphorylates these residues, G protein-coupled receptor kinase 2 (GRK2), in experiments aimed at determining whether GRK2 inhibition would alter ethanol-motivated behaviors as a function of chronic (CORT exposure during adolescence) and/or acute (injection of yohimbine) stress exposure (Experiment 2). While we only uncovered statistical trends for the ability of chronic adolescent CORT exposure to increase ethanol self-administration in Experiment 2, we found that inhibition of GRK2 in the BLA significantly attenuated yohimbine-induced increases in ethanol-motivated behavior, regardless of adolescent experience. These findings suggest that GRK2 inhibition is a promising target for reducing stress-induced increases in ethanol-motivated behaviors.

Prolonged stress exposure during adolescence has been shown to increase the vulnerability of developing psychiatric disorders, including alcoholism later in life [5,6]. Prior studies have shown that chronic CORT exposure in adolescence subsequently increases impulsivity [10] in adulthood. This indicates that there are long-lasting effects of elevated glucocorticoid levels during adolescence that could increase the risk of maladaptive behavior. However, while some studies have shown that post-weaning isolation stress increases ethanol-motivated behaviors [12,14,36], the present and previous studies utilizing the chronic CORT model in adolescence [15] showed inconsistent effects in the ability of CORT to significantly augment ethanol-motivated behaviors. The significant increase in ethanol self-administration during the final three days of training in Experiment 1 is paralleled by consistent trends for prior CORT exposure to augment responses for ethanol in Experiment 2, and cue-induced reinstatement in female rats [15]. This disparity could be due to environmental differences in the two facilities in which the present experiments were conducted. Indeed, large differences in behavioral outcomes have been documented, even when experimental conditions (other than facility) are held constant [37,38]. It is possible that these unavoidable changes in husbandry could have resulted in the diminished CORT effects on drinking in Experiment 2, potentially leading to a decreased sensitivity to stress. For example, other studies have identified differences in ingestive behavior [39] and stress/anxiety-related responses [40] in rodents housed in open-style (like Experiment 1) versus individually ventilated (like Experiment 2) cages, potentially leading to the lack of a robust effect of CORT on subsequent ethanol intake in the latter experiment. Our laboratory has conducted experiments to directly determine if light cycle phase or degree of food restriction during ethanol self-administration was responsible for the differential effects of adolescent CORT exposure between Experiments 1 and 2, and neither of these factors was found to consistently influence our results. Future studies could test exposures to higher concentrations of CORT to potentially overcome any stress-buffering effects that the current facility may have, to improve the replicability of these results.

Regardless of the sensitivity of prior chronic CORT exposure in adolescence to alter subsequent ethanol self-administration, the proteomics analysis identified a number of differentially phosphorylated proteins in the amygdala of the CORT-exposed rats that did show greater ethanol self-administration in Experiment 1, and these could potentially be targeted to treat alcohol use disorders. The effect of adolescent CORT exposure was greater than ethanol self-administration experience alone, suggesting that elevated glucocorticoids in adolescence may produce a long-lasting vulnerability that is not substantially exacerbated by ethanol intake. In addition, while there were a few phosphopeptides that exhibited significant interactions between adolescent treatment and the reinforcer that was self-administered, almost all of these phosphopeptides were of the highest abundance in the H_2O–sucrose group, with ethanol self-administration bringing the abundance in the H_2O group down to the level of the CORT exposed rats. These data are intriguing and suggestive that three weeks of 1 hour daily ethanol exposure may shift the molecular activity of the amygdala of control rats to a state that is more similar to rats that were exposed to chronic CORT in adolescence.

Importantly, the majority of significantly regulated phosphopeptides were observed in the adolescent CORT group independent of the reinforcer self-administered in adulthood, again suggesting that adolescent CORT exposure produces profound effects on the amygdala phosphoproteome, including proteins that are associated with alcohol use disorders, which may indicate heightened vulnerability to the effects of alcohol. In particular, increased phosphorylation of four serine residues in the third intracellular loop of the α2AR, which are a GRK2 substrate, was of particular interest, due to evidence pointing to the potent role of adrenergic signaling, particularly in the amygdala, in ethanol drinking and seeking. It has long been recognized that the noradrenergic system plays a critical role in the development of alcohol use disorders, but only recently has interest been revitalized in targeting this system with respect to AUD treatment [34]. Preclinical studies have shown that downregulation of noradrenergic signaling, via α_{2A}R agonism (clonidine; [35]), or antagonism of α1AR (prazosin; [41]) or βAR (propranolol; [42]), reduces ethanol drinking and seeking in high-consuming animals. Similar treatment approaches have been undertaken to treat comorbid post-traumatic stress disorder (PTSD) and AUD [33,43], as these disorders frequently co-occur, lead to significant disability, and are difficult to treat effectively [44].

Further, early life stress reduces norepinephrine in the amygdala [45] and α_{2A}R (*adra2a*) gene expression in the hypothalamus [46], which leads to increased anxiety-like behavior and ethanol drinking, respectively, and enhances ethanol-induced norepinephrine levels in the BLA [47]. Taken together, these studies suggest that early life stress alters noradrenergic functioning, consistent with the enhanced phosphorylation, and likely subsequent internalization of the autoreceptor α_{2A}R in CORT-exposed rats. Moreover, ethanol exposure can lead to a heightened noradrenergic response, consistent with the enhanced yohimbine-related ethanol-motivated behavior that was mitigated by blocking GRK2-induced α_{2A}R phosphorylation in both H_2O- and CORT-exposed rats. Our results add to a small, but growing number of studies examining the role of GRK-mediated phosphorylation of metabotropic receptors, such as serotonin 5-HT2A [48], dopamine D1/D2 [49,50], cannabinoid [51], and mu opioid receptors [52], in models of substance use disorders, and suggest that GRKs should be further studied as potential targets for novel treatment development. Finally, the proteomics results also identified novel potential targets for treating alcohol use disorders, such as gap junction signaling through connexin43, and provide further support for the development of mGluR5 modulators as treatments for substance use and other disorders [25].

5. Conclusions

The present studies expand on previous findings of enhanced vulnerability to maladaptive behavior following exposure to persistently elevated glucocorticoid levels by demonstrating that chronic adolescent CORT exposure can lead to heightened ethanol drinking, enhancement of yohimbine stress-induced ethanol seeking, and increased phosphorylation of α2ARs at residues that are the substrates for GRK2-mediated receptor desensitization/internalization. We then showed that blocking these reductions in α2AR function by inhibiting GRK in the BLA blocks yohimbine stress-induced ethanol seeking, regardless of prior CORT exposure. These results suggest that altering GRK activity, and/or facilitating noradrenergic autoinhibition, are promising targets for reducing stress-related alcohol use.

Supplementary Materials: The following are available online at http://www.mdpi.com/2227-7382/6/4/41/s1. Table S1: Protein List, Table S2: Two-way ANOVA Interaction Table.

Author Contributions: Conceptualization, M.L.B., J.T. and M.T.; Methodology, M.L.B., K.S., T.T.L. and M.T.; Software, S.B. and W.W.; Validation, K.S. and T.T.L.; Formal Analysis, M.L.B., S.B., W.W. and M.M.T..; Investigation, M.L.B., K.S. and M.M.T.; Resources, T.T.L., W.W., A.C.N., J.R.T. and M.M.T.; Data Curation, M.L.B., T.T.L., S.B., W.W. and M.M.T.; Writing-Original Draft Preparation, M.L.B., T.T.L., A.C.N., J.R.T. and M.M.T.; Writing-Review & Editing, M.L.B., T.T.L., S.B., W.W., A.C.N., J.R.T. and M.M.T.; Supervision, T.T.L., W.W., A.C.N., J.R.T. and M.M.T.; Funding Acquisition, T.T.L., A.C.N., J.R.T. and M.M.T.

Funding: This work was supported by the Interdisciplinary Research Consortium on Stress, Self-control and Addiction (UL1-DE19586 (R, Sinha)) and the NIH Roadmap for Medical Research/Common Fund, PHS DA018343 (A.C.N., T.T.L.), P50AA017537 (J.R.T.), R01MH066172 (J.R.T.), K01DA031745 (M.M.T.), AA025547 (MMT), the Connecticut Department of Mental Health and Addiction Services, and the Pennsylvania Department of Health.

Conflicts of Interest: The authors have no conflicts of interest to disclose.

References

1. Penza, K.M.; Heim, C.; Nemeroff, C.B. Neurobiological effects of childhood abuse: Implications for the pathophysiology of depression and anxiety. *Arch. Womens. Ment. Health* **2003**, *6*, 15–22. [CrossRef] [PubMed]
2. Clarke, T.-K.; Laucht, M.; Ridinger, M.; Wodarz, N.; Rietschel, M.; Maier, W.; Lathrop, M.; Lourdusamy, A.; Zimmermann, U.S.; Desrivieres, S.; et al. KCNJ6 is associated with adult alcohol dependence and involved in gene × early life stress interactions in adolescent alcohol drinking. *Neuropsychopharmacology* **2011**, *36*, 1142–1148. [CrossRef] [PubMed]
3. Crews, F.; He, J.; Hodge, C. Adolescent cortical development: A critical period of vulnerability for addiction. *Pharmacol. Biochem. Behav.* **2007**, *86*, 189–199. [CrossRef] [PubMed]
4. Burke, A.R.; Miczek, K.A. Stress in adolescence and drugs of abuse in rodent models: Role of dopamine, CRF, and HPA axis. *Psychopharmacology* **2014**, *231*, 1557–1580. [CrossRef] [PubMed]
5. Eiland, L.; Romeo, R.D. Stress and the developing adolescent brain. *Neuroscience* **2013**, *249*, 162–171. [CrossRef] [PubMed]
6. Pilowsky, D.J.; Keyes, K.M.; Hasin, D.S. Adverse childhood events and lifetime alcohol dependence. *Am. J. Public Health* **2009**, *99*, 258–263. [CrossRef] [PubMed]
7. Lloyd, D.A.; Turner, R.J. Cumulative lifetime adversities and alcohol dependence in adolescence and young adulthood. *Drug Alcohol Depend.* **2008**, *93*, 217–226. [CrossRef] [PubMed]
8. Gourley, S.L.; Wu, F.J.; Kiraly, D.D.; Ploski, J.E.; Kedves, A.T.; Duman, R.S.; Taylor, J.R. Regionally specific regulation of ERK MAP kinase in a model of antidepressant-sensitive chronic depression. *Biol. Psychiatry* **2008**, *63*, 353–359. [CrossRef] [PubMed]

9. Gourley, S.L.; Taylor, J.R. Recapitulation and reversal of a persistent depression-like syndrome in rodents. *Curr. Protoc. Neurosci.* **2009**. Chapter 9. [CrossRef] [PubMed]
10. Torregrossa, M.M.; Xie, M.; Taylor, J.R. Chronic Corticosterone Exposure during Adolescence Reduces Impulsive Action but Increases Impulsive Choice and Sensitivity to Yohimbine in Male Sprague-Dawley Rats. *Neuropsychopharmacology* **2012**, *37*, 1656–1670. [CrossRef] [PubMed]
11. Mitchell, S.H. The genetic basis of delay discounting and its genetic relationship to alcohol dependence. *Behav. Process.* **2011**, *87*, 10–17. [CrossRef] [PubMed]
12. McCool, B.A.; Chappell, A.M. Early social isolation in male Long-Evans rats alters both appetitive and consummatory behaviors expressed during operant ethanol self-administration. *Alcohol. Clin. Exp. Res.* **2009**, *33*, 273–282. [CrossRef] [PubMed]
13. Skelly, M.J.; Chappell, A.E.; Carter, E.; Weiner, J.L. Adolescent social isolation increases anxiety-like behavior and ethanol intake and impairs fear extinction in adulthood: Possible role of disrupted noradrenergic signaling. *Neuropharmacology* **2015**, *97*, 149–159. [CrossRef] [PubMed]
14. Deehan, G.A.; Cain, M.E.; Kiefer, S.W. Differential rearing conditions alter operant responding for ethanol in outbred rats. *Alcohol. Clin. Exp. Res.* **2007**, *31*, 1692–1698. [CrossRef] [PubMed]
15. Bertholomey, M.L.; Nagarajan, V.; Torregrossa, M.M. Sex differences in reinstatement of alcohol seeking in response to cues and yohimbine in rats with and without a history of adolescent corticosterone exposure. *Psychopharmacology* **2016**, *233*, 2277–2287. [CrossRef] [PubMed]
16. Gourley, S.L.; Kedves, A.T.; Olausson, P.; Taylor, J.R. A history of corticosterone exposure regulates fear extinction and cortical NR2B, GluR2/3, and BDNF. *Neuropsychopharmacology* **2009**, *34*, 707–716. [CrossRef] [PubMed]
17. Hunsucker, S.W.; Solomon, B.; Gawryluk, J.; Geiger, J.D.; Vacano, G.N.; Duncan, M.W.; Patterson, D. Assessment of post-mortem-induced changes to the mouse brain proteome. *J. Neurochem.* **2008**, *105*, 725–737. [CrossRef] [PubMed]
18. Rich, M.T.; Abbott, T.B.; Chung, L.; Gulcicek, E.E.; Stone, K.L.; Colangelo, C.M.; Lam, T.T.; Nairn, A.C.; Taylor, J.R.; Torregrossa, M.M. Phosphoproteomic analysis reveals a novel mechanism of CaMKIIα regulation inversely induced by cocaine memory extinction versus reconsolidation. *J. Neurosci.* **2016**, *36*, 7613–7627. [CrossRef] [PubMed]
19. Eason, M.G.; Moreira, S.P.; Liggett, S.B. Four consecutive serines in the third intracellular loop are the sites for beta-adrenergic receptor kinase-mediated phosphorylation and desensitization of the α2A-adrenergic receptor. *J. Biol. Chem.* **1995**, *270*, 4681–4688. [CrossRef] [PubMed]
20. Bertholomey, M.L.; Verplaetse, T.L.; Czachowski, C.L. Alterations in ethanol seeking and self-administration following yohimbine in selectively bred alcohol-preferring (P) and high alcohol drinking (HAD-2) rats. *Behav. Brain Res.* **2013**, *238*, 252–258. [CrossRef] [PubMed]
21. Feltenstein, M.W.; Henderson, A.R.; See, R.E. Enhancement of cue-induced reinstatement of cocaine-seeking in rats by yohimbine: Sex differences and the role of the estrous cycle. *Psychopharmacology* **2011**, *216*, 53–62. [CrossRef] [PubMed]
22. Taus, T.; Kocher, T.; Pichler, P.; Paschke, C.; Schmidt, A.; Henrich, C.; Mechtler, K. Universal and confident phosphorylation site localization using PhosphoRS. *J. Proteome Res.* **2011**, *10*, 5354–5362. [CrossRef] [PubMed]
23. Hirosawa, M.; Hoshida, M.; Ishikawa, M.; Toya, T. MASCOT: Multiple alignment system for protein sequences based on three-way dynamic programming. *Comput. Appl. Biosci.* **1993**, *9*, 161–167. [CrossRef] [PubMed]
24. Akkus, F.; Mihov, Y.; Treyer, V.; Ametamey, S.M.; Johayem, A.; Senn, S.; Rösner, S.; Buck, A.; Hasler, G. Metabotropic glutamate receptor 5 binding in male patients with alcohol use disorder. *Transl. Psychiatry* **2018**, *8*, 17. [CrossRef] [PubMed]
25. Olive, M.F. Metabotropic glutamate receptor ligands as potential therapeutics for addiction. *Curr. Drug Abuse Rev.* **2009**, *2*, 83–98. [CrossRef] [PubMed]
26. Cozzoli, D.K.; Courson, J.; Wroten, M.G.; Greentree, D.I.; Lum, E.N.; Campbell, R.R.; Thompson, A.B.; Maliniak, D.; Worley, P.F.; Jonquieres, G.; et al. Binge Alcohol Drinking by Mice Requires Intact Group1 Metabotropic Glutamate Receptor Signaling Within the Central Nucleus of the Amygdale. *Neuropsychopharmacology* **2014**, *39*, 435–444. [CrossRef] [PubMed]

27. Besheer, J.; Grondin, J.J.M.; Cannady, R.; Sharko, A.C.; Faccidomo, S.; Hodge, C.W. Metabotropic Glutamate Receptor 5 Activity in the Nucleus Accumbens Is Required for the Maintenance of Ethanol Self-Administration in a Rat Genetic Model of High Alcohol Intake. *Biol. Psychiatry* **2010**, *67*, 812–822. [CrossRef] [PubMed]
28. Sinclair, C.M.; Cleva, R.M.; Hood, L.E.; Olive, M.F.; Gass, J.T. mGluR5 receptors in the basolateral amygdala and nucleus accumbens regulate cue-induced reinstatement of ethanol-seeking behavior. *Pharmacol. Biochem. Behav.* **2012**, *101*, 329–335. [CrossRef] [PubMed]
29. Márquez-Rosado, L.; Solan, J.L.; Dunn, C.A.; Norris, R.P.; Lampe, P.D. Connexin43 phosphorylation in brain, cardiac, endothelial and epithelial tissues. *Biochim. Biophys. Acta* **2012**, *1818*, 1985–1992. [CrossRef] [PubMed]
30. Bao, X.; Reuss, L.; Altenberg, G.A. Regulation of purified and reconstituted connexin 43 hemichannels by protein kinase C-mediated phosphorylation of Serine 368. *J. Biol. Chem.* **2004**, *279*, 20058–20066. [CrossRef] [PubMed]
31. Solan, J.L.; Lampe, P.D. Connexin43 phosphorylation: Structural changes and biological effects. *Biochem. J.* **2009**, *419*, 261–272. [CrossRef] [PubMed]
32. Wang, Q.; Limbird, L.E. Regulation of alpha2AR trafficking and signaling by interacting proteins. *Biochem. Pharmacol.* **2007**, *73*, 1135–1145. [CrossRef] [PubMed]
33. Petrakis, I.L.; Ralevski, E.; Desai, N.; Trevisan, L.; Gueorguieva, R.; Rounsaville, B.; Krystal, J.H. Noradrenergic vs. Serotonergic Antidepressant with or without Naltrexone for Veterans with PTSD and Comorbid Alcohol Dependence. *Neuropsychopharmacology* **2012**, *37*, 996–1004. [CrossRef] [PubMed]
34. Haass-Koffler, C.L.; Swift, R.M.; Leggio, L. Noradrenergic targets for the treatment of alcohol use disorder. *Psychopharmacology* **2018**, *235*, 1625–1634. [CrossRef] [PubMed]
35. Rasmussen, D.D.; Alexander, L.; Malone, J.; Federoff, D.; Froehlich, J.C. The α2-adrenergic receptor agonist, clonidine, reduces alcohol drinking in alcohol-preferring (P) rats. *Alcohol* **2014**, *48*, 543–549. [CrossRef] [PubMed]
36. Roeckner, A.R.; Bowling, A.; Butler, T.R. Chronic social instability increases anxiety-like behavior and ethanol preference in male Long Evans rats. *Physiol. Behav.* **2017**, *173*, 179–187. [CrossRef] [PubMed]
37. Crabbe, J.C.; Wahlsten, D.; Dudek, B.C. Genetics of mouse behavior: Interactions with laboratory environment. *Science* **1999**, *284*, 1670–1672. [CrossRef] [PubMed]
38. Wahlsten, D.; Metten, P.; Phillips, T.J.; Boehm, S.L.; Burkhart-Kasch, S.; Dorow, J.; Doerksen, S.; Downing, C.; Fogarty, J.; Rodd-Henricks, K.; et al. Different data from different labs: Lessons from studies of gene-environment interaction. *J. Neurobiol.* **2003**, *54*, 283–311. [CrossRef] [PubMed]
39. Nicolaus, M.L.; Bergdall, V.K.; Davis, I.C.; Hickman-Davis, J.M. Effect of Ventilated Caging on Water Intake and Loss in 4 Strains of Laboratory Mice. *J. Am. Assoc. Lab. Anim. Sci.* **2016**, *55*, 525–533. [PubMed]
40. Shan, L.; Schipper, P.; Nonkes, L.J.P.; Homberg, J.R. Impaired fear extinction as displayed by serotonin transporter knockout rats housed in open cages is disrupted by IVC cage housing. *PLoS ONE* **2014**, *9*, e91472. [CrossRef] [PubMed]
41. Verplaetse, T.L.; Rasmussen, D.D.; Froehlich, J.C.; Czachowski, C.L. Effects of Prazosin, an α1-Adrenergic Receptor Antagonist, on the Seeking and Intake of Alcohol and Sucrose in Alcohol-Preferring (P) Rats. *Alcohol. Clin. Exp. Res.* **2012**, *36*, 881–886. [CrossRef] [PubMed]
42. Gilpin, N.W.; Koob, G.F. Effects of β-adrenoceptor antagonists on alcohol drinking by alcohol-dependent rats. *Psychopharmacology* **2010**, *212*, 431–439. [CrossRef] [PubMed]
43. Sofuoglu, M.; Rosenheck, R.; Petrakis, I. Pharmacological treatment of comorbid PTSD and substance use disorder: Recent progress. *Addict. Behav.* **2014**, *39*, 428–433. [CrossRef] [PubMed]
44. Gilpin, N.W.; Weiner, J.L. Neurobiology of comorbid post-traumatic stress disorder and alcohol-use disorder. *Genes Brain Behav.* **2017**, *16*, 15–43. [CrossRef] [PubMed]
45. Luo, X.-M.; Yuan, S.-N.; Guan, X.-T.; Xie, X.; Shao, F.; Wang, W.-W. Juvenile stress affects anxiety-like behavior and limbic monoamines in adult rats. *Physiol. Behav.* **2014**, *135*, 7–16. [CrossRef] [PubMed]
46. Comasco, E.; Todkar, A.; Granholm, L.; Nilsson, K.; Nylander, I. α2a-Adrenoceptor Gene Expression and Early Life Stress-Mediated Propensity to Alcohol Drinking in Outbred Rats. *Int. J. Environ. Res. Public Health* **2015**, *12*, 7154–7171. [CrossRef] [PubMed]
47. Karkhanis, A.N.; Alexander, N.J.; McCool, B.A.; Weiner, J.L.; Jones, S.R. Chronic social isolation during adolescence augments catecholamine response to acute ethanol in the basolateral amygdala. *Synapse* **2015**, *69*, 385–395. [CrossRef] [PubMed]

48. Franklin, J.M.; Carrasco, G.A. Cocaine Potentiates Multiple 5-HT2A Receptor Signaling Pathways and Is Associated with Decreased Phosphorylation of 5-HT2A Receptors In Vivo. *J. Mol. Neurosci.* **2015**, *55*, 770–777. [CrossRef] [PubMed]
49. Nimitvilai, S.; McElvain, M.A.; Brodie, M.S. Reversal of Dopamine D2 Agonist-Induced Inhibition of Ventral Tegmental Area Neurons by Gq-Linked Neurotransmitters Is Dependent on Protein Kinase C, G Protein-Coupled Receptor Kinase, and Dynamin. *J. Pharmacol. Exp. Ther.* **2013**, *344*, 253–263. [CrossRef] [PubMed]
50. Verma, V.; Hasbi, A.; O'Dowd, B.F.; George, S.R. Dopamine D1-D2 Receptor Heteromer-mediated Calcium Release Is Desensitized by D1 Receptor Occupancy with or without Signal Activation. *J. Biol. Chem.* **2010**, *285*, 35092–35103. [CrossRef] [PubMed]
51. Álvaro-Bartolomé, M.; García-Sevilla, J.A. Dysregulation of cannabinoid CB1 receptor and associated signaling networks in brains of cocaine addicts and cocaine-treated rodents. *Neuroscience* **2013**, *247*, 294–308. [CrossRef] [PubMed]
52. Zhang, J.; Ferguson, S.S.; Barak, L.S.; Bodduluri, S.R.; Laporte, S.A.; Law, P.Y.; Caron, M.G. Role for G protein-coupled receptor kinase in agonist-specific regulation of mu-opioid receptor responsiveness. *Proc. Natl. Acad. Sci. USA* **1998**, *95*, 7157–7162. [CrossRef] [PubMed]

proteomes

MDPI

Article

Evaluation of the Phosphoproteome of Mouse Alpha 4/Beta 2-Containing Nicotinic Acetylcholine Receptors In Vitro and In Vivo

Megan B. Miller [1,†], Rashaun S. Wilson [2,†], TuKiet T. Lam [2,3,4], Angus C. Nairn [1,2] and Marina R. Picciotto [1,*]

1 Department of Psychiatry, Yale University School of Medicine, 34 Park Street, 3rd Floor Research, New Haven, CT 06508, USA; miller.meganb@gmail.com (M.B.M.); Angus.Nairn@yale.edu (A.C.N.)
2 Yale/NIDA Neuroproteomics Center, 300 George Street, New Haven, CT 06509, USA; rashaun.wilson@yale.edu (R.S.W.); TuKiet.Lam@yale.edu (T.T.L.)
3 W.M. Keck Biotechnology Resource Laboratory, Yale University School of Medicine, 300 George Street, New Haven, CT 06509, USA
4 Department of Molecular Biophysics and Biochemistry, Yale University, 266 Whitney Avenue, New Haven, CT 06520, USA
* Correspondence: marina.picciotto@yale.edu; Tel.: +1-203-737-2041
† These authors contributed equally to this work.

Academic Editors: Angus C. Nairn and Kenneth R. Williams
Received: 12 September 2018; Accepted: 11 October 2018; Published: 15 October 2018

Abstract: Activation of nicotinic acetylcholine receptors containing α4 and β2 subunits (α4/β2* nAChRs) in the mammalian brain is necessary for nicotine reinforcement and addiction. We previously identified interactions between α4/β2* nAChRs and calcium/calmodulin-dependent protein kinase II (CaMKII) in mouse and human brain tissue. Following co-expression of α4/β2 nAChR subunits with CaMKII in HEK cells, mass spectrometry identified 8 phosphorylation sites in the α4 subunit. One of these sites and an additional site were identified when isolated α4/β2* nAChRs were dephosphorylated and subsequently incubated with CaMKII in vitro, while 3 phosphorylation sites were identified following incubation with protein kinase A (PKA) in vitro. We then isolated native α4/β2* nAChRs from mouse brain following acute or chronic exposure to nicotine. Two CaMKII sites identified in HEK cells were phosphorylated, and 1 PKA site was dephosphorylated following acute nicotine administration in vivo, whereas phosphorylation of the PKA site was increased back to baseline levels following repeated nicotine exposure. Significant changes in β2 nAChR subunit phosphorylation were not observed under these conditions, but 2 novel sites were identified on this subunit, 1 in HEK cells and 1 in vitro. These experiments identified putative CaMKII and PKA sites on α4/β2* nAChRs and novel nicotine-induced phosphorylation sites in mouse brain that can be explored for their consequences on receptor function.

Keywords: nicotinic receptor; CaMKII; PKA; quantitative phosphoproteomics; mouse; phosphorylation; nicotine

1. Introduction

High-affinity nicotinic acetylcholine receptors containing the α4 and β2 subunits (α4/β2* nAChRs, where * denotes other, potentially unidentified, subunits) are essential for the rewarding and reinforcing properties of nicotine in the mouse [1–4]. α4/β2* nAChRs are intrinsic ion channel-containing proteins that flux positive ions, including calcium, in response to nicotine or the endogenous neurotransmitter acetylcholine. Activation of α4/β2* nAChRs depolarizes neurons on which they are expressed, leading to changes in intracellular signaling, such as activation of calcium-dependent kinases [5].

In addition to initiating calcium signaling, nicotine also increases the number of nAChRs and can alter the associated proteome of α4/β2* nAChRs in mouse and human brain [6]. Biochemical studies have identified a number of interacting proteins that regulate assembly, trafficking, and function of α4/β2* nAChRs. For example, the chaperone 14-3-3 has been identified as an α4 nAChR subunit interactor in multiple studies [6–8], and this interaction can alter the physiological properties of α4/β2* nAChRs [7,9,10]. Interestingly, the association between 14-3-3 and the α4 nAChR subunit is regulated by protein kinase A (PKA), and is critical for regulating the desensitization kinetics of the receptor [7,10–12]. Other kinases can also regulate nAChR function. For example, phosphorylation of the α4 nAChR subunit by the calcium-dependent protein kinase PKC, as well as dephosphorylation by the calcium-dependent phosphatase calcineurin, also regulate desensitization of α4/β2* nAChRs in response to prolonged nicotine exposure [13–15]. Thus, biochemical studies have established an important role for nAChR phosphorylation in regulation of nicotine signaling through its receptors.

Several studies have now evaluated the α4/β2* nAChR-associated proteome, and these studies have identified several proteins that co-immunoprecipitate with the receptor from mouse and human brain [6,8,16]. Interestingly, a quantitative interaction between α4/β2* nAChRs and multiple isoforms of calcium/calmodulin-dependent protein kinase II (CaMKII) has been identified in mouse and human brain tissue [6,8]. This is of particular interest because activation of nAChRs by nicotine could activate associated CaMKII directly, leading to phosphorylation of nAChR subunits or of downstream targets. In mice, acute nicotine exposure activates CaMKII in the spinal cord [17] and brain [18], whereas chronic exposure increases CaMKII activity in the nucleus accumbens [19], all of which require α4/β2* nAChRs. CaMKII is also required for development of anxiety-like behaviors during nicotine withdrawal [20]. Taken together, these studies suggest that nAChR-mediated activation of CaMKII is important for at least a subset of the behavioral effects of nicotine related to addiction, and the direct interaction between α4/β2* nAChRs and CaMKII isoforms provides the rationale for determining whether nAChR subunits are substrates for phosphorylation.

In the current set of experiments, we used mass spectrometry to identify the residues phosphorylated on the α4 and β2 nAChR subunits when co-expressed with CaMKIIα in HEK cells, when dephosphorylated and subjected to phosphorylation in vitro with CaMKIIα or PKA, and when isolated from mouse brain at baseline, or following exposure to acute or repeated nicotine in vivo. These studies were designed to determine whether CaMKII can phosphorylate the α4 and β2 nAChR subunits in cells that do not normally express these proteins, and whether sites identified in the cellular assay were recapitulated when purified nAChRs were incubated with CaMKIIα or a kinase that is endogenously expressed in HEK cells (PKA) in vitro. The in vitro study also allowed us to determine whether previously identified PKA sites [7,10–12] could be identified in our studies. Finally, we provide the first evidence of nAChR phosphorylation in mouse brain, at baseline and following nicotine exposure in vivo.

2. Materials and Methods

2.1. Animals

Adult male C3H mice (approval number: 2016-07895) were obtained from Jackson laboratories and housed in groups of no more than 5 individuals per cage, maintained on a 12:12 h light/dark cycle, and given ad libitum access to food and water. All procedures involving animals were approved by the Yale University Institutional Animal Care and Use Committee and conformed to the standards for animal care and use set by the National Institutes of Health.

2.2. Cell Culture

HEK-293 (HEK) cells (ATCC) were grown and maintained in a humidified incubator at 37 °C and 5% CO_2. DMEM cell culture medium supplemented with 10% fetal bovine serum (FBS, Gibco) and antibiotic/antimycotic (Gibco), according to established ATCC protocols. Prior to transient transfection,

HEK cells were split and plated at medium-high density on plastic 10 cm dishes which were pre-treated with 0.05 mg/mL poly-D-lysine (PDL) in water.

2.3. Cell Transfection

Transient expression of nAChR subunits, mRuby, and CaMKII-mRuby were performed in serum-free medium (SFM) using Lipofectamine 2000 (Invitrogen, Carlsbad, CA, USA) at 2.5 μL per μg DNA. For cell-based phosphorylation assays, the following combinations of plasmids were used, each in triplicate: (1) untagged α4-nAChR + β2-nAChR-YFP + mRuby, (2) untagged α4-nAChR + β2-nAChR-YFP + CaMKII-mRuby, (3) α4-nAChR-YFP + Untagged β2-nAChR + mRuby, and (4) α4-nAChR-YFP + Untagged β2-nAChR + CaMKII. For in vitro phosphorylation assays, HEK cells were transfected with either untagged α4-nAChR + β2-nAChR-YFP or with α4-nAChR-YFP + Untagged β2-nAChR. For all experiments, transfection suspensions were prepared by combining DNA and Lipofectamine in a small volume of SFM and incubating at room temperature for 30 min. Transfection suspensions were added to cells with an additional volume of SFM, and returned to the incubator for 24 h prior to harvesting.

2.4. Plasmids

All nAChR plasmids were generous gifts from Henry Lester [21], and can be procured from Addgene (Cambridge, MA, USA): nAChR alpha4 WT (Addgene plasmid #24271), nAChR alpha4-YFP (Addgene plasmid #15245), nAChR beta2 WT (Addgene plasmid #24272), and nAChR Beta2-YFP (Addgene plasmid #15107). CaMKIIα-mRuby2 was created from full length Camui-CR (a gift from Michael Lin; Addgene plasmid #40256; [22]) using the *Nhe*1/*Apa*1 cloning site. mRuby2-C1 was purchased from Addgene (plasmid #54768; [22]).

2.5. HEK Cell Harvest and Protein Extraction

24 h after transfection, HEK transients were harvested by scraping cells into ice-cold membrane extraction buffer (MEB; 50 mM Tris-HCl (pH 7), 120 mM NaCl, 5 mM KCl, 1 mM $MgCl_2$) containing 2% Triton X-100, protease inhibitors (PMSF and Sigma protease inhibitor cocktail), and phosphatase inhibitors (5 mM sodium fluoride, 0.1 mM sodium orthovanadate, and Sigma phosphatase inhibitor cocktails 2 and 3). For in vitro phosphorylation assays, only the baseline samples were collected in the presence of phosphatase inhibitors; separate samples for the lambda phosphatase, PKA, and CaMKIIα conditions were collected in ice-cold MEB without phosphatase inhibitors. Harvested cells underwent 2 rounds of sonication/vortex cycles on ice, and were then allowed to incubate on ice for ~30 min to facilitate solubilization. Insoluble material was removed by brief centrifugation, and the resulting supernatants were used for further experiments.

2.6. In Vivo Nicotine Treatment

The nicotine treatment paradigm used was essentially as we have used previously [23,24]. Adult C3H mice were randomly assigned to each of three groups containing 5 animals each: Control, Acute, and Chronic. Animals were given ad libitum access to food and water containing either 200 μg/mL nicotine hydrogen tartrate (calculated as free base) in 2% (*w*/*v*) saccharin (chronic condition only) or 2% saccharin with molar-matched tartaric acid (Sigma-Aldrich, St. Louis, MO, USA; Acute and Control conditions) for 14 days. Animals were housed 3–4 to a cage, and pairings were set up at least 5 days prior to introduction of experimental water to allow for acclimation. Water was stored in darkened bottles to protect from light.

At the end of two weeks, animals were treated with a single, subcutaneous dose of either nicotine (0.5 mg/kg, acute condition), or saline (chronic and control conditions). Experimental drinking water was removed from cages 1 h prior to dosing, and animals were sacrificed by cervical dislocation 15 min after dosing. Whole brains were immediately removed on ice, then flash-frozen and stored at −80 °C until processing.

2.7. Brain Tissue Processing for Immunoprecipitation

On the day of tissue preparation, frozen brains were thawed on ice and homogenized in 10 volumes of ice-cold tissue homogenization buffer (10 mM HEPES [pH 7.4], 320 mM sucrose, 2 mM EDTA) containing protease inhibitors (PMSF and Sigma protease inhibitor cocktail) and a panel of phosphatase inhibitors (5 mM sodium fluoride, 0.1 mM sodium orthovanadate, and Sigma phosphatase inhibitor cocktails 2 and 3). Lysates were subjected to two rounds of sonication and vortexing, incubated on ice for 30 min, then spun at 1000× *g* for 10 min. Pellets were discarded, and "S1" supernatants were transferred to clean ultracentrifuge tubes and spun for 1 h at 100,000× *g* and 4 °C using a Beckman 70.1 Ti rotor. The resulting supernatants were removed and the pellets ("P2") were resuspended in 0.3× their initial volume in membrane extraction buffer (MEB; 50 mM Tris-HCl (pH 7), 120 mM NaCl, 5 mM KCl, 1 mM $MgCl_2$) containing 2% Triton X-100, protease inhibitors (PMSF and Sigma protease inhibitor cocktail), and phosphatase inhibitors (5 mM sodium fluoride, 0.1 mM sodium orthovanadate, and Sigma phosphatase inhibitor cocktails 2 and 3). Samples were vortexed thoroughly, and allowed to incubate on ice for ~2 h to facilitate solubilization. P2 homogenates were transferred to microfuge tubes and spun at 1000× *g* and 4 °C in for 10 min. The supernatant (S3; Triton-soluble crude membrane fraction) was used for immunoprecipitation of nAChRs.

2.8. Immunoprecipitation

Immunoprecipitation (IP) of HEK transients was conducted using a magnetic GFP-nAb resin from Allele, essentially according to manufacturer protocols. Briefly, GFP-nAb resin was washed 3× with binding buffer (10 mM Tris-HCl, 150 mM NaCl, pH 7.5), then resuspended and aliquot into microfuge tubes (20 µL resin/IP). While preparing resin, HEK cell lysates were thawed on ice and diluted in binding buffer such that the detergent concentration was not higher than 0.6% (*v*/*v*). Wash buffer was completely removed from resin using a magnetic stand, and diluted HEK lysates were added to washed resin. IP was conducted by tumbling overnight at 4 °C.

The following day, bound material was separated from the unbound supernatant by placing resin on a magnetic stand for at least 2 min. Supernatants were removed and resin was washed 1× in binding buffer and 2× in wash buffer (high salt Tris Buffered Saline). Bound fractions of HEK lysates from co-expression experiments (co-expression of nAChRs with CaMKII or mRuby in HEKs) and baseline samples from in vitro phosphorylation experiments were eluted by boiling resin in in 60 µL 1× Laemmli sample buffer (Bio-Rad, Hercules, CA, USA) with SDS (prepared in ultrapure water for MS). Bound fractions of HEK lysates, intended for in vitro phosphorylation experiments, were treated according to the dephosphorylation/phosphorylation protocol(s) detailed below, prior to eluting.

Immunoprecipitation of α4/β2-containing nAChRs from mouse brain lysates was done using M-270 Epoxy magnetic Dynabeads (Invitrogen, Carlsbad, CA, USA) linked to purified nAChR antiserum. Approximately 600 µg of rat-anti-α4 (mAb299; Lindstrom) was combined with 175 µg of rat-anti-beta2 (mAb270; Lindstrom) and linked to surface-activated M-270 Epoxy Dynabeads (Invitrogen; 5 mg Dynabeads /100 µg mAb) according to the manufacturer's instructions. mAb299 and mAb270 antisera were a generous gift from Jon Lindstrom, and were characterized previously [25]. Antibody linking was conducted overnight (~23 h) at room temperature with gentle agitation. The following day, linked resin was separated on a magnetic stand, rinsed with Phosphate Buffered Saline (PBS) + 0.1% BSA (*w*/*v*), then resuspended in PBS and stored at 4 °C until use. On the day of IP, linked resin was rinsed once in PBS and then equally divided into each of 15 clearly labeled microfuge tubes (5 samples × 3 treatment groups). Processed whole brain samples (S3 fraction) were added to the prepared resin, and IP was conducted by tumbling overnight at 4 °C. The next day, beads and bound proteins were separated on a magnetic stand and rinsed 3× with PBS and 1× with PBS containing 0.1% BSA (*w*/*v*) and 0.1% Triton X-100 (*v*/*v*). The final wash was then removed, and the bound fractions were eluted by boiling resin in 50 µL 1× Laemmli sample buffer (Bio-Rad) with SDS. Once cooled, the eluate was immediately removed from the resin and stored at −20 °C until gels were run.

2.9. In Vitro Phosphorylation

Transfected HEK cells for in vitro phosphorylation experiments were separated into four groups: baseline, lambda phosphatase, PKA, and CaMKII. Baseline nAChR transients were collected in the presence of phosphatase inhibitors, and eluted immediately following IP, as described above. The remaining three groups were harvested in the absence of phosphatase inhibitors, and immunoprecipitated receptors were subject to in vitro dephosphorylation with purified lambda phosphatase (New England Biolabs, Ipswich, MA, USA; all but baseline group received this treatment).

2.9.1. Dephosphorylation with Lambda Phosphatase

Immunoprecipitated receptors were dephosphorylated in vitro using recombinant lambda phosphatase (New England Biolabs), according to manufacturer instructions. Briefly, immunoprecipitated receptors (still bound to nAb-GFP resin) were rinsed once with 200 µL PMP phosphatase buffer (manufacturer supplied) to equilibrate. The following phosphatase mixture was then added to each tube: 100 µL PMP buffer, 10 µL $MnCl_2$ (manufacturer supplied), 1 µL each of aprotinin and PMSF, and 2 µL of lambda phosphatase. Dephosphorylation continued for 1 h at 30 °C with gentle agitation. After incubation, resins were separated from dephosphorylation mixture by placing on a magnetic stand, washed with PMP, and dephosphorylation was conducted a second time. Following the second treatment, 50 µL of 500 mM EDTA was added to each sample and suspensions incubated on ice for 5 min to stop phosphatase activity. Supernatants were then removed, and resin was washed 2× in wash buffer. Bound receptors in the lambda phosphatase group were eluted here in 40 µL 1× Laemli Sample Buffer (LSB) prepared in ultrapure water. Remaining samples continued on to rephosphorylation with PKA or CaMKIIα.

2.9.2. Phosphorylation with PKA

Following dephosphorylation, samples in the PKA group were phosphorylated with purified PKA (New England Biolabs) according to the manufacturer's instructions. Briefly, bound receptors were rinsed in 1× PK buffer (manufacturer supplied), and then incubated in the following PKA phosphorylation mixture: 100 µL 1× PK buffer, 1 µL of sigma protease inhibitor cocktail, 2 µL of 10 mM ATP stock (200 µM total), and 2 µL of PKA enzyme. Phosphorylation continued for 1–2 h at 30 °C with gentle agitation. Following incubation, supernatants were removed, and resin was washed 2× in wash buffer. Bound receptors were then eluted in 40 µL 1× LSB prepared in ultrapure water.

2.9.3. Phosphorylation with CaMKIIα

Following dephosphorylation, samples in the CaMKIIα group were phosphorylated with purified CaMKIIα (New England Biolabs) according to manufacturer instructions. CaMKIIα enzyme was activated by combining 2 µL of the purified kinase with 100 µL buffer, 200 µM ATP, 1.2 µL calmodulin, and 2 mM $CaCl_2$, all provided by the manufacturer. The activation solution incubated for 10 min at 30 °C. Meanwhile, bound receptors were washed 2× in supplied kinase buffer and then incubated with activated CaMKII phosphorylation mix (+A/PMSF) for 45 min at 30 °C with gentle agitation. After incubation, the kinase suspension was removed, and resins were washed 2× in wash buffer. Bound receptors were then eluted in 40 µL 1× LSB prepared in ultrapure water.

2.10. Protein Gels

All IP samples were separated on Bio-Rad Mini-Protean TGX precast gels (Bio-Rad, Hercules, CA, USA) using established protocols. Separated protein eluates were then stained using Simply Blue Safe Stain (Invitrogen). Samples were run alongside a protein molecular weight marker, and bands of the appropriate molecular weights were excised from the gel, transferred to microfuge tubes, and stored at −20 °C until processing for proteomics. Approximate molecular weights for the nAChR subunits and variants are as follows: α4-nAChR ~75 kDa (HEK transients and from mouse brain), β2-nAChR-YFP

~75 kDa, α4-nAChR-YFP ~100 kDa, and β2-nAChR ~55 kDa (HEK transients and from mouse brain). All processing was done in a clean environment and ultrapure water was used to prepare all buffers and reagents.

2.11. Protein Digestion for LC-MS/MS

Gel bands were first cut into small pieces and subjected to the following washes with agitation: 50% (*v*/*v*) acetonitrile (5 min), 50% (*v*/*v*) acetonitrile/10 mM NH_4HCO_3 (30 min). Gel pieces were dried with a speed vacuum, resuspended in 30 μL of 10 mM NH_4HCO_3/0.2 μg digestion grade trypsin (Promega), and incubated for 16 h at 37 °C. Peptides were acidified with 0.1% (*v*/*v*) trifluoroacetic acid (TFA) prior to mass spectrometry analysis.

2.12. Protein Identification by LC-MS/MS

Reverse phase liquid chromatography tandem mass spectrometry (RP-LC-MS/MS) was performed using a NanoACQUITY (Waters Corporation, Milford, MA) ultra-performance liquid chromatography (UPLC) coupled to a Q Exactive Plus Hybrid Quadrupole-Orbitrap (ThermoFisher Scientific, San Jose, CA, USA) mass spectrometer. Peptides were loaded onto a nanoACQUITY (Waters Corporation, Milford, MA, USA) UPLC Symmetry C18 trapping column (180 μm × 20 mm) at a flowrate of 5 μL/min prior to separation on a nanoACQUITY (Waters Corporation, Milford, MA, USA) Peptide BEH C18 column (75 μm × 250 mm). Mobile phase A and B compositions were 0.1% (*v*/*v*) formic acid in water and 0.1% (*v*/*v*) formic acid in acetonitrile, respectively. Peptides were eluted over 120 min with a mobile phase B gradient (6–20%) at a column temperature of 37 °C and a flow rate of 300 nL/min. Precursor mass scans (300 to 1500 m/z range, target value: 3×10^6, maximum ion injection times: 45 ms) were acquired and followed by HCD-based fragmentation (normalized collision energy: 28). A resolution of 70,000 at m/z 200 was used for MS1 scans, and up to 20 dynamically chosen, most abundant precursor ions were fragmented (isolation window: 1.7 m/z). The tandem MS/MS scans were acquired at a resolution of 17,500 at m/z 200 (target value: 1×10^5, maximum ion injection times: 100 ms). Mass spectrometry raw spectra were searched against the Mascot algorithm (Matrix Science, London, UK) using Proteome Discoverer software (v 2.2.0.388, ThermoFisher Scientific, San Jose, CA, USA). The search criteria were the following: precursor mass tolerance, 10.0 ppm; fragment mass tolerance, 0.020 Da; enzyme, trypsin; maximum missed cleavage sites, 2; variable modifications, carbamidomethyl (C), oxidation (M), phosphorylation (STY), propionamide (C).

2.13. Quantitative Data Analysis

Searched data was imported into Scaffold (v 4.8.7, Proteome Software Inc., Portland, OR, USA) software for validation of peptide and protein identifications. Peptide and protein identifications were accepted above a 95% and 99% probability threshold, respectively. Proteins containing less than two peptides per protein were filtered out, and proteins sharing redundant peptides were grouped. Peptide and protein probabilities were calculated by the Scaffold Local FDR algorithm and Protein Prophet algorithm [26], respectively. For label-free quantitative analysis, the Scaffold Q+ (v 4.8.7, Proteome Software Inc., Portland, OR, USA) function was used. Median-normalization of precursor ion intensities was performed across samples, which were then log-transformed and weighted by an adaptive intensity weighting algorithm. After removal of peptides not meeting the threshold criteria the following number of spectra were used for label-free quantitation: Experiment 1 (66%), 77,423 (quantitative)/117,463 (total); Experiment 2 (69%), 108,557 (quantitative)/158,253 (total); Experiment 3 (57%), 114,821 (quantitative)/200,002 (total). Analysis settings were specified for the following categories: Analysis type, Intensity-based; Experiment type, Between subjects (Independent Groups). Quantitation Preferences were selected as follows: (1) Minimum Value Preference; Use Minimum Absolute Intensity, false; Minimum Absolute Intensity, 0.0; Minimum Value: 0.01; (2) Condenser Preferences; Use Intensity Weighting, true; Use Standard Deviation Estimation, true; Use Non-Exclusive Peptides, true; (3) View Preference; View Type, Log_2 Ratio;

(4) Normalization Preference; Calculation Type, Median; Blocking Level, Unique Peptides; Use Inter Experiment Normalization, true; Use Intra Sample Normalization, false; Use Peptide Normalization, false; Use Protein Average As Reference, true; Use Iterative Normalization, true; Spectrum Quality Filter, no filter. For annotation of protein PostTranslational Modification (PTM) sites, Scaffold PTM (v 4.8.7, Proteome Software Inc., Portland, OR, USA), which integrated the MS/MS results exported from Scaffold/Scaffold Q+. For differential phosphorylation analysis, phosphorylated peptide spectral counts were first normalized to the total spectral counts for each protein. This value was then normalized to the total spectral counts for the entire sample.

2.14. Statistical Analysis

For all replicates, normalized quantitative values for the "treated" group were compared to those of the "control" group. Statistical analyses were performed in GraphPad Prism v7.01 (La Jolla, CA, USA) using two-tailed Student *t*-tests. The level for significance was set at $p < 0.05$.

3. Results

3.1. Co-Expression of nAChR Subunits and CaMKII in HEK Cells

Trafficking and activity of high-affinity nAChRs can be regulated by phosphorylation [7,14,15]. The intracellular loop between the 3rd and 4th transmembrane domain of the α4 subunit (M3/M4 loop) is the longest found among all nAChR subunits and has been identified as a locus for protein–protein interactions and phosphorylation (Figure 1). Further, activation of nAChRs can increase intracellular calcium levels in the cells on which they are expressed, and α4/β2 nAChRs are physically associated with several isoforms of the calcium-dependent kinase CaMKII in mouse and human brain [6,8]. In order to determine whether the α4 or β2 nAChR subunits can be phosphorylated by CaMKII, we co-expressed untagged and YFP-tagged nAChR subunits in HEK cells (2 independent replicates per condition), with or without CaMKIIα-mRuby, and used mass spectrometry to identify phosphorylation sites on the nAChR. To control for any effects of the fluorescent tag, parallel experiments were performed using α4-YFP with untagged β2 and untagged α4 with β2-YFP. Results did not differ depending on which subunit was fluorescently tagged, so data on phosphorylation sites were pooled between the two studies for statistical evaluation.

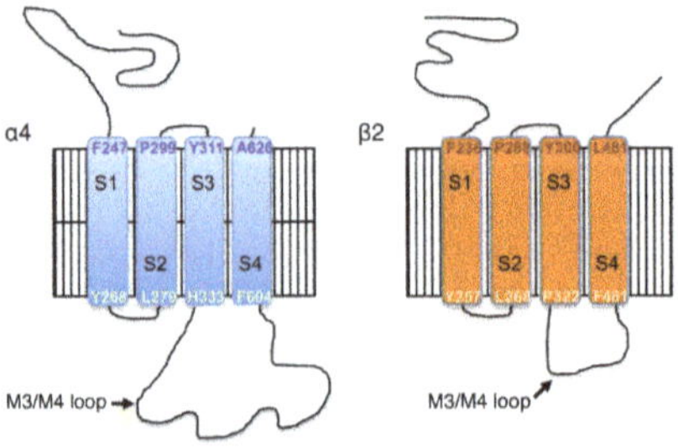

Figure 1. Amino acid structure of mouse α4/β2 nAChR subunits. Membrane topology of the mouse α4/β2 nAChR shows the boundaries of the intracellular domains. The intracellular M3/M4 loop of the α4 subunit is the longest of all the nAChR subtypes, and is the site of most identified protein–protein interactions [8,16].

Following transfection, nAChRs were immunoprecipitated and subunits were separated by gel electrophoresis. Gels were Coomassie-stained, bands of the appropriate size for the untagged and tagged α4 and β2 nAChR subunits were excised, and proteins were digested and subjected to mass spectrometry. We identified 8 serine residues on the α4 subunit with significantly increased phosphorylation following co-expression with CaMKIIα-mRuby, compared to those co-expressing mRuby alone (Figure 2, see Supplementary Materials for representative spectra). Phosphorylation of the β2 subunit on S445 was identified in the baseline condition in HEK cells, but there was no significant phosphorylation of this residue in the CaMKIIα-mRuby condition. The phosphorylated residues identified in the α4 subunit were all in the intracellular M3/M4 loop of the protein (see Figure 1) and include serine 444 (S444), S448, S468, S470, S530, S540, S543, and S563 (Figure 2, Table 1), as was the S445 phosphorylation site in the β2 subunit (Table 1).

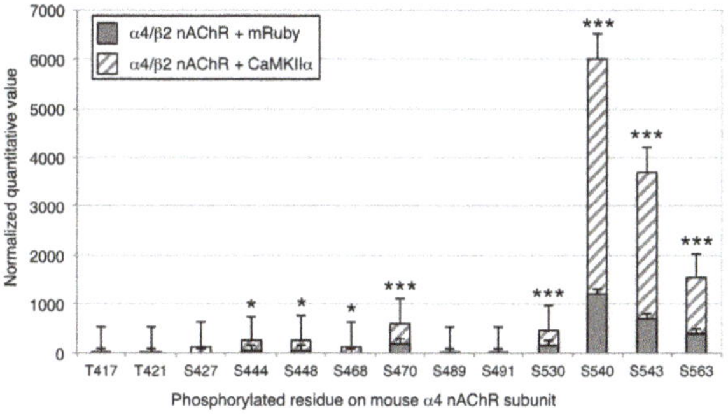

Figure 2. Co-expression of α4/β2 nAChRs and CaMKII in HEK cells. Phosphorylated residues on tagged α4 and β2 nAChR subunits co-expressed with mRuby or CaMKIIα -mRuby in HEK cells were identified by mass spectrometry following immunoprecipitation and separation by gel electrophoresis. Phosphorylation level was normalized to total subunit protein. Phosphorylation of 5 serine residues on the α4 subunit (S470, S530, S540, S543, S563) could be identified in HEK cells co-expressing mRuby, and 8 serine residues showed a significant increase in phosphorylation in HEK cells co-expressing CaMKIIα (S444, S448, S468, S470, S530, S540, S543, S563). No phosphorylation of the β2 subunit was detected. * $p < 0.05$; *** $p < 0.005$. Error bars represent standard error of the mean; $n = 6$/condition.

Table 1. Sequence of phosphorylation sites identified in the mouse α4 and β2 nAChR subunits.

Site	Sequence [1]	Observed Previously [2]	Conserved in Human	Predicted CaMKII Site [3]
T417	... RMDTAVE ...	No	No	Yes
S444	... EKASP ...	No	Yes (S441)	No
S448	... PSPG ...	No	Yes (S445)	No
S468	... KARSLSVQH ...	No	Yes (S464)	No
S470	... KARSLSVQH ...	Yes	Yes (S467)	Yes
S491	... RSRSIQ ...	Yes	Yes (S488)	Yes
S521	... TRPSQLP ...	No	No	No
S530	... DQTSPC ...	Yes	Yes (S527)	No
S540	... KEPSPVSP ...	Yes	Yes (S538)	Yes
S543	... KEPSPVSP ...	Yes	Yes (S541)	No
S563	... LPLSPAL ...	Yes	Yes (S561)	No

[1] Phosphorylated residue is underlined. [2] Sites of phosphorylation in the α4 nAChR subunit identified in **7, 11, 12, 28–30**. [3] CaMKII sites predicted using the Phyre2 site: http://www.sbg.bio.ic.ac.uk/phyre2/.

The coverage of the α4 subunit intracellular M3/M4 loop was ~80%, suggesting that the majority of physiologically relevant sites of phosphorylation were likely identified, and only 6 serine or threonine residues in the intracellular loop were uncovered (Figure 3). However, overall coverage of the α4 subunit was 61% and of the β2 subunit was 43%, so additional sites of phosphorylation could be present, but not identified in this experiment.

Experiment 1 Coverage: (218/272) x 100=80.1%

**HHRSPRTHTMPAWVRRVFLDIVPRLLFMKRPSVVKDNCRRLIESMHKMANAPRFWPEPESEPGILG
DICNQGLSPAPTFCNRMDTAVETQPTCRSPSHKVPDLKTSEVEKASPCPSPGSCHPPNSSGAPVLIKAR
SLSVQHVPSSQEAAEGSIRCRSRSIQYCVSQDGAASLTESKPTGSPASLKTRPSQLPVSDQTSPCKCTCK
EPSPVSPITVLKAGGTKAPPQHLPLSPALTRAVEGVQYIADHLKAEDTDFSVKEDWKYVAMVIDRIF**

Experiment 2 Coverage: (230/272) x 100=84.6%

**HHRSPRTHTMPAWVRRVFLDIVPRLLFMKRPSVVKDNCRRLIESMHKMANAPRFWPEPESEPGILG
DICNQGLSPAPTFCNRMDTAVETQPTCRSPSHKVPDLKTSEVEKASPCPSPGSCHPPNSSGAPVLIKAR
SLSVQHVPSSQEAAEGSIRCRSRSIQYCVSQDGAASLTESKPTGSPASLKTRPSQLPVSDQTSPCKCTCK
EPSPVSPITVLKAGGTKAPPQHLPLSPALTRAVEGVQYIADHLKAEDTDFSVKEDWKYVAMVIDRIF**

Experiment 3 Coverage: (203/272) x 100=74.6%

**HHRSPRTHTMPAWVRRVFLDIVPRLLFMKRPSVVKDNCRRLIESMHKMANAPRFWPEPESEPGILG
DICNQGLSPAPTFCNRMDTAVETQPTCRSPSHKVPDLKTSEVEKASPCPSPGSCHPPNSSGAPVLIKAR
SLSVQHVPSSQEAAEGSIRCRSRSIQYCVSQDGAASLTESKPTGSPASLKTRPSQLPVSDQTSPCKCTCK
EPSPVSPITVLKAGGTKAPPQHLPLSPALTRAVEGVQYIADHLKAEDTDFSVKEDWKYVAMVIDRIF**

Identified amino acids in the in each experiment in the a4 nAChR subunit M3/M4 intracellular loop

Identified phosphorylation sites

Potentially missed phosphosite IDs due to lack of coverage

Figure 3. Coverage of the α4 nAChR subunit M3/M4 intracellular loop across experiments. Coverage of the large intracellular loop of the α4 subunit is diagrammed for each experiment for direct comparison. Yellow: identified amino acids; Green: identified phosphorylation sites; Red: serine or threonine residue that was not covered and might represent a missed phosphorylation site.

3.2. CaMKII and PKA Can Phosphorylate nAChRs In Vitro

Since multiple kinases are found in all cell types, phosphorylation of the nAChR subunits may have resulted directly through CaMKII activity, or indirectly through activation of other kinases expressed in HEK cells. We, therefore, performed an in vitro phosphorylation experiment using α4/β2 nAChRs immunoprecipitated from HEK cells following dephosphorylation using lambda phosphatase. Following immunoprecipitation and dephosphorylation, tagged or untagged nAChRs were incubated with CaMKIIα in the presence of calcium and calmodulin or PKA. nAChRs were then separated by gel electrophoresis, and bands were excised for evaluation by mass spectrometry. As above, parallel experiments were performed using α4-YFP with untagged β2 and untagged α4 with β2-YFP, and no differences were found, so data for tagged and untagged subunits were pooled. At baseline, four highly phosphorylated serine residues were identified on the α4 subunit, S470, S530, S540, S543 (Figure 4, see Supplementary Materials for representative spectra). Except for S470, phosphorylation of these sites was decreased or nearly eliminated following incubation with lambda phosphatase (Figure 4a). Incubation with CaMKIIα resulted in phosphorylation of T417 and S468 above the phosphatase condition, whereas phosphorylation of S470 and S540 were detected, but were not higher than the phosphatase condition (Figure 4b, Table 1). Incubation with PKA resulted in phosphorylation of S470, S491, and S521 above the phosphatase condition (Figure 4c, Table 1). Of these sites, S468, S470, and S540 were detected when nAChR subunits and CaMKII were co-transfected into HEK cells as described above (Figure 2). The coverage of the α4 subunit intracellular M3/M4 loop was ~85%, suggesting that the majority of physiologically relevant phosphorylation sites were likely identified, but 6 serine or threonine residues in the intracellular loop were uncovered in this experiment (Figure 3). No significant changes in phosphorylation were identified on the β2 subunit, although phosphorylation of T375 was detected in both the CaMKII and PKA conditions, and not at baseline (Table 1). Coverage of the α4 and β2 subunits was 55% and 38%, respectively, so additional sites of phosphorylation could be present, but not identified in this experiment. These findings identify distinct phosphorylation sites on the α4 subunit for PKA and CaMKIIα, as well as sites that may be phosphorylated at baseline by these or other kinases.

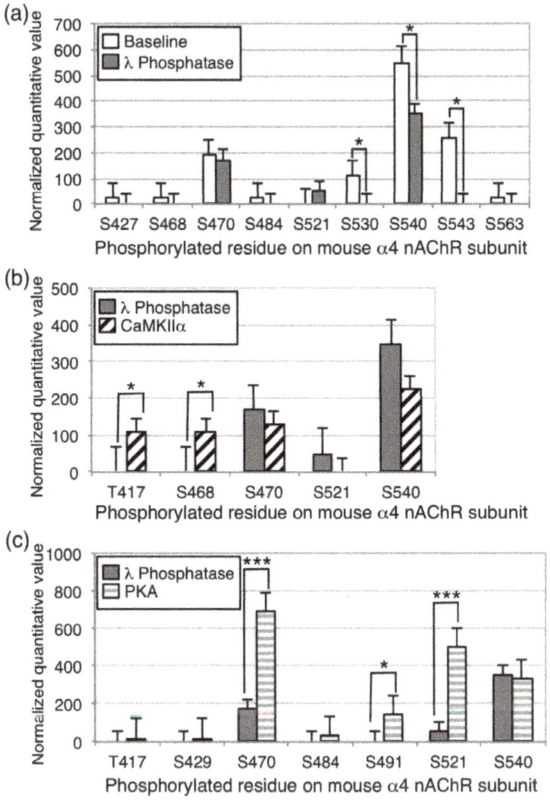

Figure 4. In vitro phosphorylation of α4/β2 nAChRs by CaMKII or PKA. The α4 and β2 nAChR subunits were co-expressed in HEK cells, isolated by immunoprecipitation, and subjected to mass spectrometry. Phosphorylation level was normalized to total subunit protein. (**a**) At baseline, there was a high level of phosphorylation of S470, S530, and S540 on the α4 subunit, and incubation with lambda phosphatase dephosphorylated S540 and S543 to undetectable levels. (**b**) Incubation with CaMKIIα in the presence of calcium and calmodulin increased phosphorylation of T417 and S468 on the α4 subunit significantly. (**c**) Incubation with PKA in the presence of cyclic AMP increased phosphorylation of S470, S491, and S521 significantly. * $p < 0.05$; *** $p < 0.005$. Error bars represent standard error of the mean; $n = 6$/condition.

3.3. Phosphorylation of nAChRs In Vivo

We next determined whether the phosphorylation sites identified as CaMKII or PKA targets in vitro were also phosphorylated in vivo under conditions in which nAChRs could be activated by nicotine. Using monoclonal antibodies raised against the α4 subunit, we immunoprecipitated native α4/β2* nAChRs from mouse brain following saline administration, a single nicotine dose in a novel environment (0.5 mg/kg), or chronic nicotine in the drinking water, a regimen known to increase locomotor activity in a dopamine-dependent manner [23]. In mice that had been handled and placed in a novel environment following saline administration, we once again identified phosphorylation of S470, S491, and S543 in the α4 subunit (Figure 5, see Supplementary Materials for representative spectra). Acute nicotine administration resulted in a significant increase in phosphorylation of S444 and S448, whereas repeated nicotine exposure resulted in no significant differences over baseline, but detectable phosphorylation of S444, S448, S470, S491, S543, and S563 (Figure 5). Increases in S406 and S563 were observed, but did not reach significance. Coverage of the M3/M4 loop of the α4 subunit was ~75% in this experiment, but 10 serine or threonine residues in the intracellular loop were uncovered in this experiment, including S521 and S530 which were identified as a potential PKA site and in HEK cells, respectively (Figure 3). Interestingly, S470 in the α4 subunit was significantly phosphorylated following saline administration, and phosphorylation was reduced to undetectable levels following acute nicotine administration, then returned to baseline levels following repeated nicotine exposure. No sites of phosphorylation were identified on the β2 subunit. Coverage of the α4 subunit was 53%, and of the β2 subunit was 47%, so additional sites of phosphorylation could be present, but not identified in this experiment.

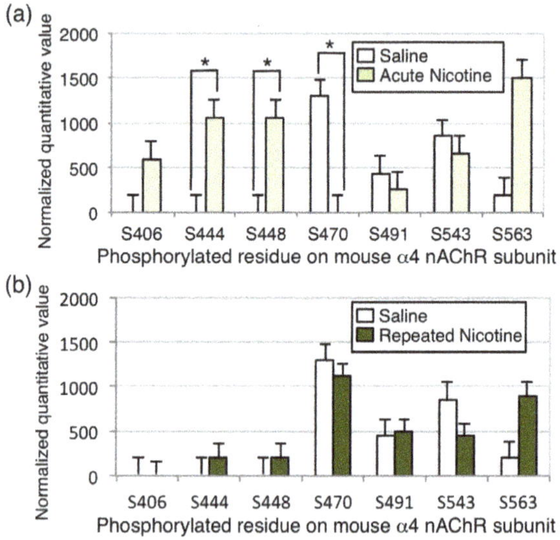

Figure 5. Phosphorylation of α4/β2 nAChRs in vivo following nicotine exposure. nAChRs were immunoprecipitated from mouse brain homogenates using a monoclonal antibody raised against the α4 subunit, isolated by gel electrophoresis, and bands corresponding to the α4 and β2 subunits were excised and subjected to mass spectrometry. Phosphorylation level was normalized to total subunit protein. Phosphorylation of S491, S543, and S563 on the α4 subunit was detected in brain homogenates from saline treated mice. (**a**) Following acute nicotine exposure in vivo, phosphorylation of S444 and S448 was significantly increased, whereas phosphorylation of S470 was significantly decreased to undetectable levels. (**b**) Following chronic exposure to nicotine, no significant differences from baseline phosphorylation were observed in the α4 subunit. No phosphorylation of the β2 subunit was detected. * $p < 0.05$. Error bars represent standard error of the mean; $n = 10$/condition.

4. Discussion

These experiments identify previously described and novel sites on the mouse α4 nAChR subunit, and the first report of specific residues on the β2 nAChR subunit, that can be phosphorylated in cells after heterologous expression with CaMKIIα, or by CaMKIIα or PKA in vitro (summarized in Figure 6). In addition, we report the first identification of in vivo nAChR phosphorylation at baseline and in response to nicotine exposure in mouse brain (Figure 5). Despite phosphorylation of the mouse β2 nAChR subunit on S445 in HEK cells and T375 in vitro by CaMKIIα or PKA, no phosphorylation in vivo at baseline or following nicotine exposure was detected, although coverage of the subunit was not complete, so additional sites may not have been revealed.Co-transfection of the α4 and β2 nAChR subunits with CaMKIIα on the α4 subunit in HEK cells induced significant phosphorylation of 8 sites on the mouse α4 subunit, all of which are conserved in the human α4 nAChR subunit and 3 of which (S444, S448 and S468) have not been reported previously. Of these, 2 sites (S470 and S540) match the minimal requirements for phosphorylation by CaMKII (RXXS/T, where R is arginine and T is threonine; [27]), however, incubation of isolated α4/β2 nAChR with CaMKIIα in vitro did not result in significant phosphorylation of these residues. Instead, in vitro CaMKIIα phosphorylated one site identified in the HEK cell experiment (S468) and a novel site not identified previously (T417). T417 conforms to the minimal consensus sequence for CaMKII phosphorylation, but S468 has a different basic residue (K (lysine) rather than R) at the −3 position.

HEK Cells		*In Vitro*			*In Vivo*		
Basal	CaMKII	PPase resistant	CaMKII	PKA	Saline	Acute Nic	Repeat Nic
			T417*				
	S444*					S444*	
	S448*					S448*	
	S468*		S468***				
S470	S470***	S470		S470***		*S470*	*S470*^{A}
				S491*	S491		
		S521		S521***			
S530	S530***						
S540	S540***	S540					
S543	S543***				S543		
S563	S563***				S563		

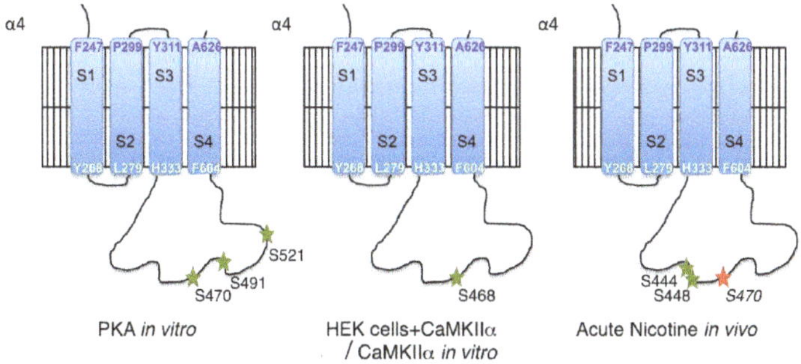

Figure 6. Summary of α4 nAChR subunit phosphorylation sites. Comparison of the sites detected in the α4 subunit across experiments and topological site of phosphorylation sites identified in vitro and in vivo. Green: sites with increased phosphorylation; Red/italics: site with decreased phosphorylation; A: no difference from baseline; * $p < 0.05$; *** $p < 0.005$.

The in vitro experiment identified two additional residues that were phosphorylated by PKA (S491 and S521) that have not been described as substrates for this kinase previously, along with highly significant phosphorylation of S470. A number of studies have identified S470 as an important site of phosphorylation on the α4 nAChR subunit [7,11,12,28–30]. These studies show that S470 can be phosphorylated by both PKA and PKC in vitro and after co-transfection in cultured cells. Our in vitro experiments confirm that S470 is a substrate for PKA, and suggest it is not a direct substrate for CaMKIIα, although activation of CaMKII appears to lead to increased phosphorylation of this residue in HEK cells, possibly through indirect activation of another kinase or decreased activity of a phosphatase. Phosphorylation of S470 on the α4 subunit by PKA is necessary for recruitment of the scaffolding protein 14-3-3, and this association increases stability of the α4/β2 nAChR and contributes to upregulation following nicotine exposure [7]. In addition, phosphorylation of the α4 subunit by PKC increases activity of the α4/β2 nAChR by enhancing recovery from desensitization following agonist exposure [28,31]. We observed baseline phosphorylation of S470 on the α4 nAChR subunit in cells and in mouse brain across experiments, however, acute nicotine exposure in vivo significantly decreased the phosphorylation of S470, whereas chronic exposure returned the phosphorylation state to baseline levels in the mouse brain. Thus, the decreased phosphorylation of S470 following acute nicotine exposure, observed here, is likely to result in decreased activity of the receptor, whereas the recovery to baseline following chronic exposure could be important for nAChR upregulation, which is observed in mouse brain following the chronic nicotine exposure regimen used here [23,32]. Note that, in all experiments, phosphorylation level was normalized to total subunit protein.

The decrease in phosphorylation of the S470 site on the mouse α4 subunit following acute nicotine treatment in vivo suggests that stimulation of nAChRs may result in activation of a protein phosphatase that dephosphorylates this residue. This observation is consistent with experiments showing that nicotine acting through nAChRs can activate the calcium-dependent phosphatase calcineurin in cultured cortical neurons [33]. In addition, activity of calcineurin is required for nicotine-induced locomotor sensitization in rats [34], suggesting that this decrease in nAChR phosphorylation could be behaviorally relevant.

The consistent association between α4/β2 nAChRs and several CaMKII isoforms in mouse and human brain prompted us to investigate whether these receptors were a substrate for phosphorylation by this kinase. Neither of the phosphorylation sites identified following in vitro phosphorylation of α4/β2 nAChRs with CaMKIIα were identified in vivo at baseline or following nicotine exposure, suggesting that the nAChR may not be a major substrate for CaMKII in the mouse brain at baseline or under that conditions of nicotine exposure tested here. However, S444, S448, and S563 were phosphorylated both when CaMKIIα was co-expressed with the α4 and β2 nAChR subunits in HEK cells and in mouse brain. The physical interaction between α4/β2 nAChRs and CaMKII could result in activation of the kinase and phosphorylation of other protein targets. Therefore, the increased phosphorylation of α4/β2 nAChRs, when co-transfected with CaMKIIα in HEK cells, could be the result of a protein kinase cascade that indirectly results in phosphorylation of the nAChR by PKA and other unidentified kinases. Alternatively, the association between α4/β2 nAChRs and CaMKII may be important for other cell biological functions, such as localization of the receptor to particular intracellular compartments or the plasma membrane. Interaction with scaffolding proteins, such as 14-3-3, contributes to trafficking of nAChRs in a PKA-dependent manner [7], and CaMKII can serve as a binding protein to target other proteins to specific intracellular membranous compartments, such as synaptic vesicles [35]. Thus, α4/β2 nAChRs may be regulated by association with CaMKII in mouse brain, even if they are not an efficient substrate for phosphorylation by the enzyme.

5. Conclusions

In summary, this phosphoproteomic study has identified novel phosphorylation sites on the mouse α4 nAChR and β2 subunits, and is the first instance of identification of a subset of α4 nAChR residues phosphorylated, in vivo, in mouse brain tissue. Further, acute nicotine exposure increases the phosphorylation of two residues on the α4 subunit (S444 and S448), but decreases phosphorylation of a very well-characterized residue (S470) that contributes to surface trafficking and resistance to desensitization of α4/β2 nAChRs in cultured neurons [7,10–12]. Taken together, these studies suggest that nAChR activity initiates intracellular signaling cascades that can alter receptor activity. Furthermore, although it is not yet clear whether α4/β2 nAChRs are a substrate for CaMKIIα, which interacts physically with the receptor, these results suggest that CaMKIIα can affect phosphorylation of the α4 subunit indirectly in cells. Future studies using purified enzymes in vitro, co-transfection studies in cells, or specific stimuli in vivo will be necessary to identify the kinases that phosphorylate these novel sites. The functional consequences of these phosphorylation events on receptor assembly, trafficking, and function should also be evaluated in cells and in vivo.

Supplementary Materials: The following are available online at http://www.mdpi.com/2227-7382/6/4/42/s1.

Author Contributions: Conceptualization, M.M., A.C.N. and M.R.P.; Data curation, M.M., R.W. and T.L.; Formal analysis, M.M. and R.W.; Funding acquisition, M.M., A.C.N. and M.R.P.; Investigation, M.M. and T.L.; Methodology, R.W., T.L. and A.C.N.; Project administration, A.C.N. and M.R.P.; Resources, A.C.N.; Supervision, A.C.N. and M.R.P.; Visualization, M.M., R.W. and M.R.P.; Writing—original draft, M.R.P.; Writing—review & editing, M.M., R.W., T.L., A.C.N. and M.R.P.

Funding: This research was funded by the Yale/NIDA Neuroproteomics Center at Yale University (DA018343), DA14241, MH77681 and the State of Connecticut, Department of Mental Health and Addiction Services. MBM was supported by the Basic Science Training Program (T32 MH014276) and received a Pilot Grant from DA018343. The Q Exactive Plus mass spectrometer used for mass spectral data collection was funded by NIH SIG from the Office of The Director, National Institutes of Health under Award Number (S10OD018034). The content is solely the responsibility of the authors and does not necessarily represent the official views of the National Institutes of Health.

Acknowledgments: The authors would like to thank Jon Lindstrom for the generous gift of antisera, Cali Calarco, Angela Lee, and Alan Lewis for help with, and discussion of, these studies. We also would like to thank Edward Voss and Jean Kanyo for assistance with mass spectrometry sample preparation and data collection.

Conflicts of Interest: The authors declare no conflict of interest.

References

1. Tapper, A.R.; McKinney, S.L.; Nashmi, R.; Schwarz, J.; Deshpande, P.; Labarca, C.; Whiteaker, P.; Marks, M.J.; Collins, A.C.; Lester, H.L. Nicotine activation of α4* receptors: Sufficient for reward, tolerance, and sensitization. *Science* **2004**, *306*, 1029–1032. [CrossRef] [PubMed]
2. Maskos, U.; Molles, B.E.; Pons, S.; Besson, M.; Guiard, B.P.; Guilloux, J.P.; Evrard, A.; Cazala, P.; Cormier, A.; Mameli-Engvall, M.; et al. Nicotine reinforcement and cognition restored by targeted expression of nicotinic receptors. *Nature* **2005**, *436*, 103–107. [CrossRef] [PubMed]
3. Picciotto, M.R.; Kenny, P.J. Molecular mechanisms underlying behaviors related to nicotine addiction. *Cold Spring Harb. Perspect. Med.* **2013**, *3*, a012112. [CrossRef] [PubMed]
4. Picciotto, M.R.; Zoli, M.; Rimondini, R.; Lena, C.; Marubio, L.M.; Pich, E.M.; Fuxe, K.; Changeux, J.-P. Acetylcholine receptors containing the β2 subunit are involved in the reinforcing properties of nicotine. *Nature* **1998**, *391*, 173–177. [CrossRef] [PubMed]
5. Picciotto, M.R. Nicotine-mediated activation of signal transduction pathways. In *Understanding Nicotine and Tobacco Addiction*; Bock, G.R., Goode, J.A., Eds.; John Wiley & Sons: New York, NY, USA, 2006; pp. 83–90.
6. McClure-Begley, T.D.; Esterlis, I.; Stone, K.L.; Lam, T.T.; Grady, S.R.; Colangelo, C.M.; Lindstrom, J.M.; Marks, M.J.; Picciotto, M.R. Evaluation of the Nicotinic Acetylcholine Receptor-Associated Proteome at Baseline and Following Nicotine Exposure in Human and Mouse Cortex. *eNeuro* **2016**, *3*. [CrossRef] [PubMed]
7. Jeanclos, E.M.; Lin, L.; Treuil, M.W.; Jayaraman, A.; DeCoster, M.A.; Anand, A. The chaperone protein 14-3-3η interacts with the nicotinic acetylcholine receptor α4 subunit. Evidence for a dynamic role in subunit stabilization. *J. Biol. Chem.* **2001**, *276*, 28281–28290. [CrossRef] [PubMed]
8. McClure-Begley, T.D.; Stone, K.L.; Marks, M.J.; Grady, S.R.; Colangelo, C.M.; Lindstrom, J.M.; Picciotto, M.R. Exploring the nicotinic acetylcholine receptor-associated proteome with iTRAQ and transgenic mice. *Genom. Proteom. Bioinform.* **2013**, *11*, 207–218. [CrossRef] [PubMed]
9. Bermudez, I.; Moroni, M. Phosphorylation and function of α4/β2 receptor. *J. Mol. Neurosci.* **2006**, *30*, 97–98. [CrossRef]
10. Exley, R.; Moroni, M.; Sasdelli, F.; Houlihan, L.M.; Lukas, R.J.; Sher, E.; Zwart, R.; Bermudez, I. Chaperone protein 14-3-3 and protein kinase A increase the relative abundance of low agonist sensitivity human α4/β2 nicotinic acetylcholine receptors in *Xenopus* oocytes. *J. Neurochem.* **2006**, *98*, 876–885. [CrossRef] [PubMed]
11. Pollock, V.V.; Pastoor, T.E.; Wecker, L. Cyclic AMP-dependent protein kinase (PKA) phosphorylates Ser362 and 467 and protein kinase C phosphorylates Ser550 within the M3/M4 cytoplasmic domain of human nicotinic receptor α4 subunits. *J. Neurochem.* **2007**, *103*, 456–466. [CrossRef] [PubMed]

12. Pollock, V.V.; Pastoor, T.; Katnik, C.; Cuevas, J.; Wecker, L. Cyclic AMP-dependent protein kinase A and protein kinase C phosphorylate α4/β2 nicotinic receptor subunits at distinct stages of receptor formation and maturation. *Neuroscience* **2009**, *158*, 1311–1325. [CrossRef] [PubMed]
13. Eilers, H.; Schaeffer, E.; Bickler, P.E.; Forsayeth, J.R. Functional deactivation of the major neuronal nicotinic receptor caused by nicotine and a protein kinase C-dependent mechanism. *Mol. Pharmacol.* **1997**, *52*, 1105–1112. [CrossRef] [PubMed]
14. Fenster, C.P.; Beckman, M.L.; Parker, J.C.; Sheffield, E.B.; Whitworth, T.L.; Quick, M.W.; Lester, R.A.J. Regulation of α4/β2 nicotinic receptor desensitization by calcium and protein kinase C. *Mol. Pharmacol* **1999**, *55*, 432–443. [PubMed]
15. Marszalec, W.; Yeh, J.Z.; Narahashi, T. Desensitization of nicotine acetylcholine receptors: Modulation by kinase activation and phosphatase inhibition. *Eur. J. Pharmacol.* **2005**, *514*, 83–90. [CrossRef] [PubMed]
16. Kabbani, N.; Woll, M.P.; Levenson, R.; Lindstrom, J.M.; Changeux, J.-P. Intracellular complexes of the 2 subunit of the nicotinic acetylcholine receptor in brain identified by proteomics. *Proc. Natl. Acad. Sci. USA* **2007**, *104*, 20570–20575. [CrossRef] [PubMed]
17. Damaj, M.I. Nicotinic regulation of calcium/calmodulin-dependent protein kinase II activation in the spinal cord. *J. Pharmacol. Exp. Ther.* **2007**, *320*, 244–249. [CrossRef] [PubMed]
18. Jackson, K.J.; Walters, C.L.; Damaj, M.I. β2 subunit-containing nicotinic receptors mediate acute nicotine-induced activation of calcium/calmodulin-dependent protein kinase II-dependent pathways in vivo. *J. Pharmacol. Exp. Ther.* **2009**, *330*, 541–549. [CrossRef] [PubMed]
19. Jackson, K.J.; Damaj, M.I. Beta2-containing nicotinic acetylcholine receptors mediate calcium/calmodulin-dependent protein kinase-II and synapsin I protein levels in the nucleus accumbens after nicotine withdrawal in mice. *Eur. J. Pharmacol.* **2013**, *701*, 1–6. [CrossRef] [PubMed]
20. Jackson, K.J.; Damaj, M.I. L-type calcium channels and calcium/calmodulin-dependent kinase II differentially mediate behaviors associated with nicotine withdrawal in mice. *J. Pharmacol. Exp. Ther.* **2009**, *330*, 152–161. [CrossRef] [PubMed]
21. Nashmi, R.; Dickinson, M.E.; McKinney, S.; Jareb, M.; Labarca, C.; Fraser, S.E.; Lester, H.A. Assembly of α4/β2 nicotinic acetylcholine receptors assessed with functional fluorescently labeled subunits: Effects of localization, trafficking, and nicotine-induced upregulation in clonal mammalian cells and in cultured midbrain neurons. *J. Neurosci.* **2003**, *23*, 11554–11567. [CrossRef] [PubMed]
22. Lam, A.J.; St-Pierre, F.; Gong, Y.; Marshall, J.D.; Cranfill, P.J.; Baird, M.A.; McKeown, M.R.; Wiedenmann, J.; Davidson, M.W.; Schnitzer, M.J.; et al. Improving FRET dynamic range with bright green and red fluorescent proteins. *Nat. Methods* **2012**, *9*, 1005–1012. [CrossRef] [PubMed]
23. King, S.L.; Caldarone, B.J.; Picciotto, M.R. β2-subunit-containing nicotinic acetylcholine receptors are critical for dopamine-dependent locomotor activation following repeated nicotine administration. *Neuropharmacology* **2004**, *47*, 132–139. [CrossRef] [PubMed]
24. Jung, Y.; Hsieh, L.S.; Lee, A.M.; Zhou, Z.; Coman, D.; Heath, C.J.; Hyder, F.; Mineur, Y.S.; Yuan, Q.; Goldman, D.; et al. An epigenetic mechanism mediates developmental nicotine effects on neuronal structure and behavior. *Nat. Neurosci.* **2016**, *19*, 905–914. [CrossRef] [PubMed]
25. Whiting, P.; Lindstrom, J. Purification and characterization of a nicotinic acetylcholine receptor from rat brain. *Proc. Natl. Acad. Sci. USA* **1987**, *84*, 595–599. [CrossRef] [PubMed]
26. Nesvizhskii, A.I.; Keller, A.; Kolker, E.; Aebersold, R. A statistical model for identifying proteins by tandem mass spectrometry. *Anal. Chem.* **2003**, *75*, 4646–4658. [CrossRef] [PubMed]
27. White, R.R.; Kwon, Y.G.; Taing, M.; Lawrence, D.S.; Edelman, A.M. Definition of optimal substrate recognition motifs of Ca^{2+}-calmodulin-dependent protein kinases IV and II reveals shared and distinctive features. *J. Biol. Chem.* **1998**, *273*, 3166–3172. [CrossRef] [PubMed]
28. Pacheco, M.A.; Pastoor, T.E.; Wecker, L. Phosphorylation of the α4 subunit of human α4/β2 nicotinic receptors: Role of cAMP-dependent protein kinase (PKA) and protein kinase c (PKC). *Mol. Brain Res.* **2003**, *114*, 65–72. [CrossRef]
29. Wecker, L.; Guo, X.; Rycerz, A.M.; Edwards, S.C. Cyclic AMP-dependent protein kinase (PKA) and protein kinase C phosphorylate sites in the amino acid sequence corresponding to the M3/M4 cytoplasmic domain of α4 neuronal nicotinic receptor subunits. *J. Neurochem.* **2008**, *76*, 711–720. [CrossRef]

30. Guo, X.; Wecker, L. Identification of three cAMP-dependent protein kinase (PKA) phosphorylation sites within the major intracellular domain of neuronal nicotinic receptor α4 subunits. *J. Neurochem.* **2002**, *82*, 439–447. [CrossRef] [PubMed]
31. Lee, A.M.; Wu, D.F.; Dadgar, J.; Wang, D.; McMahon, T.; Messing, R.O. PKCε phosphorylates α4/β2 nicotinic ACh receptors and promotes recovery from desensitization. *Br. J. Pharmacol.* **2015**, *172*, 4430–4441. [CrossRef] [PubMed]
32. Sparks, J.A.; Pauly, J.R. Effects of continuous oral nicotine administration on brain nicotinic receptors and responsiveness to nicotine in C57Bl/6 mice. *Psychopharmacology* **1999**, *141*, 145–153. [CrossRef] [PubMed]
33. Stevens, T.R.; Krueger, S.R.; Fitzsimonds, R.M.; Picciotto, M.R. Neuroprotection by nicotine in mouse primary cortical cultures involves activation of calcineurin and L-type calcium channel inactivation. *J. Neurosci.* **2003**, *23*, 10093–10099. [CrossRef] [PubMed]
34. Addy, N.A.; Fornasiero, E.F.; Stevens, T.R.; Taylor, J.R.; Picciotto, M.R. Role of calcineurin in nicotine-mediated locomotor sensitization. *J. Neurosci.* **2007**, *27*, 8571–8580. [CrossRef] [PubMed]
35. Benfenati, F.; Valtorta, F.; Rubenstein, J.L.; Gorelick, F.S.; Greengard, P.; Czernik, A.J. Synaptic vesicle-associated Ca^{2+}/calmodulin-dependent protein kinase II is a binding protein for synapsin I. *Nature* **1992**, *359*, 417–420. [CrossRef] [PubMed]

proteomes

MDPI

Article

Proteomic Analysis of the Spinophilin Interactome in Rodent Striatum Following Psychostimulant Sensitization

Darryl S. Watkins [1], Jason D. True [2,3], Amber L. Mosley [2] and Anthony J. Baucum II [4,5,6,*]

1. Stark Neurosciences Research Institute, Indiana University School of Medicine Medical Neuroscience Graduate Program, Indianapolis, IN 46278, USA; dswatkin@iu.edu
2. Department of Biochemistry and Molecular Biology, Indiana University School of Medicine, Indianapolis, IN 46278, USA; Jdtrue@bsu.edu (J.D.T.); almosley@iu.edu (A.L.M.)
3. Department of Biology, Ball State University, Muncie, IN 47306, USA
4. Department of Biology, Indiana University-Purdue University Indianapolis, Indianapolis, IN 46202, USA
5. Stark Neurosciences Research Institute Indianapolis, Indianapolis, IN 46202, USA
6. Department of Pharmacology and Toxicology, Indiana University School of Medicine, Indianapolis, IN 46202, USA

* Correspondence: ajbaucum@iupui.edu; Tel.: +1-317-274-0540; Fax: +1-317-274-2846

Received: 12 October 2018; Accepted: 13 December 2018; Published: 17 December 2018

Abstract: Glutamatergic projections from the cortex and dopaminergic projections from the substantia nigra or ventral tegmental area synapse on dendritic spines of specific GABAergic medium spiny neurons (MSNs) in the striatum. Direct pathway MSNs (dMSNs) are positively coupled to protein kinase A (PKA) signaling and activation of these neurons enhance specific motor programs whereas indirect pathway MSNs (iMSNs) are negatively coupled to PKA and inhibit competing motor programs. An imbalance in the activity of these two programs is observed following increased dopamine signaling associated with exposure to psychostimulant drugs of abuse. Alterations in MSN signaling are mediated by changes in MSN protein post-translational modifications, including phosphorylation. Whereas direct changes in specific kinases, such as PKA, regulate different effects observed in the two MSN populations, alterations in the specific activity of serine/threonine phosphatases, such as protein phosphatase 1 (PP1) are less well known. This lack of knowledge is due, in part, to unknown, cell-specific changes in PP1 targeting proteins. Spinophilin is the major PP1-targeting protein in striatal postsynaptic densities. Using proteomics and immunoblotting approaches along with a novel transgenic mouse expressing hemagglutainin (HA)-tagged spinophilin in dMSNs and iMSNs, we have uncovered cell-specific regulation of the spinophilin interactome following a sensitizing regimen of amphetamine. These data suggest regulation of spinophilin interactions in specific MSN cell types and may give novel insight into putative cell-specific, phosphatase-dependent signaling pathways associated with psychostimulants.

Keywords: amphetamine; spinophilin; protein phosphatase-1; dopamine; striatum

1. Introduction

Psychostimulant drug abuse is becoming increasingly popular and costly globally [1,2]. Psychostimulant drugs of abuse, such as methamphetamine, amphetamine, and cocaine, have been associated with dopamine (DA) receptor dysfunction, improper synaptic transmission, and other neuronal perturbations that may contribute to addiction pathology [3–8]. Psychostimulants drive hyper-dopaminergic signaling within the striatum by increasing DA concentrations and enhancing DA transmission [9–14]. When low doses of psychostimulants are administered chronically, response to the drug also increases, causing progressive potentiation of motor programs, a process known as behavioral sensitization [15,16]. Thus, DA plays a critical role in basal ganglia regulated motor programs [17–22].

The striatum is the largest structure within the basal ganglia and has been shown to play a role in disease states, such as Huntington and Parkinson Disease (HD and PD, respectively), and neurological disorders like obsessive-compulsive disorder (OCD) and drug addiction/abuse [23–28]. The striatum is divided into two main regions: The dorsal striatum (dStr) and the ventral striatum (vStr), which includes the nucleus accumbens (NAc) and the olfactory tubercle (OT). The dStr is innervated by dopaminergic projections arising from the substantia nigra (SN) and has been functionally described as a modulator of motor domains specifically involving action selection and initiation [29,30]. The vStr is innervated by dopaminergic projections from the ventral tegmental area (VTA) and is involved in mediating reward and motivational domains [31–33]. However, studies also suggest that there is significant overlap in motor and reward functional domains within the striatum [34,35]. Approximately 90–95% of the neuronal populations within the striatum are gamma-aminobutyric acid (GABA)-ergic medium spiny neurons (MSNs). There are two MSN subtypes within the striatum that are characteristically distinct based on physiological and structural properties, as well as differential expression of DA receptor subtypes and neuropeptide hormones [36–38]. Differential expression of DA receptors allows for differential signaling within striatal MSNs. Studies have shown that characteristics and behaviors associated with striatal specific pathological maladies can occur when there is an imbalance in the activity and/or signaling between the two MSN classes [14,39–41]. Direct pathway MSNs (dMSNs) contain the D1 class of DA receptors, which are positively coupled to PKA signaling. Activation of dMSNs enhances basal ganglia related motor programs. Conversely, indirect pathway MSNs (iMSNs) contain the D2 class of DA receptors, which are negatively coupled to PKA signaling. Activation of iMSNs inhibits inappropriate basal ganglia regulated motor function [42–45]. Thus, the opposing functions of dMSNs and iMSNs are, in part, regulated by post-synaptic responses to DA-dependent signaling.

In the post-synaptic density (PSD) of MSNs, reversible protein phosphorylation is facilitated by kinase and phosphatase activity, contributing to competent neuronal signaling, communication, and synaptic plasticity. To achieve proper signaling, serine/threonine kinases phosphorylate substrates utilizing specific consensus sites; however, serine/threonine phosphatases, such as protein phosphatase 1 (PP1), associate with targeting proteins to attain specificity [46,47]. The most abundant targeting protein for PP1 in the PSD is spinophilin [46,48–50]. Spinophilin acts as a scaffolding protein by targeting PP1 to specific substrates; however, spinophilin can also inhibit the activity of PP1, driving changes in synaptic strength and plasticity [48,51–53]. Furthermore, spinophilin is enriched in the PSD of dendritic spines, and is essential for proper dendritic spine function by regulation of critical dendrite properties [48,54–56]. Changes in MSN dendritic spine density and morphology, perturbations in synaptic transmission and concomitant aberrant dopaminergic signaling are all major contributors to striatal disease states like drug addiction and myriad others [23,26,27,31,57]. In addition, psychostimulant administration, which drives hyper-dopaminergic responses was shown to increase spinophilin expression in the striatum [58,59]. As stated above, alterations in DA levels will regulate DA receptor activity and downstream activation of kinases, such as PKA. Spinophilin is phosphorylated by PKA and PKA phosphorylation of spinophilin is known to decrease its binding to F-actin [49,60]. DA depletion, which decreases DA terminals in the striatum, modulates spinophilin interactions within the striatum. Specifically, there were increases in spinophilin binding to PP1; however, the interactions of spinophilin with a plurality of spinophilin-associated proteins (SpAPs) were decreased [61,62]. We have shown that whole-body spinophilin knockout (KO) mice do not undergo d-amphetamine-induced behavioral sensitization [63]. However, how excessive DA signaling, as occurs following psychostimulant sensitization, modulates spinophilin interactions is unclear. Here we show that in contrast to DA depletion, amphetamine sensitization increases a majority of striatal spinophilin interactions. Moreover, in our preliminary studies using a novel transgenic mouse line that allows for Cre-dependent expression of an hemagglutinin (HA)-tagged form of spinophilin, we observed both pan-MSN and putative cell-specific alterations in spinophilin interactions following amphetamine treatment. Together, these data delineate alterations in spinophilin interactions that may contribute to psychostimulant-induced pathologies.

2. Materials and Methods

2.1. Animals—HA Spinophilin Mice Generation

All animal studies were performed in accordance with the Guide for the Care and Use of Laboratory Animals as disseminated by the U.S. National Institutes of Health and were approved by Indiana University-Purdue University School of Science Animal Care and Use Committees (Approval #SC270R). A human, HA-tagged spinophilin construct [62] containing a P2A sequence and the mNeptune 3 protein were assembled into the pBigT vector between the ClaI and SacI restriction sites (Figure 1A). Gene files were assembled in SnapGene (GSL Biotech, Chicago, IL, USA) or Vector NTI (ThermoFisher Scientific, Waltham, MA, USA). The insert was then subcloned into the AscI/PacI sites on the pROSA26.PA vector (Figure 1B) for generation of targeted embryonic stem cells (ES) cells. pROSA26.PA vector was linearized with AscI and injected into SV129 ES cells by the Vanderbilt Transgenic Mouse/ESC Shared Resource. These ES cells were transferred into pseudo-pregnant C57Bl6/J females and chimeric pups were born from two of these clones (2D4 and 2E12). Chimeras were transferred from the Vanderbilt Transgenic Mouse/ESC Shared Resource to the mouse colony at IUPUI. One clone was maintained in house. Mice were backcrossed at least six generations onto a C57Bl6/J background. Mice were subsequently crossed with either the Drd1a-Cre line or onto an A2A-Cre line [64,65] that were on the C57Bl6/J background. For proteomics, mice expressing a single copy of spinophilin knocked-in to the ROSA locus were used. For immunoblotting, mice expressing HA-spinophilin knocked into one or both copies of the ROSA locus were used. For those expressing a single copy of spinophilin, the other ROSA allele was either WT or had a flox-stop tdTomato reporter sequence inserted (Jackson laboratories Stock #007914, Bar Harbor, ME, USA).

2.2. Animals—Proteomics Studies

Adult male and female mice were used for the proteomics studies (Table 1).

Table 1. Animals used for proteomics studies. Sex, genotype, weight of animals used for proteomics studies.

Eartag	Sex	Condition	Genotype	Cre	Initial Weight	Final Weight	Birth Date	Sacrifice Date
2450	M	Saline	Het/Cre+	D1	23.8	25.0	31 January 2018	30 March 2018
2452	M	Treated	Het/Cre+	D1	22.4	23	31 January 2018	30 March 2018
2453	M	Treated	tdHet/HA-Het/Cre+	D1	23.0	23.3	31 January 2018	30 March 2018
2454	M	Saline	Het/Cre+	D1	22.3	22.8	31 January 2018	30 March 2018
2390	F	Saline	Het/Cre+	A2A	22.9	22.8	3 January 2018	30 March 2018
2393	F	Treated	Het/Cre+	A2A	20.9	20.8	3 January 2018	30 March 2018
2443	F	Saline	Het/Cre+	D1	21.5	21.7	29 January 2018	30 March 2018
2444	F	Treated	Het/Cre+	D1	19.6	19.8	29 January 2018	30 March 2018

2.3. Animals—Immunoblotting Studies

Four male or female P85–P120 mice were used for immunoblotting analysis of HA spinophilin. In addition, one adult P90 WT and one adult spinophilin KO mouse [63,66] were used to validate a subset of spinophilin interactions.

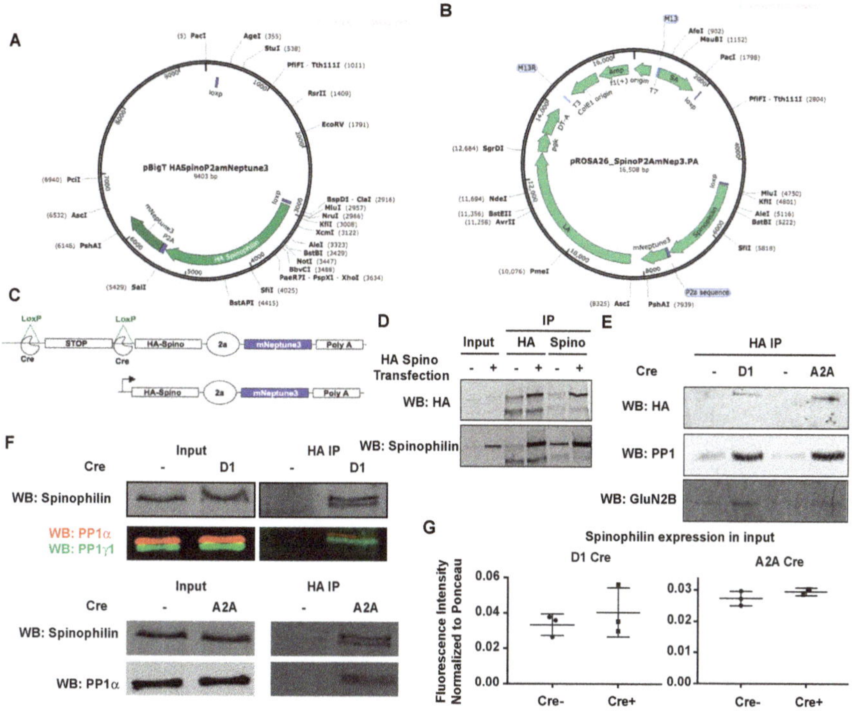

Figure 1. Generation and characterization of Cre-expressing, HA-tagged human spinophilin mice. (**A**) A construct containing DNA encoding HA-tagged human spinophilin with a P2A sequence and mNeptune3 fluorescent protein was cloned into the pBIGT vector that contains a floxed-stop sequence. (**B**) The construct encoding the floxed-stop sequence and the HA-spinophilin-P2A-mNeptune 3 sequence was subcloned into the ROSA targeting vector pROSA_26.PA. (**C**) The modified ROSA vector was used for generation of the targeted transgenic mice. (**D**) Striatal cells were transfected without or with HA-tagged human spinophilin. Lysates were immunoprecipitated with either an HA or spinophilin antibody and immunoblotted with an HA antibody or a spinophilin antibody. HA-spinophilin was selectively detected when it was overexpressed. (**E**) Mice express HA-tagged spinophilin upon crossing with Cre recombinase expressed in the direct pathway (D1) or indirect pathway (A2A) medium spiny neurons. (**F**) Spinophilin and protein phosphatase 1 immunoblots of inputs and HA-immunoprecipitates from HA spinophilin mice crossed with D1 or A2A Cre-recombinase-expressing mice. (**G**) Mice expressing HA-spinophilin had non-significant increases in total spinophilin expression.

2.4. d-Amphetamine Sensitization

Mice received daily intraperitoneal (i.p) injections of d-amphetamine at 3.0 mg/kg (Sigma-Aldrich, St. Louis, MO, USA) or saline (10 mL/kg) for five consecutive days. Mice were then sacrificed and striata dissected 72 h after the last injection.

2.5. Brain Tissue Lysis

Mouse striatum (including both dorsal and ventral (accumbens) striatum or olfactory tubercle) was homogenized and sonicated in 1 mL in a low-ionic strength Tris buffer containing 2 mM Tris-HCl, 1 mM DTT, 2 mM EDTA, 1% Triton X-100, and 1X protease inhibitor cocktail (Bimake, Houston, TX, USA), phosphatase inhibitors (20 mM sodium fluoride, 20 mM sodium orthovanadate, 20 mM β-glycerophosphate, and 10 mM sodium pyrophosphate; Sigma-Aldrich or ThermoFisher Scientific

(Waltham, MA, USA)). Homogenates were incubated for 15 min at 4 °C and then centrifuged at 13,600× *g* for 10 min. The cleared lysate was mixed with Laemmli sample buffer to generate the input or subjected to immunoprecipitation.

2.6. Transfections

Mouse STHdhQ7/7 striatal cell line (a kind gift from Dr. Gunnar Kwakye, Oberlin College, Oberlin, OH, USA) were cultured in Dulbecco's modified Eagle's medium (DMEM) that contained 10% fetal bovine serum, 1% GlutaMAX^{TM} (ThermoFisher Scientific), 400 µg/mL G418-Sulfate (Geneticin) (ThermoFisher Scientific), 100 U/mL penicillin, and 100 µg/mL streptomycin. Culture plates were incubated at a constant 33 °C and 5% CO_2 in myTemp Mini CO_2 digital incubator (Benchmark Scientific; Edison, NJ, USA). Cells were transfected overnight with 2 µg of HA-tagged human spinophilin and PolyJet reagent (SignaGen Laboratories, Gaithersburg, MD, USA) per the manufacturers' instructions. Cells were lysed in the low-ionic strength Tris buffer.

2.7. Immunoprecipitations

Striatal lysates were immunoprecipitated with an HA-epitope antibody or spinophilin antibody. 3 µg of goat HA polyclonal antibody (Bethyl Laboratories, Montgomery, TX, USA, A190-238A) or 5 µg goat spinophilin polyclonal antibody (Santa Cruz Biotechnology, Dallas, TX, USA, SC14774) were incubated at 4 °C with 750–800 µL (75–80%) of total striatal lysate overnight. Striatal cell lysates were immunoprecipitated with 1.6 µg of a sheep spinophilin antibody (ThermoFisher Scientific). The following day, protein G magnetic beads (DynaBeads, ThermoFisher Scientific) were added, and the mixture was incubated for 2 h. Beads were washed three times by magnetic separation in an immunoprecipitation wash buffer (150 mM NaCl, 50 mM Tris-HCl pH 7.5, 0.5% (v/v) Triton X-100). For immunoblotting, beads were resuspended in 2X sample buffer. For Tandem Mass Tag (TMT) labeling, beads were subsequently washed three times in PBS by centrifugation. Washed beads were submitted for tryptic digestion and each sample was labeled with an isobaric tandem mass tag to allow for quantitation.

2.8. TMT Labeling and Mass Spectrometry

Following washes, immunoprecipitated samples on beads were reduced with 5 mM tris(2-carboxyethyl)phosphine hydrochloride (TCEP) and alkylated with 10 mM chloroacetamide (CAM). Beads were then incubated with Trypsin Gold (Promega) at 37 °C overnight. Digested samples were cleaned up using a Waters Sep-Pak C18 plate per manufacturer's instructions.

For TMT labeling, tryptic peptides from each individual condition were labeled with eight different isobaric TMT tags using 8 of a 10-plex TMT kit (ThermoFisherScientific) and following manufacturer's instructions. Following individual labeling, samples were mixed and separated by HPLC and subjected to mass spectrometry.

For HPLC and mass spectrometry analysis, digested peptides were loaded onto an Acclaim PepMap C18 trapping column and eluted on a PepMap C18 analytical column with a linear gradient from 3% to 35% acetonitrile (in water with 0.1% formic acid) over 120 min in-line with an Orbitrap Fusion Lumos Mass Spectrometer (ThermoFisher Scientific). Raw files generated from the run were analyzed using Thermo Proteome Discoverer (PD) 2.2. SEQUEST HT (as a node in PD 2.2) was utilized to perform database searches as previously described [67] with a few modifications: Trypsin digestion, two maximum missed cleavages, precursor mass tolerance of 10 ppm, fragment mass tolerance of 0.8 Da, fixed modifications of +57.021 Da on cysteine, +229.163 on lysines and peptide N-terminii, and a variable modification of +15.995 Da on methionine. The spectral false discovery rate (FDR) was set to $\leq$1% as previously described [68]. The FASTA database used was a mouse proteome downloaded from Uniprot on January 9, 2017 with the addition of 72 common contaminants. Results and quantitative information from the TMT were exported to an Excel Spreadsheet (Tables S1 and S2, Microsoft, Seattle, WA, USA) and tables were generated from these data. A total intensity from the TMT labels, derived from all of the tryptic peptides matching to a specific protein, is given. These intensity data were used for comparing the abundance of different proteins. Examples of these TMT peaks are shown in Figures S2–S5.

2.9. Immunoblotting

Striatal lysates were immunoblotted with the appropriate primary antibody (see below). For protein detection the following primary antibodies were used: HA rabbit antibody (Bethyl Laboratories A190-208A), mouse Clathrin Heavy Chain monoclonal antibody (Santa Cruz, CA, USA, sc12734), rabbit SAP102 monoclonal antibody (Cell Signaling Technology, Danvers, MA, USA; 47421S), rabbit SNIP/p140Cap (SRCIN1) polyclonal antibody (Cell Signaling Technology 3757) and goat or sheep spinophilin polyclonal antibody (as above) or rabbit spinophilin antibody (Cell Signaling Technology 14136S). Antibody dilutions were used at 1:500–1:2000 for immunoblotting. Following overnight incubation at 4 °C, the following secondary antibodies were used for fluorescence detection: Donkey anti-rabbit (H+L) Alexa Fluor 790 (Jackson Immunoresearch, West Grove, PA, USA #711-655-152, 1:50,000 dilution), Donkey anti Goat Alexa Fluor 680 (ThermoFisher Scientific A21084; 1:10,000 dilution), Donkey anti-mouse (H+L) Alexa Fluor 680 (ThermoFisher Scientific, A10038; 1:10,000 dilution), Donkey anti-mouse (H+L) Alexa Fluor 790 (Jackson Immunoresearch #715-655-151, 1:50,000). Imaging was performed on an Odyssey CLx system (LI-COR Biosciences, Lincoln, NE, USA).

2.10. Pathway Analysis

Proteins that were increased in HA-spinophilin immunoprecipitates isolated from amphetamine treated animals were analyzed using the Database for Annotation, Visualization and Integrated Discovery (DAVID) Bioinformatics Resource (version 6.8; https://david.ncifcrf.gov/, National Cancer Instutitue at Frederick, MD, USA [69,70]). Proteins were analyzed using the Kyoto encyclopedia of genes and genomes (KEGG) and gene ontology (GO) pathway databases.

2.11. Statistics

To compare saline to amphetamine treatment across all groups combined, a *t*-test was performed to compare spinophilin abundance in the amphetamine vs. saline treated samples. A non-adjusted *t*-test and a *t*-test adjusted for multiple comparisons (using the Holm-Sidak method) were performed to compare the abundance of the proteins isolated from amphetamine compared to saline-treated samples (both non-normalized (Table S1) and normalized (Table S2). For normalization for PCA analysis and protein abundance, we divide the abundance of the individual protein in the individual sample by the total peptide or total spinophilin abundance detected in that sample. Given that there was an N of 1 in some of the sub-categories (e.g., A2A male and D1 female) the study was not powered nor intended to make statistical conclusions and these results are a qualitative display of sex- and cell-specific protein interactions.

3. Results

3.1. Generation of HA-Tagged Spinophilin Mice

We created constructs that encoded HA-tagged human spinophilin [62] along with a P2A sequence and a far-red fluorescent protein (mNeptune3 [71]). Mice were generated from these constructs by the Vanderbilt Transgenic Mouse/ESC Shared Resource (see methods). When crossed with Cre-expressing mice, these animals express HA-tagged human spinophilin under control of the ROSA promoter (Figure 1C). When crossed with mice expressing Cre recombinase under control of the *Drd1a* gene or the *Adora2a* gene, we were able to detect HA signal in HA immunoprecipitates by immunoblotting (Figure 1D,E). Furthermore, we were able to detect known spinophilin interacting proteins PP1 and GluN2B [48,50,72,73] in the HA immunoprecipitates isolated from Cre-expressing lines (Figure 1E). Less PP1 and GluN2B co-precipitated from the Cre-negative animals (Figure 1E).

Moreover, a spinophilin antibody also detected a band in the HA immunoprecipitates isolated from Cre-expressing, but not Cre-negative mice (Figure 1D,F). Of note, we detected a doublet in the HA-immunoprecipitates when immunoblotting for spinophilin isolated from the HA-spinophilin expressing mice, as well as a striatal cell line transfected with an HA-spinophilin construct (Figure 1F). This is not surprising as spinophilin is thought to homo-dimerize and this suggests that the human HA-spinophilin is complexing with the endogenous, mouse spinophilin. However, there was no significant difference in the amount of total spinophilin in the mice expressing HA-spinophilin, suggesting a low overexpression of spinophilin (Figure 1G). Moreover, given the low expression of epitope tagged spinophilin and fluorescent protein, we were unable to detect either HA-tagged protein or fluorescent protein by immunohistochemistry (data not shown).

3.2. Amphetamine Modulates Spinophilin Expression and Interactions

DA signaling within the striatum modulates MSN activity and signaling [74,75]. Amphetamine increases the release of DA at dopaminergic terminals synapsing on MSNs [9,11,12]. Our previous studies show that DA depletion alters the spinophilin interactome [62] and that spinophilin KO mice do not undergo amphetamine-induced locomotor sensitization [63]. However, how spinophilin normally contributes to synaptic changes associated with amphetamine-dependent striatal changes is unclear. As spinophilin targets PP1 to regulate synaptic protein phosphorylation, in order to identify potential spinophilin-dependent synaptic protein targets that are regulated by spinophilin following amphetamine sensitization, we utilized our HA-tagged spinophilin mice (Figure 1) to measure spinophilin interactions in saline- or amphetamine-treated mice expressing spinophilin in D1 DA or A2A adenosine-receptor containing neurons of the striatum. Mice were injected with 3 mg/kg amphetamine every day for five days and sacrificed 72 h after the final amphetamine treatment. Striatal lysates were immunoprecipitated with an HA antibody, digested with trypsin, labeled with TMTs, and analyzed by mass spectrometry (Figure 2A). A total TMT abundance for spinophilin was detected in all conditions. As human and mouse spinophilin differ by fewer than 30 amino acids (Figure S1) and as spinophilin homo-dimerizes we searched only the mouse database. Forty-eight spectral counts matching spinophilin were detected across all eight samples. While the human construct and mouse spinophilin are highly homologous, there are 28 (out of 817) different amino acids between the two species that lead to the generation of ~12 different potential tryptic fragments (Figure S1). We validated two MS/MS spectra generated from tryptic peptides that were predicted to be different between mouse and human spinophilin (Figures S2–S5). This further validates the expression of our HA-tagged human spinophilin construct. Based on the unnormalized abundance of the TMT tag from the different samples, we observed more spinophilin in the amphetamine-treated compared to control treated samples (Figure 2B). A principal component analysis from all proteins (1454) of the individual samples, normalized to total peptide amount, revealed that 46.2% of the total variability is due to amphetamine treatment (Figure 2C). We next evaluated changes in the TMT tag abundance ratios (amphetamine/saline) of the spinophilin interacting proteins. For this, we eliminated all contaminant proteins and only included those proteins detected in all eight samples (e.g., with all eight tags). This led to the detection of 984 total proteins in the HA-spinophilin immunoprecipitates (Table S1). We plotted these unnormalized values using a volcano plot with Log2 abundance ratio on the X-axis and −log10 *p*-value (*t*-test, non-adjusted) on the Y-axis (Figure 2D). All but two proteins showed an increased abundance in spinophilin immunoprecipitates isolated from amphetamine compared to saline treated samples.

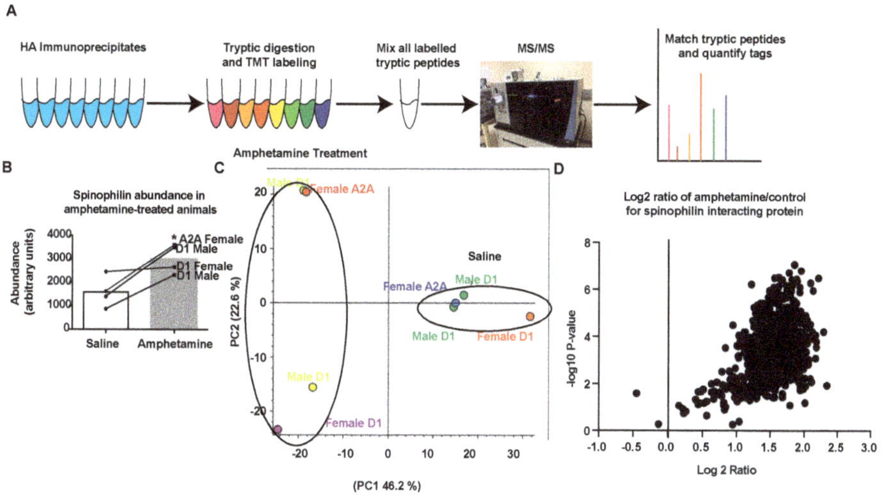

Figure 2. Quantitation of spinophilin complexes isolated from dMSNs and iMSNs using tandem mass tag (TMT) analysis. (**A**) Striatal lysates isolated from male or female mice expressing HA spinophilin under the control of D1 or A2A promoters and treated with saline or amphetamine were immunoprecipitated with an HA antibody, digested with trypsin, labelled with eight different TMT tags, mixed and analyzed by mass spectrometry (MS/MS). (**B**) A higher intensity of TMT reporter abundance matching spinophilin was observed in amphetamine-treated compared to saline treated animals (*t*-test; * $p < 0.05$). (**C**) Principal component analysis of individual samples normalized to total peptide amount within each sample. (**D**) A volcano plot showing a majority of the protein have increased abundance in HA immunoprecipitates isolated from amphetamine treated animals.

3.3. Regulation of Spinophilin Interactions by Amphetamine

Given that spinophilin abundance was increased in immunoprecipitates isolated from amphetamine treatment compared to saline-treated samples, to determine if the increased association of spinophilin with interacting proteins was due exclusively to increased spinophilin levels, we normalized the abundance of each individual interacting protein to the abundance of spinophilin in the corresponding sample (Table S2A). Moreover, in Table S2, we show only those proteins that contained at least eight peptide spectral matches (PSMs) as these would average 2 PSMs per condition. We detected 423 total proteins across all conditions that met these criteria. Of these proteins, 134 were unchanged (Log2 ratio −0.5 to +0.5), three had a decreased association (<Log2 Ratio −0.5), and 286 had an increased association (>Log2 Ratio +0.5) with spinophilin. Those proteins with a decreased interaction ratio of <−0.5 and increased interaction ratio of >1.0 are shown in Table 2. We performed a second PCA analysis of the data normalized to spinophilin for those peptides having eight PSMs or more (Figure 3A). This mode of analysis decreased the variability within the amphetamine treatment group but increased the variability in the saline treatment group. We plotted these normalized values using a volcano plot with Log2 abundance ratio on the X-axis and −log10 P-value (*t*-test, non-adjusted) on the Y-axis (Figure 3B). Therefore, even when normalized to spinophilin there is a higher number of proteins with an enhanced association with spinophilin compared to a decreased or no change in association.

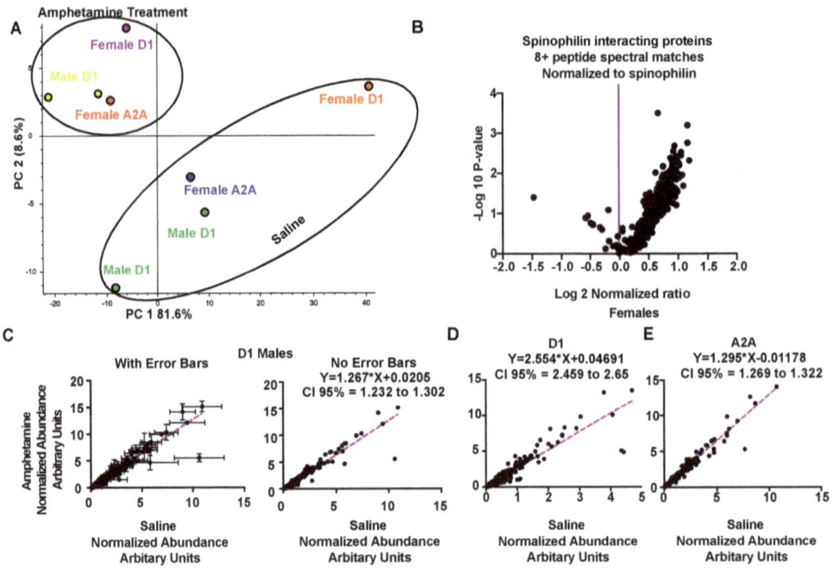

Figure 3. Greater abundance of spinophilin interacting proteins in amphetamine-treated animals occurs across both sexes and cell types. (**A**) Principal component analysis of individual samples normalized to spinophilin abundance within each sample and filtered for eight or more PSMs. (**B**) A volcano plot showing a majority of the proteins have increased abundance in HA immunoprecipitates isolated from amphetamine treated animals when normalized to the amphetamine-dependent increase in spinophilin abundance. (**C**) A plot of the abundance of spinophilin interacting proteins isolated from male, D1 Cre expressing animals and normalized for spinophilin expression (Table S2) and quantified from treated (Y-axis) or control (X-axis) samples. Left panel shows mean ± standard deviation, the right panel just shows the mean of the two values. (**D**) A plot of the abundance of spinophilin interacting proteins isolated from female, D1 Cre expressing animals and normalized for spinophilin expression (Table S2) and quantified from treated (Y-axis) or control (X-axis) samples. (**E**) A plot of the abundance of spinophilin interacting proteins isolated from female, A2A-Cre expressing animals and normalized for spinophilin expression (Table S2) and quantified from treated (Y-axis) or control (X-axis) samples.

We next evaluated the spinophilin interacting proteins from the different cell types and sexes. It is important to note that evaluation of these sub-categories (sex and genotype) are qualitative and no statistical inference can be made; however, these preliminary studies denote the importance of evaluating different sexes and cell types. We plotted the abundance of the amphetamine treated samples on the Y-axis and the abundance of the saline-treated samples isolated from D1 males (Figure 3C), D1 females (Figure 3D) and A2A females (Figure 3E). All data were normalized to spinophilin abundance. All three sets had a slope greater than 1, suggesting that there was greater abundance in the amphetamine treatment compared to the saline treatment. Of note, the D1 females had the greatest slope (M = 2.554), suggesting the greatest increased association in this group compared to the D1 males (M = 1.267) or A2A females (M = 1.269). Together, our data suggest that amphetamine treatment enhances synaptic protein interactions with spinophilin across multiple sexes and cell types.

To begin to delineate amphetamine-dependent regulation of specific spinophilin interactors in the different Cre lines, we generated a ratio of the abundance ratios from the D1 Cre animals to the A2A Cre animals (Table S2B). As stated above, it is important to note that, given the N of 1 in the A2A animals, these ratios are qualitative and no statistical inference can be obtained; however, these data will inform novel lines of inquiry in future cell-specific studies. Those proteins with this ratio of ratios greater than 2 are shown in Table 3. No interactions with a ratio of less than 0.5 (2-fold decrease) were detected. All ratios are shown in Table S2B.

Table 2. Spinophilin interacting proteins that had altered abundance following amphetamine treatment.

Description	# PSMs	Normalized Abundance Ratio (Treatment)/(Control)	Normalized Abundance Ratio (log2): (Treatment)/(Control)
		Decreased Interactions	
E3 ubiquitin-protein ligase XIAP	22	0.69	−0.54
Disks large homolog 3	143	0.67	−0.58
Granulins	10	0.36	−1.47
		Increased Interactions	
Myelin proteolipid protein	38	2.28	1.19
Hemoglobin subunit alpha	17	2.24	1.16
ADP/ATP translocase 2	36	2.23	1.16
Clathrin light chain A	17	2.13	1.09
Adenylyl cyclase-associated protein 2	8	2.10	1.07
MCG10343, isoform CRA_b	35	2.08	1.05
Tubulin alpha-1B chain	54	2.07	1.05
Tubulin alpha chain (Fragment)	54	2.03	1.02
Cytochrome c oxidase subunit NDUFA4	13	2.03	1.02
Profilin-2	14	2.02	1.02
Reticulon (Fragment)	20	2.02	1.01
Myelin-oligodendrocyte glycoprotein	18	2.01	1.01
Serine/threonine-protein phosphatase 2A 65 kDa regulatory subunit A alpha isoform	28	2.00	1.00

HA spinophilin was immunoprecipitated from saline and amphetamine-treated D1 Cre and A2A Cre mice. The abundance of the individual proteins was normalized to the abundance of spinophilin. A ratio of the abundance of proteins isolated from the amphetamine treated over the saline treated mice was generated. A subset of spinophilin interacting proteins that had at least eight spectral counts (PSMs) and had a decreased (<0.05) or increased (≥1.00) log2 ratio is shown. A complete list of interacting proteins (without contaminants) and their abundance ratios are shown in Table S2A.

Table 3. Spinophilin interacting proteins that had altered abundance ratios in D1 Cre animals compared to A2A Cre animals.

Description	PSMs	Female A2A Ratios	Female D1 Ratios	Male D1 Avg Ratios	Avg D1/A2A Ratios
Endophilin-A2	12	0.50	1.38	1.28	2.63
Cofilin-1	14	1.13	4.14	1.57	2.52
Malate dehydrogenase, mitochondrial	27	0.92	3.22	1.38	2.51
Calreticulin	8	0.87	3.15	1.11	2.46
Alpha-synuclein	21	1.06	3.56	1.61	2.43
Malate dehydrogenase, cytoplasmic	41	1.05	3.59	1.38	2.36
Glutamate dehydrogenase 1, mitochondrial	23	1.05	3.58	1.39	2.36
Fructose-bisphosphate aldolase	87	1.14	3.93	1.41	2.35
Protein disulfide-isomerase A3	9	0.93	3.19	1.13	2.31
Citrate synthase, mitochondrial	32	1.37	4.58	1.55	2.24
Synapsin-1	42	1.11	3.51	1.45	2.23
Tubulin polymerization-promoting protein	9	1.08	3.24	1.56	2.23
Nucleoside diphosphate kinase	12	1.20	3.92	1.39	2.21
Fascin	26	0.89	2.78	1.13	2.20
Cytochrome b-c1 complex subunit 1, mitochondrial	19	0.99	3.09	1.26	2.19
Carbonic anhydrase 2	14	1.22	3.75	1.54	2.18
Rab GDP dissociation inhibitor alpha	22	1.12	3.58	1.26	2.16
Protein kinase C and casein kinase substrate in neurons protein 1	18	1.10	3.36	1.30	2.12
Endophilin-A1	23	1.11	3.37	1.33	2.12
Cytochrome c oxidase subunit 5B, mitochondrial	13	1.17	3.52	1.43	2.12

Table 3. *Cont.*

Description	PSMs	Female A2A Ratios	Female D1 Ratios	Male D1 Avg Ratios	Avg D1/A2A Ratios
Adenylyl cyclase-associated protein 2	8	1.43	4.36	1.63	2.09
Cytochrome b-c1 complex subunit 2, mitochondrial	40	1.04	3.08	1.22	2.06
Pyruvate kinase PKM	36	1.10	3.22	1.30	2.06
Profilin-2	14	1.33	3.85	1.58	2.04
Myelin proteolipid protein	38	1.58	4.56	1.89	2.04
60S ribosomal protein L17	8	1.06	3.20	1.12	2.03
40S ribosomal protein S23	11	0.89	2.40	1.23	2.03
Beta-synuclein	16	1.18	3.31	1.47	2.03
Heat shock 70 kDa protein 4	23	1.07	3.01	1.27	2.01
L-lactate dehydrogenase B chain	28	1.23	3.58	1.35	2.00

HA spinophilin was immunoprecipitated from saline and amphetamine-treated D1 Cre and A2A Cre mice. The abundance of the individual proteins was normalized to the abundance of spinophilin. A ratio of the abundance of proteins isolated from the amphetamine treated over the saline treated mice was generated. A second ratio comparing the amphetamine/saline ratios identified in the 3 D1 samples and the 1 A2A sample was generated. Those D1/A2A ratios ≥ 2.00 are shown. A complete list of interacting proteins (without contaminants) and the cell-specific abundance ratios are shown in Table S2B.

3.4. Pathway and GO Analysis of Spinophilin Interacting Proteins Enhanced by Amphetamine

Using the DAVID Bioinformatics resource [69,70], we performed the Kyoto Encyclopedia of Genes and Genomes (KEGG) pathway analysis on the 286 proteins that had an increased association with spinophilin across all samples. 283 total proteins were detected from the list. A total of 94 pathways were detected from this analysis (Table S3). 49 of the pathways were significantly enriched using a Bonferroni adjustment (Table S3 highlighted). These include pathways associated with striatal function, including amphetamine addiction. The top 10 pathways also include other disease states associated with striatal dysfunction, such as Parkinson disease and Huntington disease (Table 4).

Table 4. Top 10 Kyoto Encyclopedia of Genes and Genomes (KEGG) pathways associated with proteins that have amphetamine-dependent increases in spinophilin.

Term	Count	%	*p* Value	Bonferroni
mmu01200:Carbon metabolism	27	9.54	3.52×10^{-18}	7.04×10^{-16}
mmu05012:Parkinson's disease	29	10.25	2.34×10^{-17}	4.69×10^{-15}
mmu05016:Huntington's disease	32	11.31	9.23×10^{-17}	2.22×10^{-14}
mmu00190:Oxidative phosphorylation	26	9.19	4.42×10^{-15}	8.88×10^{-13}
mmu04721:Synaptic vesicle cycle	19	6.71	5.04×10^{-15}	9.99×10^{-13}
mmu00020:Citrate cycle (TCA cycle)	15	5.30	7.94×10^{-15}	1.60×10^{-12}
mmu05010:Alzheimer's disease	27	9.54	1.82×10^{-13}	3.65×10^{-11}
mmu01130:Biosynthesis of antibiotics	29	10.25	3.85×10^{-13}	7.70×10^{-11}
mmu04961:Endocrine and other factor-regulated calcium reabsorption	15	5.30	1.92×10^{-11}	3.84×10^{-9}
mmu00010:Glycolysis/Gluconeogenesis	16	5.65	5.11×10^{-11}	1.02×10^{-8}

The 286 proteins that had an increased association with spinophilin (log2 ratio ≥ 0.5; Table S2A) were input into the DAVID Bioinformatics resource and analyzed using KEGG pathway analysis. The top 10 enriched pathways are shown. All pathways are given in Table S3.

We next evaluated these increased interactions in DAVID using gene ontology terms (GO). We evaluated Biological Processes (BP), Cellular Components (CC), and Molecular Function (MF). We detected 247 BPs, 154 CCs, and 127 different MFs (Table S4) in the increased spinophilin interactors. Of these, 33, 55, and 31, respectively, were significantly enriched. For BPs, we observed a large number of metabolic processes, including ATP and NADH metabolism. We also observed vesicle trafficking processes, including synaptic vesicle endocytosis, vesicle-mediated transport, and endocytosis. For CCs, we observed known areas where spinophilin is enriched, including membrane, postsynaptic density, dendrite, dendritic spine, and cytoskeleton. In addition, we matched other locations where spinophilin has been implicated, such as synaptic vesicle membrane [76]. For MFs, we observed known roles for spinophilin as a scaffold, including protein binding, protein complex binding, protein kinase binding, and actin filament binding. In addition, we observed novel putative roles for spinophilin, including GTPase binding, ATP binding, and syntaxin-1 binding. The top 10 pathways for BP, CC, and MF are shown in Table 5.

Table 5. Top 10 GO, BP, CC, and MF pathways associated with proteins that have amphetamine-dependent increases in spinophilin. The 286 proteins that had an increased association with spinophilin (log 2 ratio $\geq$ 0.05; Table S2A) were input into the DAVID Bioinformatics resource and analyzed using GO BP, CC, and MF pathway analyses. The top 10 enriched pathways are shown. All pathways are given in Table S4.

Term	Count	%	PValue	Bonferroni
Biological Process				
GO:0006099~tricarboxylic acid cycle	13	4.59	3.85×10^{-15}	6.78×10^{-12}
GO:0006810~transport	68	24.03	1.55×10^{-12}	2.70×10^{-9}
GO:0006096~glycolytic process	11	3.89	8.12×10^{-11}	1.42×10^{-7}
GO:0006734~NADH metabolic process	8	2.83	1.11×10^{-10}	1.93×10^{-7}
GO:0046034~ATP metabolic process	11	3.89	2.57×10^{-10}	4.48×10^{-7}
GO:0015992~proton transport	12	4.24	7.90×10^{-10}	1.38×10^{-6}
GO:0015991~ATP hydrolysis coupled proton transport	10	3.53	8.81×10^{-10}	1.54×10^{-6}
GO:0050821~protein stabilization	15	5.30	7.11×10^{-9}	1.24×10^{-5}
GO:0015986~ATP synthesis coupled proton transport	8	2.83	2.98×10^{-8}	5.20×10^{-5}
GO:1904871~positive regulation of protein localization to Cajal body	6	2.12	3.79×10^{-8}	6.61×10^{-5}
Cellular Compartment				
GO:0043209~myelin sheath	92	32.51	3.70×10^{-120}	1.53×10^{-117}
GO:0070062~extracellular exosome	159	56.18	2.16×10^{-65}	8.91×10^{-63}
GO:0005739~mitochondrion	88	31.10	3.66×10^{-27}	1.51×10^{-24}
GO:0005829~cytosol	88	31.10	4.60×10^{-26}	1.90×10^{-23}
GO:0005737~cytoplasm	173	61.13	3.24×10^{-22}	1.34×10^{-19}
GO:0016020~membrane	178	62.90	6.47×10^{-22}	2.66×10^{-19}
GO:0014069~postsynaptic density	32	11.31	5.02×10^{-21}	2.07×10^{-18}
GO:0043005~neuron projection	40	14.13	6.70×10^{-21}	2.76×10^{-18}
GO:0043234~protein complex	46	16.25	1.63×10^{-19}	6.71×10^{-17}
GO:0005743~mitochondrial inner membrane	37	13.07	2.23×10^{-19}	9.20×10^{-17}
Molecular Function				
GO:0005515~protein binding	138	48.8	1.12×10^{-22}	6.18×10^{-20}
GO:0032403~protein complex binding	36	12.7	2.54×10^{-18}	1.40×10^{-15}
GO:0019901~protein kinase binding	36	12.7	1.24×10^{-15}	6.74×10^{-13}
GO:0044822~poly(A) RNA binding	55	19.4	4.52×10^{-14}	2.49×10^{-11}
GO:0000166~nucleotide binding	74	26.1	2.72×10^{-13}	1.50×10^{-10}
GO:0019904~protein domain specific binding	27	9.5	5.67×10^{-13}	3.13×10^{-10}
GO:0098641~cadherin binding involved in cell-cell adhesion	26	9.2	1.77×10^{-12}	9.79×10^{-10}
GO:0008022~protein C-terminus binding	22	7.8	1.30×10^{-11}	7.15×10^{-9}
GO:0005516~calmodulin binding	19	6.7	5.09×10^{-10}	2.81×10^{-7}
GO:0003779~actin binding	25	8.8	6.35×10^{-10}	3.51×10^{-7}

3.5. Interactome Analysis of Spinophilin Interacting Proteins Enhanced by Amphetamine

We next wanted to organize spinophilin interacting proteins that were enhanced by amphetamine based on interactions and functional classifications. To do this, we utilized the string-db program [77,78]. This allows for the pictorial representation of proteins and their interactions. To reduce the complexity of the submitted proteins, we used a high stringency confidence score (0.900) and removed any proteins that were not connected. We next grouped proteins into 12 categories based on known function and these interactions (Figure 4). These categories are: Metabolism, ATPases, vesicle trafficking, synaptic signaling, cytoskeleton, ribosomal and nuclear, scaffolding, heatshock, G-proteins and GTPases, semaphorin signaling, BBSome, and other.

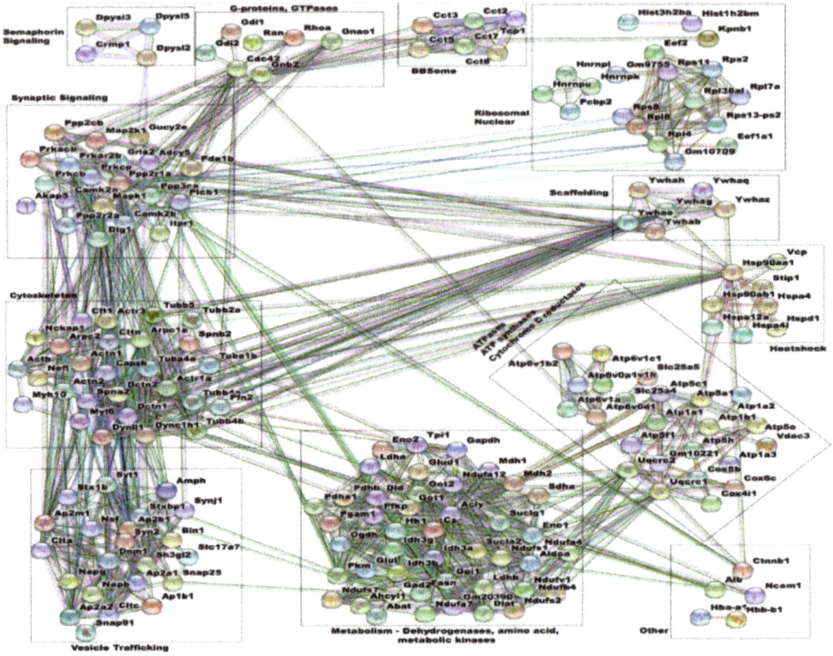

Figure 4. Graphical representation of interactors input into the string-db from protein complexes that had greater abundance in the HA immunoprecipitates isolated from amphetamine-treated animals. Proteins with greater abundance in HA immunoprecipitates isolated from amphetamine-treated animals were input into the string-db program (www.string-db.org) and separated by hand based on function. To reduce the complexity of this map, only those proteins that had at least 1 interaction at a confidence of 0.9 (highest confidence).

3.6. Alteration and Validation of Novel Spinophilin Interacting Proteins

While some of the proteins isolated using the HA spinophilin antibody are known spinophilin interactors (e.g., PP1, glutamate receptors, actin), some have not been previously validated. Therefore, we wanted to validate some of the proteins that had an altered association with spinophilin that are involved in different processes. We chose SAP102 (*Dlg3*), src kinase inhibitor protein 1 (*Srcin1*), and clathrin heavy chain (*Cltc*) as proteins involved in synaptic scaffolding, signaling, and vesicle trafficking, respectively. To determine if these proteins interact with spinophilin in a specific manner, we dissected out total (dorsal and ventral (e.g., accumbens)) striatum (Str) and olfactory tubercle (OT), a further ventral portion of the striatum from wildtype and whole-body spinophilin KO animals. We chose these regions based on their roles in psychostimulant sensitization [31–33]. We immunoprecipitated striatal and tubercle lysates with a spinophilin antibody and immunoblotted for spinophilin, SAP102, Srcin1, and clathrin heavy chain. Spinophilin was only detected in WT and not KO samples (Figure 5). Moreover, spinophilin interactions were only detected in WT and not KO animals. These data suggest that these proteins are specific interactors with spinophilin.

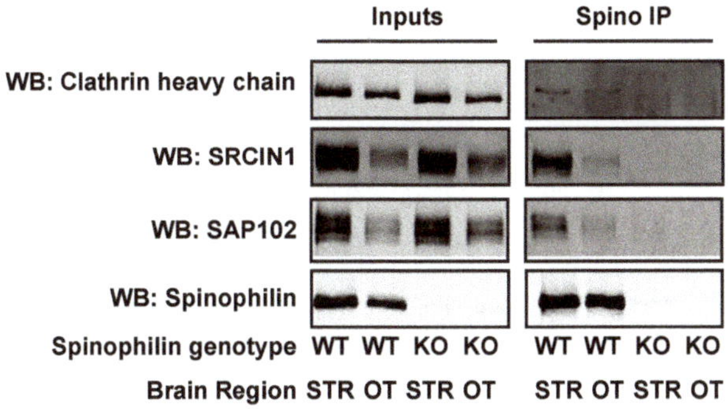

Figure 5. Validation of spinophilin interactions. WT or spinophilin KO striatal (STR) or olfactory tubercle (OT) lysates were immunoprecipitated with a spinophilin antibody. Lysates or immunoprecipitates were immunoblotted for spinophilin and three interacting proteins that were detected in the HA immunoprecipitates that had a decreased (SAP102) or increased (Clathrin heavy chain and SRCIN1) interaction with spinophilin in amphetamine-treated animals. Spinophilin and all associated proteins were detected in the spinophilin immunoprecipitates from WT animals, but were absent in immunoprecipitates isolated from KO animals.

4. Discussion

4.1. Spinophilin Functional Localization

Spinophilin is a highly abundant spine-enriched protein. We have previously found that loss of spinophilin modulates striatal behaviors. Specifically, loss of spinophilin decreases motor performance and motor learning on a rotarod apparatus [66]. Moreover, we previously reported that spinophilin KO mice, in contrast to WT mice, do not undergo amphetamine-induced locomotor sensitization [63]. In addition, others have observed alterations in the response of spinophilin KO mice in rotarod behaviors and following cocaine treatment [79,80]. Together, these data suggest that spinophilin is important in striatal based behaviors. However, how spinophilin contributes to these behaviors is unclear. As the major postsynaptic density-enriched PP1-interacting protein, spinophilin's functional regulation of the above striatal behaviors may be due to its targeting of PP1 to synaptic proteins. Moreover, while spinophilin, as its name implies, is enriched in dendritic spines, it is also present in dendrites, presynaptic terminals, and glial cells [54,55]. Therefore, spinophilin may have functions beyond just dendritic spines.

4.2. Regulation of the Spinophilin Interactome Following Amphetamine Treatment

To begin to identify cell-type specific spinophilin interactions that may be important in behavioral changes observed following psychostimulant abuse, we created mice that Cre-dependently overexpress an epitope-tagged (HA) form of human spinophilin. Using HA immunoprecipitation from different cell types and TMT labeling, we probed the spinophilin interactome in the striatum of saline compared to amphetamine treated animals. As previously observed in rats [58,59], the amount of spinophilin detected (based on labeled peptide abundance) was increased following amphetamine treatment. Interestingly, across all samples the abundance of the different proteins was greater in the amphetamine-treated compared to the control treated lysates, even when normalizing to the increased spinophilin expression. This is in contrast to what we observed previously in DA-depleted striatum (an animal model of Parkinson disease) [62]. In that previous study, 60 proteins were decreased whereas 31 total proteins were increased across two different fractions. In contrast, in the current study, we only observed three proteins that were decreased and 286 proteins that were increased following amphetamine treatment. These data suggest that spinophilin interactions are decreased by DA depletion and increased by hyperdopaminergic signaling. However, it is critical to note that while specific, there are low levels of expression of the HA-tagged, human spinophilin and while we validated expression of the HA-tagged form of the human protein by WB and MS/MS, as well as the interactor, PP1, additional interacting proteins may be non-specifically interacting with the beads. Therefore, future studies will need to follow-up on these studies to delineate those interactions that are real and that are modulated by amphetamine.

Regulation of DA leads to alterations in striatal medium spiny neuron spine density. Specifically, loss of DA decreases spine density [81–83] and psychostimulant treatment increases spine density [84,85]. Spinophilin is also known to regulate spine density, with acute knockdown of spinophilin decreasing spine density in hippocampal cultures [86]. Whole-body spinophilin KO animals do not have loss of dendritic spines in adulthood (and have a paradoxical increase in young animals) [56]. This lack of an effect may be due to compensatory changes, such as decreases in expression of PP1 [80]. However, data showing amphetamine-dependent increases in spinophilin expression [58,59] and our data showing a lack of amphetamine-dependent locomotor sensitization [66] and the data presented in this paper may suggest that changes in the spinophilin expression and/or interactions are critical for normal psychostimulant-induced behaviors. However, it is currently unclear if amphetamine-induced increases in spine density are also abrogated or modulated in spinophilin KO mice.

4.3. Classes and Specific Spinophilin Protein Interactions that are Modulated by Amphetamine

Using GO and KEGG analysis along with hand annotation of altered interactions in string-db we detailed different classes of spinophilin interacting proteins that have an increased interaction with amphetamine. These protein classes co-purify with spinophilin, including cytoskeletal and vesicle trafficking proteins [62,73,87,88]. Moreover, many of the pathways identified associate with protein binding, striatal function and diseases, and synaptic/postsynaptic protein organization. In addition to these known functions of spinophilin and striatum-dependent regulation by amphetamine, we observed novel/less well-characterized spinophilin interactions. Some of these may be non-specific. For instance, myelin sheath was one of the most abundant cellular components. Myelin sheath components may be non-specifically sticky and may not be specific interactions. Indeed, myelin sheath components are associated with the CRAPome [89], a list of non-specific interactions that may non-specifically co-precipitate. However, even though some of these interactions may be non-specific, these pathways may be regulated by amphetamines. For instance, psychostimulant abuse in humans may be associated with altered neuron myelination [90]. Therefore, while additional studies will need to detail specific spinophilin interactions, our data may delineate alterations in protein expression following amphetamine treatment.

Another major class of altered spinophilin interacting proteins were vesicle trafficking proteins. For instance, the synaptic vesicle cycle was identified as being enriched in the KEGG pathway and we delineated multiple vesicle trafficking proteins. While not much is known about the role of spinophilin in vesicle trafficking, one study has detailed spinophilin as a regulator of presynaptic vesicle function [76]. We also validated a vesicle trafficking protein, clathrin heavy chain, as a specific interactor with spinophilin. Spinophilin may play a role in vesicle trafficking on glutamatergic (or dopaminergic) presynaptic terminals or postsynaptic MSNs; however, given the enrichment of endogenous spinophilin in spines and a lack of detected presynaptic vesicle proteins (e.g., syntaxin, SNAP-25, synaptobrevin, etc.), we posit that this role is more postsynaptic; however, we cannot rule out a presynaptic role for spinophilin in striatum.

We observed many metabolic proteins, including lactate and succinate dehydrogenases, and ATPases, that were increased in spinophilin immunoprecipitates following amphetamine treatment. Moreover, some of the pathways with altered protein expression included glycolytic process, ATP metabolism/hydrolysis/synthesis, tricarboxylic acid cycle, and Citrate (TCA) cycle. Spinophilin has been shown to associate with and regulate the membrane localization of the Na^+-K^+-ATPase [91]. Moreover, amphetamines may modify the activity of the Kreb's cycle [92,93]. However, how alterations in spinophilin interactions with metabolic proteins modulate response to amphetamines is an unexplored area.

4.4. Direct and Indirect Pathway Striatal MSNs and Spinophilin Interactions

By using mice expressing spinophilin Cre dependently, we were able to isolate complexes from dMSNs and iMSNs. We used a D1 DA receptor Cre line and an A2A adenosine receptor Cre line. These Cre lines were created as part of the GENSAT project [64,65]. While D1 and A2A are enriched in striatal MSNs, there may be expression in other cell types and other brain regions. We observed some HA spinophilin expression in other brain regions, including the hippocampus and prefrontal cortex (data not shown). While these Cre lines only minimally express in these other regions, it may be sufficient for driving expression of the HA spinophilin.

Previous studies have observed persistent psychostimulant-dependent increases in spinophilin density in dMSNs (months), whereas in iMSNs these changes are more transient (days) [84]. We found that there was a greater association of spinophilin with multiple proteins 72 h following amphetamine treatment and that this occurred in both Cre-driver lines. This suggests that amphetamine may be regulating interactions in both cell types. While our preliminary study suggests the level of increase was greater overall in the dMSNs compared to the iMSNs, it is unclear if these changes will persist in both populations and future studies will need to evaluate the persistence of these changes and the link between spinophilin and modulation of dendritic spine density.

While most protein interaction changes were similar in the two cell types, the magnitude of the effect was different between the different cell types. However, there was a cell-specific effect of amphetamine treatment in two proteins that are known to be involved in striatal pathologies. We detected both α-synuclein (*SNCA*) and tau (*MAPT*) proteins in the HA-spinophilin immunoprecipitates. While additional work needs to determine if endogenous spinophilin associates with these proteins, it was interesting that amphetamine increased the association of spinophilin with both of these proteins, but this increase was only in the D1 Cre containing animals. These proteins were enriched 2.45-fold (α-synuclein) and 1.85-fold (Tau) in the dMSNs compared to the iMSNs. These proteins play major roles in PD and AD, respectively, and amphetamine is known to increase α-synuclein and tau protein levels [94,95]. Moreover, phosphorylation of these proteins is important in modulating their function and aggregation potential, but if spinophilin plays a role in modulating this aggregation has, to our knowledge, not been evaluated.

5. Conclusions

Our data identify novel putative spinophilin interactions that are modulated by amphetamine. As a whole, amphetamine increased spinophilin expression and enhanced spinophilin interactions. While these changes occur in both MSN cell types, this preliminary study suggests dMSNs appear to be more influenced by amphetamine. Future studies need to validate these interactions, use additional approaches to enhance cell specific expression and interactions (e.g., viral transduction of Cre-dependent epitope tagged spinophilin), and evaluate long-term amphetamine changes (e.g., 1-month) in the different striatal MSN subtypes. However, the current proteomics study is the first to outline potential pathways of spinophilin interactors that are modulated by amphetamine. Moreover, we have begun to uncover differences in amphetamine-dependent spinophilin interactions in the different striatal cell types. This knowledge will enhance our understanding of amphetamine-dependent regulation of cell-specific striatal biology.

Supplementary Materials: The Supplementary Materials are available online at http://www.mdpi.com/2227-7382/6/4/53/s1.

Author Contributions: Conceptualization, D.S.W., A.J.B.II, A.L.M.; Formal Analysis, D.S.W., J.D.T., A.J.B.II; Funding Acquisition, A.L.M. and A.J.B.II; Methodology, D.S.W., A.L.M., J.D.T., A.J.B.II; Resources, A.L.M. and A.J.B.II; Writing—Original Draft Preparation, D.S.W. and A.J.B.II; Writing—Review and Editing, D.S.W., A.L.M., J.D.T., A.J.B.II; All authors approved of the final submitted version of the manuscript.

Funding: Acquisition of the IUSM Proteomics core instrumentation used for this project was provided by the Indiana University Precision Health Initiative. Transgenic mice were generated at the Vanderbilt Transgenic animal core with support from the Cancer Center Support Grant (CA68485) and the Vanderbilt Diabetes Research and Training Center (DK020593). AJB was supported by K01NS073700, R21/R33DA04187, Department of Biology Start-up funds, and Bridge Funding. DSW was supported by a supplement R33DA04187-03S1.

Acknowledgments: Mass spectrometry was provided by the Indiana University School of Medicine Proteomics Core Facility and we acknowledge the help of Guihong Qi and Aruna Wijeratne for mass spectrometry technical assistance and data analysis.

Conflicts of Interest: The authors have no conflicts of interest to declare.

References

1. Degenhardt, L.; Hall, W. Extent of illicit drug use and dependence, and their contribution to the global burden of disease. *Lancet* **2012**, *379*, 55–70. [CrossRef]
2. Ashok, A.H.; Mizuno, Y.; Volkow, N.D.; Howes, O.D. Association of Stimulants With Dopaminergic Alterations in Users of Cocaine, Amphetamine, and Methamphetamine: A Systematic Review and Meta-analysis. *JAMA Psychiatry* **2017**, *74*, 511–519. [CrossRef] [PubMed]
3. Burke, A.R.; Miczek, K.A. Stress in adolescence and drugs of abuse in rodent models: Role of dopamine, CRF, and HPA axis. *Psychopharmacology* **2014**, *231*, 1557–1580. [CrossRef] [PubMed]
4. Gerdeman, G.L.; Partridge, J.G.; Lupica, C.R.; Lovinger, D.M. It could be habit forming: Drugs of abuse and striatal synaptic plasticity. *Trends Neurosci.* **2003**, *26*, 184–192. [CrossRef]
5. Sanchez-Ramos, J. Chapter Seven–Neurologic Complications of Psychomotor Stimulant Abuse. In *International Review of Neurobiology*; Taba, P., Lees, A., Sikk, K., Eds.; Academic Press: Amsterdam, The Netherlands, 2015; Volume 120, pp. 131–160.
6. Chiodi, V.; Mallozzi, C.; Ferrante, A.; Chen, J.F.; Lombroso, P.J.; Di Stasi, A.M.M.; Popoli, P.; Domenici, M.R. Cocaine-Induced Changes of Synaptic Transmission in the Striatum are Modulated by Adenosine A(2A) Receptors and Involve the Tyrosine Phosphatase STEP. *Neuropsychopharmacology* **2014**, *39*, 569–578. [CrossRef] [PubMed]
7. Centonze, D.; Picconi, B.; Baunez, C.; Borrelli, E.; Pisani, A.; Bernardi, G.; Calabresi, P. Cocaine and Amphetamine Depress Striatal GABAergic Synaptic Transmission through D2 Dopamine Receptors. *Neuropsychopharmacology* **2002**, *26*, 164. [CrossRef]
8. Ungless, M.A.; Whistler, J.L.; Malenka, R.C.; Bonci, A. Single cocaine exposure in vivo induces long-term potentiation in dopamine neurons. *Nature* **2001**, *411*, 583. [CrossRef]
9. Di Chiara, G.; Imperato, A. Drugs abused by humans preferentially increase synaptic dopamine concentrations in the mesolimbic system of freely moving rats. *Proc. Natl. Acad. Sci. USA* **1988**, *85*, 5274–5278. [CrossRef]
10. Volkow, N.D.; Wang, G.-J.; Fowler, J.S.; Logan, J.; Gatley, S.J.; Wong, C.; Hitzemann, R.; Pappas, N.R. Reinforcing Effects of Psychostimulants in Humans Are Associated with Increases in Brain Dopamine and Occupancy of D2 Receptors. *J. Pharmacol. Exp. Ther.* **1999**, *291*, 409–415.
11. Siviy, S.M.; McDowell, L.S.; Eck, S.R.; Turano, A.; Akopian, G.; Walsh, J.P. Effects of amphetamine on striatal dopamine release, open-field activity, and play in Fischer 344 and Sprague–Dawley rats. *Behav. Pharmacol.* **2015**, *26*, 720–732. [CrossRef]
12. dela Peña, I.; Gevorkiana, R.; Shi, W.-X. Psychostimulants affect dopamine transmission through both dopamine transporter-dependent and independent mechanisms. *Eur. J. Pharmacol.* **2015**, *764*, 562–570. [CrossRef]
13. Cass, W.A.; Gerhardt, G.A.; Mayfield, R.D.; Curella, P.; Zahniser, N.R. Differences in Dopamine Clearance and Diffusion in Rat Striatum and Nucleus Accumbens Following Systemic Cocaine Administration. *J. Neurochem.* **1992**, *59*, 259–266. [CrossRef] [PubMed]
14. Bertran-Gonzalez, J.; Bosch, C.; Maroteaux, M.; Matamales, M.; Hervé, D.; Valjent, E.; Girault, J.-A. Opposing patterns of signaling activation in dopamine D1 and D2 receptor-expressing striatal neurons in response to cocaine and haloperidol. *J. Neurosci.* **2008**, *28*, 5671. [CrossRef] [PubMed]
15. Kalivas, P.W.; Stewart, J. Dopamine transmission in the initiation and expression of drug- and stress-induced sensitization of motor activity. *Brain Res. Rev.* **1991**, *16*, 223–244. [CrossRef]

16. Robinson, T.E.; Becker, J.B. Enduring changes in brain and behavior produced by chronic amphetamine administration: A review and evaluation of animal models of amphetamine psychosis. *Brain Res. Rev.* **1986**, *11*, 157–198. [CrossRef]
17. Joshua, M.; Adler, A.; Bergman, H. The dynamics of dopamine in control of motor behavior. *Curr. Opin. Neurobiol.* **2009**, *19*, 615–620. [CrossRef] [PubMed]
18. Ishiguro, A.; Inagaki, M.; Kaga, M. Stereotypic circling behavior in mice with vestibular dysfunction: Asymmetrical effects of intrastriatal microinjection of a dopamine agonist. *Int. J. Neurosci.* **2007**, *117*, 1049–1064. [CrossRef] [PubMed]
19. Ferro, M.M.; Bellissimo, M.I.; Anselmo-Franci, J.A.; Angellucci, M.E.M.; Canteras, N.S.; Da Cunha, C. Comparison of bilaterally 6-OHDA- and MPTP-lesioned rats as models of the early phase of Parkinson's disease: Histological, neurochemical, motor and memory alterations. *J. Neurosci. Method.* **2005**, *148*, 78–87. [CrossRef]
20. Archer, T.; Danysz, W.; Fredriksson, A.; Jonsson, G.; Luthman, J.; Sundström, E.; Teiling, A. Neonatal 6-hydroxydopamine-induced dopamine depletions: Motor activity and performance in maze learning. *Pharmacol. Biochem. Behav.* **1988**, *31*, 357–364. [CrossRef]
21. Herbin, M.; Simonis, C.; Revéret, L.; Hackert, R.; Libourel, P.-A.; Eugène, D.; Diaz, J.; de Waele, C.; Vidal, P.-P. Dopamine Modulates Motor Control in a Specific Plane Related to Support. *PLoS ONE* **2016**, *11*, e0155058. [CrossRef]
22. Wong, L.S.; Eshel, G.; Dreher, J.; Ong, J.; Jackson, D.M. Role of dopamine and GABA in the control of motor activity elicited from the rat nucleus accumbens. *Pharmacol. Biochem. Behav.* **1991**, *38*, 829–835. [CrossRef]
23. Rikani, A.A.; Choudhry, Z.; Choudhry, A.M.; Rizvi, N.; Ikram, H.; Mobassarah, N.J.; Tuli, S. The mechanism of degeneration of striatal neuronal subtypes in Huntington disease. *Annal. Neurosci.* **2014**, *21*, 112–114. [CrossRef] [PubMed]
24. Belluscio, M.A.; Escande, M.V.; Keifman, E.; Riquelme, L.A.; Murer, M.G.; Zold, C.L. Oscillations in the basal ganglia in Parkinson's disease: Role of the striatum. *Basal Ganglia* **2014**, *3*, 203–212. [CrossRef]
25. Greenberg, B.D.; Gabriels, L.A.; Malone, D.A., Jr.; Rezai, A.R.; Friehs, G.M.; Okun, M.S.; Shapira, N.A.; Foote, K.D.; Cosyns, P.R.; Kubu, C.S.; et al. Deep brain stimulation of the ventral internal capsule/ventral striatum for obsessive-compulsive disorder: Worldwide experience. *Mol. Psychiatry* **2008**, *15*, 64. [CrossRef] [PubMed]
26. Yip, S.W.; Worhunsky, P.D.; Rogers, R.D.; Goodwin, G.M. Hypoactivation of the Ventral and Dorsal Striatum During Reward and Loss Anticipation in Antipsychotic and Mood Stabilizer-Naive Bipolar Disorder. *Neuropsychopharmacology* **2014**, *40*, 658. [CrossRef] [PubMed]
27. Yager, L.M.; Garcia, A.F.; Wunsch, A.M.; Ferguson, S.M. The ins and outs of the striatum: Role in drug addiction. *Neuroscience* **2015**, *301*, 529–541. [CrossRef] [PubMed]
28. Burguière, E.; Monteiro, P.; Mallet, L.; Feng, G.; Graybiel, A.M. Striatal circuits, habits, and implications for obsessive-compulsive disorder. *Curr. Opin. Neurobiol.* **2015**, *0*, 59–65. [CrossRef] [PubMed]
29. Graybiel, A.M.; Grafton, S.T. The Striatum: Where Skills and Habits Meet. *CSH Perspect. Biol.* **2015**, *7*, a021691. [CrossRef] [PubMed]
30. Balleine, B.W.; O'Doherty, J.P. Human and Rodent Homologies in Action Control: Corticostriatal Determinants of Goal-Directed and Habitual Action. *Neuropsychopharmacology* **2010**, *35*, 48–69. [CrossRef] [PubMed]
31. Durieux, P.F.; Bearzatto, B.; Guiducci, S.; Buch, T.; Waisman, A.; Zoli, M.; Schiffmann, S.N.; de Kerchove d'Exaerde, A. D2R striatopallidal neurons inhibit both locomotor and drug reward processes. *Nat. Neurosci.* **2009**, *12*, 393. [CrossRef] [PubMed]
32. Flanigan, M.; LeClair, K. Shared Motivational Functions of Ventral Striatum D1 and D2 Medium Spiny Neurons. *J. Neurosci.* **2017**, *37*, 6177. [CrossRef] [PubMed]
33. Ikemoto, S. Dopamine reward circuitry: Two projection systems from the ventral midbrain to the nucleus accumbens-olfactory tubercle complex. *Brain Res. Rev.* **2007**, *56*, 27–78. [CrossRef] [PubMed]
34. Belin, D.; Everitt, B.J. Cocaine Seeking Habits Depend upon Dopamine-Dependent Serial Connectivity Linking the Ventral with the Dorsal Striatum. *Neuron* **2008**, *57*, 432–441. [CrossRef] [PubMed]

35. Kravitz, A.V.; Kreitzer, A.C. Striatal Mechanisms Underlying Movement, Reinforcement, and Punishment. *Physiology* **2012**, *27*. [CrossRef] [PubMed]
36. Gerfen, C.R.; Engber, T.M.; Mahan, L.C.; Susel, Z.; Chase, T.N.; Monsma, F.J.; Sibley, D.R. D1 and D2 dopamine receptor-regulated gene expression of striatonigral and striatopallidal neurons. *Science* **1990**, *250*, 1429. [CrossRef] [PubMed]
37. Gerfen, C.R.; Scott Young, W. Distribution of striatonigral and striatopallidal peptidergic neurons in both patch and matrix compartments: An in situ hybridization histochemistry and fluorescent retrograde tracing study. *Brain Res.* **1988**, *460*, 161–167. [CrossRef]
38. Lanciego, J.L.; Luquin, N.; Obeso, J.A. Functional Neuroanatomy of the Basal Ganglia. *CSH Perspect. Med.* **2012**, *2*, a009621. [CrossRef] [PubMed]
39. Burns, R.S.; LeWitt, P.A.; Ebert, M.H.; Pakkenberg, H.; Kopin, I.J. The Clinical Syndrome of Striatal Dopamine Deficiency. *N. Eng. J. Med.* **1985**, *312*, 1418–1421. [CrossRef] [PubMed]
40. Chen, J.Y.; Wang, E.A.; Cepeda, C.; Levine, M.S. Dopamine imbalance in Huntington's disease: A mechanism for the lack of behavioral flexibility. *Front. Neurosci.* **2013**, *7*, 114. [CrossRef]
41. Surmeier, D.J.; Graves, S.M.; Shen, W. Dopaminergic modulation of striatal networks in health and Parkinson's disease. *Curr. Opin. Neurobiol.* **2014**, *29*, 109–117. [CrossRef] [PubMed]
42. Goto, A.; Nakahara, I.; Yamaguchi, T.; Kamioka, Y.; Sumiyama, K.; Matsuda, M.; Nakanishi, S.; Funabiki, K. Circuit-dependent striatal PKA and ERK signaling underlies rapid behavioral shift in mating reaction of male mice. *Proc. Natl. Acad. Sci. USA* **2015**, *112*, 6718–6723. [CrossRef] [PubMed]
43. Flores-Hernández, J.; Cepeda, C.; Hernández-Echeagaray, E.; Calvert, C.R.; Jokel, E.S.; Fienberg, A.A.; Greengard, P.; Levine, M.S. Dopamine Enhancement of NMDA Currents in Dissociated Medium-Sized Striatal Neurons: Role of D1 Receptors and DARPP-32. *J. Neurophysiol.* **2002**, *88*, 3010–3020. [CrossRef]
44. Surmeier, D.J.; Ding, J.; Day, M.; Wang, Z.; Shen, W. D1 and D2 dopamine-receptor modulation of striatal glutamatergic signaling in striatal medium spiny neurons. *Trends Neurosci.* **2007**, *30*, 228–235. [CrossRef]
45. Bahuguna, J.; Weidel, P.; Morrison, A. Exploring the role of striatal D1 and D2 medium spiny neurons in action selection using a virtual robotic framework. *Eur. J. Neurosci.* **2018**. [CrossRef] [PubMed]
46. Peti, W.; Nairn, A.C.; Page, R. Structural Basis for Protein Phosphatase 1 Regulation and Specificity. *FEBS J.* **2013**, *280*, 596–611. [CrossRef] [PubMed]
47. Woolfrey, K.M.; Dell'Acqua, M.L. Coordination of Protein Phosphorylation and Dephosphorylation in Synaptic Plasticity. *J. Biol. Chem.* **2015**, *290*, 28604–28612. [CrossRef] [PubMed]
48. Allen, P.B.; Ouimet, C.C.; Greengard, P. Spinophilin, a novel protein phosphatase 1 binding protein localized to dendritic spines. *Proc. Natl. Acad. Sci. USA* **1997**, *94*, 9956–9961. [CrossRef]
49. Hsieh-Wilson, L.C.; Allen, P.B.; Watanabe, T.; Nairn, A.C.; Greengard, P. Characterization of the Neuronal Targeting Protein Spinophilin and Its Interactions with Protein Phosphatase-1. *Biochemistry* **1999**, *38*, 4365–4373. [CrossRef]
50. Colbran, R.J.; Bass, M.A.; McNeill, R.B.; Bollen, M.; Zhao, S.; Wadzinski, B.E.; Strack, S. Association of brain protein phosphatase 1 with cytoskeletal targeting/regulatory subunits. *J. Neurochem.* **1997**, *69*, 920–929. [CrossRef]
51. Morishita, W.; Connor, J.H.; Xia, H.; Quinlan, E.M.; Shenolikar, S.; Malenka, R.C. Regulation of Synaptic Strength by Protein Phosphatase 1. *Neuron* **2001**, *32*, 1133–1148. [CrossRef]
52. Ragusa, M.J.; Dancheck, B.; Critton, D.A.; Nairn, A.C.; Page, R.; Peti, W. Spinophilin directs Protein Phosphatase 1 specificity by blocking substrate binding sites. *Nat. Struct. Mol. Biol.* **2010**, *17*, 459–464. [CrossRef] [PubMed]
53. Hu, X.D.; Huang, Q.; Roadcap, D.W.; Shenolikar, S.S.; Xia, H. Actin-associated neurabin–protein phosphatase-1 complex regulates hippocampal plasticity. *J. Neurochem.* **2006**, *98*, 1841–1851. [CrossRef] [PubMed]
54. Muly, E.C.; Smith, Y.; Allen, P.; Greengard, P. Subcellular distribution of spinophilin immunolabeling in primate prefrontal cortex: Localization to and within dendritic spines. *J. Comp. Neurol.* **2004**, *469*, 185–197. [CrossRef] [PubMed]

55. Muly, E.C.; Allen, P.; Mazloom, M.; Aranbayeva, Z.; Greenfield, A.T.; Greengard, P. Subcellular Distribution of Neurabin Immunolabeling in Primate Prefrontal Cortex: Comparison with Spinophilin. *Cereb. Cortex* **2004**, *14*, 1398–1407. [CrossRef] [PubMed]
56. Feng, J.; Yan, Z.; Ferreira, A.; Tomizawa, K.; Liauw, J.A.; Zhuo, M.; Allen, P.B.; Ouimet, C.C.; Greengard, P. Spinophilin regulates the formation and function of dendritic spines. *Proc. Natl. Acad. Sci. USA* **2000**, *97*, 9287–9292. [CrossRef]
57. Roze, E.; Cahill, E.; Martin, E.; Bonnet, C.; Vanhoutte, P.; Betuing, S.; Caboche, J. Huntington's Disease and Striatal Signaling. *Front. Neuroanat.* **2011**, *5*, 55. [CrossRef] [PubMed]
58. Boikess, S.R.; O'Dell, S.J.; Marshall, J.F. A sensitizing d-amphetamine dose regimen induces long-lasting spinophilin and VGLUT1 protein upregulation in the rat diencephalon. *Neurosci. Lett.* **2010**, *469*, 49–54. [CrossRef]
59. Boikess, S.R.; Marshall, J.F. A sensitizing d-amphetamine regimen induces long-lasting spinophilin protein upregulation in the rat striatum and limbic forebrain. *Eur. J. Neurosci.* **2008**, *28*, 2099–2107. [CrossRef]
60. Xu, J.; Chen, Y.; Lu, R.; Cottingham, C.; Jiao, K.; Wang, Q. Protein Kinase A Phosphorylation of Spinophilin Modulates Its Interaction with the α2A-Adrenergic Receptor (AR) and Alters Temporal Properties of α2AAR Internalization. *J. Biol. Chem.* **2008**, *283*, 14516–14523. [CrossRef]
61. Brown, A.M.; Baucum, A.J.; Bass, M.A.; Colbran, R.J. Association of protein phosphatase 1 gamma 1 with spinophilin suppresses phosphatase activity in a Parkinson disease model. *J. Biol. Chem.* **2008**, *283*, 14286–14294. [CrossRef]
62. Hiday, A.C.; Edler, M.C.; Salek, A.B.; Morris, C.W.; Thang, M.; Rentz, T.J.; Rose, K.L.; Jones, L.M.; Baucum, A.J. Mechanisms and Consequences of Dopamine Depletion-Induced Attenuation of the Spinophilin/Neurofilament Medium Interaction. *Neural Plast.* **2017**. [CrossRef] [PubMed]
63. Morris, C.W.; Watkins, D.S.; Salek, A.B.; Edler, M.C.; Baucum, A.J., II. The association of spinophilin with disks large-associated protein 3 (SAPAP3) is regulated by metabotropic glutamate receptor (mGluR) 5. *Mol. Cell. Neurosci.* **2018**, *90*, 60–69. [CrossRef] [PubMed]
64. Gerfen, C.R.; Paletzki, R.; Heintz, N. GENSAT BAC cre-recombinase driver lines to study the functional organization of cerebral cortical and basal ganglia circuits. *Neuron* **2013**, *80*, 1368–1383. [CrossRef] [PubMed]
65. Gong, S.; Doughty, M.; Harbaugh, C.R.; Cummins, A.; Hatten, M.E.; Heintz, N.; Gerfen, C.R. Targeting Cre recombinase to specific neuron populations with bacterial artificial chromosome constructs. *J. Neurosci.* **2007**, *27*, 9817–9823. [CrossRef] [PubMed]
66. Edler, M.C.; Salek, A.B.; Watkins, D.S.; Kaur, H.; Morris, C.W.; Yamamoto, B.K.; Baucum, A.J. Mechanisms Regulating the Association of Protein Phosphatase 1 with Spinophilin and Neurabin. *ACS Chem. Neurosci.* **2018**. [CrossRef] [PubMed]
67. Smith-Kinnaman, W.R.; Berna, M.J.; Hunter, G.O.; True, J.D.; Hsu, P.; Cabello, G.I.; Fox, M.J.; Varani, G.; Mosley, A.L. The interactome of the atypical phosphatase Rtr1 in Saccharomyces cerevisiae. *Mol. Biosyst.* **2014**, *10*, 1730–1741. [CrossRef] [PubMed]
68. Mosley, A.L.; Sardiu, M.E.; Pattenden, S.G.; Workman, J.L.; Florens, L.; Washburn, M.P. Highly reproducible label free quantitative proteomic analysis of RNA polymerase complexes. *Mol. Cell. Proteom.* **2011**, *10*, M110000687. [CrossRef]
69. Huang da, W.; Sherman, B.T.; Lempicki, R.A. Systematic and integrative analysis of large gene lists using DAVID bioinformatics resources. *Nat. Protoc.* **2009**, *4*, 44–57. [CrossRef]
70. Huang da, W.; Sherman, B.T.; Lempicki, R.A. Bioinformatics enrichment tools: Paths toward the comprehensive functional analysis of large gene lists. *Nucleic Acids Res.* **2009**, *37*, 1–13. [CrossRef]
71. Chu, J.; Haynes, R.D.; Corbel, S.Y.; Li, P.; Gonzalez-Gonzalez, E.; Burg, J.S.; Ataie, N.J.; Lam, A.J.; Cranfill, P.J.; Baird, M.A.; et al. Non-invasive intravital imaging of cellular differentiation with a bright red-excitable fluorescent protein. *Nat. Methods* **2014**, *11*, 572–578. [CrossRef]
72. Baucum, A.J.; Brown, A.M.; Colbran, R.J. Differential association of postsynaptic signaling protein complexes in striatum and hippocampus. *J. Neurochem.* **2013**, *124*, 490–501. [CrossRef]
73. Baucum, A.J.; Jalan-Sakrikar, N.; Jiao, Y.; Gustin, R.M.; Carmody, L.C.; Tabb, D.L.; Ham, A.J.; Colbran, R.J. Identification and validation of novel spinophilin-associated proteins in rodent striatum using an enhanced ex vivo shotgun proteomics approach. *Mol. Cell. Proteom.* **2010**, *9*, 1243–1259. [CrossRef]

74. Gerfen, C.R.; Surmeier, D.J. Modulation of striatal projection systems by dopamine. *Annu. Rev. Neurosci.* **2011**, *34*, 441–466. [CrossRef]
75. Hu, X.T.; Wang, R.Y. Comparison of effects of D-1 and D-2 dopamine receptor agonists on neurons in the rat caudate putamen: An electrophysiological study. *J. Neurosci.* **1988**, *8*, 4340. [CrossRef]
76. Muhammad, K.; Reddy-Alla, S.; Driller, J.H.; Schreiner, D.; Rey, U.; Bohme, M.A.; Hollmann, C.; Ramesh, N.; Depner, H.; Lutzkendorf, J.; et al. Presynaptic spinophilin tunes neurexin signalling to control active zone architecture and function. *Nat. Commun.* **2015**, *6*, 8362. [CrossRef]
77. Szklarczyk, D.; Franceschini, A.; Wyder, S.; Forslund, K.; Heller, D.; Huerta-Cepas, J.; Simonovic, M.; Roth, A.; Santos, A.; Tsafou, K.P.; et al. STRING v10: Protein-protein interaction networks, integrated over the tree of life. *Nucleic Acids Res.* **2015**, *43*, D447–D452. [CrossRef]
78. Szklarczyk, D.; Morris, J.H.; Cook, H.; Kuhn, M.; Wyder, S.; Simonovic, M.; Santos, A.; Doncheva, N.T.; Roth, A.; Bork, P.; et al. The STRING database in 2017: Quality-controlled protein-protein association networks, made broadly accessible. *Nucleic Acids Res.* **2017**, *45*, D362–D368. [CrossRef]
79. Allen, P.B.; Zachariou, V.; Svenningsson, P.; Lepore, A.C.; Centonze, D.; Costa, C.; Rossi, S.; Bender, G.; Chen, G.; Feng, J.; et al. Distinct roles for spinophilin and neurabin in dopamine-mediated plasticity. *Neuroscience* **2006**, *140*, 897–911. [CrossRef]
80. Lu, R.; Chen, Y.; Cottingham, C.; Peng, N.; Jiao, K.; Limbird, L.E.; Wyss, J.M.; Wang, Q. Enhanced hypotensive, bradycardic, and hypnotic responses to alpha2-adrenergic agonists in spinophilin-null mice are accompanied by increased G protein coupling to the alpha2A-adrenergic receptor. *Mol. Pharmacol.* **2010**, *78*, 279–286. [CrossRef]
81. Ingham, C.A.; Hood, S.H.; Arbuthnott, G.W. Spine density on neostriatal neurones changes with 6-hydroxydopamine lesions and with age. *Brain Res.* **1989**, *503*, 334–338. [CrossRef]
82. Ingham, C.A.; Hood, S.H.; van Maldegem, B.; Weenink, A.; Arbuthnott, G.W. Morphological changes in the rat neostriatum after unilateral 6-hydroxydopamine injections into the nigrostriatal pathway. *Exp. Brain Res.* **1993**, *93*, 17–27. [CrossRef]
83. Day, M.; Wang, Z.; Ding, J.; An, X.; Ingham, C.A.; Shering, A.F.; Wokosin, D.; Ilijic, E.; Sun, Z.; Sampson, A.R.; et al. Selective elimination of glutamatergic synapses on striatopallidal neurons in Parkinson disease models. *Nat. Neurosci.* **2006**, *9*, 251–259. [CrossRef]
84. Lee, K.W.; Kim, Y.; Kim, A.M.; Helmin, K.; Nairn, A.C.; Greengard, P. Cocaine-induced dendritic spine formation in D1 and D2 dopamine receptor-containing medium spiny neurons in nucleus accumbens. *Proc. Natl. Acad. Sci. USA* **2006**, *103*, 3399–3404. [CrossRef]
85. Li, Y.; Kolb, B.; Robinson, T.E. The location of persistent amphetamine-induced changes in the density of dendritic spines on medium spiny neurons in the nucleus accumbens and caudate-putamen. *Neuropsychopharmacology* **2003**, *28*, 1082–1085. [CrossRef]
86. Evans, J.C.; Robinson, C.M.; Shi, M.; Webb, D.J. The Guanine Nucleotide Exchange Factor (GEF) Asef2 Promotes Dendritic Spine Formation via Rac Activation and Spinophilin-dependent Targeting. *J. Biol. Chem.* **2015**, *290*, 10295–10308. [CrossRef]
87. Hsieh-Wilson, L.C.; Benfenati, F.; Snyder, G.L.; Allen, P.B.; Nairn, A.C.; Greengard, P. Phosphorylation of spinophilin modulates its interaction with actin filaments. *J. Biol. Chem.* **2003**, *278*, 1186–1194. [CrossRef]
88. Satoh, A.; Nakanishi, H.; Obaishi, H.; Wada, M.; Takahashi, K.; Satoh, K.; Hirao, K.; Nishioka, H.; Hata, Y.; Mizoguchi, A.; et al. Neurabin-II/spinophilin. An actin filament-binding protein with one pdz domain localized at cadherin-based cell-cell adhesion sites. *J. Biol. Chem.* **1998**, *273*, 3470–3475. [CrossRef]
89. Mellacheruvu, D.; Wright, Z.; Couzens, A.L.; Lambert, J.P.; St-Denis, N.A.; Li, T.; Miteva, Y.V.; Hauri, S.; Sardiu, M.E.; Low, T.Y.; et al. The CRAPome: A contaminant repository for affinity purification-mass spectrometry data. *Nat. Methods* **2013**, *10*, 730–736. [CrossRef]
90. Berman, S.; O'Neill, J.; Fears, S.; Bartzokis, G.; London, E.D. Abuse of amphetamines and structural abnormalities in the brain. *Ann. N. Y. Acad. Sci.* **2008**, *1141*, 195–220. [CrossRef]
91. Kimura, T.; Allen, P.B.; Nairn, A.C.; Caplan, M.J. Arrestins and spinophilin competitively regulate Na+,K+-ATPase trafficking through association with a large cytoplasmic loop of the Na+,K+-ATPase. *Mol. Biol. Cell* **2007**, *18*, 4508–4518. [CrossRef]
92. King, L.J.; Carl, J.L.; Lao, L. Cocaine and amphetamine modification of cerebral energy metabolism in vivo. *Psychopharmacologia* **1975**, *44*, 43–45. [CrossRef]

93. Valvassori, S.S.; Calixto, K.V.; Budni, J.; Resende, W.R.; Varela, R.B.; de Freitas, K.V.; Goncalves, C.L.; Streck, E.L.; Quevedo, J. Sodium butyrate reverses the inhibition of Krebs cycle enzymes induced by amphetamine in the rat brain. *J. Neural Transm.* **2013**, *120*, 1737–1742. [CrossRef]
94. Klongpanichapak, S.; Phansuwan-Pujito, P.; Ebadi, M.; Govitrapong, P. Melatonin protects SK-N-SH neuroblastoma cells from amphetamine-induced neurotoxicity. *J. Pineal Res.* **2007**, *43*, 65–73. [CrossRef]
95. Straiko, M.M.; Coolen, L.M.; Zemlan, F.P.; Gudelsky, G.A. The effect of amphetamine analogs on cleaved microtubule-associated protein-tau formation in the rat brain. *Neuroscience* **2007**, *144*, 223–231. [CrossRef]

proteomes

MDPI

Correction

Correction: Baucum II, Anthony J. et al. Proteomic Analysis of the Spinophilin Interactome in Rodent Striatum Following Psychostimulant Sensitization. *Proteomes* 2018, 6, 53

Darryl S. Watkins [1], Jason D. True [2,3], Amber L. Mosley [2] and Anthony J. Baucum II [4,5,6,*]

1 Stark Neurosciences Research Institute, Indiana University School of Medicine Medical Neuroscience Graduate Program, Indianapolis, IN 46278, USA; dswatkin@iu.edu
2 Department of Biochemistry and Molecular Biology, Indiana University School of Medicine, Indianapolis, IN 46278, USA; Jdtrue@bsu.edu (J.D.T.); almosley@iu.edu (A.L.M.)
3 Department of Biology, Ball State University, Muncie, IN 47306, USA
4 Department of Biology, Indiana University-Purdue University Indianapolis, Indianapolis, IN 46202, USA
5 Stark Neurosciences Research Institute Indianapolis, Indianapolis, IN 46202, USA
6 Department of Pharmacology and Toxicology, Indiana University School of Medicine, Indianapolis, IN 46202, USA
* Correspondence: ajbaucum@iupui.edu; Tel.: +1-317-274-0540; Fax: +1-317-274-2846

Received: 16 January 2019; Accepted: 17 January 2019; Published: 13 February 2019

The author wishes to make the following corrections to the methods section of their paper [1]:

The corrections include: (1) Replace the sentence "Mouse striatum (including both dorsal and ventral (accumbens) striatum or olfactory tubercle) was homogenized and sonicated in 1 mL in a low-ionic strength Tris buffer containing 20 mM Tris, 10 mM DTT, 2 mM 0.5 M EDTA, 10% Triton X-100, 1% protease inhibitor cocktail (Bimake, Houston, TX, USA)" with "Mouse striatum (including both dorsal and ventral (accumbens) striatum or olfactory tubercle) was homogenized and sonicated in 1 mL in a low-ionic strength Tris buffer containing 2 mM Tris-HCl, 1 mM DTT, 2 mM EDTA, 1% Triton X-100, and 1X protease inhibitor cocktail (Bimake, Houston, TX, USA)"; (2) replace the sentence "magnetic separation in an immunoprecipitation wash buffer (50 mM NaCl, 50 mM Tris-HCl pH 7.5, 0.5% (*v*/*v*) Triton X-100)" with "magnetic separation in an immunoprecipitation wash buffer (150 mM NaCl, 50 mM Tris-HCl pH 7.5, 0.5% (*v*/*v*) Triton X-100)".

These corrections do not affect the conclusions of the study. We apologize for the typographical errors, and the authors would like to apologize for any inconvenience caused to the readers by these changes. The manuscript will be updated, and the original manuscript will remain online on the article webpage.

References

1. Watkins, D.S.; True, J.D.; Mosley, A.L.; Baucum, A.J., II. Proteomic Analysis of the Spinophilin Interactome in Rodent Striatum Following Psychostimulant Sensitization. *Proteomes* **2018**, *6*, 53. [CrossRef] [PubMed]

proteomes

MDPI

Article

Proteases Shape the *Chlamydomonas* Secretome: Comparison to Classical Neuropeptide Processing Machinery

Raj Luxmi [1], Crysten Blaby-Haas [2], Dhivya Kumar [3,†], Navin Rauniyar [4], Stephen M. King [3], Richard E. Mains [1] and Betty A. Eipper [1,3,*]

1. Department of Neuroscience, University of Connecticut Health Center, Farmington, CT 06030-3401, USA; luxmi@uchc.edu (R.L.); mains@uchc.edu (R.E.M.)
2. Department of Biology, Brookhaven National Laboratory, Upton, NY 11973-5000, USA; cblaby@bnl.gov
3. Department of Molecular Biology and Biophysics, University of Connecticut Health Center, Farmington, CT 06030-3305, USA; dhivya.kumar@ucsf.edu (D.K.); king@uchc.edu (S.M.K.)
4. W.M. Keck Biotechnology Resource Laboratory, Yale University, New Haven, CT 06511-6624, USA; navin.rauniyar@yale.edu

* Correspondence: eipper@uchc.edu; Tel.: +1-860-679-8898

† Current address: Department of Biochemistry and Biophysics, University of California, San Francisco, San Francisco, CA 94143-0525, USA

Received: 30 August 2018; Accepted: 20 September 2018; Published: 23 September 2018

Abstract: The recent identification of catalytically active peptidylglycine α-amidating monooxygenase (PAM) in *Chlamydomonas reinhardtii*, a unicellular green alga, suggested the presence of a PAM-like gene and peptidergic signaling in the last eukaryotic common ancestor (LECA). We identified prototypical neuropeptide precursors and essential peptide processing enzymes (subtilisin-like prohormone convertases and carboxypeptidase B-like enzymes) in the *C. reinhardtii* genome. Reasoning that sexual reproduction by *C. reinhardtii* requires extensive communication between cells, we used mass spectrometry to identify proteins recovered from the soluble secretome of mating gametes, and searched for evidence that the putative peptidergic processing enzymes were functional. After fractionation by SDS-PAGE, signal peptide-containing proteins that remained intact, and those that had been subjected to cleavage, were identified. The *C. reinhardtii* mating secretome contained multiple matrix metalloproteinases, cysteine endopeptidases, and serine carboxypeptidases, along with one subtilisin-like proteinase. Published transcriptomic studies support a role for these proteases in sexual reproduction. Multiple extracellular matrix proteins (ECM) were identified in the secretome. Several pherophorins, ECM glycoproteins homologous to the *Volvox* sex-inducing pheromone, were present; most contained typical peptide processing sites, and many had been cleaved, generating stable N- or C-terminal fragments. Our data suggest that subtilisin endoproteases and matrix metalloproteinases similar to those important in vertebrate peptidergic and growth factor signaling play an important role in stage transitions during the life cycle of *C. reinhardtii*.

Keywords: peptidylglycine α-amidating monooxygenase; cilia; mating; signal peptide; prohormone convertase; carboxypeptidase; matrix metalloproteinase; subtilisin; pherophorin

1. Introduction

Identification of the enkephalins as endogenous ligands for opioid receptors led to the successful description of hundreds of additional bioactive peptides in the nervous systems of species as diverse as *Drosophila*, *Caenorhabditis elegans*, and *Hydra* [1–3]. Like proinsulin and proopiomelanocortin, the precursors to these neuropeptides have N-terminal signal sequences, multiple potential paired basic amino acid endoproteolytic cleavage sites, potential amidation sites, generally lack recognized domains, and often contain multiple copies of similar peptides (Figure 1). Bioinformatic criteria for the identification of potential neuropeptide precursors have been successfully applied to many systems [4,5].

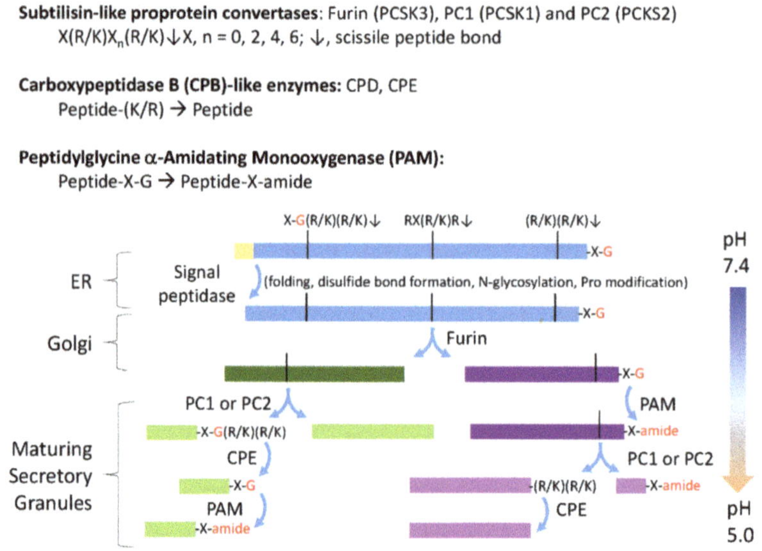

Figure 1. Classical neuropeptide precursors and processing enzymes. The biosynthesis and post-translational processing of neuropeptide precursors from organisms as diverse as human and *Hydra* employs a common set of subcellular organelles and processing enzymes. The reactions catalyzed by subtilisin-like prohormone convertases, carboxypeptidase B (CPB)-like enzymes and peptidylglycine α-amidating monooxygenase (PAM) are shown. Endoplasmic reticulum (ER) entry requires an N-terminal signal peptide, which is quickly removed. As for other secreted proteins, N-glycosylation, disulfide bond formation, proline hydroxylation, and proline isomerization are accomplished before transit through the Golgi complex. A family of subtilisin-like endoproteases, referred to as prohormone convertases (PCs), catalyze a series of ordered endoproteolytic cleavages, with furin (PCSK3), PC1 (PCSK1), and PC2 (PCSK2) playing especially important roles in many neurons and endocrine cells. Endoproteolytic cleavage is controlled, in large part, by the pH of the luminal compartment, with furin active in the *trans*-Golgi network and endocytic compartments, and PC1 and PC2 more active in the low pH environment encountered in immature and mature secretory granules. CPB-like enzymes (CPE and CPD) remove the C-terminal Lys and Arg residues produced by furin, PC1, and PC2. The amidating enzyme, PAM, requires only a C-terminal Gly residue to amidate the penultimate residue (–X–amide); in the presence of adequate copper, ascorbate, and molecular oxygen, PAM can function throughout the biosynthetic pathway [6].

With the availability of genomic and transcriptomic data from a diverse array of organisms, it quickly became clear that "neuropeptide precursors" were quite prevalent in species lacking recognizable neurons or endocrine cells. *Trichoplax*, a basal, multicellular animal, lacks muscles and neurons [7–9]. Despite this, its genome encodes many of the proteins that define the nervous system: candidate voltage-gated ion channels, SNARE proteins, and neuropeptide precursors are present. *Trichoplax* use the beating of ventrally located cilia to glide over surfaces, pausing to secrete the enzymes needed to feed on algae. Ciliated gland cells produce peptide-containing secretory granules that may be used to control locomotion and digestion. Single celled eukaryotes exhibit behaviors such as phototaxis and chemotaxis, leading to the suggestion that the first steps of nervous system evolution occurred in a ciliated organism, with neural circuits evolving to control locomotor cilia [10]. For example, amidated peptides synthesized in sensory neurons regulate the swimming of *Platynereis dumerilii* larvae by controlling ciliary beat frequency [11].

The enzymes involved in converting neuropeptide precursors into bioactive peptide products are highly conserved (Figure 1). All must function within the secretory pathway, where luminal pH plays a key role in controlling precursor cleavage and product storage in secretory granules. Like other secreted proteins, neuropeptide precursors often contain essential disulfide bonds, are modified by N- and O-linked glycosylation, and are phosphorylated. A set of calcium-dependent subtilisin-like endoproteases, CPB-like enzymes, and PAM, are generally regarded as reliable markers for neuropeptide-producing cells [12–15]. The discovery of a fully functional PAM protein in *Chlamydomonas reinhardtii* suggested the presence of "neuropeptides" in *C. reinhardtii*, and in the last eukaryotic common ancestor. Secretory granules have not been observed in *C. reinhardtii* and CrPAM is localized to the Golgi and to the ciliary membrane [16]. Cells in which expression of CrPAM was reduced were unable to assemble cilia that extended beyond the transition zone [17]. Using a bioinformatics approach, we searched for and found putative prepro neuropeptides, and a complete set of peptide-processing enzymes encoded by the *C. reinhardtii* genome.

Sexual reproduction, which began in the last eukaryotic common ancestor [18], is triggered in *C. reinhardtii* by nutrient restriction. Based on studies in multiple systems, this process is likely to involve peptidergic signaling [2,4] and regulated secretion [19]. Many of the interactions and signaling pathways involved in *C. reinhardtii* mating have been well characterized, and detailed transcriptomic data for specific stages are available [20,21]. Therefore, to evaluate the functional role(s) of "neuropeptide" processing machinery in a unicellular eukaryote, we undertook an analysis of the mating secretome of *C. reinhardtii*. Using differential centrifugation, we separated mating ectosomes, small vesicles derived from the cilia, from the soluble mating secretome. We focus here on our analysis of the proteases and cleaved products identified in the soluble mating secretome.

2. Material and Methods

2.1. Strains and Growth Conditions

CC124 (mating type *minus*) and CC125 (mating type *plus*) *C. reinhardtii* strains (Chlamydomonas Resource Center) were cultured in R-medium aerated with 95% air and 5% CO_2 under a 12 h light/12 h dark cycle at 22 °C. Vegetative cells were grown for 5 days under these culture conditions. To induce gametogenesis, vegetative cells were harvested, washed, and resuspended in nitrogen-deficient minimal medium (M-N/5 medium) for 16–20 h [22]. The mating competency of the gametes was assessed by mixing mating type *minus* and *plus* gametes and microscopically verifying >80% agglutination.

2.2. Preparation of Soluble Mating Secretome

In preliminary experiments, CC124 and CC125 gametes checked for high mating efficiency were mixed together for 1 h with gentle aeration at 22 °C. Following removal of cells (1600× *g* for 5 min) and debris (20,000× *g* for 10 min), the mating medium was filtered through a 0.22 µm filter to remove particulate material. The filtrate was concentrated 7-fold using a tangential flow filter system with a 2 kDa cutoff (Sartorius Vivaflow 200, Hydrosart, 2K; Thermo Fisher Scientific, Waltham, MA, USA). The concentrated mating medium was subjected to SDS-PAGE; after visualization by silver staining (Pierce, SilverSnap kit; Thermo Fisher Scientific, Rockford, IL, USA), the gel lane was excised. Gel fragments containing high and low molecular weight proteins were separately subjected to in-gel trypsin digestion, essentially as described below. Tandem mass spectrometry was carried out at the Proteomics and Mass Spectrometry Facility at the University of Massachusetts Medical School. The sample analyzed in this preliminary experiment contained both ectosomes and soluble proteins; in subsequent analyses, additional centrifugation steps yielded ectosomes and the soluble secretome, which were analyzed separately.

Gametes of both mating types (*plus* and *minus*) were washed and resuspended in 10 mL of fresh nitrogen-free medium at a density of 8–10 × 10^6 cells/mL. An equal number of mating type *plus* and *minus* gametes were mixed; after a 1 h incubation, the cultures were centrifuged at 1600× *g* as described above. These supernatants were then centrifuged at 20,000× *g* for 30 min at 4 °C to pellet cell debris. The resulting supernatants were next centrifuged at 200,000× *g* for 60 min at 4 °C to sediment all particulate material, including extracellular vesicles. These final supernatants are referred to as the soluble mating secretome. A protease inhibitor cocktail (Roche, cOmplete ULTRA Tablets, Cat. No. 05 892 791 001) and phenylmethylsulfonyl fluoride (final concentration 0.3 mg/mL) were added to each sample, which was dispensed in aliquots that were stored at −80 °C. A total of six samples prepared at two different times were subjected to analysis, ultimately yielding Dataset 1 (Samples A, B and C) and Dataset 2 (Samples D, E and F). SDS-PAGE fractionation of 40 μL of the soluble mating secretome yielded bands that were readily visualized using silver staining (Silver Stain for Mass Spectrometry; Thermo Fisher Scientific, Rockford, IL, USA). The low speed pellet (1600× *g* for 5 min) was resuspended in 0.30 mL 20 mM 2-[tris(hydroxymethyl)-methylamino]-ethanesulfonic acid (TES), 10 mM mannitol, pH 7.4 containing 1% TX-100 and a protease inhibitor cocktail; aliquots were assayed for protein content using the bicinchoninic acid assay (Thermo Fisher Scientific, Rockford, IL, USA).

2.3. Fractionation for Mass Spectroscopy

Aliquots (40 μL) of each soluble mating secretome were prepared for SDS-PAGE by mixing with 4× Laemmli sample buffer (Bio-Rad, Hercules, CA, USA) and denaturation at 55 °C for 5 min. Each sample was then fractionated on a Criterion TGX 4–15% gradient gel (Bio-Rad, Hercules, CA, USA). Electrophoresis of the first set of samples was stopped when the dye band had traveled 3 cm, and the gel was stained with QC Colloidal Coomassie (Bio-Rad, Hercules, CA, USA), according to the manufacturer's protocol. Based on molecular weight standards analyzed at the same time, each lane was cut into 10 slices that covered material migrating from the dye band to the top of the gel. Electrophoresis of the second set of samples was stopped when the dye band had traveled 2.5 cm; after staining with QC Colloidal Coomassie, each lane was cut into 4 slices covering the entire molecular weight range. Gel slices were stored frozen in microfuge tubes before preparation for LC-MS/MS analysis after in-gel digestion with trypsin.

2.4. Preparation of Vegetative Secretome

Vegetative cells of both mating types (*plus* and *minus*) were grown for 5 days in R-media (500 mL cultures). Cells were washed and resuspended in 5 mL of fresh R-medium. Cells were incubated for 4 h under continuous light with gentle aeration. The soluble vegetative secretome was prepared as described above for the soluble mating secretome. For comparing the vegetative and mating secretomes, the volume of soluble secretome analyzed by gel electrophoresis was adjusted to represent secretion by 200 (vegetative) or 50 (mating) μg of cell protein.

2.5. Mass Spectrometry

Gel bands were cut into small pieces, washed with 250 μL of 50% acetonitrile for 5 min with rocking, then washed with 50% acetonitrile/50 mM NH_4HCO_3 for 30 min on a tilt-table. After a final 30 min wash with 50% acetonitrile/10 mM NH_4HCO_3, gel fragments were dried using a speed vacuum. Each sample was suspended in 30 μL of 10 mM NH_4HCO_3 containing 0.20 μg of digestion grade trypsin (Promega, V5111) and incubated at 37 °C for 16 h. The digestion supernatant was acidified and placed into a vial for LC-MS/MS analysis (5 μL injected).

Data acquisition was conducted on an Orbitrap Fusion Tribrid mass spectrometer. Reversed phase (RP)-LC-MS/MS was performed using a nanoACQUITY UPLC system (Waters Corporation, Milford, MA, USA) connected to an Orbitrap Fusion Tribrid (Thermo Fisher Scientific, San Jose, CA, USA) mass spectrometer. After injection, samples were loaded into a trapping column (nanoACQUITY UPLC Symmetry C18 Trap column, 180 μm × 20 mm) at a flowrate of 5 μL/min and separated using a C18 column (nanoACQUITY column Peptide BEH C18, 75 μm × 250 mm). The compositions of mobile phases

A and B were 0.1% formic acid in water and 0.1% formic acid in acetonitrile, respectively. Peptides were eluted with a gradient extending from 3% to 20% mobile phase B in 85 min, and then to 35% mobile phase B in another 35 min at a flowrate of 300 nL/min and a column temperature of 37 °C. The data were acquired with the mass spectrometer operating in a top speed data-dependent acquisition (DDA) mode. An Orbitrap full MS scan was performed in the range of 300–1500 m/z at a resolution setting of 120,000 with an automatic gain control (AGC) target value of 4×10^5. Iterative isolation (with a 1.6 Thomson unit isolation window and minimum intensity threshold of 5×10^4) and fragmentation by higher-energy collisional dissociation of ions were carried out after the Orbitrap full MS scan. Ions were injected with a maximum injection time of 110 ms and an AGC target of 1×10^5.

Raw data were processed using Proteome Discoverer software (version 2.1, Thermo Fisher Scientific, San Jose, CA, USA). MS2 spectra were searched using Mascot (Matrix Science, London, UK), which was set up to search against the *Chlamydomonas reinhardtii* database (Creinhardtii_281_v5.5). The search criteria included 10 ppm precursor mass tolerance, 0.02 Da fragment mass tolerance, trypsin enzyme, and maximum missed cleavage sites of two. Dynamic modifications included propionamide on cysteine, oxidation on methionine, deamidation on asparagine and glutamine, and Gly-loss+Amide on C-terminal glycine. Peptide spectral match (PSM) error rates were determined using the target-decoy strategy coupled to Percolator modeling of true and false matches [23,24]. The mass spectrometry proteomics data have been deposited to the ProteomeXchange Consortium via the PRIDE partner repository with the dataset identifier PXD010945.

Scaffold (version Scaffold_4.8.4, Proteome Software Inc., Portland, OR, USA) was used to validate MS/MS-based peptide and protein identifications. Peptide identifications were accepted if they could be established at greater than 95.0% probability by the Scaffold Local FDR algorithm. Protein identifications were accepted if they could be established at greater than 99% probability and contained at least 1 identified peptide. Proteins that contained similar peptides and could not be differentiated based on MS/MS analysis alone were grouped to satisfy the principles of parsimony. Proteins sharing significant peptide evidence were grouped into clusters.

A total of 1291 proteins were identified in Dataset 1 and in Dataset 2. Proteins recognized in less than four of the six samples were eliminated, yielding a Merged Dataset with 1216 proteins. We used PredAlgo (https://giavap-genomes.ibpc.fr/predalgo/) to predict the subcellular localization of each identified protein. We used SignalP 4.1 (www.cbs.dtu.dk/services/SignalP/) to identify proteins categorized as O (Other) or NA (Not Assigned) but predicted to contain a signal peptide (SP). The merged dataset contained 102 signal peptide-containing proteins, which we refer to as the soluble mating secretome.

When possible, proteins in the soluble mating secretome were grouped on the basis of function, as assigned by Phytozome v12.1 (https://phytozome.jgi.doe.gov), Uniprot (https://uniprot.org), and literature searches. Where indicated, mammalian homologues of these proteins were identified using BLASTp (https://blast.ncbi.nlm.nih.gov) and the Uniprot database. NeuroPred (stagbeetle.animal.uiuc.edu/cgi-bin/neuropred.py) [5] and SMART (smart.embl-heidelberg.de) were used to predict cleavage sites and protein domains. For the analysis of proteases, and we utilized the hierarchical, structure-based classification system of Families and Clans available in the MEROPS database, Release 12.0 (https://www.ebi.ac.uk/merops/).

2.6. Bioinformatic Analyses

To search for any neuropeptide-like precursors encoded by the *C. reinhardtii* genome, we identified primary transcripts (Phytozome v5.5; a total of 17,741) encoding proteins with a predicted signal peptide (SignalP v4.1) and no transmembrane helices (TMHMM v2.0), resulting in a list of 771 proteins. Using a custom script, these protein sequences were searched for potential prohormone convertase cleavage sites ((K/R)X_n(K/R), where n = 0, 2, 4 or 6); likely furin cleavage sites (RX(K/R)R) were screened for separately. The cleavage sites identified were then screened for the presence of potential amidation sites (G(K/R)(K/R), (K/R)X_nG(K/R) where n = 1 or 3, and RG(K/R)R). In addition, NeuroPred [5] was used with "known motifs" to predict amidated product peptides that could be produced from these proteins; based on cleavages observed in *Aplysia californica*, RK, KXXK, and KXXR sites were excluded from "known motifs". Multiple sequence alignments utilized T-COFFEE, Version 11.00 [25].

3. Results

3.1. The *C. reinhardtii* Genome Encodes Multiple Proteins with the Characteristics of Neuropeptide Precursors

We screened the *C. reinhardtii* transcriptome for the presence of proteins that fit the generally accepted criteria for neuropeptide precursors. We limited our search to the 771 soluble proteins predicted to contain an N-terminal signal peptide (Figure 2A) (next page). Screening this set of proteins for consensus subtilisin-like prohormone convertase cleavage sites ((R/K)X_n(R/K)↓, where n = 0 or 2, and ↓ identifies the cleavage site) [26], yielded 756 proteins. Potential amidation sites were identified in 331 proteins with G(K/R)(K/R) sites and in 298 proteins with (K/R)XG(K/R) sites. The prohormone convertases that recognize paired basic cleavage sites function optimally in the low pH environment of secretory granules; since furin does not require an acidic environment, we screened for furin-like cleavage sites (RX(K/R)R), which were found in 224 proteins; 33 could yield amidated peptides. Proteins with a C-terminal –Gly or –Gly–(Lys/Arg)$_n$ can be amidated without prior endoproteolytic cleavage [27,28]; 49 of the 771 *C. reinhardtii* proteins were predicted to be secretory, terminate with –Gly, and 24 could be converted into PAM substrates by the action of a CPB-like enzyme. The full dataset is provided in Supplemental Table S1.

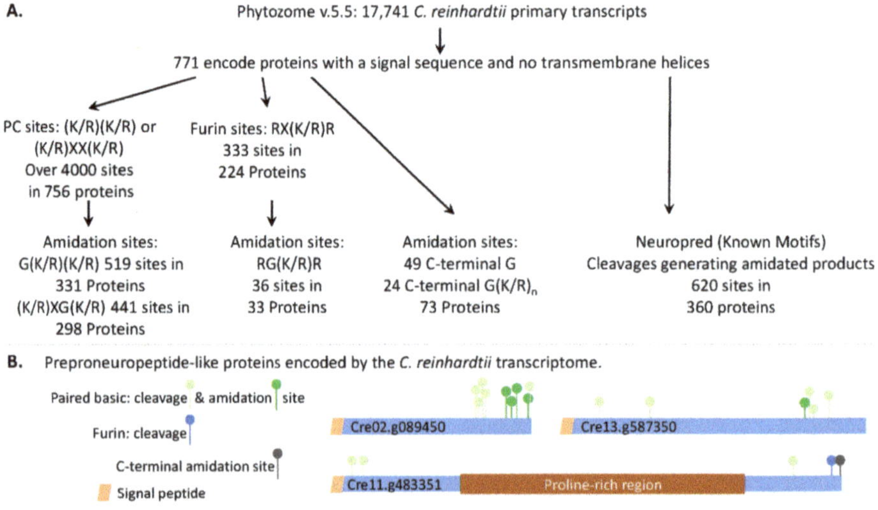

Figure 2. The *C. reinhardtii* transcriptome encodes multiple proteins that resemble classical neuropeptide precursors. (**A**) The strategies used to identify prohormone convertase (PC) and furin cleavage sites, and the amidated products that could be produced from signal peptide-containing soluble *C. reinhardtii* proteins, are outlined. Potential amidation sites that do not require the action of an endoprotease were also identified. The *C. reinhardtii* transcriptome was also analyzed using Neuropred (Known Motifs) [5] to identify potential amidated peptides and their lengths. (**B**) Diagrams illustrate the location of paired basic cleavage sites, furin sites, and amidation sites in three *C. reinhardtii* proteins that resemble classical neuropeptide precursors.

We also used Neuropred [5] to identify *C. reinhardtii* proteins that could generate amidated peptides following cleavages at "Known Motifs" (Figure 2A); 360 proteins were identified. A total of 620 amidated peptides, ranging in size from 2 to over 100 amino acids, were identified (Supplemental Table S2). While 60% of these proteins could yield only one amidated peptide, 30 could yield 3 amidated peptides and 4 could each yield 8 amidated peptides. All 20 amino acids were identified as potential products; Gly-amide, the predicted C-terminus of 94 of the amidated peptides, was the most prevalent.

Examples of precursors that could yield multiple amidated peptides, utilize furin and paired basic cleavage sites, and undergo C-terminal amidation, are shown in Figure 2B. The *C. reinhardtii* transcriptome encodes multiple proteins that could be acted upon by the classical neuropeptide processing machinery.

3.2. The *C. reinhardtii* Genome Encodes Enzymes that Resemble the PCs and CBP-Like Enzymes Essential for Neuropeptide Production

The first step unique to neuropeptide precursor processing involves limited endoproteolytic cleavage by subtilisin-like prohormone convertases as the newly synthesized proteins move through the secretory pathway lumen (Figure 1). When we searched for homologs of human furin, PC1, and PC2 in the *C. reinhardtii* genome using BLASTp, five subtilisin-like *C. reinhardtii* proteins were identified (Figure 3A). Like furin, Cre14.g628800 is a Type I integral membrane protein, while Cre04.g213400 is a soluble protein, as are PC1 and PC2. Strikingly, the other 3 homologs are predicted to be Type II integral membrane enzymes. Functional data are available only for Cre01.g049950, which encodes VLE1 (vegetative lytic enzyme, also known as sporangin), the endoprotease essential for the hatching of daughter cells from the mother cell wall [29].

The catalytic core (S8 domain) of subtilisin-like endoproteases includes an essential D, H, S catalytic triad; spacing of the active site residues in Cre01.g049950 (VLE1), Cre16.g685250, Cre14.g628800, and Cre04.g213400 resembles the spacing in furin, PC1, and PC2 (Figure 3 and Table 1).

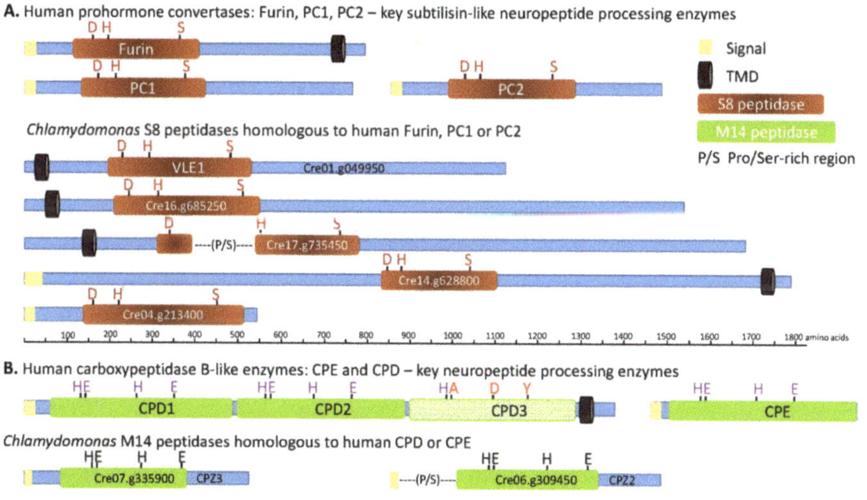

Figure 3. Subtilisin- and carboxypeptidase B-like proteins encoded by the *C. reinhardtii* transcriptome. The closest homologs of human PC1 (BAA11133.1), PC2 (AAB32656.1), furin (NP_002560.1), CPE (P16870.1), and CPD (AAH51702.1) were identified using NCBI BLASTp and v5.5 of the *C. reinhardtii* proteome (Phytozome 12). The subtilisin (S8) and CPB (M14) catalytic cores, signal peptide, transmembrane domains, and other features were identified using SMART and the MEROPS data base. Diagrams are drawn to scale, with active site residues indicated: D, H, and S form the catalytic *triad of S8 family proteases; the M14 peptidases rely on zinc binding to an H, E, H* motif, and an active site E. The third catalytic domain of CPD is not active; altered residues are shown in brown. P/S, indicates a Pro/Ser-rich region.

Table 1. S8 domain-containing subtilisin-like proteins in the *C. reinhardtii* genome.

Protein ID	Length	Signal	TM Helix	S8 Location	Most Closely Related Human Protein (*GENE*)
Secreted					
Cre02.g076950	1355	1–22	no	189–992	
Cre04.g213400	539	1–23	no	171–521	PC1 (*PCSK1*)
Cre07.g329500	945	1–29	no	601–847	
Cre10.g459450	866	1–30	no	536–836	
Cre05.g242100	1264	1–26	no	428–991	
Cre05.g242750	1301	1–23	no	511–1053	
Cre19.g750447	1141	1–26	no	700–1140	
Type I membrane					
Cre14.g628800	1787	1–47	1719–1741	837–1103	SKI-1 (*MBTPS1*)
Type II membrane					
Cre01.g049950 VLE1, sporangin	1117	no	37–59	210–539	PC7 (*PCSK7*)
Cre03.g145827	1512	no	45–67	219–552	
Cre16.g685250	1532	no	55–77	233–562	PC4 (*PCSK4*)
Cre17.g708400	1794	no	451–473	767–1163	
Cre17.g735450	1674	no	138–160	333–774	PC2 (*PCSK2*) & PACE4 (*PCSK6*)
Cre03.g190250	1229	no	1149–1171	298–792	
Other					
Cre13.g585800	809	no	no	154–425	
Cre16.g675350	1492	no	no	119–599	
Cre17.g713600	1982	no	no	764–1116	
Cre03.g170300	1374	no	no	Split	
Cre05.g242700	777	no	no	26–383	
Cre05.g242856	1419	no	no	585–1129	
Cre09.g406700	1890	no	no	482–1018	

For each S8 domain-containing protein, SMART was used to determine its total number of amino acids, the presence and length of any signal peptide (Signal), the location of potential transmembrane helices (TM Helix), and the position of the S8 domain (S8 location). Proteins highlighted in gray were identified based on screening for homologs to full-length human furin, PC1, and PC2. Based solely on their S8 domains, the most closely related hPCSK is identified; the preferred protein name is shown, with the corresponding gene name in parenthesis. Cre01.g0499450 was previously identified as sporangin (also known as vegetative lytic enzyme (VLE1)) [29].

The catalytic cores of Cre01.g049950 (VLE1), Cre16.g685250, and Cre04.g213400 are most similar to those of human PC7, PC4, and PC1, respectively; the human enzymes each cleave secretory products after basic residues [13]. The catalytic core of Cre17.g735450, which is interrupted by a region rich in Pro and Ser (P/S), has short segments homologous to PC2 and PACE4, which also cleave after basic amino acids but, overall, is most homologous to a putative human subtilisin (SJM30502.1) that lacks a signal peptide. The catalytic core of Cre14.g628800 most closely resembles that of human SKI-1/S1P, which cleaves after non-basic sites in membrane-bound transcription factors [13].

Searching the Phytozome 12 (v5.5) database identified 21 *C. reinhardtii* proteins that contain the S8 peptidase domain (Table 1). SMART analysis predicts that 7 of the 21 proteins are secreted, 6 adopt a Type II topology, 1 has a Type I topology, and 7 may be cytosolic. A phylogenetic analysis identified a cluster containing the four *C. reinhardtii* S8 domain proteins most closely related to human furin, PC1, and PC2 (Supplemental Figure S2). The oxyanion asparagine in PC2 has been replaced by aspartic acid; this same substitution occurs in one of the 21 *C. reinhardtii* S8 domain proteins (Cre17.g713600). As expected, the S8 catalytic domain is well conserved, with evolutionary divergence reflected in features outside of the catalytic core [30].

To identify *C. reinhardtii* CPB homologs, we used the amino acid sequences of human CPE and CPD. Two CPB-like proteins were identified in *C. reinhardtii* (Figure 3B) (Supplementary Figure S3). Both contain a catalytic core (M14 peptidase; MEROPS database) that includes the three essential zinc-binding residues (H, E, H) and the active site Glu. The catalytic core of CPZ2 (Cre06.g309450) is preceded by a Pro/Ser-rich region. While human CPD is a Type 1 integral membrane protein, both CPZ2 (Cre06.g309450) and CPZ3 (Cre07.g335900) lack a transmembrane domain and, thus, resemble CPE.

3.3. Expression of Transcripts Encoding *C. reinhardtii* Neuropeptide Processing Enzymes Is Regulated During Sexual Reproduction

Neuropeptide processing enzyme expression often varies with neuropeptide expression [31–33]. Published transcriptomic data for *C. reinhardtii* reveal dramatic changes in expression of the *C. reinhardtii* genes that encode the proteases most closely related to human furin, PC1, PC2, CPD, and CPE [20] (Supplemental Figure S4). Transcript levels were reported for asynchronous vegetative cells and for cells synchronized using a light/dark cycle, resting gametes (*plus* and *minus*) and gametes of both mating types treated with lysin to remove the cell wall or with dibutyryl-cAMP to mimic flagellar activation of the mating signaling pathway.

Expression of Cre04.g213400, the protein most homologous to PC1, was not reported in this dataset, but each of the four other subtilisin-like enzymes exhibits a unique pattern [20]. VLE1 (Cre01.g049950) is most highly expressed in vegetative cells. Expression of Cre16.g685250 is especially sensitive to lysin treatment, while expression of Cre17.g735450 increases in response to dibutyryl-cAMP, and expression of Cre14.g628800 drops in response to dibutyryl-cAMP. CPZ3 (Cre07.g335900), a CPB-like enzyme, is highly expressed in vegetative cells and resting gametes, dropping to low levels in lysin- or dibutyryl-cAMP-treated gametes (Supplemental Figure S4) [20]. By contrast, expression of CPZ2 (Cre06.g309450), another CPB-like enzyme, rises in response to Lysin-treatment and is higher after dibutyryl-cAMP treatment than in resting gametes. Different subtilisin-like and CPB-like *C. reinhardtii* enzymes appear to be used to perform distinct functions. The *C. reinhardtii* genome encodes a single PAM protein (Cre03.g152850) [16,34]; its expression drops after dibutyryl-cAMP-treatment.

3.4. Preparation and Analysis of the Soluble Mating Secretome

In addition to vesicle-mediated secretion, *C. reinhardtii* release bioactive ectosomes from their cilia [35–37]; we used differential centrifugation to separate ectosomes from the soluble secretome. Vegetative cells and gametes of both mating types were prepared (Figure 4A). A series of differential centrifugation steps yielded an ectosome-rich pellet and the soluble secretome. The soluble vegetative secretome and the soluble mating secretome were visualized after SDS-PAGE (Figure 4B). The major proteins in the soluble vegetative secretomes varied with mating type. Furthermore, mating gametes released substantially more protein per unit time than vegetative cells.

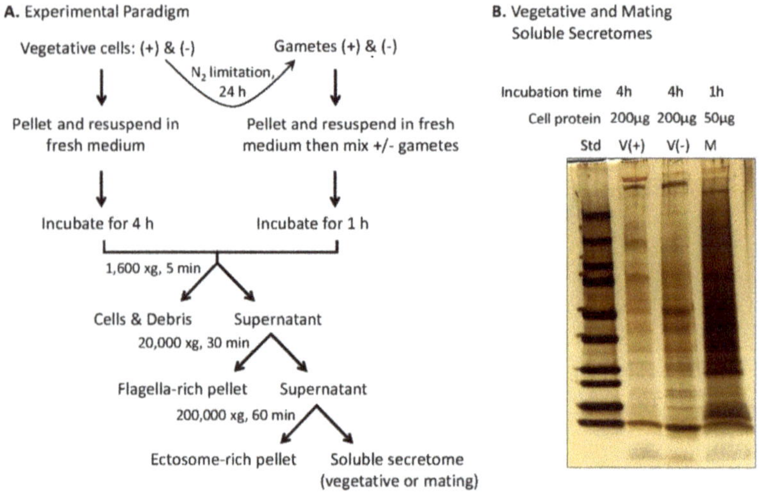

Figure 4. Preparation of the soluble mating secretome. (**A**) Experimental paradigm. Equal numbers of vegetative cells (*minus* and *plus*) (10–12 $\times$ 10^6 cells/mL) suspended in fresh medium were grown in constant light for 4 h; the indicated series of centrifugation steps yielded the vegetative ectosome-rich pellet, and the soluble vegetative secretome. For preparation of mating ectosomes and the soluble mating secretome, gametes were prepared by incubation of vegetative cells in nitrogen-limited medium for 24 h; after mixing an equal number of *plus* and *minus* gametes (1–2 $\times$ 10^8 cells of each mating type) in 10 mL nitrogen-free medium), mating was allowed to proceed for 1 h in constant light with gentle aeration. The mating medium was processed, as described, for the vegetative medium, yielding the mating ectosome enriched pellet and the soluble mating secretome. (**B**) SDS-PAGE analysis (4 to 15% polyacrylamide gradient gels) of the soluble vegetative secretome (*plus* and *minus*) and the soluble mating secretome. For vegetative cells, the volume of soluble secretome loaded came from 200 μg of cell protein; for mating cells, the volume of soluble secretome loaded came from 50 μg of cell protein.

We fractionated soluble mating secretome samples by SDS-PAGE; a total of six independent samples were analyzed (Figure 5A). Proteins identified in at least 4 of the 6 samples are listed in Supplemental Table S3. Almost half were assigned to the category that includes cytosolic, endosomal and vacuolar proteins (Other), with 32% identified as chloroplast proteins (Figure 5B). The selective destruction of *minus* gamete chloroplast nucleoids and *plus* gamete mitochondrial DNA that accompanies sexual reproduction may contribute to the prevalence of proteins associated with organelles in the soluble mating secretome [21]. Signal peptide-containing proteins accounted for 8% of the secretome proteins; the complete list appears in Supplemental Table S4. ER and cell wall proteins each accounted for about one-fifth of the signal peptide-containing proteins, with proteases accounting for one-eighth (Figure 5C).

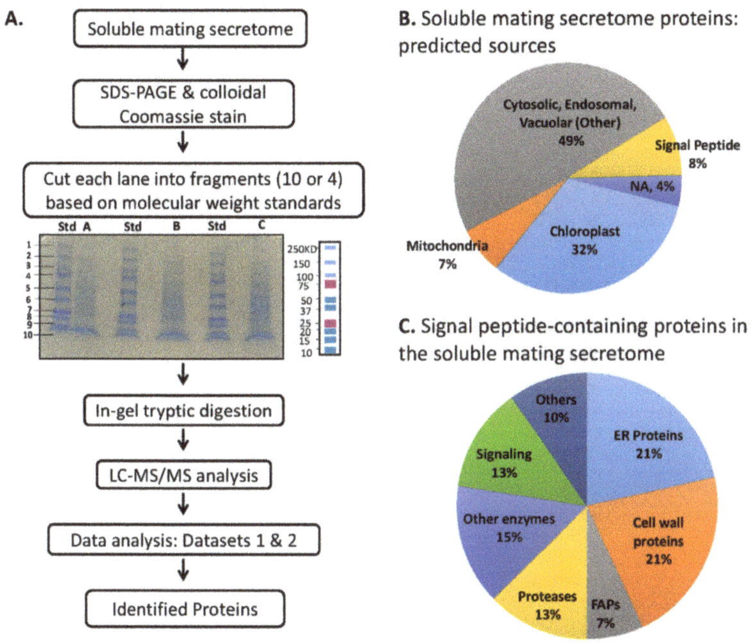

Figure 5. Analysis of soluble mating secretome. (**A**) Separate preparations of gametes were used on two different occasions to generate a total of six soluble mating secretome samples. For the first set of three samples, gels were sliced into 10 fragments, generating Dataset 1 (1233 proteins). For the second set of three samples, gels were sliced into 4 fragments, generating Dataset 2 (1494 proteins). The two datasets were combined as described in Materials and Methods, yielding a merged dataset; normalized spectral counts were used to calculate the spectral count average and standard error of the mean. Spectral count data, predicted location, signal peptide presence, number of transmembrane helices, functional information and data for individual gel slices appear in Supplemental Table S3. (**B**) Predalgo predictions for the subcellular localization of all proteins in the merged dataset were used to generate the pie chart. (**C**) For soluble mating secretome proteins predicted to contain a signal peptide, functional predictions were made using Phytozome and literature analyses (the complete list appears in Supplemental Table S4).

ER chaperones (4) and cell wall proteins (12) were abundant, as expected. Further work will be required to determine whether the identification of Cre11.g477950, an ADP-ribosylglycohydrolase, as the most prevalent component of the secretome, indicates a prominent role for the reversible ADP-ribosylation of chaperone proteins in controlling protein folding and secretion in *C. reinhardtii* [38,39]. The prevalence of importin β, importin β-3 homolog, and Ran GTPase-activating protein in the secretome, is consistent with the suggestion that proteins involved in controlling nuclear pore traffic also play a role in controlling ciliary protein trafficking [40,41]. Three putative tRNA synthetases (Ala, Glu, Thr) were identified in the soluble mating secretome, with alanyl- and glutamyl-tRNA synthetase among the 30 most prevalent proteins (Table 2). These ancient enzymes, which catalyze the ATP-dependent attachment of a specific amino acid to its tRNA, are known to perform additional functions in other organisms, and several human tRNA synthetases are secreted [42–44].

Table 2. Thirty most prevalent components of the soluble mating secretome.

Rank	Accession Number	Mol Mass	Avg Tot Spectral Counts	SEM/ Avg	Description
1	Cre11.g477950.t1.1	94 kDa	189.1	0.13	ADP-ribosylglycohydrolase
2	Cre02.g088200.t1.2	58 kDa	148.4	0.09	Protein disulfide isomerase 1, RB60
3	Cre02.g143200.t1.1	122 kDa	71.7	0.05	Alanine tRNA ligase
4	Cre02.g080700.t1.2	72 kDa	66.9	0.10	ER associated Hsp70 protein
5	Cre01.g038400.t1.2	47 kDa	56.4	0.15	Calreticulin 2, calcium-binding protein
6	Cre14.g633750.t1.1	122 kDa	51.7	0.07	Importin β-3 homolog
7	Cre06.g298650.t1.2	53 kDa	51.0	0.43	Translation initiation factor 4A
8	Cre02.g080650.t1.2	93 kDa	41.4	0.09	ER associated heat shock protein 90B
9	Cre09.g394200.t1.1	156 kDa	37.7	0.34	Flagellar associated protein
10	Cre01.g034000.t1.2	97 kDa	37.3	0.26	Importin β
11	Cre14.g620600.t1.2	52 kDa	34.7	0.24	Pherophorin, PHC2
12	Cre09.g406600.t1.1	38 kDa	33.7	0.35	ChlamyFPv5, 2 KCl peptides
13	Cre06.g258800.t1.1	132 kDa	26.7	0.24	OH-Pro-rich glycoprotein, GP2 (FAP3)
14	Cre10.g431800.t1.2	70 kDa	25.2	0.37	Arylsulfatase
15	Cre07.g330200.t1.2	30 kDa	23.2	0.12	Radial spoke protein 9
16	Cre02.g089500.t1.2	43 kDa	22.1	0.21	Proline rich extensin signature
17	Cre03.g144564.t1.1	81 kDa	22.0	0.30	Matrix metalloproteinase, MMP13
18	Cre07.g321400.t1.1	199 kDa	20.5	0.13	Flagellar associated protein
19	Cre02.g089450.t1.2	38 kDa	20.4	0.33	Proline rich extensin signature, HRP5
20	Cre02.g077850.t1.2	83 kDa	18.6	0.34	Flagellar associated protein, FAP212
21	Cre09.g407700.t1.2	54 kDa	17.3	0.08	Cysteine endopeptidase, CEP1
22	Cre09.g401900.t1.2	132 kDa	16.9	0.37	Proline rich extensin signature
23	Cre12.g487700.t1.2	77 kDa	15.9	0.13	Serine/threonine protein kinase
24	Cre09.g393700.t1.1	71 kDa	15.7	0.16	Matrix metalloproteinase, MMP3
25	Cre11.g467547.t1.1	83 kDa	13.0	0.26	Glutamyl/glutaminyl-tRNA synthetase
26	Cre02.g077800.t1.2	86 kDa	12.5	0.45	Proline rich extensin signature (FAP310)
27	Cre02.g102050.t1.1	92 kDa	12.4	0.20	Proline rich extensin signature (FAP328)
28	Cre11.g479250.t1.2	54 kDa	12.1	0.11	Ran GTPase-activating protein,
29	Cre06.g304500.t1.2	40 kDa	10.9	0.18	Zygote-specific protein
30	Cre12.g533100.t1.1	21 kDa	10.6	0.41	CHRD domain, PF07452

The signal peptide-containing proteins identified in the merged dataset were sorted by average total spectral counts (Avg Tot Spectral Counts); the entire list, grouped by function, is provided in Supplemental Table S4. The 30 most prevalent proteins, which account for 72% of the total spectral counts, are shown here.

Four of the seven *C. reinhardtii* proteins, identified as homologs of mammalian neuropeptide processing enzymes (Figure 3), were present in our analysis of mating medium that contained both ectosomes and soluble proteins. Only VLE1 (Cre01.g049950) was identified in the soluble mating secretome. VLE1 is released in vegetative ciliary ectosomes, which can cleave the mother cell wall [35]. It is not clear how VLE1 activation, which presumably involves an autoproteolytic cleavage, is triggered, or the subcellular site at which it occurs. It is possible that the other neuropeptide processing enzyme homologs identified (Cre16.g685250, Cre17.g735450 and Cre06.g309450 (CPZ2)), along with PAM (Cre03.g152850), will be found in mating ectosomes.

3.5. Multiple Signal Peptide-Containing Proteases Are Present in the Soluble Mating Secretome

Proteases representative of several different classes were identified in the soluble mating secretome (Figure 6 and Supplemental Table S5). Cell wall removal during mating is accomplished by gametolysin, a zinc-containing matrix metalloprotease (MMP) [45,46]. Gametolysin is stored in the periplasm as an inactive 65 kDa proenzyme (progametolysin); its cleavage and activation are triggered by the increase in intracellular cAMP that occurs during mating [19,21]. A serine protease (p-lysinase) released in response to flagellar agglutination, cleaves and activates gametolysin [19]. With multiple candidate proteases, the genes encoding the proteases that account for these activities have not yet been identified.

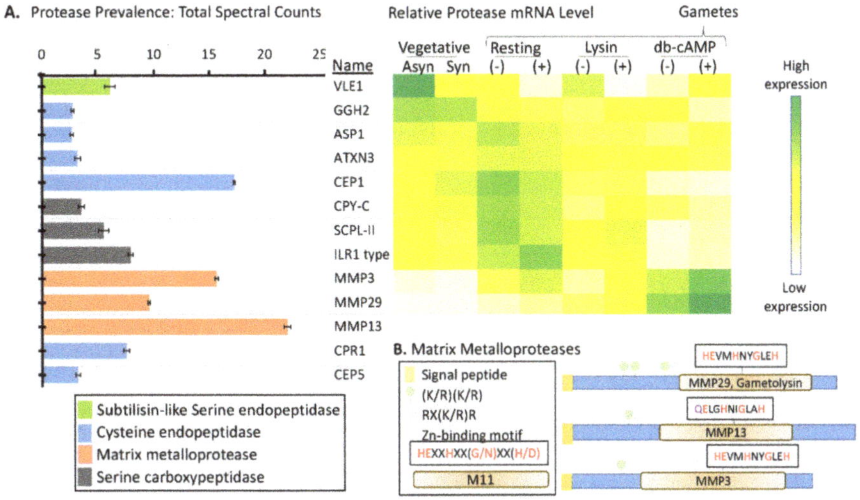

Figure 6. Signal peptide-containing proteases in the soluble mating secretome. (**A**) Average normalized total spectral counts ($n = 6$; SEM) are shown for each of the signal peptide-containing proteases identified in the merged dataset. The heatmap used transcriptomic data from Ning et al. [20] to identify stages during which mRNAs encoding many of these proteases are most highly expressed; transcriptomic data for MMP13, CPR1, and CEP5 were not available in the cited study. (**B**) Diagrams illustrating key features of the three MMPs identified. Furin recognition motif: RX(R/K)R; Zn-binding motif: HEXXHXX(G/N)XX(H/D) [47].

The only Ser endoprotease identified in the soluble mating secretome was VLE1, a type 2 integral membrane protein (Figure 6A). VLE1 is stored in cells as an inactive 127 kDa proenzyme; after hatching, active 125 kDa VLE1 can be recovered from the culture medium, presumably reflecting the occurrence of an endoproteolytic cleavage that separates active enzyme from the transmembrane domain [29]. Biochemical studies indicate some specificity of VLE1 for basic amino acids and two candidate sites (R^{136}KR and R^{168}R) precede the catalytic domain; the 9 tryptic peptides identified are consistent with autocatalytic cleavage at either of these sites. As expected, expression of VLE1 mRNA is highest in vegetative cells [20,29] (Figure 6). Lysin-catalyzed cell wall removal from *minus* gametes increases VLE1 mRNA levels, perhaps accounting for the presence of this protease in the soluble mating secretome.

Three matrix metalloproteinases (MMPs), MMP3, MMP29, and MMP13, were identified. Based on spectral counts, the MMPs, which specialize in the degradation and remodeling of ECM and in the extracellular release of signaling proteins, were the most highly expressed proteases. Proteases more associated with lysosomal degradation (serine and metallo-carboxypeptidases and cysteine endoproteases) were also present. Expression of transcripts encoding MMP3 and MMP29 peak following db-cAMP treatment of both (*plus*) and (*minus*) gametes, consistent with a role for each of these enzymes in gametic cell wall removal. Expression of the remaining proteases peaked in vegetative cells or in resting gametes.

MMPs, ancient enzymes found in all kingdoms of life, are synthesized as inactive zymogens, with their signal peptide followed by a prodomain [47]. In plants, they are involved in remodeling the ECM and in cell–cell communication and signaling [48]. Each of the MMPs identified in the *C. reinhardtii* soluble mating secretome is predicted to be a soluble protein (Figure 6B). Based on homology to better characterized MMPs, activation will require an endoproteolytic cleavage that separates the prodomain from the catalytic domain. MMP activation is frequently catalyzed by subtilisin-like enzymes [49], and consensus furin-cleavage sites occur in the prodomains that precede the M11 catalytic domains in MMP29, MMP13, and MMP3 (Figure 6B). Based on transcript expression patterns, VLE1, Cre16.g685250, or Cre17.g735450 could be involved in activating MMPs (Supplemental Figure S3).

The serine carboxypeptidases and cysteine and aspartyl endopeptidases identified are cathepsin-like, suggesting that they function in a degradative pathway. The human homologs of CEP1, CEP5, and CPR1 play a role in the degradation of extracellular matrix material. Transcripts encoding CEP1 and both serine carboxypeptidases (SCPL-II and CPY-C) are highest in resting gametes of both mating types, falling dramatically after lysin treatment or db-cAMP stimulation [20].

3.6. Many Cell Wall Pherophorins Recovered from the Soluble Mating Secretome Are Cleaved, While Hydroxyproline-Rich Proteins Remain Intact

Carefully controlled removal of the *C. reinhardtii* cell wall is required during vegetative growth, mating, and zygospore activation. Different sets of hydroxyproline-rich glycoproteins appear in the vegetative/gametic vs zygotic cell wall [21]. *Volvox* and *C. reinhardtii* cell wall proteins, termed pherophorins, resemble the subset of higher plant extensins referred to as solanaceous lectins, suggesting that their globular domains bind carbohydrates [50]. The pherophorins have a hydroxyproline-rich rod-like domain that separates globular N- and C-terminal domains [50]. Pherophorins are prevalent in the vegetative/gametic cell wall, but absent from early zygote-specific gene clusters [20,21]. The vegetative hatching enzyme (VLE1) must be specific enough to cleave the mother cell wall without attacking the vegetative cell wall. By contrast, the gametic lytic enzyme (gametolysin) can cleave the cell wall at all stages of the *C. reinhardtii* life cycle, except the zygospore [51]. Each of these proteases is produced as an inactive zymogen, with endoproteolytic removal of its prodomain, an essential part of the activation process.

A total of 21 cell wall proteins were identified in the soluble mating secretome (Figure 7A). Three pherophorins were identified as highly expressed in vegetative cells (PHC5) or resting gametes (PHC2, PHC4) [20] (Figure 7B). Three pherophorins (PHC1, PHC21, and PHC51) and four hydroxyproline-rich proteins (VSP3, HRP3, HRP5, and CWP2), present in the soluble mating secretome, fell into the group of cell wall proteins whose transcript levels increased in response to g-lysin treatment of both *plus* and *minus* gametes [20,21]. Only two of the cell wall proteins identified (CWP2 and PHC28) fell into the group of transcripts whose levels responded to both g-lysin and dibutyryl-cAMP treatment of gametes.

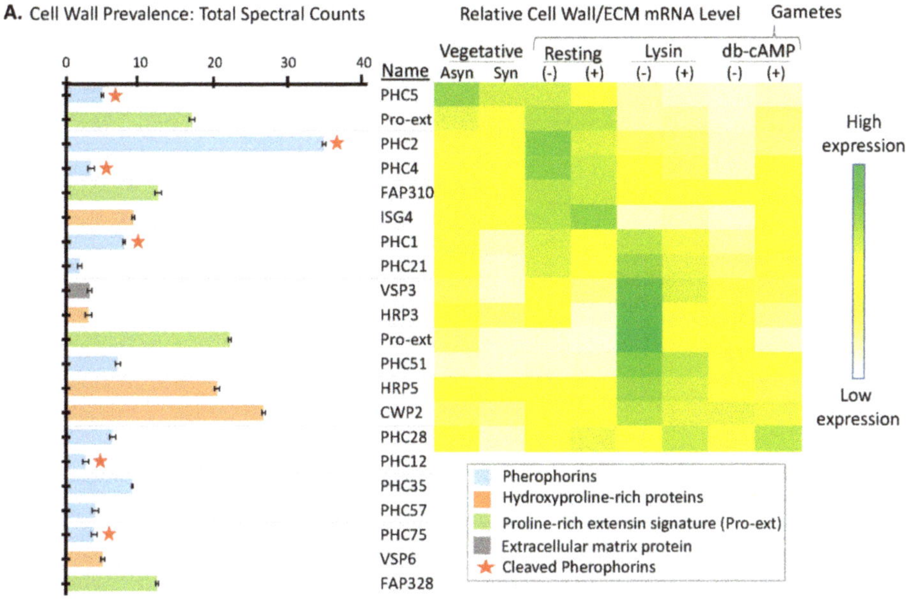

Figure 7. Signal peptide-containing cell wall proteins in the soluble mating secretome. (**A**) Average normalized total spectral counts are reported for cell wall proteins identified in the soluble mating secretome, as described in Figure 6. Cell wall proteins with a pherophorin domain were more prevalent than hydroxyproline-rich proteins. (**B**) As described in Figure 6, the heatmap uses transcriptomic data from Ning et al. [20]; PHC12, PHC35, PHC57, PHC75, VSP6, and FAP328 were not included in the cited analysis.

In *Volvox*, cleavage of a cell wall protein releases a globular pherophorin domain that plays a role as a sexual inducer or pheromone [50]. By analyzing individual gel slices from both datasets (Supplemental Table S3), we looked for pherophorins or hydroxyproline rich proteins that had undergone endoproteolytic cleavage, generating smaller stable products (Figure 8).

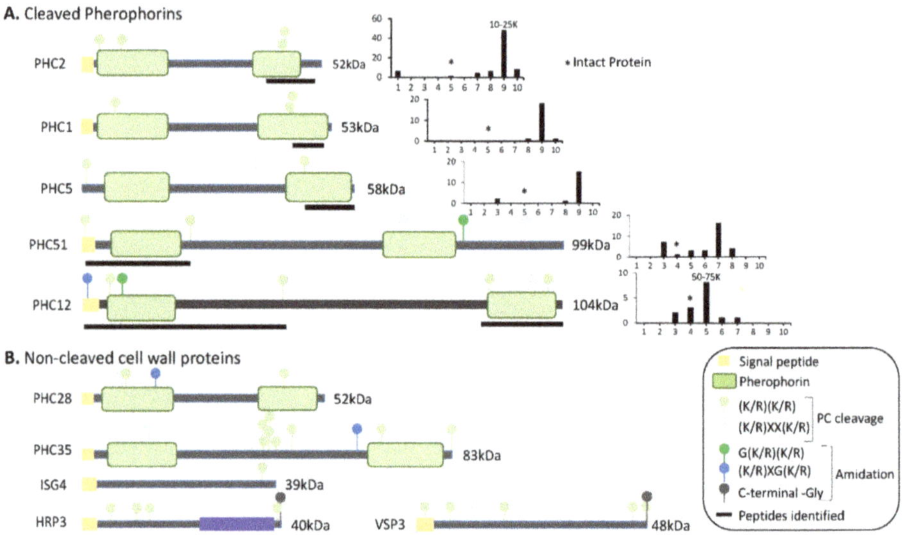

Figure 8. Predicted domain structures of cell wall proteins. (**A**) SMART was used to predict domain structures for cell wall proteins whose tryptic peptides were identified in gel slices containing proteins smaller than the intact protein. Signal peptides, pherophorin domains, and potential prohormone convertase cleavage sites are shown. To the right of each diagram, total spectral counts in each gel slice (from Dataset 1) are shown; * marks the gel slice in which that intact protein would be located; the mass range of the slice containing the highest spectral counts is indicated above the bar. Black bars under each diagram identify the region from which the tryptic peptides came. (**B**) The domain structures predicted for cell wall proteins, whose tryptic peptides were identified only in gel slices containing proteins at least as large as the intact protein, are shown.

For seven of the pherophorins, smaller fragments that contained peptides derived from one or both pherophorin domains were detected (Figure 8A). PHC2, the most abundant pherophorin, contains N- and C-terminal pherophorin domains. While intact PHC2 (52 kDa) was not detected, peptides from its C-terminal pherophorin domain were identified in the 10–18 kDa region of the gel. PHC1 and PHC5 also appeared to undergo cleavages that generated a stable C-terminal pherophorin domain (Figure 8A). Stable N- and C-terminal pherophorin domains were generated from PHC12. Seven of the eleven pherophorins identified in the soluble mating secretome had undergone cleavage; by contrast, the six hydroxyproline-rich cell wall proteins appeared to remain intact. While more detailed analyses will be required to identify the cleavage sites used, prohormone convertase-like cleavage of the pherophorins could generate the products observed (Figure 8A).

4. Conclusions

In neurons, the biosynthesis of neuropeptides, which are stored in secretory granules, and many growth factors, which are associated with the extracellular matrix, is orchestrated by the controlled cleavage of inactive precursors. Activation of the subtilisin-like endoproteases that produce most of the peptides stored in secretory granules is controlled by luminal pH, which declines as secretory products move from the ER, through the Golgi complex, and into secretory granules (Figure 9). Although secretory granules have not yet been identified in *C. reinhardtii*, their transcriptome encodes candidate preproneuropeptides. VLE1 (sporangin), a subtilisin-like enzyme resembling furin, PC1, and PC2, plays an essential role in vegetative growth, while gametolysin, an MMP-like enzyme resembling those responsible for the extracellular cleavage of proTGFβ family members, is essential for sexual reproduction.

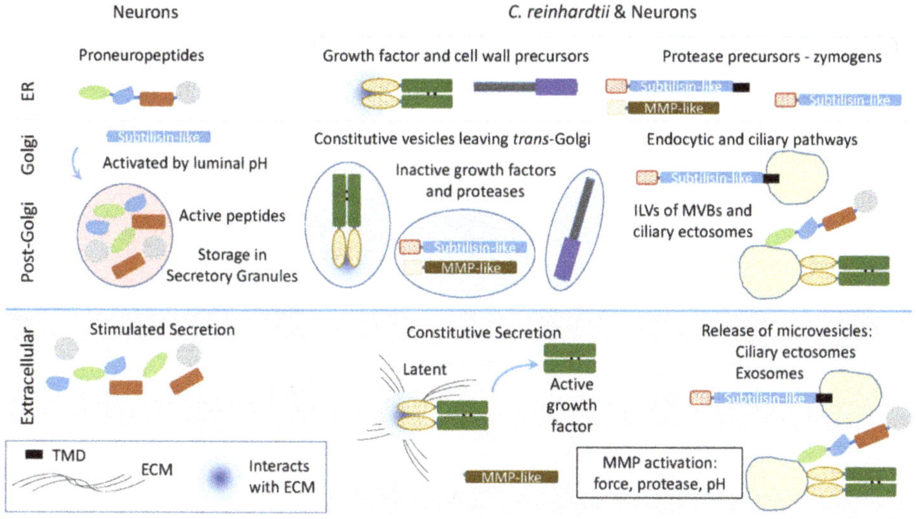

Figure 9. Using subtilisin- and MMP-like enzymes in *C. reinhardtii* and secretory granule-containing cells. The *C. reinhardtii* genome encodes subtilisin-like proteins resembling the enzymes that cleave proneuropeptides and generate the peptides stored in secretory granules (Table 1). MMP-like proteins resembling those that cleave latent (inactive) growth factor precursors extracellularly are also encoded by the *C. reinhardtii* genome. Ectosomes released from the cilia of hatching vegetative cells contain active VLE1, the subtilisin-like enzyme that degrades the mother cell wall. Latent growth factor activation in vertebrates involves extracellular activation of MMPs, along with the interaction of proteases and growth factors with the extracellular matrix and plasma membrane. The presence of secretory granules allows control of zymogen activation by declining luminal pH and storage of active peptides for release in response to secretagogues.

In addition to constitutive secretion of vesicles exiting the *trans*-Golgi, *C. reinhardtii* use ciliary ectosomes to deliver essential cargo to the appropriate target. The endomembrane system is well developed in *C. reinhardtii*, with multivesicular bodies (MVBs), contractile vacuoles, and acidocalcisomes. Topologically, the formation of ciliary ectosomes resembles the formation of intraluminal vesicles in MVBs. Upon MVB fusion with the plasma membrane, exosomes are released. Vertebrate PAM and furin function in the endocytic pathway, with PAM identified in exosomes from several different sources [52,53]. Our data suggest that the controlled endoproteolytic activation of proneuropeptides and growth factors had their molecular and enzymatic origins in unicellular organisms. The complex endomembrane system thought to be present in the last eukaryotic common ancestor presumably supported the evolution of the prepronenropeptides and growth factors essential for nervous system development and adult nervous system function well before the appearance of neurons.

Supplementary Materials: The following are available online at http://www.mdpi.com/2227-7382/6/4/36/s1, Figure S1: Alignment of five S8 subtilisin-like *C. reinhardtii* proteins with human homologs; Figure S2: Alignment of S8 domains of 21 *C. reinhardtii* family members; Figure S3: Alignment of two M14 CPB-like *C. reinhardtii* proteins with human proteins homologs; Figure S4: Expression of *C. reinhardtii* subtilisin-like, CPB-like and PAM transcripts; Table S1: Prepronenropeptides encoded by the *C. reinhardtii* transcriptome; Table S2: Neuropred analysis of *C. reinhardtii* transcriptome; Table S3: Mass spectrometric analysis of soluble mating secretome proteins. SolMatSecretome MergedDatasets; Table S4: Signal peptide-containing proteins; Table S5: Proteases identified in the soluble mating secretome.

Author Contributions: Conceptualization, R.L., D.K., S.M.K., R.E.M., B.A.E.; Methodology, R.L., S.M.K., N.R.; Formal Analysis, R.L., N.R., C.B.-H., S.M.K., R.E.M., B.A.E.; Investigation, R.L., D.K., N.R.; Resources, C.B.-H., S.M.K., B.A.E.; Data Curation, R.L., C.B.-H., N.R.; Writing—Original Draft Preparation, R.L., B.A.E.; Writing—Review & Editing, R.L., D.K., S.M.K., C.B.-H., R.E.M., B.A.E.

Funding: This research was supported by the Janice and Rodney Reynolds Endowment, grants from the National Institutes of Health [DK032949 (BAE) and GM051293 (SMK)], the U.S. Department of Energy, Office of Science and Office of Biological and Environmental Research (CB) and by the Yale/NIDA Neuroproteomics Center (DA018343). Mass spectral data were collected on an Orbitrap Fusion Tribrid mass spectrometer funded by a National Institutes of Health SIG grant (1S10ODOD018034).

Acknowledgments: These studies would not have been possible without the assistance of Ramila Patel-King, Kathryn Powers, Jean Kanyo and TuKiet Lam. Thanks to Iris Lindberg for her help in understanding prohormone convertases.

Conflicts of Interest: The authors declare no conflict of interest. The funding sponsors had no role in study design, data collection, analysis and interpretation or in manuscript writing and the decision to publish the results.

References

1. Taghert, P.H.; Nitabach, M.H. Peptide neuromodulation in invertebrate model systems. *Neuron* **2012**, *76*, 82–97. [CrossRef] [PubMed]
2. Hauser, F.; Grimmelikhuijzen, C.J.P. Evolution of the AKH/corazonin/ACP/GnRH receptor superfamily and their ligands in the *Protostomia*. *Gen. Comp. Endocrinol.* **2014**, *209*, 35–49. [CrossRef] [PubMed]
3. Hansen, G.N.; Williamson, M.; Grimmelikhuijzen, C.J. A new case of neuropeptide coexpression (rgamide and lw amides) in *hydra*, found by whole-mount, two-color double-labeling in situ hybridization. *Cell Tissue Res.* **2002**, *308*, 157–165. [CrossRef] [PubMed]
4. Takeda, N.; Kon, Y.; Artigas, G.Q.; Lapebie, P.; Barreau, C.; Kizumi, O.; Kishimoto, T.; Tachibana, K.; Houliston, E.; Deguchi, R. Identification of jellyfish neuropeptides that act directly as oocyte maturation-inducing hormones. *Development* **2018**, *145*, dev156786. [CrossRef] [PubMed]
5. Southey, B.R.; Amare, A.; Zimmerman, T.A.; Rodriguez-Zas, S.L.; Sweedler, J.V. Neuropred: A tool to predict cleavage sites in neuropeptide precursors and provide the masses of resulting peptides. *Nucleic Acids Res.* **2006**, *34*, W267–W272. [CrossRef] [PubMed]
6. Carlson, K.; Pomerantz, S.C.; Vafa, O.; Naso, M.; Strohl, W.; Mains, R.E.; Eipper, B.A. Optimizing production of fc-amidated peptides by chinese hamster ovary cells. *BMC Biotechnol.* **2015**, *15*, 95. [CrossRef] [PubMed]
7. Jorgensen, E.M. Animal evolution: Looking for the first nervous system. *Curr. Biol.* **2016**, *24*, R655–R658. [CrossRef] [PubMed]
8. Smith, C.L.; Varoqueaux, F.; Kittelmann, M.; Azzam, R.N.; Cooper, B.; Winters, C.A.; Eitel, M.; Fasshauer, D.; Reese, T.S. Novel cell types, neurosecretory cells, and body plan of the early-diverging metazoan *Trichoplax adhaerens*. *Curr. Biol.* **2014**, *24*, 1565–1572. [CrossRef] [PubMed]
9. Senatore, A.; Reese, T.S.; Smith, C.L. Neuropeptidergic integration of behavior in *Trichoplax adhaerens*, an animal without synapses. *J. Exp. Biol.* **2017**, *220*, 3381–3390. [CrossRef] [PubMed]
10. Jekely, G. Global view of the evolution and diversity of metazoan neuropeptide signaling. *Proc. Natl. Acad. Sci. USA* **2013**, *110*, 8702–8707. [CrossRef] [PubMed]
11. Conzelmann, M.; Offenburger, S.L.; Asadulina, A.; Keller, T.; Munch, T.A.; Jekely, G. Neuropeptides regulate swimming depth of *Platynereis* larvae. *Proc. Natl. Acad. Sci. USA* **2011**, *108*, E1174–E1183. [CrossRef] [PubMed]
12. Kumar, D.; Mains, R.E.; Eipper, B.A. 60 years of POMC: From POMC and α-msh to PAM, molecular oxygen, copper, and vitamin C. *J. Mol. Endocrinol.* **2016**, *56*, T63–T76. [CrossRef] [PubMed]
13. Seidah, N.G.; Abifadel, M.; Prost, S.; Boileau, C.; Prat, A. The proprotein convertases in hypercholesterolemia and cardiovascular diseases: Emphasis on proprotein convertase subtilisin/kexin 9. *Pharmacol. Rev.* **2017**, *69*, 33–52. [CrossRef] [PubMed]
14. Rouille, Y.; Duguay, S.J.; Lund, K.; Furuta, M.; Gong, Q.; Lipkind, G.; Oliva, A.A.; Chan, S.J.; Steiner, D.F. Proteolytic processing mechanisms in the biosynthesis of neuroendocrine peptides. *Front. Neuro Endocrinol.* **1995**, *16*, 322–361.
15. Thacker, C.; Rose, A.M. A look at the *Caenorhabditis elegans* kex2/subtilisin-like proprotein convertase family. *Bioessays* **2000**, *22*, 545–553. [CrossRef]
16. Kumar, D.; Blaby-Haas, C.; Merchant, S.; King, S.; Mains, R.E.; Eipper, B.A. Early eukaryotic origins for cilia-associated bioactive peptide-amidating activity. *J. Cell Sci.* **2016**, *129*, 943–956. [CrossRef] [PubMed]
17. Kumar, D.; Strenkert, D.; Patel-King, R.S.; Leonard, M.T.; Merchant, S.S.; Mains, R.E.; King, S.M.; Eipper, B.A. A bioactive peptide amidating enzyme is required for ciliogenesis. *eLife* **2017**, *6*, e25728. [CrossRef] [PubMed]
18. Goodenough, U.; Heitman, J. Origins of eukaryotic sexual reproduction. *Cold Spring Harbor Perspect. Biol.* **2014**, *6*, a016154. [CrossRef] [PubMed]

19. Snell, W.J.; Eskue, W.A.; Buchanan, M.J. Regulated secretion of a serine protease that activates an extracellular matrix-degrading metalloprotease during fertilization in *Chlamydomonas*. *J. Cell Biol.* **1989**, *109*, 1689–1694. [CrossRef] [PubMed]
20. Ning, J.; Otto, T.D.; Pfander, C.; Schwach, F.; Brochet, M.; Bushell, E.; Goulding, D.; Sanders, M.; Lefebvre, P.A.; Pei, J.; et al. Comparative genomics in *Chlamydomonas* and *Plasmodium* identifies an ancient nuclear envelope protein family essential for sexual reproduction in protists, fungi, plants and vertebrates. *Genes Dev.* **2013**, *27*, 1198–1215. [CrossRef] [PubMed]
21. Joo, S.; Nishimura, Y.; Cronmiller, E.; Hong, R.H.; Kariyawasam, T.; Wang, M.H.; Shao, N.C.; El Akkad, S.E.D.; Suzuki, T.; Higashiyama, T.; et al. Gene regulatory networks for the haploid-to-diploid transition of *Chlamydomonas reinhardtii*. *Plant Physiol.* **2017**, *175*, 314–332. [CrossRef] [PubMed]
22. Iomini, C.; Till, J.E.; Dutcher, S.K. Genetic and phenotypic analysis of flagellar assembly mutants in *Chlamydomonas reinhardtii*. *Mtds. Cell Biol.* **2009**, *93*, 121–143.
23. Elias, J.E.; Gygi, S.P. Target-decoy search strategy for increased confidence in large-scale protein identifications by mass spectrometry. *Nat. Methods* **2007**, *4*, 207–214. [CrossRef] [PubMed]
24. Kall, L.; Canterbury, J.D.; Weston, J.; Noble, W.S.; MacCoss, M.J. Semi-supervised learning for peptide identification from shotgun proteomics datasets. *Nat. Methods* **2007**, *7*, 923–925. [CrossRef] [PubMed]
25. Chang, J.M.; Di Tommaso, P.; Notredame, C. TCS: A new multiple sequence alignment reliability measure to estimate alignment accuracy and improve phylogenetic tree reconstruction. *Mol. Biol. Evol.* **2014**, *31*, 1625–1637. [CrossRef] [PubMed]
26. Dahms, S.O.; Hardes, K.; Steinmetzer, T.; Than, M.E. X-ray structures of the proprotein convertase furin bound with substrate analogue inhibitors reveal substrate specificity determinants beyond the s4 pocket. *Biochemistry* **2018**, *57*, 925–934. [CrossRef] [PubMed]
27. Simpson, P.D.; Eipper, B.A.; Katz, M.J.; Gandara, L.; Wappner, P.; Fischer, R.; Hodson, E.J.; Ratcliffe, P.J.; Masson, N. Striking oxygen sensitivity of the peptidylglycine alpha-amidating monooxygenase (pam) in neuroendocrine cells. *J. Biol. Chem.* **2015**, *290*, 24891–24901. [CrossRef] [PubMed]
28. Skulj, M.; Pezdirec, D.; Gaser, D.; Kreft, M.; Zorec, R. Reduction in c-terminal amidated species of recombinant monoclonal antibodies by genetic modification of cho cells. *BMC Biotechnol.* **2014**, *14*, 76. [CrossRef] [PubMed]
29. Kubo, T.; Kaida, S.; Abe, J.; Saito, T.; Fukuzawa, H.; Matsuda, Y. The clamydomonas hatching enzyme, sporangin, is expressed in specific phases of the cell cycle and is localized to the flagella of daughter cells within the sporangial cell wall. *Plant Cell Physiol.* **2009**, *50*, 572–583. [CrossRef] [PubMed]
30. Laskar, A.; Rodger, E.J.; Chatterjee, A.; Mandal, C. Modeling and structural analysis of evolutionarily diverse s8 family serine proteases. *Bioinformation* **2011**, *7*, 239–245. [CrossRef] [PubMed]
31. Oyarce, A.M.; Hand, T.A.; Mains, R.E.; Eipper, B.A. Dopaminergic regulation of secretory granule-associated proteins in rat intermediate pituitary. *J. Neurochem.* **1996**, *67*, 229–241. [CrossRef] [PubMed]
32. Ayoubi, T.A.Y.; Jenks, B.G.; Roubos, E.W.; Martens, G.J.M. Transcriptional and posttranscriptional regulation of the proopiomelanocortin gene in the pars intermedia of the pituitary gland of *Xenopus laevis*. *Endocrinology* **1992**, *130*, 3560–3566. [CrossRef] [PubMed]
33. Dotman, C.H.; van Herp, F.; Martens, G.J.M.; Jenks, B.G.; Roubos, E.W. Dynamics of pomc and pc2 gene expression in xenopus melanotrope cells during long-term background adaptation. *J. Endocrinol.* **1998**, *159*, 281–286. [CrossRef] [PubMed]
34. Attenborough, R.M.; Hayward, D.C.; Kitahara, M.V.; Miller, D.J.; Ball, E.E. A "neural" enzyme in nonbilaterian animals and algae: Preneural origins for peptidylglycine a-amidating monooxygenase. *Mol. Biol. Evol.* **2012**, *29*, 3095–3109. [CrossRef] [PubMed]
35. Wood, C.R.; Huang, K.; Diener, D.R.; Rosenbaum, J.L. The cilium secretes bioactive ectosomes. *Curr. Biol.* **2013**, *23*, 906–911. [CrossRef] [PubMed]
36. Avasthi, P.; Marshall, W. Ciliary secretion: Switching the cellular antenna to transmit. *Curr. Biol.* **2013**, *23*, R471–R473. [CrossRef] [PubMed]
37. Cao, M.; Ning, J.; Hernandez-Lara, C.I.; Belzile, O.; Wang, Q.; Dutcher, S.K.; Liu, Y.; Snell, W.J. Uni-directional ciliary membrane trafficking by a cytoplasmic retrograde ift motor and ciliary ectosome shedding. *eLife* **2015**, e05242. [CrossRef] [PubMed]
38. Chambers, J.E.; Petrova, K.; Tomba, G.; Vendruscolo, M.; Ron, D. Adp ribosylation adapts an er chaperone response to short-term fluctuations in unfolded protein load. *J. Cell Biol.* **2012**, *198*, 371–385. [CrossRef] [PubMed]

39. Feijs, K.L.; Verheugd, P.; Lüscher, B. Expanding functions of intracellular resident mono-adp-ribosylation in cell physiology. *FEBS J.* **2013**, *280*, 3519–3529. [CrossRef] [PubMed]
40. Lu, L.; Madugula, V. Mechanisms of ciliary targeting: Entering importins and rabs. *Cell. Mol. Life Sci.* **2018**, *75*, 597–606. [CrossRef] [PubMed]
41. Cavazza, T.; Vernos, I. The rangtp pathway: From nucleo-cytoplasmic transport to spindle assembly and beyond. *Front. Cell Dev. Biol.* **2016**, *3*, 82. [CrossRef] [PubMed]
42. Park, S.G.; Schimmel, P.; Kim, S. Aminoacyl trna synthetases and their connections to disease. *Proc. Natl. Acad. Sci. USA* **2008**, *105*, 11043–11049. [CrossRef] [PubMed]
43. Lee, M.S.; Kwon, H.; Nguyen, L.T.; Lee, E.Y.; Lee, C.Y.; Choi, S.H.; Kim, M.H. Shiga toxins trigger the secretion of lysyl-trna synthetase to enhance proinflammatory responses. *J. Microbiol. Biotechnol.* **2016**, *26*, 432–439. [CrossRef] [PubMed]
44. Williams, T.F.; Mirando, A.C.; Wilkinson, B.; Francklyn, C.S.; Lounsbury, K.M. Secreted threonyl-trna synthetase stimulates endothelial cell migration and angiogenesis. *Sci. Rep.* **2013**, *3*, 1317. [CrossRef] [PubMed]
45. Abe, J.; Kubo, T.; Takagi, Y.; Saito, T.; Miura, K.; Fukuzawa, H.; Matsuda, Y. The transcriptional program of synchronous gametogenesis in *Chlamydomonas reinhardtii*. *Curr. Genet.* **2004**, *46*, 304–315. [CrossRef] [PubMed]
46. Kinoshita, T.; Fukuzawa, H.; Shimada, T.; Saito, T.; Matsuda, Y. Primary structure and expression of a gamete lytic enzyme in *Chlamydomonas reinhardtii*: Similarity of functional domains to matrix metalloproteases. *Proc. Natl. Acad. Sci. USA* **1992**, *89*, 4693–4697. [CrossRef] [PubMed]
47. Marino-Puertas, L.; Goulas, T.; Gomis-Ruth, F.X. Matrix metalloproteinases outside vertebrates. *Biophys. Biochim. Acta Mol. Cell Res.* **2017**, *1864*, 2026–2036. [CrossRef] [PubMed]
48. Marino, G.; Funk, C. Matrix metalloproteinases in plants: A brief overview. *Physiol. Plant* **2012**, *145*, 196–202. [CrossRef] [PubMed]
49. Fanjul-Fernandez, M.; Folgueras, A.R.; Cabrera, S.; Lopez-Otin, C. Matrix metalloproteinases: Evolution, gene regulation and functional analysis in mouse models. *Biochim. Biophys. Acta* **2010**, *1803*, 3–19. [CrossRef] [PubMed]
50. Hallmann, A. The pherophorins: Common, versatile building blocks in the evolution of extracellular matrix architecture in *Volvocales*. *Plant J.* **2006**, *45*, 292–307. [CrossRef] [PubMed]
51. Harris, E.H. *Cell architecture. The Chlamydomonas Sourcebook*, 2nd ed.; Harris, E.H., Ed.; Elsevier: New York, NY, USA, 2009; Volume 1, pp. 25–63.
52. Bäck, N.; Kanerva, K.; Vishwanatha, K.S.; Yanik, A.; Ikonen, E.; Mains, R.E.; Eipper, B.A. The endocytic pathways of a secretory granule membrane protein in HEK293 cells: PAM and EGF traverse a dynamic multivesicular body network together. *Eur. J. Cell Biol.* **2017**, *86*, 407–417. [CrossRef] [PubMed]
53. Sinha, S.; Hoshino, D.; Hong, N.H.; Kirkbride, K.C.; Grega-Larson, N.E.; Seiki, M.; Tyska, M.J.; Weaver, A.M. Cortactin promotes exosome secretion by controlling branched actin dynamics. *J. Cell Biol.* **2016**, *214*, 197–213. [CrossRef] [PubMed]

proteomes

MDPI

Article

Sex-Specific Proteomic Changes Induced by Genetic Deletion of Fibroblast Growth Factor 14 (FGF14), a Regulator of Neuronal Ion Channels

Mark L. Sowers [1,2,†], Jessica Di Re [2,3,†], Paul A. Wadsworth [1,2,4], Alexander S. Shavkunov [2], Cheryl Lichti [2], Kangling Zhang [2] and Fernanda Laezza [2,*]

1 UTMB MD/PhD Combined Degree Program, University of Texas Medical Branch, Galveston, TX 77555, USA; mlsowers@utmb.edu (M.L.S.); pawadswo@utmb.edu (P.A.W.)
2 Department of Pharmacology and Toxicology, University of Texas Medical Branch, Galveston, TX 77555, USA; jedire@utmb.edu (J.D.R.); asshavku@utmb.edu (A.S.S.); clichti@wustl.edu (C.L.); kazhang@utmb.edu (K.Z.)
3 Neuroscience Graduate Program, University of Texas Medical Branch, Galveston, TX 77555, USA
4 Biochemistry and Molecular Biology Graduate Program, University of Texas Medical Branch, Galveston, TX 77555, USA
* Correspondence: felaezza@utmb.edu; Tel.: +1-409-772-9672
† These authors contributed equally.

Received: 16 November 2018; Accepted: 17 January 2019; Published: 23 January 2019

Abstract: Fibroblast growth factor 14 (FGF14) is a member of the intracellular FGFs, which is a group of proteins involved in neuronal ion channel regulation and synaptic transmission. We previously demonstrated that male *Fgf14*$^{-/-}$ mice recapitulate the salient endophenotypes of synaptic dysfunction and behaviors that are associated with schizophrenia (SZ). As the underlying etiology of SZ and its sex-specific onset remain elusive, the *Fgf14*$^{-/-}$ model may provide a valuable tool to interrogate pathways related to disease mechanisms. Here, we performed label-free quantitative proteomics to identify enriched pathways in both male and female hippocampi from *Fgf14*$^{+/+}$ and *Fgf14*$^{-/-}$ mice. We discovered that all of the differentially expressed proteins measured in *Fgf14*$^{-/-}$ animals, relative to their same-sex wildtype counterparts, are associated with SZ based on genome-wide association data. In addition, measured changes in the proteome were predominantly sex-specific, with the male *Fgf14*$^{-/-}$ mice distinctly enriched for pathways associated with neuropsychiatric disorders. In the male Fgf14$^{-/-}$ mouse, we found molecular characteristics that, in part, may explain a previously described neurotransmission and behavioral phenotype. This includes decreased levels of ALDH1A1 and protein kinase A (PRKAR2B). ALDH1A1 has been shown to mediate an alternative pathway for gamma-aminobutyric acid (GABA) synthesis, while PRKAR2B is essential for dopamine 2 receptor signaling, which is the basis of current antipsychotics. Collectively, our results provide new insights in the role of FGF14 and support the use of the *Fgf14*$^{-/-}$ mouse as a useful preclinical model of SZ for generating hypotheses on disease mechanisms, sex-specific manifestation, and therapy.

Keywords: mass spectroscopy; bioinformatics; FGF14; voltage gated channels; schizophrenia; autism; Alzheimer's Disease; sex-specific differences; synaptic plasticity; cognitive impairment; excitatory/inhibitory tone

1. Introduction

Originally identified as the genetic locus of missense mutations leading to spinocerebellar ataxia type 27 [1–7], fibroblast growth factor 14 (FGF14) is an emerging risk factor for neuropsychiatric disorders [8]. Unlike canonical secreted FGFs, which act through the activation of FGF receptor signaling, FGF14 is retained intracellularly, where it has been shown to regulate ion channel function [9–13]. Much evidence indicates that FGF14 within neurons binds directly to and regulates the voltage-gated sodium (Nav) channel, targeting the axonal initial segment (AIS) and biophysical properties [9–21]. Other reported functions of FGF14 suggest a much more complex role within the brain, including the regulation of presynaptic glutamate and gamma-aminobutyric acid (GABA) release, and calcium signaling [18–21]. Studies focused on signaling pathways demonstrated that FGF14 is also a hub for regulatory kinases [11,22], including glycogen synthase kinase 3 [15], which is an enzyme that is linked to depression, bipolar disorder, and schizophrenia (SZ) [8,23–25].

Given the variety of key cellular functions associated with FGF14, it is not surprising that the deletion of the gene results in disrupted function and behavior associated with complex brain disorders. Recent studies have shown that male mice lacking *Fgf14* (*Fgf14*$^{-/-}$) recapitulate key features of SZ endophenotypes. Namely, male *Fgf14*$^{-/-}$ mice present with the loss of parvalbumin positive GABAergic interneurons in the hippocampus, disrupted gamma frequency, and reduced working memory, all of which are hallmarks of cognitive impairment in SZ animal models and post-mortem studies [21,26]. Concomitant changes in these mice are found at the glutamatergic synapses with reduced presynaptic release and long-term potentiation [20,27], which may be the common underlying pathology of SZ and other neurodevelopmental disorders [28]. Additional evidence of disease endophenotypes is brought by studies reporting disrupted adult neurogenesis in the dentate gyrus (DG) of *Fgf14*$^{-/-}$ mice that is consistent with an immature dentate gyrus [21,29] and is another hallmark of SZ and other neuropsychiatric disorders [30].

In addition to reduced working memory, male *Fgf14*$^{-/-}$ mice exhibit behavioral deficits that align with disrupted dopamine signaling, including altered aggressive and reproductive behavior, and blunt response to cocaine and methamphetamine [26,31].

Taken together, these findings indicate that the male *Fgf14*$^{-/-}$ mouse recapitulates the endophenotypes of SZ, including changes in GABA and glutamatergic synaptic signaling, leading to perturbations of the excitatory/inhibitory (E/I) tone of the brain [32–36], impaired neurogenesis in the DG, and disruption of dopamine signaling, which are all functional nodes in SZ pathophysiology.

Although many lines of evidence converge to suggest that male *Fgf14*$^{-/-}$ mice are useful animals for the study of SZ, little is known about how these complex phenotypes develop, how they relate to other neurodevelopmental diseases, or whether sex-specific differences exist in female *Fgf14*$^{-/-}$ animals.

We chose to investigate this potentially useful animal model to gain further insight into the etiology of SZ and related disorders. We performed label-free proteomic mass spectrometry and a variety of bioinformatic approaches on isolated hippocampi from male and female wild-type (WT) and *Fgf14*$^{-/-}$ mice to determine the molecular pathways disrupted in this model. As a result, we found evidence that this animal model recapitulates the molecular aspects found in patients afflicted with SZ. Our results will aid in the generation of new hypotheses about neuropsychiatric diseases, and are expected to elucidate several gender-specific differences in the etiology of SZ, such as the age of diagnosis, symptom clustering, premorbid function, treatment response, and prognosis [37–41].

2. Materials and Methods

2.1. Hippocampal Tissue Preparation

Fgf14$^{-/-}$ and *Fgf14*$^{+/+}$ male and female mice are maintained on an inbred C57/BL6J background with greater than 10 generations of backcrossing to C57/BL6J. Animals were bred in the University of Texas Medical Branch animal care facility: either heterozygous *Fgf14*$^{+/-}$ males and females or, in a few cases, homozygotes (*Fgf14*$^{-/-}$ males with *Fgf14*$^{+/-}$ females); *Fgf14*$^{+/+}$ WT mice served as control. Both male and female mice were used in this study at four to six months of age, unless otherwise stated. The University of Texas Medical Branch operates in compliance with the United States Department of Agriculture Animal Welfare Act, the Guide for the Care and Use of Laboratory

Animals, and Institutional Animal Care and Use Committee approved protocols (0904029C). Mice were housed, $n \leq 5$ per cage, and kept under a 12-h light/12-h dark cycle with sterile food and water *ad libitum*. All of the genotypes described were confirmed by genotyping of the progeny using DNA extraction and PCR amplification following established protocols or conducted at Transnetyx Inc. (Cordova, TN, USA).

Both hippocampi were dissected from each mouse brain of *Fgf14*$^{-/-}$ and *Fgf14*$^{+/+}$ male and female adult mice. A total of three biological replicates were in each group. Biological replicates were combined to maximize the amount of total protein. Protein extraction was done on these combined samples and analyzed three times for a total of three technical replicates. Tissue was homogenized in RIPA buffer (Thermo Fisher Scientific, Rockford, IL, 25 mM of TrisHCl pH 7.6, 150 mM of NaCl, 1% NP-40, 1% sodium deoxycholate, 0.1% SDS) containing Halt protease and phosphatase EDTA-free inhibitor cocktail (Thermo Fisher Scientific, Rockford, IL, USA) and one mM of phenylmethylsulfonyl fluoride. Mechanical homogenization was performed using Polytron™ PT 10/35 GT Homogenizer (Kinematica, Bohemia, NY, USA), 20 s × three pulses, at 10,000 rpm. After homogenization, Pierce universal nuclease (Thermo Fisher Scientific, Rockford, IL) was added to samples (25 units per one mL of tissue lysate) and incubated on ice for 30 min. Protein concentration was determined using a BCA Protein Assay Kit (Pierce). Then, 100 µg aliquots of total protein were reduced and alkylated. 5 µL of 200 mM of tris (2-carboxyethyl) phosphine (TCEP) buffered with 50 mM of triethylammonium bicarbonate (TEAB) were added to each sample (final TCEP concentration: 10 mM) and incubated at 55 °C for 1 h. 5 µL of 375 mM of iodoacetamide (buffered with 50 mM of TEAB) were added and incubated in the dark for 30 min. Proteins were precipitated in four volumes (440 µL) of ice-cold acetone overnight at −20 °C. Samples were centrifuged at 10,000× *g* for 30 min (4 °C), after which the supernatants were removed and discarded. Protein pellets were delipidated and incubated in one mL of ice-cold tri-n-butylphosphate/acetone/methanol (1:12:1 by volume), followed by centrifugation (Eppendorf 5415D, Hamburg, Germany) at 2800× *g* for 15 min at 4 °C, and sequential incubations in ice-cold tri-n-butyl phosphate, acetone, and methanol, for 15 min each [42]. Pellets were air-dried and resuspended in 12.5 µL of eight M of urea. Trypsin (4 µg in 87.5 µL of TEAB buffer) was added, and the samples were incubated for 24 h at 37 °C. A final sample clean-up and removal of urea were performed using Mark C18 Sep-Pak® Vac 1cc cartridges (Waters, Milford, MA, USA) attached to a vacuum manifold. Cartridges were pre-equilibrated with 3 × 1 mL of acetonitrile and washed with 3 × 1 mL of 0.25% trifluoroacetic acid (flow rate ~ 2 mL/min); digested samples were loaded onto the cartridges after adding trifluoroacetic acid to 1% final concentration, washed with 4 × 1 mL of 0.25% trifluoroacetic acid, eluted in one mL of 80% acetonitrile/0.1% formic acid, and dried in the CentriVamp Concentrator (Labconco, Kansas City, MO, USA).

2.2. Mass Spectrometry and Chromatography

Chromatographic separation and mass spectrometric analysis were performed with a nano-LC chromatography system (Easy-nLC 1000, Thermo Scientific) coupled online to a hybrid linear ion trap-Orbitrap mass spectrometer (Orbitrap Elite, Thermo Scientific) through a Nano-Flex II nanospray ion source (Thermo Scientific). Mobile phases were 0.1% formic acid in water (A) and 0.1% formic acid in acetonitrile (ACN, B). After equilibrating the column in 95% solvent A and 5% solvent B, the samples (5 µL in 5% *v*/*v* ACN/0.1% (*v*/*v*) formic acid in water, corresponding to 1 µg of tissue protein digest) were injected onto a trap column (C18, 100 µm ID × 2 cm) and subsequently eluted (250 nL/min) by gradient elution onto a C18 column (10 cm × 75 µm ID, 15 µm tip, ProteoPep II, 5 µm, 300 Å, New Objective). The gradient was as follows: isocratic flow at 5% Solvent B for 5 min, 5% to 35% Solvent B for 89 min, and 35% to 95% Buffer B for 16 min followed by isocratic flow at 95% Buffer B for 10 min.

All of the LC-MS/MS data were acquired using XCalibur, version 2.7 SP1 (Thermo Fisher Scientific). The survey scans (m/z 350–1650) (MS) were acquired in the Orbitrap at 60,000 resolution (at m/z = 400) in profile mode, followed by top five higher energy collisional dissociation (HCD) fragmentation centroid MS/MS spectra, acquired at 15 K resolution in data-dependent analyses (DDA) mode. The automatic gain control targets for the Orbitrap were 1×10^6 for the MS scans and 5×10^4 for MS/MS scans. The maximum injection times for the MS1 and MS/MS scans in the Orbitrap were both 500 ms. For MS/MS acquisition, the following settings were used: parent threshold = 10,000; isolation width = 4.0 Da; normalized collision energy = 30%; and activation time = 10 ms. zMonoisotopic precursor selection, charge-state screening, and charge-state rejection were enabled, with the rejection of singly charged and unassigned charge states. Dynamic exclusion was used to remove selected precursor ions (±10 ppm) for 90 s after MS/MS acquisition. A repeat count of one, and a maximum exclusion list size of 500, were used. The following ion source parameters were used: capillary temperature 275 °C, source voltage 2.1 kV, source current 100 µA, and S-lens RF level 40%. Each sample was analyzed in triplicate, and the order of runs was block-randomized.

2.3. *Quantification of Peptides and Proteins*

Maxquant version 1.6.1.0 was used to process raw files [43,44]. Default settings were used unless otherwise specified. Briefly, peptide spectrum match and protein false discovery rate (FDR) were set to 1% and a minimum of one unique peptide for identification. Fixed modifications were set to carbamidomethyl for cysteine, and variable modifications were set to methionine oxidation and N-terminal acetylation. Matches between runs were enabled with a default match time window of 0.7 min and alignment window of 20 min. The mouse Uniprot reference proteome was downloaded on 18 September 2018, last updated 28 July 2018, with canonical and isoform sequences. For label-free quantification (LFQ), MS2 was required, while a minimum of one peptide was required for quantification across samples, including both razor and unique peptides.

2.4. *Statistical Analysis*

Statistical analysis was performed with Perseus 1.6.0.7 [45]. LFQ intensity values were $\log_2$ transformed to render the data normally distributed. Proteins identified by site, reverse, and potential contaminants were filtered prior to analysis. Proteins with missing values in any sample, including replicates, were filtered. Differentially expressed proteins were determined using a moderated *t* test statistic with the FDR controlled at 5% and the s0 parameter set to 0.1. Multiple test correction was done using a permutation-based randomization procedure where values are randomly shuffled to generate a "null" distribution to estimate the random type one error, or false detection rate, with 250 randomizations. This is the preferred procedure in Perseus.

3. Results

To efficiently detect specific changes in the cellular proteome, it is important to limit the biological complexity of the subject of study. Whole brain proteome analysis is likely to miss or downplay prominent changes of protein expression in particular brain regions. Therefore, the proteomic analysis of isolated brain structures is preferred. Previous studies, including our own, have largely focused on the role of FGF14 in the hippocampus. The hippocampus is part of the limbic system, which is critically involved in cognition and a primary site of FGF14 expression [13,20,31]. FGF14 knockout causes pronounced changes in the synaptic transmission [20,21] and cellular composition [21,29] of the hippocampus, which correlate with changes in electrophysiology and behavioral deficits [20,27,31]. Given the documented role of hippocampal pathology in cognitive impairment in SZ [46–49], we hypothesize that these gross changes play a critical part in the development of SZ-related endophenotypes in *Fgf14*$^{-/-}$ mice. An additional advantage is that the hippocampus can be readily isolated from adjacent brain structures, which makes sample preparation more robust and reproducible. The workflow of our study is presented in Figure 1.

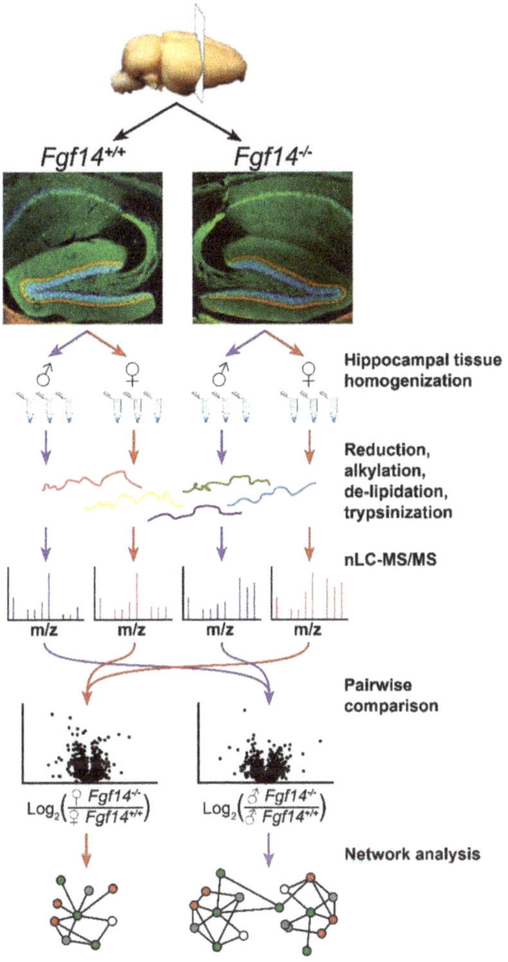

Figure 1. Overview of label-free proteomics workflow and analysis. Workflow outlining experimental procedures and LC-MS/MS data acquisition for the analysis of hippocampal brain tissue [50–52] from male and female *fgf14*$^{+/+}$ and *Fgf14*$^{-/-}$ mice, as detailed in the text. Representative confocal images of triple staining of the entire hippocampus from *Fgf14*$^{+/+}$ (**left**) and *Fgf14*$^{-/-}$ (**right**) mice representing calbindin (green), calretinin (red), and Topro-3 nuclear staining (blue) at low magnification of the dentate gyrus (DG).

While having many advantages, primarily ease of use, label-free proteomics chromatography conditions must be standardized and assessed for reproducibility and overall data quality. As shown in Figure 2, the various samples and their technical replicates are highly reproducible after appropriate filtering (see methods). Furthermore, Maxquant quantifies protein intensity using MS2 spectra. However, if an MS2 spectra, which is needed for peptide sequencing/identification, is missing in one run due to the stochastic nature of data-dependent acquisition, Maxquant has the match between run (MBR) feature. MBR allows for quantification by imputing the estimated MS2 intensity by using the mass and retention time alignment of the corresponding MS1 peak [43,44].

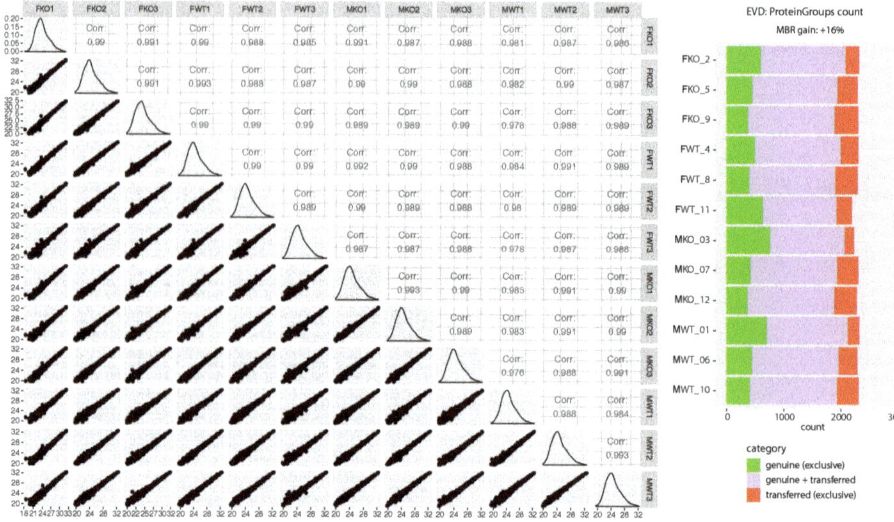

Figure 2. Quality control of label-free quantitative proteomics. The scatter matrix shows pairwise Pearson correlations between animal groups and their technical replicates, histograms of $\log_2$ label-free quantification (LFQ intensity) distributions, and their respective scatter plots. On the right is a quality control figure showing ~2300 proteins identified in each sample after applying match-between-runs (MBR) in Maxquant. Most proteins were identified by a combination of MS1 matching from other samples, where peptide identification was successful, as well as directly by MS2 (purple). A smaller subset of proteins could be identified exclusively by MS2 in a given run (green), and proteins identified only after the retention time and the *m/z* alignment of MS1 peaks in comparison to other runs resulted in a 16% gain of quantified proteins (red). The quality control figure was prepared using R programming language and Proteomics Quality Control (PTXQC) [53].

3.1. *Differentially Expressed Proteins in Fgf14$^{-/-}$ Mice and their Implications*

After log transforming and filtering, we analyzed ~1500 proteins whose distribution was approximately normal across all of the samples. Then, we compared male and female *Fgf14*$^{-/-}$ mice to their respective wild-type counterparts using statistical analysis of microarrays, which is a moderated t-test statistic (Figure 3, Table S1). We chose to investigate both male and female homozygous knockouts of FGF14, as we had previously shown that male knockouts have SZ-like dysfunction, while female mice for this model had not been previously investigated.

In the female *Fgf14*$^{-/-}$ mice, we found *Snap25* and *Mtatp6* upregulated. SNAP25 is part of the SNARE complex, which mediates neurotransmitter–vesicle fusion and controls receptor trafficking at post-synaptic sites of glutamatergic and GABAergic synapses [54]. It is unclear how the genetic deletion of FGF14 causes a change in SNAP25 expression, but previous proteomic studies have shown that FGF14 immunoprecipitates with SNAP25 [10]. Thus, the genetic deletion of FGF14 could lead to SNAP25 loss of function, which in animal models is considered a mechanism leading to SZ endophenotypes [55].

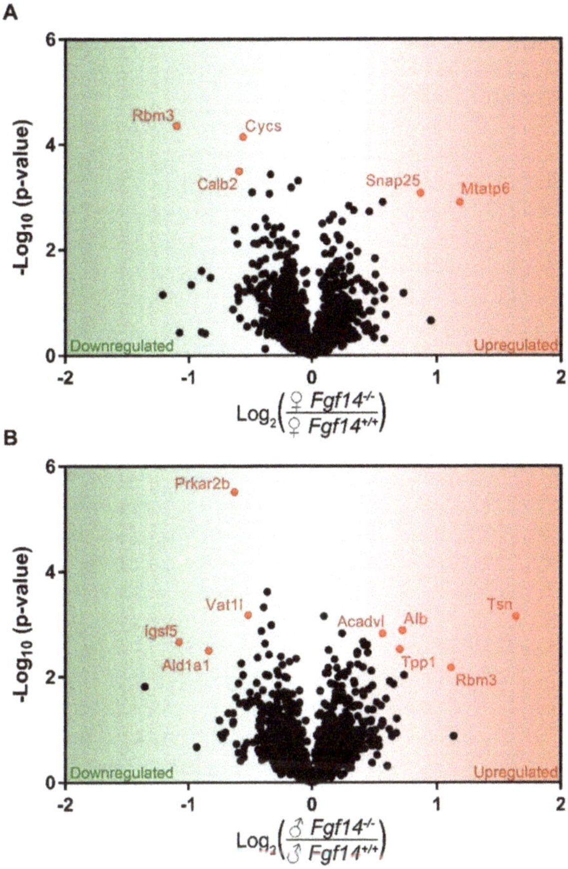

Figure 3. Volcano plot of mass spectrometry results. Proteins that are significantly upregulated in sex by genotype conditions: (**a**) shows the proteins significantly upregulated in female *Fgf14*$^{-/-}$ compared to *Fgf14*$^{+/+}$ mice; and (**b**) shows the proteins significantly upregulated in male *Fgf14*$^{-/-}$ compared to *Fgf14*$^{+/+}$ mice. The Y-axis represents negative log10 (p-value) based on the test statistic, and the X-axis shows proteins with a positive $\log_2$ fold change (FC) as upregulated (red), and negative values as downregulated (green) proteins in the *Fgf14*$^{-/-}$ mice, respectively.

MTATP6, or ATP synthase/Complex V, has been associated with SZ as either decreased mRNA levels or as genetic polymorphisms [56,57]. Thus both SNAP25 and MTATP6, which have been shown either knocked down or decreased in association with SZ, are upregulated in female knockouts; this is a possible mechanism of resistance to the genetic deletion of FGF14. SNAP25 and MTATP6, while upregulated in female mice, were not differentially expressed in male mice after multiple hypothesis test correction. This suggests that the cognitive deficits seen in male *Fgf14*$^{-/-}$ mice may be a consequence of reduced energy production, while their female *Fgf14*$^{-/-}$ counterparts may be able to compensate. Despite the upregulation of these two proteins, we would expect some dysfunction in mitochondrial energy production as well as GABA-ergic signaling in the female knockout mice, as represented by decreased Calretinin (CalB2) and Cytochrome C (Cycs).

CYCS is an essential component of oxidative phosphorylation that is a major source of energy, particularly in neurons. Mitochondrial dysfunction is also believed to be one of the potential risk factors of SZ [56]. CALB2 is a calcium-buffering protein that is predominantly expressed in calretinin positive interneurons, which is a subtype of cells expressed in the hippocampus [58]. This suggests that there may be a decrease in calretinin-positive interneurons in the hippocampus in female *Fgf14*$^{-/-}$ mice. This is in direct opposition to the increase in these interneurons and immature dentate gyrus that were previously reported in *Fgf14*$^{-/-}$ males [29].

Differentially expressed proteins were almost entirely different between female *Fgf14*$^{-/-}$ and male *Fgf14*$^{-/-}$ compared to their respective *Fgf14*$^{+/+}$ controls, with the exception of RBM3. RBM3 is a cold inducible protein that is believed to be protective against neurodegeneration and mediate structural plasticity [59]. While believed to aid in translation, RBM3 has also been reported as two alternatively spliced isoforms, with the variant lacking arginine more highly expressed in the dendritic spines of mature neurons [60]. Smart et al. also reported in the same study that both RBM3 isoforms are post-translationally modified. Thus, the difference in RBM3 expression between male and female *Fgf14*$^{-/-}$ mice could be due, in part, to the lack of quantitation of some peptides, since only unmodified peptides were quantified. While RBM3 was found to be upregulated in males, which is generally thought to be protective, it is unclear if this is due to the stress response, lack of post-translational modifications, or some combination of the two. Additionally, we found that RBM3 expression was lower in male versus female WT groups (Figure S1). This might suggest that females and males have either differential expression in the hippocampus; alternatively, again, sex-specific post-translational modifications could also play a role. Furthermore, differential expression may be a consequence of different dendritic morphology and branching [61]. There are known differences in male and female C57BL/6J mice, as RBM3 is enriched in dendritic spines. However, further targeted investigation would be needed in order to determine the effect of sex and FGF14 on RBM3 expression and post-translational modifications.

Although we focused our studies primarily on male knockouts (see Discussion), as they displayed the cognitive and synaptic functions of interest, we identified key differences between normal male and female hippocampi. Namely, most of the proteins that were differentially expressed were related to the "neuron part" cellular component of the gene ontology (GO) term (Figure S1). As mentioned previously, dendritic morphology has sex-specific differences. Estrogen may also play a role in the hippocampal neuronal spine shape and long-term potentiation [62]. Our results support that there are sex-dependent differences in proteins that are important for spine formation and dendrite morphology in the hippocampus. Proteins overexpressed in the females, relative to other WT mice that were male, were related to calcium signaling (CAMK2A) and calcium regulation (ANXA6). The former is most strongly implicated in the early phases of long-term potentiation [63]. Copine 6 was also upregulated, which is a calcium-binding protein that is believed to be responsible for translating calcium signals into morphological changes at the level of synaptic spines [64].

Interestingly, two differentially expressed proteins that were found downregulated in *Fgf14*$^{-/-}$ male mice than male WT, IGSF5 and VAT1L, were also found to be more abundantly expressed in the male WT than female WT. Not only do these proteins appear to be differentially regulated by loss FGF14 in only the male mice, they are more abundantly expressed in male WT than female. This suggests they may be of central importance in the male hippocampus.

3.2. *Differentially Expressed Proteins Highly Associated with Schizophrenia and/or Autism*

Interestingly, we discovered that in an analysis of various genome-wide association studies (GWAS), all of our differentially expressed proteins, with the exception of MTATP6, were identified to be statistically associated with either autism and/or SZ, suggesting that the *Fgf14* knockout mouse might be a valuable model for a wider range of neuropsychiatric and neurodevelopmental disorders. Protein level p-values were determined by Seyfried et al. using the MAGMA tool, which controls for various confounders to determine the p-values for each protein coding gene [65,66] (Figure 4, Table S2). MTATP6 may have been missing in this dataset due to being coded on the mitochondrial genome.

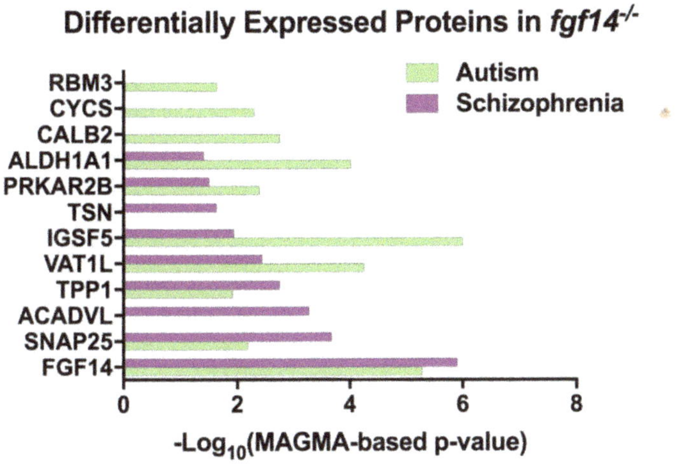

Figure 4. Differentially expressed proteins in *Fgf14*$^{-/-}$ mice are associated with autism and schizophrenia. The proteins that were identified in our study were found in an analysis of genome-wide association study (GWAS) data [65] (Table S10) using the MAGMA tool [66].

3.3. Central Role of ALDH1A1 and SNAP25 in Pathophysiology of Fgf14$^{-/-}$ *Mice*

Experimental protein–protein interaction networks were constructed with the differentially expressed proteins for males and females, separately, using OmicsNet, which identifies known interactors [67] (Figure 5). Interactions were based only on high-confidence STRING interactions with experimental evidence. The networks were imported into Cytoscape for visual purposes. Three-dimensional predicted protein–protein interaction networks were constructed with the differentially expressed proteins for males and females, separately (Figure 5). The network construction did not generate any connections to other significant proteins other than *Snap25* and *Aldh1a1* for females and males, respectively. Although protein–protein interaction data are far from complete, this suggests that Snap25 and Aldh1a1 may be key players in the pathogenesis observed in *Fgf14*$^{-/-}$ mice and perhaps SZ and/or autism [68,69].

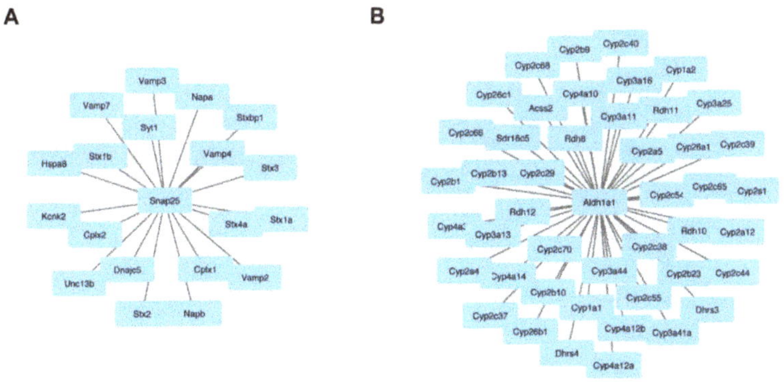

Figure 5. Central node proteins networks. OmicsNet was used to generate protein–protein networks with differentially expressed proteins and known experimental interactors. The networks that were created based on input gene names are shown for both the male and female *Fgf14*$^{-/-}$ mice. In the center are the input genes, and the connected genes are known interactors. This analysis identified *snap25* (**A**) and in female *Fgf14*$^{-/-}$ and a*ldh1a1* (**B**) male *Fgf14*$^{-/-}$ mice as central interactors.

3.4. Hierarchical Clustering Reveals Subtype-Specific Clusters

We performed the hierarchical clustering of quantified proteins and sample groups using Euclidean distance metric with average linkage as well as preprocessing with k-means for data reduction purposes, prior to the generation of the heatmap shown in Figure 6. Sample replicates were median-averaged, and the measured proteins were Z-score normalized across sample groups prior to clustering. Both Z-scoring and clustering were done in the Perseus bioinformatics suite (default clustering settings) [45]. Our analysis identified four protein clusters of interest, because they were upregulated in each of the respective animals. We submitted these group-specific clusters to the STRING protein–protein network database using only the highest confidence interactions based on all of the data types, and identified the positively enriched pathways for each animal-specific protein cluster (Figure 7, Table S3).

Of particular note are the enriched pathways in male *Fgf14*$^{-/-}$mice, which includes alcoholism, drug addiction, and related pathologies. These pathways may explain the endophenotype of male *Fgf14*$^{-/-}$ mice. Furthermore, all of the animal groups had enriched terms related to vesicle, membrane-bound vesicle, or vesicle-mediated transport, likely indicating their important roles in the mouse hippocampus. This also suggests that both sex and the presence of FGF14 may affect different aspects of neurotransmission given that these terms are positively enriched in all of the protein clusters.

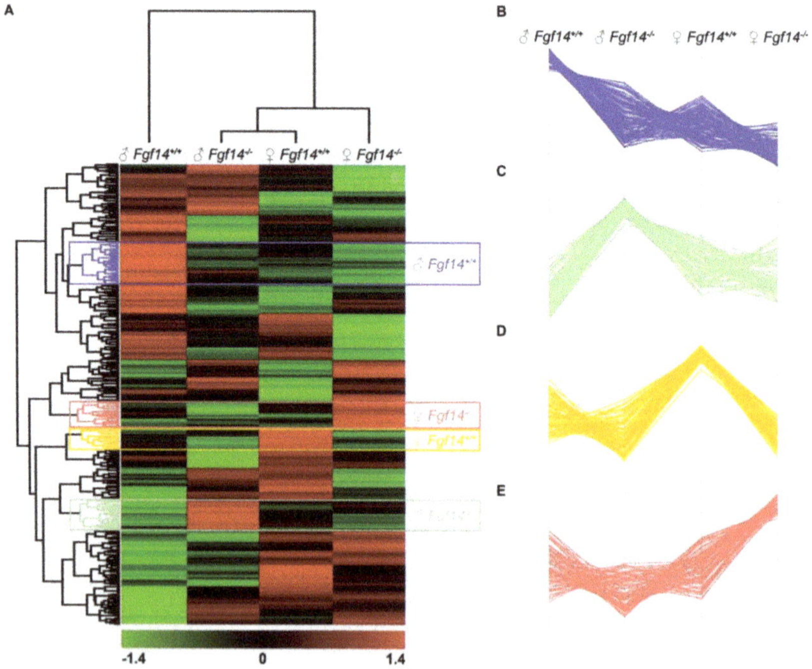

Figure 6. Hierarchical clustering, heatmap, and cluster analysis of differentially expressed proteins. (**A**) Heatmap of differentially expressed proteins in male and female *Fgf14*$^{+/+}$ and *Fgf14*$^{-/-}$ mice. LFQ intensities were averaged for technical replicates, and averages across animal groups were Z-scored prior to Euclidean distance-based hierarchical clustering with Perseus (**B–E**). Protein clusters specific to each animal group, male *Fgf14*$^{+/+}$, female *Fgf14*$^{-/-}$, female *Fgf14*$^{+/+}$, and male *Fgf14*$^{-/-}$, respectively.

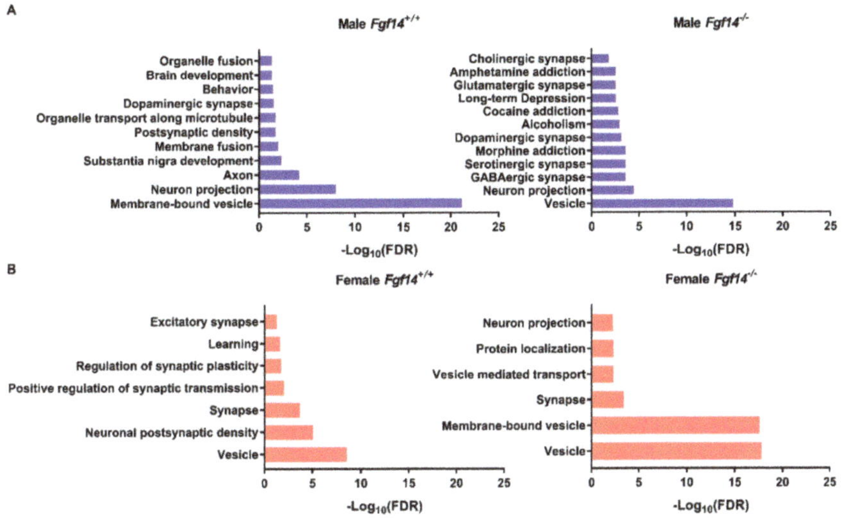

Figure 7. Protein–protein interaction and pathway enrichment for animal-specific clusters. (**a**) Enriched pathways and their adjusted *p*-values (FDR) were obtained from the STRING database after inputting cluster-specific gene names. Male *Fgf14*$^{-/-}$ display protein expression changes broadly associated with neurotransmitter-based synaptic activation, drug addiction, and alcoholism. This was unique to the male *Fgf14*$^{-/-}$ specific cluster. (**b**) Female *Fgf14*$^{-/-}$ mice display protein expression changes broadly associated with synaptic vesicles, synaptic transport, and protein localization; this is an important function of FGF14.

4. Discussion

Using a label-free proteomic approach and bioinformatics, we analyzed sex-specific differences in the hippocampi of *Fgf14*$^{-/-}$ mice relative to their sex-specific controls. Previous work has demonstrated that male *Fgf14*$^{-/-}$ mice present with cognitive deficits and changes in neuronal function that mimic the endophenotypes of SZ and other neuropsychiatric disorders [21,29]. However, the results presented in this study provide a biological context as to which specific pathways might be disrupted. Importantly, we found that many of the proteins differentially expressed in male *Fgf14*$^{-/-}$ mice have previously been linked to neuropsychiatric disorders with cognitive impairment, such as SZ and autism (Figure 4). In fact, a network analysis of proteomic data from the brains of Alzheimer's disease (AD) patients has shown that synaptic transmission, synaptic membrane, and mitochondrion pathways are disrupted [65]. Perhaps this indicates a general mechanism for cognitive impairment that may be related to SZ, autism, and even the cognitive aspects of AD.

Importantly, many of the proteins with significantly altered expression in male *Fgf14*$^{-/-}$ mice, including ALDH1A1, PRKAR2B, and VAT1L, have previously been linked to SZ and other neuropsychiatric disorders within the domain of cognitive symptoms (Figure 3) [70–73]. These results also further support the role of FGF14 in synaptic signaling [18,21]. For example, it is known that *Fgf14*$^{-/-}$ male mice present with changes in GABA-ergic signaling in the hippocampus [18,21]. Our results here support that FGF14 may regulate the composition of GABA-ergic synapses both presynaptically and postsynaptically through SNAP25 (Figure 3) and synaptic function (Figure 7).

Alterations in the dopaminergic signaling of male *Fgf14*$^{-/-}$ mice may be due to changes in ALDH1A1. ALDH1A1 is not only an important enzyme for the breakdown of alcohol, it also defines a subpopulation of dopaminergic neurons in the rodent and human substantia nigra pars compacta, which are sensitive to α-synuclein cytotoxicity [71]. As shown in Liu et al., the deletion of ALDH1A1 exacerbates dopaminergic neurodegeneration in a mouse model of Parkinson's disease. This effect may be mediated through changes in the E/I tone of the brain, as retinoic acid, which is synthesized by ALDH1A1, regulates synaptic scaling at glutamatergic synapses by regulating AMPA receptor trafficking [74]. Furthermore, ALDH1A1 is part of a highly conserved pathway that provides an alternative method of GABA synthesis through putrescine [70]. Disrupting this pathway through decreasing ALDH1A1 might cause the deprivation of alternative pathways to synthesize GABA, which could in turn reduce inhibitory transmission and disrupt the E/I tone. These findings support others who have shown that male *Fgf14*$^{-/-}$ mice exhibit changes in synaptic function as well as in their response to drugs of abuse, such as cocaine and methamphetamine [31].

Changes in dopaminergic and GABAergic signaling in male *Fgf14*$^{-/-}$ mice can also be attributed to a decreased level of protein kinase A (PRKAR2B). PRKAR2B has been linked to GABA receptor breakdown by the endothelial gene claudin-5 in the prefrontal cortex of patients with SZ [75], and may partially underlie the mechanism of action of several antipsychotics through the increase of GABA receptors [72,76]. PRKAR2B also plays a role in dopaminergic neuromodulation, although this has typically been shown in the nucleus accumbens for D2 receptor signaling and in neuronal firing in medium spiny neurons [75,77–80]. Probes that target the interface between FGF14 and the voltage-gated Na^+ channel 1.6 (Nav1.6) have been shown to disrupt medium spiny neuron firing, which is a phenotype found in the same neuron subtype in male *Fgf14*$^{-/-}$ mice [14]. Therefore, it is plausible that protein kinase A (PKA) and FGF14 provide a regulatory mechanism of medium spiny neuron firing that contributes to maintaining dopaminergic tone in the nucleus accumbens.

Other proteins with altered expression in the male *Fgf14*$^{-/-}$ mice are members of pathways altered in neuropsychiatric disorders. Translin (TSN) is an RNA binding protein that regulates the dendritic trafficking of brain-derived neurotrophic factor (BDNF) [81]. BDNF is tied to synaptic transmission, plasticity, and homeostasis, and decreased serum levels of BDNF and mutations in the BDNF receptor, tyrosine receptor kinase B, have also been linked to SZ [82,83]. Although little is known about the effects of increased TSN [84], its altered expression in male *Fgf14*$^{-/-}$ mice, along with altered levels of expression of other SZ associated proteins, support additional findings that the male *Fgf14*$^{-/-}$ mouse model may be a new model of SZ and other disorders with a disrupted cognition component.

Overall, these findings support that *Fgf14*$^{-/-}$ male mice have several key features constituting an endophenotype of SZ [21]. As there are currently no pharmacotherapies for the treatment of the cognitive symptoms of SZ, this animal model may be a powerful tool in the discovery and testing of new disease treatments.

Of similar importance is the striking finding that the differentially expressed proteome of female *Fgf14*$^{-/-}$ mice is different from their male counterparts. This is especially critical given the gender differences in several domains of neuropsychiatric disorders, including the age of diagnosis, premorbid functioning, and symptom clustering [37–41]. These results indicate that there is a need to study behavioral changes, if any, in female *Fgf14*$^{-/-}$. In fact, the upregulation of SNAP25 and MTATP6 not only indicate that female *Fgf14*$^{-/-}$ mice have a unique proteomic signature, but that they may have a mechanism of resilience that compensates for changes in the synaptic functions seen in male mice of the same age.

SNAP25 is an important member of the SNAP/SNARE complex, which is necessary for the proper release of vesicles at the synapse [85] and has previously been shown to co-immunoprecipitate with FGF14 [10]. Not only is the deletion of SNAP25 linked to an increase in E/I tone through increased glutamatergic neurotransmission [55], but the deletion of SNAP25 has also been linked to improper neurogenesis in the adult mammalian brain, which is an important endophenotype of several neuropsychiatric disorders, including SZ and bipolar disorder [86,87]. Previously, it has been shown that male *Fgf14*$^{-/-}$ mice also show traits of an immature DG [29]. These findings suggest that this could be mediated through SNAP25, although more research is needed to determine whether neurogenesis is also altered in the brain of adult female *Fgf14*$^{-/-}$ mice. Decreases in both SNAP25 and MTATP6 have been seen in patients with SZ, as well as in potential animal models of neuropsychiatric diseases [55,57,85,88,89]. Furthermore, genetic variants of SNAP25 leading to low protein expression levels have been associated with hyperactivity and/or with low cognitive scores in autistic patients [69,90], corroborating our results linking differentially expressed proteins in *Fgf14*$^{-/-}$ mice with autism (Figure 4). Not only does our study highlight the importance of sex-specific research in basic science, it lays the groundwork for further investigations on the mechanisms of potential resilience to neuropsychiatric disorders in females in preclinical models as well as in humans.

Supplementary Materials: The following are available online at http://www.mdpi.com/2227-7382/7/1/5/s1, Figure S1: Male vs. Female WT volcano plot; Table S1: Protein fold changes, statistics, and proteomic results data from proteingroups.txt output from Maxquant. This file corresponds to actual data presented in Figure 3; Table S2: Autism and Schizophrenia associated genes, actual data for Figure 4; Table S3: Group-specific protein clusters identified from hierarchical clustering, complete protein lists used to prepare Figure 7.

Author Contributions: Conceptualization, F.L., J.D.R., and M.L.S.; Validation, M.L.S.; Formal Analysis, M.L.S. and C.L.; Investigation, A.S.S.; Resources, F.L., K.Z., and C.L.; Data Curation, A.S.S.; Writing—Original Draft Preparation, J.D.R., M.L.S.; Writing—Review & Editing, F.L.; Visualization, P.A.W.; Supervision, F.L.; Project Administration, F.L. and K.Z.; Funding Acquisition, F.L. and K.Z.

Funding: This research was funded by NIH R01MH111107 (F.L.), R01MH095995 (F.L.), R01DA047102 (F.L.), R01CA184097 (K.Z.), NIA T32 AG051131 (P.A.W.), and University of Texas Medical Branch Jeanne B. Kempner Scholarship (J.D.R.).

Acknowledgments: We thank Heather Lander for critical reading and editing of the manuscript.

Conflicts of Interest: The authors declare no conflict of interest.

References

1. Brusse, E.; de Koning, I.; Maat-Kievit, A.; Oostra, B.A.; Heutink, P.; van Swieten, J.C. Spinocerebellar Ataxia Associated with a Mutation in the Fibroblast Growth Factor 14 Gene (SCA27): A New Phenotype. *Mov. Disord.* **2006**. [CrossRef]
2. Pablo, J.L.; Pitt, G.S. Fibroblast Growth Factor Homologous Factors. *Neuroscientist* **2016**, *22*, 19–25. [CrossRef]
3. Groth, C.L.; Berman, B.D. Spinocerebellar Ataxia 27: A Review and Characterization of an Evolving Phenotype. *Tremor Other Hyperkinet. Mov.* **2018**, *8*, 534. [CrossRef]
4. Hoxha, E.; Tempia, F.; Lippiello, P.; Miniaci, M.C. Modulation, Plasticity and Pathophysiology of the Parallel Fiber-Purkinje Cell Synapse. *Front. Synaptic Neurosci.* **2016**, *8*, 35. [CrossRef]
5. Choquet, K.; La Piana, R.; Brais, B. A Novel Frameshift Mutation in FGF14 Causes an Autosomal Dominant Episodic Ataxia. *Neurogenetics* **2015**, *16*, 233–236. [CrossRef]
6. Hoxha, E.; Balbo, I.; Miniaci, M.C.; Tempia, F. Purkinje Cell Signaling Deficits in Animal Models of Ataxia. *Front. Synaptic Neurosci.* **2018**, *10*, 6. [CrossRef]
7. Ornitz, D.M.; Itoh, N. The Fibroblast Growth Factor Signaling Pathway. *Wiley Interdiscip. Rev. Dev. Biol.* **2015**, *4*, 215–266. [CrossRef]
8. Di Re, J.; Wadsworth, P.A.; Laezza, F. Intracellular Fibroblast Growth Factor 14: Emerging Risk Factor for Brain Disorders. *Front. Cell. Neurosci.* **2017**, *11*, 1–7. [CrossRef]
9. Ali, S.R.; Singh, A.K.; Laezza, F. Identification of Amino Acid Residues in Fibroblast Growth Factor 14 (FGF14) Required for Structure-Function Interactions with Voltage-Gated Sodium Channel Nav1.6. *J. Biol. Chem.* **2016**, *291*, 11268–11284. [CrossRef]
10. Bosch, M.K.; Nerbonne, J.M.; Townsend, R.R.; Miyazaki, H.; Nukina, N.; Ornitz, D.M.; Marionneau, C. Proteomic Analysis of Native Cerebellar IFGF14 Complexes. *Channels* **2016**. [CrossRef]
11. Hsu, W.C.J.; Scala, F.; Nenov, M.N.; Wildburger, N.C.; Elferink, H.; Singh, A.K.; Chesson, C.B.; Buzhdygan, T.; Sohail, M.; Shavkunov, A.S.; et al. CK2 Activity Is Required for the Interaction of FGF14 with Voltage-Gated Sodium Channels and Neuronal Excitability. *FASEB J.* **2016**. [CrossRef] [PubMed]
12. Laezza, F.; Gerber, B.R.; Lou, J.-Y.; Kozel, M.A.; Hartman, H.; Craig, A.M.; Ornitz, D.M.; Nerbonne, J.M. The FGF14(F145S) Mutation Disrupts the Interaction of FGF14 with Voltage- Gated Na_+ Channels and Impairs Neuronal Excitability. *J. Neurosci.* **2007**, *27*, 12033–12044. [CrossRef] [PubMed]
13. Lou, J.-Y.; Laezza, F.; Gerber, B.R.; Xiao, M.; Yamada, K.A.; Hartmann, H.; Craig, A.M.; Nerbonne, J.M.; Ornitz, D.M. Fibroblast Growth Factor 14 Is an Intracellular Modulator of Voltage-Gated Sodium Channels. *J. Physiol.* **2005**, *5691*, 179–193. [CrossRef] [PubMed]

14. Ali, S.R.; Liu, Z.; Nenov, M.N.; Folorunso, O.; Singh, A.K.; Scala, F.; Chen, H.; James, T.F.; Alshammari, M.; Panova-Elektronova, N.I.; et al. Functional Modulation of Voltage-Gated Sodium Channels by a FGF14-Based Peptidomimetic. *ACS Chem. Neurosci.* **2018**. [CrossRef] [PubMed]
15. Shavkunov, A.S.; Wildburger, N.C.; Nenov, M.N.; James, T.F.; Buzhdygan, T.P.; Panova-Elektronova, N.I.; Green, T.A.; Veselenak, R.L.; Bourne, N.; Laezza, F. The Fibroblast Growth Factor 14??Voltage-Gated Sodium Channel Complex Is a New Target of Glycogen Synthase Kinase 3 (GSK3). *J. Biol. Chem.* **2013**, *288*, 19370–19385. [CrossRef] [PubMed]
16. Goldfarb, M.; Schoorlemmer, J.; Williams, A.; Diwakar, S.; Wang, Q.; Huang, X.; Giza, J.; Tchetchik, D.; Kelley, K.; Vega, A.; et al. Fibroblast Growth Factor Homologous Factors Control Neuronal Excitability through Modulation of Voltage-Gated Sodium Channels. *Neuron* **2007**. [CrossRef] [PubMed]
17. Goldfarb, M. Voltage-Gated Sodium Channel-Associated Proteins and Alternative Mechanisms of Inactivation and Block. *Cell. Mol. Life Sci.* **2012**. [CrossRef] [PubMed]
18. Tempia, F.; Hoxha, E.; Negro, G.; Alshammari, M.A.; Alshammari, T.K.; Panova-Elektronova, N.; Laezza, F. Parallel Fiber to Purkinje Cell Synaptic Impairment in a Mouse Model of Spinocerebellar Ataxia Type 27. *Front. Cell. Neurosci.* **2015**, *9*, 205. [CrossRef]
19. Yan, H.; Pablo, J.L.; Pitt, G.S. FGF14 Regulates Presynaptic Ca2+ Channels and Synaptic Transmission. *Cell Rep.* **2013**. [CrossRef]
20. Xiao, M.; Xu, L.; Laezza, F.; Yamada, K.; Feng, S.; Ornitz, D.M. Impaired Hippocampal Synaptic Transmission and Plasticity in Mice Lacking Fibroblast Growth Factor 14. *Mol. Cell. Neurosci.* **2007**, *34*, 366–377. [CrossRef]
21. Alshammari, T.; Alshammari, M.; Nenov, M.; Hoxha, E.; Cambiaghi, M.; Marcinno, A.; James, T.; Singh, P.; Labate, D.; Li, J.; et al. Genetic Deletion of Fibroblast Growth Factor 14 Recapitulates Phenotypic Alterations Underlying Cognitive Impairment Associated with Schizophrenia. *Transl. Psychiatry* **2016**, *666*. [CrossRef] [PubMed]
22. Hsu, W.-C.; Nenov, M.N.; Shavkunov, A.; Panova, N.; Zhan, M.; Laezza, F. Identifying a Kinase Network Regulating FGF14:Nav1.6 Complex Assembly Using Split-Luciferase Complementation. *PLoS ONE* **2015**, *10*, e0117246. [CrossRef] [PubMed]
23. Hsu, W.-C.J.; Nilsson, C.L.; Laezza, F. Role of the Axonal Initial Segment in Psychiatric Disorders: Function, Dysfunction, and Intervention. *Front. Psychiatry* **2014**, *5*, 109. [CrossRef] [PubMed]
24. Wildburger, N.C.; Laezza, F. Control of Neuronal Ion Channel Function by Glycogen Synthase Kinase-3: New Prospective for an Old Kinase. *Front. Mol. Neurosci.* **2012**, *5*, 80. [CrossRef] [PubMed]
25. Scala, F.; Nenov, M.N.; Crofton, E.J.; Singh, A.K.; Folorunso, O.; Zhang, Y.; Chesson, B.C.; Wildburger, N.C.; James, T.F.; Alshammari, M.A.; et al. Environmental Enrichment and Social Isolation Mediate Neuroplasticity of Medium Spiny Neurons through the GSK3 Pathway. *Cell Rep.* **2018**, *23*, 555–567. [CrossRef] [PubMed]
26. Hoxha, E.; Marcinnò, A.; Montarolo, F.; Masante, L.; Balbo, I.; Ravera, F.; Laezza, F.; Tempia, F. Emerging Roles of Fgf14 in Behavioral Control. *Behav. Brain Res.* **2019**, *356*, 257–265. [CrossRef] [PubMed]
27. Wozniak, D.F.; Xiao, M.; Xu, L.; Yamada, K.A.; Ornitz, D.M. Impaired Spatial Learning and Defective Theta Burst Induced LTP in Mice Lacking Fibroblast Growth Factor 14. *Neurobiol. Dis.* **2007**. [CrossRef] [PubMed]
28. Volk, L.; Chiu, S.-L.; Sharma, K.; Huganir, R.L. Glutamate Synapses in Human Cognitive Disorders. *Annu. Rev. Neurosci.* **2015**, *38*, 127–149. [CrossRef]
29. Alshammari, M.A.; Alshammari, T.K.; Nenov, M.N.; Scala, F.; Laezza, F. Fibroblast Growth Factor 14 Modulates the Neurogenesis of Granule Neurons in the Adult Dentate Gyrus. *Mol. Neurobiol.* **2016**. [CrossRef]
30. Sacco, R.; Cacci, E.; Novarino, G. Neural Stem Cells in Neuropsychiatric Disorders. *Curr. Opin. Neurobiol.* **2018**, *48*, 131–138. [CrossRef]
31. Wang, Q.; Bardgett, M.E.; Wong, M.; Wozniak, D.F.; Lou, J.; McNeil, B.D.; Chen, C.; Nardi, A.; Reid, D.C.; Yamada, K.; et al. Ataxia and Paroxysmal Dyskinesia in Mice Lacking Axonally Transported FGF14. *Neuron* **2002**. [CrossRef]
32. Savanthrapadian, S.; Wolff, A.R.; Logan, B.J.; Eckert, M.J.; Bilkey, D.K.; Abraham, W.C. Enhanced Hippocampal Neuronal Excitability and LTP Persistence Associated with Reduced Behavioral Flexibility in the Maternal Immune Activation Model of Schizophrenia. *Hippocampus* **2013**, *23*, 1395–1409. [CrossRef] [PubMed]
33. Chen, C.M.A.; Stanford, A.D.; Mao, X.; Abi-Dargham, A.; Shungu, D.C.; Lisanby, S.H.; Schroeder, C.E.; Kegeles, L.S. GABA Level, Gamma Oscillation, and Working Memory Performance in Schizophrenia. *NeuroImage Clin.* **2014**, *4*, 531–539. [CrossRef] [PubMed]

34. Kantrowitz, J.T.; Epstein, M.L.; Beggel, O.; Rohrig, S.; Lehrfeld, J.M.; Revheim, N.; Lehrfeld, N.P.; Reep, J.; Parker, E.; Silipo, G.; et al. Neurophysiological Mechanisms of Cortical Plasticity Impairments in Schizophrenia and Modulation by the NMDA Receptor Agonist D-Serine. *Brain* **2016**, *139*, 3281–3295. [CrossRef] [PubMed]
35. Falkenberg, L.E.; Westerhausen, R.; Craven, A.R.; Johnsen, E.; Kroken, R.A.; LØberg, E.M.; Specht, K.; Hugdahl, K. Impact of Glutamate Levels on Neuronal Response and Cognitive Abilities in Schizophrenia. *NeuroImage Clin.* **2014**, *4*, 576–584. [CrossRef] [PubMed]
36. Frankle, W.G.; Cho, R.Y.; Prasad, K.M.; Mason, N.S.; Paris, J.; Himes, M.L.; Walker, C.; Lewis, D.A.; Narendran, R. In Vivo Measurement of GABA Transmission in Healthy Subjects and Schizophrenia Patients. *Am. J. Psychiatry* **2015**, *172*, 1148–1159. [CrossRef] [PubMed]
37. McGlashan, T.H.; Bardenstein, K.K. Gender Differences in Affective, Schizoaffective, and Schizophrenic Disorders. *Schizophr. Bull.* **1990**, *16*, 319–329. [CrossRef] [PubMed]
38. Morgan, V.A.; Castle, D.J.; Jablensky, A.V. Do Women Express and Experience Psychosis Differently from Men? Epidemiological Evidence from the Australian National Study of Low Prevalence (Psychotic) Disorders. *Aust. N. Z. J. Psychiatry* **2008**, *42*, 74–82. [CrossRef] [PubMed]
39. Shtasel, D.L.; Gur, R.E.; Gallacher, F.; Heimberg, C.; Gur, R.C. Gender Differences in the Clinical Expression of Schizophrenia. *Schizophr. Res.* **1992**, *7*, 225–231. [CrossRef]
40. Ochoa, S.; Usall, J.; Cobo, J.; Labad, X.; Kulkarni, J. Gender Differences in Schizophrenia and First- Episode Psychosis: A Comprehensive Literature Review. *Schizophr. Res. Treat.* **2012**, *2012*, 1–9. [CrossRef]
41. Pinares-Garcia, P.; Stratikopoulos, M.; Zagato, A.; Loke, H.; Lee, J. Sex: A Significant Risk Factor for Neurodevelopmental and Neurodegenerative Disorders. *Brain Sci.* **2018**, *8*, 154. [CrossRef] [PubMed]
42. Mastro, R.; Hall, M. Protein Delipidation and Precipitation by Tri-n-Butylphosphate, Acetone, and Methanol Treatment for Isoelectric Focusing and Two-Dimensional Gel Electrophoresis. *Anal. Biochem.* **1999**, *273*, 313–315. [CrossRef] [PubMed]
43. Cox, J.; Mann, M. MaxQuant Enables High Peptide Identification Rates, Individualized p.p.b.-Range Mass Accuracies and Proteome-Wide Protein Quantification. *Nat. Biotechnol.* **2008**, *26*, 1367–1372. [CrossRef] [PubMed]
44. Tyanova, S.; Temu, T.; Cox, J. The MaxQuant Computational Platform for Mass Spectrometry-based Shotgun Proteomics. *Nat. Protoc.* **2016**, *11*, 2301–2319. [CrossRef] [PubMed]
45. Tyanova, S.; Temu, T.; Sinitcyn, P.; Carlson, A.; Hein, M.Y.; Geiger, T.; Mann, M.; Cox, J. The Perseus Computational Platform for Comprehensive Analysis of (Prote)Omics Data. *Nat. Methods* **2016**, *13*, 731–740. [CrossRef] [PubMed]
46. Bähner, F.; Meyer-Lindenberg, A. Hippocampal–prefrontal Connectivity as a Translational Phenotype for Schizophrenia. *Eur. Neuropsychopharmacol.* **2017**, *27*, 93–106. [CrossRef] [PubMed]
47. Chevaleyre, V.; Piskorowski, R.A. Hippocampal Area CA2: An Overlooked but PromisingTherapeutic Target. *Trends Mol. Med.* **2016**, *22*, 645–655. [CrossRef]
48. Kang, E.; Wen, Z.; Song, H.; Christian, K.M.; Ming, G. Adult Neurogenesis and Psychiatric Disorders. *Cold Spring Harb. Perspect. Biol.* **2016**, *8*, a019026. [CrossRef]
49. Nakahara, S.; Adachi, M.; Ito, H.; Matsumoto, M.; Tajinda, K.; van Erp, T.G.M. Hippocampal Pathophysiology: Commonality Shared by Temporal Lobe Epilepsy and Psychiatric Disorders. *Neurosci. J.* **2018**, *2018*, 1–9. [CrossRef]
50. Papp, E.A.; Leergaard, T.B.; Calabrese, E.; Johnson, G.A.; Bjaalie, J.G. Waxholm Space Atlas of the Sprague Dawley Rat Brain. *Neuroimage* **2014**, *97*, 374–386. [CrossRef]
51. Sergejeva, M.; Papp, E.A.; Bakker, R.; Gaudnek, M.A.; Okamura-Oho, Y.; Boline, J.; Bjaalie, J.G.; Hess, A. Anatomical Landmarks for Registration of Experimental Image Data to Volumetric Rodent Brain Atlasing Templates. *J. Neurosci. Methods* **2015**, *240*, 161–169. [CrossRef] [PubMed]
52. Kjonigsen, L.J.; Lillehaug, S.; Bjaalie, J.G.; Witter, M.P.; Leergaard, T.B. Waxholm Space Atlas of the Rat Brain Hippocampal Region: Three-Dimensional Delineations Based on Magnetic Resonance and Diffusion Tensor Imaging. *Neuroimage* **2015**, *108*, 441–449. [CrossRef] [PubMed]
53. Bielow, C.; Mastrobuoni, G.; Kempa, S. Proteomics Quality Control: Quality Control Software for MaxQuant Results. *J. Proteome Res.* **2016**, *15*, 777–787. [CrossRef] [PubMed]
54. Gu, Y.; Chiu, S.-L.; Liu, B.; Wu, P.-H.; Delannoy, M.; Lin, D.-T.; Wirtz, D.; Huganir, R.L. Differential Vesicular Sorting of AMPA and GABA A Receptors. *Proc. Natl. Acad. Sci. USA* **2016**, *113*, E922–E931. [CrossRef] [PubMed]

55. Yang, H.; Zhang, M.; Shi, J.; Zhou, Y.; Wan, Z.; Wang, Y.; Wan, Y.; Li, J.; Wang, Z.; Fei, J. Brain-Specific SNAP-25 Deletion Leads to Elevated Extracellular Glutamate Level and Schizophrenia-Like Behavior in Mice. *Neural Plast.* **2017**, *2017*. [CrossRef]
56. Hjelm, B.E.; Rollins, B.; Mamdani, F.; Lauterborn, J.C.; Kirov, G.; Lynch, G.; Gall, C.M.; Sequeira, A.; Vawter, M.P. Evidence of Mitochondrial Dysfunction within the Complex Genetic Etiology of Schizophrenia. *Mol. Neuropsychiatry* **2015**, *1*, 201–219. [CrossRef]
57. Ueno, H.; Nishigaki, Y.; Kong, Q.-P.; Fuku, N.; Kojima, S.; Iwata, N.; Ozaki, N.; Tanaka, M. Analysis of Mitochondrial DNA Variants in Japanese Patients with Schizophrenia. *Mitochondrion* **2009**, *9*, 385–393. [CrossRef]
58. Brisch, R.; Bielau, H.; Saniotis, A.; Wolf, R.; Bogerts, B.; Krell, D.; Steiner, J.; Braun, K.; Krzyżanowska, M.; Krzyżanowski, M.; et al. Calretinin and Parvalbumin in Schizophrenia and Affective Disorders: A Mini-Review, a Perspective on the Evolutionary Role of Calretinin in Schizophrenia, and a Preliminary Post-Mortem Study of Calretinin in the Septal Nuclei. *Front. Cell. Neurosci.* **2015**, *9*, 393. [CrossRef]
59. Peretti, D.; Bastide, A.; Radford, H.; Verity, N.; Molloy, C.; Martin, M.G.; Moreno, J.A.; Steinert, J.R.; Smith, T.; Dinsdale, D.; et al. RBM3 Mediates Structural Plasticity and Protective Effects of Cooling in Neurodegeneration. *Nature* **2015**, *518*, 236–239. [CrossRef]
60. Smart, F.; Aschrafi, A.; Atkins, A.; Owens, G.C.; Pilotte, J.; Cunningham, B.A.; Vanderklish, P.W. Two Isoforms of the Cold-Inducible MRNA-Binding Protein RBM3 Localize to Dendrites and Promote Translation. *J. Neurochem.* **2007**, *101*, 1367–1379. [CrossRef]
61. Keil, K.P.; Sethi, S.; Wilson, M.D.; Chen, H.; Lein, P.J. In Vivo and in Vitro Sex Differences in the Dendritic Morphology of Developing Murine Hippocampal and Cortical Neurons. *Sci. Rep.* **2017**, *7*, 1–15. [CrossRef] [PubMed]
62. Li, C.; Brake, W.G.; Romeo, R.D.; Dunlop, J.C.; Gordon, M.; Buzescu, R.; Magarinos, A.M.; Allen, P.B.; Greengard, P.; Luine, V.; et al. Estrogen Alters Hippocampal Dendritic Spine Shape and Enhances Synaptic Protein Immunoreactivity and Spatial Memory in Female Mice. *Proc. Natl. Acad. Sci. USA* **2004**, *101*, 2185–2190. [CrossRef] [PubMed]
63. Lisman, J.; Yasuda, R.; Raghavachari, S. Mechanisms of CaMKII Action in Long-Term Potentiation. *Nat. Rev. Neurosci.* **2012**, *13*, 169–182. [CrossRef] [PubMed]
64. Reinhard, J.R.; Kriz, A.; Galic, M.; Angliker, N.; Rajalu, M.; Vogt, K.E.; Ruegg, M.A. The Calcium Sensor Copine-6 Regulates Spine Structural Plasticity and Learning and Memory. *Nat. Commun.* **2016**, *7*, 1–14. [CrossRef] [PubMed]
65. Seyfried, N.T.; Dammer, E.B.; Swarup, V.; Nandakumar, D.; Duong, D.M.; Yin, L.; Deng, Q.; Nguyen, T.; Hales, C.M.; Wingo, T.; et al. A Multi-Network Approach Identifies Protein-Specific Co-Expression in Asymptomatic and Symptomatic Alzheimer's Disease. *Cell Syst.* **2017**, *4*, 60–72. [CrossRef] [PubMed]
66. De Leeuw, C.A.; Mooij, J.M.; Heskes, T.; Posthuma, D. MAGMA: Generalized Gene-Set Analysis of GWAS Data. *PLoS Comput. Biol.* **2015**, *11*, e1004219. [CrossRef] [PubMed]
67. Zhou, G.; Xia, J. OmicsNet: A Web-Based Tool for Creation and Visual Analysis of Biological Networks in 3D Space. *Nucleic Acids Res.* **2018**, *46*, W514–W522. [CrossRef] [PubMed]
68. Braida, D.; Ponzoni, L.; Matteoli, M.; Sala, M.M. Different Attentional Abilities among Inbred Mice Strains Using Virtual Object Recognition Task (VORT): SNAP25$^{+/-}$ Mice as a Model of Attentional Deficit. *Behav. Brain Res.* **2016**, *296*, 393–400. [CrossRef] [PubMed]
69. Braida, D.; Guerini, F.R.; Ponzoni, L.; Corradini, I.; De Astis, S.; Pattini, L.; Bolognesi, E.; Benfante, R.; Fornasari, D.; Chiappedi, M.; et al. Association between SNAP-25 Gene Polymorphisms and Cognition in Autism: Functional Consequences and Potential Therapeutic Strategies. *Transl. Psychiatry* **2015**, *5*, e500-11. [CrossRef]
70. Kim, J.-I.; Ganesan, S.; Luo, S.X.; Wu, Y.-W.; Park, E.; Huang, E.J.; Chen, L.; Ding, J.B. Aldehyde Dehydrogenase 1a1 Mediates a GABA Synthesis Pathway in Midbrain Dopaminergic Neurons. *Science* **2015**, *350*, 102–106. [CrossRef]
71. Liu, G.; Yu, J.; Ding, J.; Xie, C.; Sun, L.; Rudenko, I.; Zheng, W.; Sastry, N.; Luo, J.; Rudow, G.; et al. Aldehyde Dehydrogenase 1 Defines and Protects a Nigrostriatal Dopaminergic Neuron Subpopulation. *J. Clin. Investig.* **2014**, *124*, 3032–3046. [CrossRef] [PubMed]
72. Adams, M.R.; Brandon, E.P.; Chartoff, E.H.; Idzerda, R.L.; Dorsa, D.M.; McKnight, G.S. Loss of Haloperidol Induced Gene Expression and Catalepsy in Protein Kinase A-Deficient Mice. *Proc. Natl. Acad. Sci. USA* **1997**, *94*, 12157–12161. [CrossRef] [PubMed]

73. Chang, S.; Fang, K.; Zhang, K.; Wang, J. Network-Based Analysis of Schizophrenia Genome-Wide Association Data to Detect the Joint Functional Association Signals. *PLoS ONE* **2015**, *10*, e0133404. [CrossRef] [PubMed]
74. Aoto, J.; Nam, C.I.; Poon, M.M.; Ting, P.; Chen, L. Synaptic Signaling by All-Trans Retinoic Acid in Homeostatic Synaptic Plasticity. *Neuron* **2008**, *60*, 308–320. [CrossRef] [PubMed]
75. Nishiura, K.; Ichikawa-Tomikawa, N.; Sugimoto, K.; Kunii, Y.; Kashiwagi, K.; Tanaka, M.; Yokoyama, Y.; Hino, M.; Sugino, T.; Yabe, H.; et al. PKA Activation and Endothelial Claudin-5 Breakdown in the Schizophrenic Prefrontal Cortex. *Oncotarget* **2017**, *8*, 93382–93391. [CrossRef] [PubMed]
76. Pan, B.; Lian, J.; Huang, X.-F.; Deng, C. Aripiprazole Increases the PKA Signalling and Expression of the GABAA Receptor and CREB1 in the Nucleus Accumbens of Rats. *J. Mol. Neurosci.* **2016**, *59*, 36–47. [CrossRef] [PubMed]
77. Chen, Y.; Yu, F.H.; Surmeier, D.J.; Scheuer, T.; Catterall, W.A. Neuromodulation of Na^+ Channel Slow Inactivation via CAMP-Dependent Protein Kinase and Protein Kinase C. *Neuron* **2006**, *49*, 409–420. [CrossRef]
78. Maurice, N.; Tkatch, T.; Meisler, M.; Sprunger, L.K.; Surmeier, D.J. D1/D5 Dopamine Receptor Activation Differentially Modulates Rapidly Inactivating and Persistent Sodium Currents in Prefrontal Cortex Pyramidal Neurons. *J. Neurosci.* **2001**, *21*, 2268–2277. [CrossRef]
79. Flores-Hernandez, J.; Hernandez, S.; Snyder, G.L.; Yan, Z.; Fienberg, A.A.; Moss, S.J.; Greengard, P.; Surmeier, D.J. D(1) Dopamine Receptor Activation Reduces GABA(A) Receptor Currents in Neostriatal Neurons through a PKA/DARPP-32/PP1 Signaling Cascade. *J. Neurophysiol.* **2000**, *83*, 2996–3004. [CrossRef]
80. Surmeier, D.J.; Bargas, J.; Hemmings, H.C.; Nairn, A.C.; Greengard, P. Modulation of Calcium Currents by a D1 Dopaminergic Protein Kinase/Phosphatase Cascade in Rat Neostriatal Neurons. *Neuron* **1995**, *14*, 385–397. [CrossRef]
81. Wu, Y.-C.; Williamson, R.; Li, Z.; Vicario, A.; Xu, J.; Kasai, M.; Chern, Y.; Tongiorgi, E.; Baraban, J.M. Dendritic Trafficking of Brain-Derived Neurotrophic Factor MRNA: Regulation by Translin- Dependent and -Independent Mechanisms. *J. Neurochem.* **2011**, *116*, 1112–1121. [CrossRef] [PubMed]
82. Libman-Sokołowska, M.; Drozdowicz, E.; Nasierowski, T. BDNF as a Biomarker in the Course and Treatment of Schizophrenia. *Psychiatr. Pol.* **2015**, *49*, 1149–1158. [CrossRef] [PubMed]
83. Kheirollahi, M.; Kazemi, E.; Ashouri, S. Brain-Derived Neurotrophic Factor Gene Val66Met Polymorphism and Risk of Schizophrenia: A Meta-Analysis of Case-Control Studies. *Cell. Mol. Neurobiol.* **2016**, *36*, 1–10. [CrossRef] [PubMed]
84. Ishida, R.; Okado, H.; Sato, H.; Shionoiri, C.; Aoki, K.; Kasai, M. A Role for the Octameric Ring Protein, Translin, in Mitotic Cell Division. *FEBS Lett.* **2002**, *525*, 105–110. [CrossRef]
85. Antonucci, F.; Corradini, I.; Fossati, G.; Tomasoni, R.; Menna, E.; Matteoli, M. SNAP-25, a Known Presynaptic Protein with Emerging Postsynaptic Functions. *Front. Synaptic Neurosci.* **2016**, *8*. [CrossRef] [PubMed]
86. Hagihara, H.; Takao, K.; Walton, N.M.; Matsumoto, M.; Miyakawa, T. Immature Dentate Gyrus: An Endophenotype of Neuropsychiatric Disorders. *Neural Plast.* **2013**, *2013*, 318596. [CrossRef] [PubMed]
87. Yamasaki, N.; Maekawa, M.; Kobayashi, K.; Kajii, Y.; Maeda, J.; Soma, M.; Takao, K.; Tanda, K.; Ohira, K.; Toyama, K.; et al. Alpha-CaMKII Deficiency Causes Immature Dentate Gyrus, a Novel Candidate Endophenotype of Psychiatric Disorders. *Mol. Brain* **2008**, *1*, 6. [CrossRef]
88. Thompson, P.M.; Egbufoama, S.; Vawter, M.P. SNAP-25 Reduction in the Hippocampus of Patients with Schizophrenia. *Prog. Neuro-Psychopharmacol. Biol. Psychiatry* **2003**, *27*, 411–417. [CrossRef]
89. Etain, B.; Dumaine, A.; Mathieu, F.; Chevalier, F.; Henry, C.; Kahn, J.P.; Deshommes, J.; Bellivier, F.; Leboyer, M.; Jamain, S. A SNAP25 Promoter Variant Is Associated with Early-Onset Bipolar Disorder and a High Expression Level in Brain. *Mol. Psychiatry* **2010**, *15*, 748–755. [CrossRef]
90. Safari, M.R.; Omrani, M.D.; Noroozi, R.; Sayad, A.; Sarrafzadeh, S.; Komaki, A.; Manjili, F.A.; Mazdeh, M.; Ghaleiha, A.; Taheri, M. Synaptosome-Associated Protein 25 (SNAP25) Gene Association Analysis Revealed Risk Variants for ASD, in Iranian Population. *J. Mol. Neurosci.* **2017**, *61*, 305–311. [CrossRef]

MDPI
St. Alban-Anlage 66
4052 Basel
Switzerland
Tel. +41 61 683 77 34
Fax +41 61 302 89 18
www.mdpi.com

Proteomes Editorial Office
E-mail: proteomes@mdpi.com
www.mdpi.com/journal/proteomes

www.ingramcontent.com/pod-product-compliance
Lightning Source LLC
LaVergne TN
LVHW071937080526
838202LV00064B/6627
9783039281060